Annals of Mathematics Studies

Number 167

The Hypoelliptic Laplacian and Ray-Singer Metrics

Jean-Michel Bismut
Gilles Lebeau

PRINCETON UNIVERSITY PRESS
PRINCETON AND OXFORD
2008

Published by Princeton University Press
41 William Street, Princeton, New Jersey 08540

In the United Kingdom: Princeton University Press
6 Oxford Street, Woodstock, Oxfordshire, OX20 1TW

Library of Congress Cataloging-in-Publication Data

Bismut, Jean-Michel.
The hypoelliptic Laplacian and Ray-Singer metrics / Jean-Michel Bismut, Gilles Lebeau.
p. cm.
Includes bibliographical references and index.
ISBN-13: 978-0-691-13731-5 (alk. paper)
ISBN-13: 978-0-691-13732-2 (pbk. : alk. paper)
1. Differential equations, Hypoelliptic. 2. Laplacian operator. 3. Metric spaces. I. Lebeau, Gilles. II. Title.
QA377 .B674 2008
515′.7242–dc22 2008062103

British Library Cataloging-in-Publication Data is available

This book has been composed in LaTeX

The publisher would like to acknowledge the authors of this volume for providing the camera-ready copy from which this book was printed.

Printed on acid-free paper. ∞

press.princeton.edu

Printed in the United States of America

10 9 8 7 6 5 4 3 2 1

Contents

Introduction 1

Chapter 1. Elliptic Riemann-Roch-Grothendieck and flat vector bundles 11

1.1 The Clifford algebra 11
1.2 The standard Hodge theory 12
1.3 The Levi-Civita superconnection 14
1.4 Superconnections and Poincaré duality 15
1.5 A group action 16
1.6 The Lefschetz formula 16
1.7 The Riemann-Roch-Grothendieck theorem 17
1.8 The elliptic analytic torsion forms 19
1.9 The Chern analytic torsion forms 21
1.10 Analytic torsion forms and Poincaré duality 22
1.11 The secondary classes for two metrics 22
1.12 Determinant bundle and Ray-Singer metric 23

Chapter 2. The hypoelliptic Laplacian on the cotangent bundle 25

2.1 A deformation of Hodge theory 25
2.2 The hypoelliptic Weitzenböck formulas 29
2.3 Hypoelliptic Laplacian and standard Laplacian 30
2.4 A deformation of Hodge theory in families 33
2.5 Weitzenböck formulas for the curvature 35
2.6 $\mathfrak{F}^{\mathcal{M}}_{\phi_b,\pm\mathcal{H}-b\omega^H}, \mathfrak{E}^{\mathcal{M},2}_{\phi_b,\pm\mathcal{H}-b\omega^H}$ and Levi-Civita superconnection 40
2.7 The superconnection $A^{\mathcal{M}}_{\phi,\mathcal{H}-\omega^H}$ and Poincaré duality 40
2.8 A **2**-parameter rescaling 41
2.9 A group action 43

Chapter 3. Hodge theory, the hypoelliptic Laplacian and its heat kernel 44

3.1 The cohomology of $\mathbf{T}^*\mathbf{X}$ and the Thom isomorphism 44
3.2 The Hodge theory of the hypoelliptic Laplacian 45
3.3 The heat kernel for $\mathfrak{A}^{\mathbf{2}}_{\phi,\mathcal{H}^c}$ 50
3.4 Uniform convergence of the heat kernel as $\mathbf{b} \to \mathbf{0}$ 53
3.5 The spectrum of $\mathfrak{A}'^{\mathbf{2}}_{\phi_{\mathbf{b}},\pm\mathcal{H}}$ as $\mathbf{b} \to \mathbf{0}$ 55
3.6 The Hodge condition 58
3.7 The hypoelliptic curvature 60

Chapter 4. Hypoelliptic Laplacians and odd Chern forms 62

4.1 The Berezin integral 63

4.2 The even Chern forms 64
4.3 The odd Chern forms and a 1-form on $\mathbf{R}^{*2}$ 65
4.4 The limit as $\mathbf{t} \to \mathbf{0}$ of the forms $\mathbf{u_{b,t}}, \mathbf{v_{b,t}}, \mathbf{w_{b,t}}$ 68
4.5 A fundamental identity 68
4.6 A rescaling along the fibers of $\mathbf{T^*X}$ 69
4.7 Localization of the problem 70
4.8 Replacing $\mathbf{T^*X}$ by $\mathbf{T_xX} \oplus \mathbf{T_x^*X}$ and the rescaling of Clifford variables on $\mathbf{T^*X}$ 76
4.9 The limit as $t \to 0$ of the rescaled operator 80
4.10 The limit of the rescaled heat kernel 82
4.11 Evaluation of the heat kernel for $\frac{\mathbf{\Delta^V}}{\mathbf{4}} + \mathbf{a}\nabla_{\mathbf{p}}$ 87
4.12 An evaluation of certain supertraces 91
4.13 A proof of Theorems 4.2.1 and 4.4.1 92

Chapter 5. The limit as $t \to +\infty$ and $b \to 0$ of the superconnection forms 98
5.1 The definition of the limit forms 98
5.2 The convergence results 101
5.3 A contour integral 102
5.4 A proof of Theorem 5.3.1 104
5.5 A proof of Theorem 5.3.2 104
5.6 A proof of the first equations in (5.2.1) and (5.2.2) 109

Chapter 6. Hypoelliptic torsion and the hypoelliptic Ray-Singer metrics 113
6.1 The hypoelliptic torsion forms 113
6.2 Hypoelliptic torsion forms and Poincaré duality 115
6.3 A generalized Ray-Singer metric on the determinant of the cohomology 116
6.4 Truncation of the spectrum and Ray-Singer metrics 120
6.5 A smooth generalized metric on the determinant bundle 122
6.6 The equivariant determinant 123
6.7 A variation formula 125
6.8 A simple identity 126
6.9 The projected connections 126
6.10 A proof of Theorem 6.7.2 127

Chapter 7. The hypoelliptic torsion forms of a vector bundle 131
7.1 The function $\tau(\mathbf{c}, \eta, \mathbf{x})$ 131
7.2 Hypoelliptic curvature for a vector bundle 133
7.3 Translation invariance of the curvature 134
7.4 An automorphism of E 135
7.5 The von Neumann supertrace of $\exp\left(-\mathfrak{L}_{\mathbf{c}}^{\mathbf{E}}\right)$ 136
7.6 A probabilistic expression for $\mathbf{Q}_{\mathbf{c}}'$ 138
7.7 Finite dimensional supertraces and infinite determinants 139
7.8 The evaluation of the form $\mathrm{Tr_s}\left[\mathbf{g}\exp\left(-\mathfrak{L}_{\mathbf{c}}^{\mathbf{E}}\right)\right]$ 148
7.9 Some extra computations 152
7.10 The Mellin transform of certain Fourier series 155
7.11 The hypoelliptic torsion forms for vector bundles 160

Chapter 8. Hypoelliptic and elliptic torsions: a comparison formula 162

8.1 On some secondary Chern classes 162
8.2 The main result 163
8.3 A contour integral 164
8.4 Four intermediate results 165
8.5 The asymptotics of the $\mathbf{I}_{\mathbf{k}}^{\mathbf{0}}$ 166
8.6 Matching the divergences 169
8.7 A proof of Theorem 8.2.1 170

Chapter 9. A comparison formula for the Ray-Singer metrics 171

Chapter 10. The harmonic forms for $b \to 0$ and the formal Hodge theorem 173

10.1 A proof of Theorem 8.4.2 173
10.2 The kernel of $\mathbf{A}^{\mathbf{2}}_{\phi,\mathcal{H}^c}$ as a formal power series 175
10.3 A proof of the formal Hodge Theorem 178
10.4 Taylor expansion of harmonic forms near $\mathbf{b} = \mathbf{0}$ 180

Chapter 11. A proof of equation (8.4.6) 182

11.1 The limit of the rescaled operator as $\mathbf{t} \to \mathbf{0}$ 182
11.2 The limit of the supertrace as $\mathbf{t} \to \mathbf{0}$ 187
11.3 A proof of equation (8.4.6) 189

Chapter 12. A proof of equation (8.4.8) 190

12.1 Uniform rescalings and trivializations 190
12.2 A proof of (8.4.8) 192

Chapter 13. A proof of equation (8.4.7) 194

13.1 The estimate in the range $\mathbf{t} \geq \mathbf{b}^{\beta}$ 194
13.2 Localization of the estimate near $\pi^{-1}\mathbf{X}_{\mathbf{g}}$ 196
13.3 A uniform rescaling on the creation annihilation operators 198
13.4 The limit as $\mathbf{t} \to \mathbf{0}$ of the rescaled operator 200
13.5 Replacing $\mathbf{X}$ by $\mathbf{T}_{\mathbf{x}}\mathbf{X}$ 202
13.6 A proof of (13.2.11) 205
13.7 A proof of Theorem 13.6.2 206

Chapter 14. The integration by parts formula 214

14.1 The case of Brownian motion 215
14.2 The hypoelliptic diffusion 217
14.3 Estimates on the heat kernel 219
14.4 The gradient of the heat kernel 220

Chapter 15. The hypoelliptic estimates 224

15.1 The operator $\mathfrak{A}'^{2}_{\phi_b,\pm\mathcal{H}}$ 224
15.2 A Littlewood-Paley decomposition 226
15.3 Projectivization of $\mathbf{T}^*\mathbf{X}$ and Sobolev spaces 227
15.4 The hypoelliptic estimates 229
15.5 The resolvent on the real line 238
15.6 The resolvent on $\mathbf{C}$ 240

15.7 Trace class properties of the resolvent 243

Chapter 16. Harmonic oscillator and the J_0 function 247
16.1 Fock spaces and the Bargman transform 247
16.2 The operator $\mathbf{B}(\xi)$ 249
16.3 The spectrum of $\mathbf{B}(\mathbf{i}\xi)$ 251
16.4 The function $\mathbf{J_0}(\mathbf{y}, \lambda)$ 253
16.5 The resolvent of $\mathbf{B}(\mathbf{i}\xi) + \mathbf{P}$ 261

Chapter 17. The limit of $\mathfrak{A}'^2_{\phi_b,\pm\mathcal{H}}$ as $b \to 0$ 264
17.1 Preliminaries in linear algebra 268
17.2 A matrix expression for the resolvent 268
17.3 The semiclassical Poisson bracket 270
17.4 The semiclassical Sobolev spaces 271
17.5 Uniform hypoelliptic estimates for $\mathbf{P_h}$ 272
17.6 The operator $\mathbf{P^0_h}$ and its resolvent $\mathbf{S_{h,\lambda}}$ for $\lambda \in \mathbf{R}$ 277
17.7 The resolvent $\mathbf{S_{h,\lambda}}$ for $\lambda \in \mathbf{C}$ 281
17.8 A trivialization over $\mathbf{X}$ and the symbols $\mathcal{S}^{\mathbf{d},\mathbf{k}}_{\rho,\delta,\mathbf{c}}$ 283
17.9 The symbol $\mathbf{Q^0_h}(\mathbf{x}, \xi) - \lambda$ and its inverse $\mathbf{e_{0,h,\lambda}}(\mathbf{x}, \xi)$ 289
17.10 The parametrix for $\mathbf{S_{h,\lambda}}$ 306
17.11 A localization property for $\mathbf{E_0}, \mathbf{E_1}$ 307
17.12 The operator $\mathbf{P_\pm S_{h,\lambda}}$ 308
17.13 A proof of equation (17.12.9) 309
17.14 An extension of the parametrix to $\lambda \in \mathcal{V}$ 318
17.15 Pseudodifferential estimates for $\mathbf{P_\pm S_{h,\lambda} i_\pm}$ 319
17.16 The operator $\mathbf{\Theta_{h,\lambda}}$ 323
17.17 The operator $\mathbf{T_{h,\lambda}}$ 326
17.18 The operator $(\mathbf{J_1}/\mathbf{J_0})\left(\mathbf{hD^X}/\sqrt{\mathbf{2}}, \lambda\right)$ 329
17.19 The operator $\mathbf{U_{h,\lambda}}$ 331
17.20 Estimates on the resolvent of $\mathbf{T_{h,h^2\lambda}}$ 337
17.21 The asymptotics of $(\mathbf{L_c} - \lambda)^{-1}$ 340
17.22 A localization property 348

Bibliography 353

Subject Index 359

Index of Notation 361

The Hypoelliptic Laplacian and Ray-Singer Metrics

Introduction

The purpose of this book is to develop the analytic theory of the hypoelliptic Laplacian and to establish corresponding results on the associated Ray-Singer analytic torsion. We also introduce the corresponding theory for families of hypoelliptic Laplacians, and we construct the associated analytic torsion forms. The whole setting will be equivariant with respect to the action of a compact Lie group G.

Let us put in perspective the various questions which are dealt with in this book. In [B05], one of us introduced a deformation of classical Hodge theory. Let X be a compact Riemannian manifold, let $\left(F, \nabla^F, g^F\right)$ be a complex flat Hermitian vector bundle on X. Let $\left(\Omega^{\cdot}(X,F), d^X\right)$ be the de Rham complex of smooth forms on X with coefficients in F, let d^{X*} be the formal adjoint of d^X with respect to the obvious Hermitian product on $\Omega^{\cdot}(X,F)$. Then the Laplacian $\square^X = \left[d^X, d^{X*}\right]$ is a second order nonnegative elliptic operator acting on $\Omega^{\cdot}(X,F)$. Let $\mathcal{H}^X = \ker \square^X$ be the vector space of harmonic forms. Classical Hodge theory asserts that we have a canonical isomorphism,

$$\mathcal{H}^X \simeq H^{\cdot}(X,F). \tag{0.1}$$

Let T^*X be the cotangent bundle of X, let $\left(\Omega^{\cdot}(T^*X, \pi^*F), d^{T^*X}\right)$ be the corresponding de Rham complex over T^*X. In [B05], a deformation of classical Hodge theory was constructed, which is associated to a Hamiltonian $\mathcal{H}$ on T^*X. The corresponding Laplacian is denoted by $A^2_{\phi,\mathcal{H}}$. In the case where $\mathcal{H} = \frac{|p|^2}{2}$ and $\mathcal{H}^c = c\mathcal{H}$ depends on a parameter $c = \pm 1/b^2 \in \mathbf{R}^*$, with $b \in \mathbf{R}^*_+$, an operator which is conjugate to $A^2_{\phi,\mathcal{H}^c}$, the operator $\mathfrak{A}^2_{\phi,\mathcal{H}^c}$, is given by the formula

$$\begin{aligned}\mathfrak{A}^2_{\phi,\mathcal{H}^c} = \frac{1}{4}\Bigg(&-\Delta^V + c^2|p|^2 + c\left(2\widehat{e}_i i_{\widehat{e}^i} - n\right)\\ &- \frac{1}{2}\left\langle R^{TX}(e_i,e_j)e_k, e_l\right\rangle e^i e^j i_{\widehat{e}^k} i_{\widehat{e}^l}\Bigg)\\ -\frac{1}{2}\Bigg(&cL_{Y^{\mathcal{H}}} + \frac{c}{2}\omega\left(\nabla^F, g^F\right)\left(Y^{\mathcal{H}}\right) + \frac{1}{2}e^i i_{\widehat{e}^j}\nabla^F_{e_i}\omega\left(\nabla^F, g^F\right)(e_j)\\ &+ \frac{1}{2}\omega\left(\nabla^F, g^F\right)(e_i)\nabla_{\widehat{e}^i}\Bigg).\end{aligned} \tag{0.2}$$

In (0.2), Δ^V is is the Laplacian along the fibers of T^*X, R^{TX} is the curvature tensor of the Levi-Civita connection ∇^{TX}, the $e^i, \widehat{e}_i$ are horizontal and vertical 1-forms, which produce orthonormal bases of T^*X and TX, $Y^{\mathcal{H}}$ is the Hamiltonian vector field associated to $\mathcal{H}$, i.e., the generator of the geodesic flow, $L_{Y^{\mathcal{H}}}$ is the Lie derivative operator associated to $Y^{\mathcal{H}}$, and $\omega\left(\nabla^F, g^F\right)$ is the variation of g^F with respect to ∇^F. The differential operator which appears in the first line in the right-hand side of (0.2) is a harmonic oscillator. A fundamental feature of the operator $\mathfrak{A}^2_{\phi,\mathcal{H}^c}$ is that by a theorem of Hörmander [Hör67], $\frac{\partial}{\partial u} - \mathfrak{A}^2_{\phi,\mathcal{H}^c}$ is hypoelliptic.

In [B05], algebraic arguments were given which indicated that when b varies between 0 and $+\infty$, the Laplacian $2A^2_{\phi,\mathcal{H}^c}$ interpolates in a proper sense between the Hodge Laplacian $\square^X/2$ and the operator $|p|^2/2 - L_{Y^{\mathcal{H}}}$. Moreover, $A^2_{\phi,\mathcal{H}^c}$ was shown in [B05] to be self-adjoint with respect to a Hermitian form of signature (∞,∞).

A key motivation for the construction of the Laplacian $A^2_{\phi,\mathcal{H}^c}$ is its relation to the Witten deformation of classical Hodge theory. Let us simply recall that if $f : X \to \mathbf{R}$ is a smooth function, the associated Witten Laplacian is a one parameter deformation $\square^X_T$ of the classical Laplacian $\square^X$, which coincides with $\square^X$ for $T = 0$, which also consists of elliptic self-adjoint operators for which the Hodge theorem holds. If f is a Morse function, Witten showed that as $T \to \pm\infty$, the small eigenvalue eigenspaces localize near the critical points of f. He also conjectured that the corresponding complex of small eigenvalue eigenforms can be identified with the corresponding Thom-Smale [T49, Sm61] of the gradient field $-\nabla f$, in the case where this gradient field satisfies the Thom transversality conditions [T49]. This conjecture was proved by Helffer-Sjöstrand [HeSj85]. The Witten deformation was used in [BZ92, BZ94] to give a new proof of the Cheeger-Müller theorem [C79, Mül78] on the equality of the Reidemeister torsion and of the analytic torsion for unitary flat exact vector bundles, and more generally of the Ray-Singer metric on $\lambda = \det H^{\cdot}(X,F)$, which one defines using the Ray-Singer torsion, with the so-called Reidemeister metric [Re35], which is defined combinatorially. Let us just recall here that the Ray-Singer analytic torsion can be obtained via the derivative at $s = 0$ of the zeta functions of the Laplacian $\square^X$.

Let LX be the loop space of X, i.e., the set of smooth maps $s \in S_1 \to X$, and let E be the energy functional $E = \frac{1}{2}\int_0^1 |\dot{x}|^2\, ds$. The functional integral interpretation of the Laplacian $A^2_{\phi,\mathcal{H}^c}$ is explained in detail in [B04, B05]. In particular $2A^2_{\phi,\mathcal{H}^c}$ interpolating between $\square^X/2$ and $|p|^2/2 - L_{Y^{\mathcal{H}}}$ should be thought of as a semiclassical version of the fact that the Witten Laplacian $\square^{LX}_T$ on LX associated to the energy functional E should interpolate between the Hodge Laplacian $\square^{LX}$ and the Morse theory for E, whose critical points are precisely the closed geodesics. Incidentally, let us recall that neither $\square^{LX}$ nor its Witten deformation has ever been constructed. Let us also mention that if one follows the analogy of the deformation in [B05] with the Witten Laplacian, then $c = 1/T$, so that $T = \pm b^2$.

It was also observed in [B05] that at least formally, Fried's conjecture

[F86, F88] on the relation of the Ray-Singer torsion to dynamical Ruelle's zeta functions could be thought of as a consequence of a infinite dimensional version of the Cheeger-Müller theorem, where X is replaced by LX. This conjecture by Fried has been proved by Moscovici-Stanton [MoSta91] for symmetric spaces using Selberg's trace formula.

The present book has four main purposes:

- To develop the full Hodge theory of the Laplacian $A^2_{\phi,\mathcal{H}^c}$. This means not only proving a corresponding version of the Hodge theorem, but also studying the precise properties of its resolvent and of the corresponding heat kernel. The main difficulty is related to the fact that T^*X is noncompact, and also that the operator $A^2_{\phi,\mathcal{H}^c}$ is not classically self-adjoint.
- To develop the appropriate local index theory for the associated heat kernel.
- To adapt to such Laplacians the theory of the Ray-Singer torsion [RS71] of Ray-Singer, and of the analytic torsion forms of Bismut-Lott [BLo95].
- To give an explicit formula relating the analytic torsion objects associated to the hypoelliptic Laplacian to the classical Ray-Singer torsion for the classical Laplacian $\square^X$.

To reach these above objectives, we use the following tools:

- We refine the hypoelliptic estimates of Hörmander [Hör85] in order to control hypoellipticity at infinity in the cotangent bundle. Some of the arguments we use are similar to arguments already given by Helffer-Nier [HeN05] and Hérau-Nier [HN04] in the case where $X = \mathbf{R}^n$ in their study of the return to equilibrium for Fokker-Planck equations. It is quite striking that although we view our hypoelliptic equations as coming from a degeneration of elliptic equations on LX, we end up dealing with kinetic equations on X.
- We develop the adequate theory of semiclassical pseudodifferential operators with parameter $h = b$, combined to a computation of the resolvents as $(2,2)$ matrices, by a method formally similar to a method we developed in the context of Quillen metrics in [BL91], in order to study the convergence as $b \to 0$ of the operator $A^2_{\phi,\mathcal{H}^c}$ to $\square^X$. One basic difference with respect to [BL91] is that our operators are no longer self-adjoint.
- We develop a hypoelliptic local index theory. This local index theory extends the well-known local index theory for the operator $d^X + d^{X*}$ [P71, Gi84, ABP73, G86]. Still, the fact that we work also with analytic torsion forms forces us to develop a very general machinery which will extend to the analysis of Dirac operators. The hypoelliptic local index theory is itself a deformation of classical elliptic local index theory.

- We study the deformation of the Ray-Singer metric and also the corresponding hypoelliptic analytic torsion forms by a method formally similar to the one used in [BL91] and later extended in [B97] to holomorphic torsion forms. At least at a formal level, even though we deal with essentially different objects, the proofs are formally very close, even in their intermediate steps.

- We develop the adequate probabilistic machinery which allows us to prove certain localization estimates, and also the Malliavin calculus [M78] corresponding to the hypoelliptic diffusion process. In particular we establish an integration by parts formula for a geometric hypoelliptic diffusion, which extends a corresponding formula established in [B84] for the classical Brownian motion.

Let us now elaborate on the functional integral interpretation of the above techniques, along the lines of [B04, B05]. For $c = 1/b^2$, the dynamics of the diffusion $(x_s, p_s) \in T^*X$ associated to the hypoelliptic Laplacian $2A^2_{\phi,\mathcal{H}^c}$ can be described by the stochastic differential equation

$$\dot{x} = p, \qquad \dot{p} = \left(-p + \dot{w}\right)/b^2, \tag{0.3}$$

where w is a standard Brownian motion. The first order differential system (0.3) can also be written as the second order differential equation on X,

$$\ddot{x} = \left(-\dot{x} + \dot{w}\right)/b^2. \tag{0.4}$$

When $b \to 0$, equation (0.4) degenerates to

$$\dot{x} = \dot{w}. \tag{0.5}$$

In (0.3), p is an Ornstein-Uhlenbeck process, whose trajectories are continuous, x is a so-called physical Brownian motion, and the trajectories of x are C^1. Incidentally observe that p is a Gaussian process with covariance $\exp\left(-\left|t-s\right|/b^2\right)/b^2$. In (0.5), x is a standard Brownian motion, and its trajectories are nowhere differentiable. Now Brownian motion is precisely the process corresponding to the Hodge Laplacian $\square^X/2$. The fact that equation (0.4) degenerates into (0.5) when $b \to 0$ is one of the arguments to justify the convergence of $2A^2_{\phi,\mathcal{H}^c}$ toward $\square^X/2$ when $b \to 0$ at a dynamical level.

The convergence argument of the trajectories in (0.4) to those in (0.5) can indeed be justified. In another form, it was already present in earlier work of Stroock and Varadhan [StV72], where another convergence scheme of the solution of a differential equation to the solution of a stochastic differential equation was given. Such convergence arguments provide the critical link between classical differential calculus and the Itô calculus.

But as explained in [B05], we are asking much more, since we want to understand the functional analytic behavior of the Laplacian $A^2_{\phi,\mathcal{H}^c}$ when $b \to 0$, and this in every degree. Arguments in favor of such a possibility were given in [B05], writing the operator $A^2_{\phi,\mathcal{H}^c}$ as a $(2,2)$ matrix with respect to to a natural splitting of a corresponding Hilbert space.

Our proof of the convergence of $2A^2_{\phi,\mathcal{H}^c}$ to $\square^X/2$ can be thought of as a functional analytic version of the Itô calculus. The analytic difficulties are in part revealing the tormented path connecting a C^1 dynamics for $b > 0$ to a nowhere differentiable dynamics for $b = 0$.

Let us still elaborate on this point from a formal point of view, along the lines of [B04, B05]. Indeed for $b > 0$, the path integral representation for the supertrace $\mathrm{Tr_s}\left[\exp\left(-tA^2_{\phi,\mathcal{H}^c}\right)\right]$ is given by

$$\mathrm{Tr_s}\left[\exp\left(-tA^2_{\phi,\mathcal{H}^c}\right)\right] = \int_{LX} \exp\left(-\frac{1}{2t}\int_0^1 |\dot{x}|^2\, ds - \frac{b^4}{2t^3}\int_0^1 |\ddot{x}|^2\, ds + \dots\right). \tag{0.6}$$

In (0.6), ... represents the fermionic part of the integral. One should be aware of the fact that the process x in (0.4) which corresponds to (0.6) is such that $\frac{1}{2}\int_0^1 |\ddot{x}|^2\, ds = +\infty$.

In (0.6), if we make $b = 0$, in the right-hand side, we recover the standard representation of the Brownian measure, for which $\frac{1}{2}\int_0^1 |\dot{x}|^2\, ds = +\infty$. Making $b = 0$ seems to be an innocuous operation in (0.6), which could be Taylor expanded. The opposite is true. First of all the H^1 norm of $\dot{x}$ is much "bigger" than its H^0 norm. Any perturbative expansion of (0.6) to $b = 0$ will lead to inconsistent divergences. The rigorous process through which one shows the convergence of (0.6) to the corresponding expression with $b = 0$ is much subtler and involves functional analytic arguments, which we now describe in more detail.

The arguments in [B05] show that the convergence of $A^2_{\phi,\mathcal{H}^c}$ to $\square^X/4$ should be obtained by inverting the harmonic oscillator fiberwise. However, this picture provides only the limit view, in which b has already been made equal to 0. Namely, the inverse of the harmonic oscillator should be viewed as a fiberwise pseudodifferential operator, supported on the diagonal of X. For $b > 0$ close to 0, the inverse of the relevant operator is no longer supported over the diagonal of X. A suitably defined version of this inverse can be viewed as a semiclassical pseudodifferential operator on X with semiclassical parameter $h = b$. This semiclassical description is valid only to describe the more and more chaotic behavior of the component $p \in T^*X$ as $b \to 0$ in (0.3). As explained in (0.4), (0.5), as $b \to 0$, the dynamics of x converges to a Brownian motion on X. The obvious implication is that the relevant calculus on operators which will give a precise account of the transition from the dynamics in (0.3) to the Brownian dynamics (0.5) will necessarily have two scales, a semiclassical scale with parameter $h = b$ and an ordinary scale.

Let us point out that we also study the transition from the small time asymptotics of the heat kernel in (0.3) to the corresponding small time asymptotics for the standard heat kernel corresponding to (0.5). This requires proving the required uniform localization in b as $t \to 0$, and also using a two scale pseudodifferential calculus, with semiclassical parameters t, b.

No attempt is made in this book to study the limit $b \to +\infty$, which should concentrate the analysis near the closed geodesics.

We now present three key results which are established in this book. Let $\lambda = \det H^{\cdot}(X, F)$ be the determinant of the cohomology of F, so that λ is a complex line. By proceeding as in [B05], for $b \in \mathbf{R}_+^*, c = 1/b^2$, we construct a generalized metric $\| \|_{\lambda,b}^2$ on the line λ, using in particular the Ray-Singer torsion for $A^2_{\phi,\mathcal{H}^c}$ in the sense of [RS71]. A generalized metric differs from a usual metric in the sense it may have a sign. Let $\|\|\|_{\lambda,0}^2$ be the corresponding classical Ray-Singer metric, associated to the analytic torsion for $\square^X$. The following result is established in Theorem 9.0.1.

Theorem 0.0.1. *Given $b > 0, c = 1/b^2$, we have the identity*

$$\| \|_{\lambda,b}^2 = \| \|_{\lambda,0}^2 . \tag{0.7}$$

More generally, if G is a compact Lie group acting isometrically on the above geometric objects, along the lines of [B95], we can define the logarithm of an equivariant Ray-Singer metric $\log\left(\|\|\|_{\lambda,b}^2\right)$, which one should compare with the equivariant Ray-Singer metric $\log\left(\| \|_{\lambda,0}^2\right)$. Take $g \in G$ and let $X_g \subset X$ be the fixed point manifold of X. Let $\zeta(\theta, s) = \sum_{n=1}^{+\infty} \frac{\cos(n\theta)}{n^s}, \eta(\theta, s) = \sum_{n=1}^{+\infty} \frac{\sin(n\theta)}{n^s}$ be the real and imaginary parts of the Lerch function [Le88]. Set

$$^0J(\theta) = \frac{1}{2}\left(\frac{\partial \zeta}{\partial s}(\theta, 0) - \frac{\partial \zeta}{\partial s}(0,0)\right). \tag{0.8}$$

We denote by $e(TX_g)$ the Euler class of TX_g, and by $^0J_g\left(TX|_{X_g}\right)$ the locally constant function on X_g which is associated to the splitting of $TX|_{X_g}$ using the locally constant eigenvalues of g acting on $TX|_{X_g}$. In Theorem 9.0.1, we also establish the following extension of Theorem 0.0.1.

Theorem 0.0.2. *For $g \in G, b > 0, c = 1/b^2$, we have the identity*

$$\log\left(\frac{\| \|_{\lambda,b}^2}{\| \|_{\lambda,0}^2}\right)(g) = 2\int_{X_g} e(TX_g)\, {}^0J_g\left(TX|_{X_g}\right) \mathrm{Tr}^F[g]. \tag{0.9}$$

A more general result is for the torsion forms $\mathcal{T}_{\mathrm{ch},g,b_0}\left(T^HM, g^{TX}, \nabla^F, g^F\right)$ which we define in chapter 6 as analogues in the hypoelliptic case of the analytic torsion forms of Bismut and Lott [BLo95] $\mathcal{T}_{\mathrm{ch},g,0}\left(T^HM, g^{TX}, \nabla^F, g^F\right)$, normalized as in [BG01], which were obtained in the context of standard elliptic theory. The torsion forms $\mathcal{T}_{\mathrm{ch},g,0}\left(T^HM, g^{TX}, \nabla^F, g^F\right)$ are secondary invariants which refine the theorem of Riemann-Roch-Grothendieck for flat vector bundles established in [BLo95] at the level of differential forms. They were constructed using the superconnection formalism of Quillen [Q85b]. We make here a similar construction to obtain the hypoelliptic torsion forms $\mathcal{T}_{\mathrm{ch},g,b_0}\left(T^HM, g^{TX}, \nabla^F, g^F\right)$.

Let us now explain our results on hypoelliptic torsion forms in more detail. We consider indeed a projection $p : M \to S$ with compact fiber X, the flat Hermitian vector bundle $\left(F, \nabla^F, g^F\right)$ is now defined on M, and $T^HM \subset TM$ is a horizontal vector bundle on M. The Lie group G acts along the fibers

X. The equivariant analytic torsion forms $\mathcal{T}_{\mathrm{ch},g,b_0}\left(T^HM, g^{TX}, \nabla^F, g^F\right)$ are smooth even forms on S.

Put

$$J(\theta, x) = \frac{1}{2}\left[\sum_{\substack{p\in\mathbf{N}\\ p\,\mathrm{even}}} \frac{\partial\zeta}{\partial s}(\theta, -p)\frac{x^p}{p!} + i\sum_{\substack{p\in\mathbf{N}\\ p\,\mathrm{odd}}} \frac{\partial\eta}{\partial s}(\theta, -p)\frac{x^p}{p!}\right], \tag{0.10}$$

$$^0J(\theta, x) = J(\theta, x) - J(0,0).$$

The functions $J(\theta, x)$ and $^0J(\theta, x)$ were introduced in [BG01, Definitions 4.21 and 4.25, Theorem 4.35, and Definition 7.3].

Take $g \in G$. Here $^0J_g\left(TX|_{X_g}\right)$ is now a cohomology class on $M_g \subset M$. The class $\widetilde{\mathrm{ch}}_g^{\circ}\left(\nabla^{\mathfrak{H}^{\cdot}(X,F)}, \mathfrak{h}_0^{\mathfrak{H}^{\cdot}(X,F)}, \mathfrak{h}_{b_0}^{\mathfrak{H}^{\cdot}(X,F)}\right) \in \Omega^{\cdot}(S)/d\Omega^{\cdot}(S)$ is defined in equation (8.1.1). It is a secondary class attached to a couple of generalized metrics on $\mathfrak{H}^{\cdot}(X,F) \simeq H^{\cdot}(X,F)$.

We now state a formula comparing the elliptic and the hypoelliptic torsion forms, which is established in Theorem 8.2.1.

Theorem 0.0.3. *For $b_0 > 0, c = 1/b_0^2$ and b_0 small enough, the following identity holds:*

$$\begin{aligned} &-\mathcal{T}_{\mathrm{ch},g,b_0}\left(T^HM, g^{TX}, \nabla^F, g^F\right) + \mathcal{T}_{\mathrm{ch},g,0}\left(T^HM, g^{TX}, \nabla^F, g^F\right) \\ &-\widetilde{\mathrm{ch}}_g^{\circ}\left(\nabla^{\mathfrak{H}^{\cdot}(X,F)}, \mathfrak{h}_0^{\mathfrak{H}^{\cdot}(X,F)}, \mathfrak{h}_{b_0}^{\mathfrak{H}^{\cdot}(X,F)}\right) + \int_{X_g} e(TX_g)\, {}^0J_g\left(TX|_{M_g}\right)\mathrm{Tr}^F[g] = 0 \\ &\qquad\qquad\qquad\qquad \text{in } \Omega^{\cdot}(S)/d\Omega^{\cdot}(S). \end{aligned} \tag{0.11}$$

Note that except for the restriction that b_0 has to be small, Theorems 0.0.1 and 0.0.2 follow from Theorem 0.0.3.

Let us also point out that in [BL91], given an embedding of compact complex Kähler manifolds $i : Y \to X$, and a resolution of a holomorphic vector bundle η on Y by a holomorphic complex of vector bundles (ξ, v) on X, we gave a local formula for the ratio of the Quillen metrics on the line $\det H^{0,\cdot}(Y,\eta) \simeq \det H^{0,\cdot}(X,\xi)$. This problem seems to be of a completely different nature from the one which is being considered here. In particular all the operators considered in [BL91] are self-adjoint. Still from a certain point of view, the structures of the proofs are very similar, probably because of the underlying path integrals, which are very similar in both cases.

The book is organized as follows. In chapter 1, we describe the results obtained by Bismut and Lott [BLo95] and Bismut and Goette [BG01] in the context of classical Hodge theory. In particular we recall the construction in [BLo95] of the analytic elliptic torsion forms, which are obtained by transgression of certain elliptic odd Chern forms, and we describe various properties of Ray-Singer metrics on the line $\det H^{\cdot}(X,F)$.

In chapter 2, we recall the construction given in [B05] of a deformation of classical Hodge theory on a Riemannian manifold X, whose Laplacian $A^2_{\phi,\mathcal{H}^c}$ is a hypoelliptic operator on T^*X, this theory being also developed in the context of families. Also we give the general set up which will ultimately permit us to establish the above three results.

In chapter 3, given $b > 0$, we discuss the Hodge theory for the hypoelliptic Laplacian, and we summarize the main properties of its heat kernel. We discuss in detail the spectral theory of $A^2_{\phi,\mathcal{H}^c}$ and the behavior of the spectrum as $b \to 0$. We show that for $b > 0$, the spectrum is discrete and conjugation-invariant. We prove that for $b > 0$ small enough, the results of classical Hodge theory still hold, and also that except for the 0 eigenvalue, the other eigenvalues have a positive real part and remain real at finite distance. Also we prove that the set of $b > 0$ such that the Hodge theorem does not hold is discrete. The bulk of the analytic arguments used in this chapter is taken from the key chapters 15 and 17.

In chapter 4, we construct hypoelliptic odd Chern forms, which depend on two parameters, $b > 0, t > 0$, with $c = \pm 1/b^2$. Also we show that their asymptotics as $t \to 0$ coincide with the asymptotics of the corresponding elliptic odd Chern forms. These results are obtained using a new version of the Getzler rescaling of Clifford variables [G86] in the context of hypoelliptic operators. The arguments of localization are obtained using probabilistic methods and arguments from chapter 14. Let us also point out that in [L05], one of us has studied in detail the asymptotics of the hypoelliptic heat kernel on functions, also outside the diagonal, and obtained a corresponding large deviation principle, in which the action considered in the formal representation (0.6) ultimately appears in an exponentially small term as $t \to 0$. Alternative localization techniques are given in chapters 15 and 17. These techniques will play an essential role when studying the combined asymptotics for the heat kernel as $b \to 0, t \to 0$.

In chapter 5, we study the behavior of the hypoelliptic odd Chern forms when $t \to +\infty$ or $b \to 0$. We study in particular the uniformity of the convergence.

In chapter 6, using the results of chapters 4 and 5, for $b > 0$ small enough, we construct the corresponding analytic hypoelliptic torsion forms, which are obtained by transgression of the hypoelliptic odd Chern forms, and we construct corresponding hypoelliptic Ray-Singer metrics for any b. The elliptic and hypoelliptic torsion forms verify similar transgression equations, which makes plausible Theorem 0.0.3, which asserts essentially that their difference is topological. Also we show that the hypoelliptic Ray-Singer metrics does not depend on b.

In chapter 7, we compute the hypoelliptic torsion forms which are attached to a vector bundle. This chapter is based on explicit computations involving the harmonic oscillator and Clifford variables. This computation plays a key role in the proof of our final formula.

In chapter 8, we establish our main result, which was stated as Theorem 0.0.3, where we give a formula comparing the hypoelliptic to the elliptic torsion forms. The proof is based on a series of intermediate results, whose proofs are themselves deferred to chapters 10-13.

In chapter 9, we prove Theorems 0.0.1 and 0.0.2, i.e., we give a formula comparing the elliptic and hypoelliptic Ray-Singer metrics.

In chapter 10, given a cohomology class, we calculate the asymptotic expansion of the corresponding suitably rescaled harmonic forms as $b \to 0$.

In chapter 11, we give the proof of an intermediate result associated with the smooth kernel for $\exp\left(-tA^2_{\phi,\mathcal{H}^c}\right)$ when $b \simeq \sqrt{t}$.

In chapter 12, we get uniform bounds on the heat kernel when $b \in \left[\sqrt{t}, b_0\right]$, with $t \in]0,1]$, and $b_0 > 0$.

In chapter 13, we study the heat kernel for $A^2_{\phi,\mathcal{H}^c}$ in the range $b \in]0, \sqrt{t}], t \in]0,1]$. Note here that local index methods are also developed in chapters 12 and 13.

In chapter 14, we establish an integration by parts formula for the hypoelliptic diffusion, in the context of the Malliavin calculus [M78]. Some of the objects which appear there are the concrete manifestation of the dreams described in [B05].

Chapters 15-17 contain most of the analytic machine used in the book.

In chapter 15, given a fixed $b > 0$, we develop the hypoelliptic estimates for the operator $A^2_{\phi,\mathcal{H}^c}$. The noncompactness of T^*X introduces extra difficulties with respect to Hörmander [Hör67, Hör85]. These are handled using a Littlewood-Paley decomposition of the chapters of the given vector bundles on annuli. We show that the spectrum of $A^2_{\phi,\mathcal{H}^c}$ is included in a region of $\mathbf{C}$ which is limited by a cusplike boundary. Also we study the trace class properties of adequate powers of the resolvent.

In chapter 16, we develop some of the key tools which are needed to study the limit $b \to 0$. Indeed when microlocalizing this asymptotics, we are essentially back to the case of a flat manifold. In the case of flat tori, it was shown in [B05, subsection 3.10] that the hypoelliptic operator $A^2_{\phi,\mathcal{H}^c}$ is essentially isospectral to $\square^X/4$. In particular the spectrum of $A^2_{\phi,\mathcal{H}^c}$ is real. Still the method used in chapter 17 to study the limit $b \to 0$ consists in writing our operator as a $(2,2)$ matrix. Even in the case of the torus, this method is nontrivial. The asymptotics as $b \to 0$ of the matrix component are determined by a function $J_0(y,\lambda), (y,\lambda) \in \mathbf{R} \times \mathbf{C}$, whose behavior is studied in detail. The Bargman representation of the harmonic oscillator in terms of bosonic creation and annihilation operators plays a key role in the analysis.

Finally, in chapter 17, we study the asymptotics of the resolvent of the operator $A^2_{\phi,\mathcal{H}^c}$ as $b \to 0$. This chapter is technically difficult. Its purpose is to give a detailed analysis of the behavior of the resolvent of $A^2_{\phi,\mathcal{H}^c}$ as $b \to 0$. This means that the hypoelliptic estimates of chapter 15 have to be combined with the computation of the resolvent as a $(2,2)$ matrix. Here, in the hypoelliptic analysis, as in chapter 15, we use Kohn's method of proof [Ko73] of Hörmander's theorem [Hör67] to get a global estimate with a gain of $1/4$ derivative, and a parametrix construction in which we use a subelliptic estimate with a gain of $2/3$ derivatives in appropriate function spaces. One should observe here that this subelliptic estimate is not optimal for large $|p|$, but that in the $(2,2)$ matrix calculus, projection on the kernel of the fiberwise harmonic oscillator which appears in (0.2) compensates for that. Optimal hypoelliptic estimates have been obtained by one of us in [L06]. In chapter 17, we also study the behavior of the heat kernel when $b \to 0, t \to 0$.

In the text, to make the book more readable, we often use results of chapters 15-17, referring to those chapters for the complete proofs. This is the case in particular in chapter 13. In principle, except for notation, the various chapters in the book can be read independently, with the help of the index of notation which is given at the end of the book.

In the whole book, the positive constants C which appear in our estimates can vary from line to line, even when the same notation is used for them. Also in many cases, when dependence on parameters is crucial, the parameters on which they depend are noted as subscripts.

The results contained in this book were announced in [BL05].

In the whole book, if $\mathcal{A}$ is a $\mathbf{Z}_2$-graded algebra, if $a, a' \in \mathcal{A}$, we denote by $[a, a']$ their supercommutator.

The authors would like to thank Lucy Day Werts Hobor for her kind help in the preparation of the final version of the book.

Chapter One

Elliptic Riemann-Roch-Grothendieck and flat vector bundles

The purpose of this chapter is to recall the results on elliptic analytic torsion forms obtained by Bismut and Lott [BLo95] and later extended by Bismut and Goette [BG01] to the equivariant context.

This chapter is organized as follows. In section 1.1, we state elementary results on Clifford algebras.

In section 1.2, we recall some basic results of standard Hodge theory.

In section 1.3, we give a short account of the construction of the Levi-Civita superconnection in the context of [BLo95].

In section 1.4, we review the relations of this construction to Poincaré duality.

In section 1.5, we introduce a group action on the considered manifold.

In section 1.6, we give elementary results on Lefschetz formulas.

In section 1.7, we state the the Riemann-Roch-Grothendieck theorem for flat vector bundles of [BLo95].

In section 1.8, we explain the construction of the analytic torsion forms of [BLo95].

In section 1.9, we give the relevant formulas for the Chern analytic torsion forms of [BG01], which are simple modifications of the forms in [BLo95].

In section 1.10, we describe the behavior of the analytic torsion forms under Poincaré duality.

In section 1.11, we briefly review the construction of certain secondary classes for flat vector bundles.

Finally, in section 1.12, we describe the determinant of the cohomology of a flat vector bundle and the construction of corresponding Ray-Singer metrics via the Ray-Singer analytic torsion.

1.1 THE CLIFFORD ALGEBRA

Let V be a real Euclidean vector space. We identify V and V^* by the scalar product of V. If $U \in V$, let $U^* \in V^*$ correspond to V by the metric.

Let $c(V)$ be the Clifford algebra of V. Then $c(V)$ is spanned by $1, U \in V$, with the commutation relations

$$UU' + U'U = -2\langle U, U' \rangle . \tag{1.1.1}$$

If $U \in V$, set

$$c(U) = U^* \wedge -i_U, \qquad \widehat{c}(U) = U^* \wedge +i_U. \tag{1.1.2}$$

Then $c(U), \widehat{c}(U)$ lie in $\mathrm{End}^{\mathrm{odd}}(\Lambda(V^*))$. Moreover, if $U, U' \in V$,

$$\begin{aligned} &[c(U), c(U')] = -2\langle U, U'\rangle, \qquad [\widehat{c}(U), \widehat{c}(U')] = 2\langle U, U'\rangle, \\ &[c(U), \widehat{c}(U')] = 0. \end{aligned} \tag{1.1.3}$$

By (1.1.3),

$$U^* \wedge = \frac{1}{2}(\widehat{c}(U) + \widehat{c}(U)), \qquad i_U = \frac{1}{2}(\widehat{c}(U) - c(U)). \tag{1.1.4}$$

Let $E = E_+ \oplus E_-$ be a $\mathbf{Z}_2$-graded finite dimensional vector space, and let τ be the involution defining the grading, i.e., $\tau = \pm 1$ on $E_\pm$. The algebra $\mathrm{End}(E)$ is $\mathbf{Z}_2$-graded, its even (resp. odd) elements commuting (resp. anticommuting) with τ.

If $A \in \mathrm{End}(E)$, we define its supertrace $\mathrm{Tr_s}[A]$ by the formula

$$\mathrm{Tr_s}[A] = \mathrm{Tr}[\tau A]. \tag{1.1.5}$$

Of course, the definition of the supertrace extends to the case where E is infinite dimensional, as long as A is trace class.

Let F be another vector space. Then the exterior algebra $\Lambda(F^*)$ is also a $\mathbf{Z}_2$-graded algebra. Let $\Lambda^{\cdot}(F^*)\widehat{\otimes}\mathrm{End}(E)$ be the $\mathbf{Z}_2$-graded tensor product of the algebras $\Lambda^{\cdot}(F^*)$ and $\mathrm{End}(E)$.

As in [Q85b], we extend $\mathrm{Tr_s}$ to a map from $\Lambda^{\cdot}(F^*)\widehat{\otimes}\mathrm{End}(E)$ into $\Lambda^{\cdot}(F^*)$, with the convention that if $\alpha \in \Lambda^{\cdot}(F^*), A \in \mathrm{End}(E)$,

$$\mathrm{Tr_s}[\alpha A] = \alpha \mathrm{Tr_s}[A]. \tag{1.1.6}$$

A basic fact [Q85b] is that the supertrace of a supercommutator vanishes.

1.2 THE STANDARD HODGE THEORY

Let X be a compact manifold of dimension n. Let (F, ∇^F) be a complex flat vector bundle on X, so that ∇^F is the corresponding flat connection of F. Let $(\Omega^{\cdot}(X,F), d^X)$ be the de Rham complex of smooth sections of $\Lambda^{\cdot}(T^*X)\widehat{\otimes}F$, equipped with the de Rham map d^X. Let $H^{\cdot}(X,F)$ be the cohomology of this complex. Then $H^{\cdot}(X,F)$ is a finite dimensional $\mathbf{Z}$-graded vector space.

Let g^{TX} be a Riemannian metric on X, let g^F be a Hermitian metric on F. Let dv_X be the volume on X attached to g^{TX}. Let $\langle\,\rangle_{\Lambda^{\cdot}(T^*X)\widehat{\otimes}F}$ be the Hermitian product on $\Lambda^{\cdot}(T^*X)\widehat{\otimes}F$ which is associated to g^{TX}, g^F. We equip $\Omega^{\cdot}(X,F)$ with the Hermitian product $g^{\Omega^{\cdot}(X,F)}$ defined by

$$\langle s, s'\rangle_{g^{\Omega^{\cdot}(X,F)}} = \int_X \langle s, s'\rangle_{\Lambda^{\cdot}(T^*X)\otimes F}\, dv_X. \tag{1.2.1}$$

Let d^{X*} be the formal adjoint of d^X with respect to the Hermitian product (1.2.1). Set

$$D^X = d^X + d^{X*}. \tag{1.2.2}$$

Then D^X is a Dirac type operator, and $D^{X,2} = \left[d^X, d^{X*}\right]$ is the corresponding Laplacian, which we denote $\square^X$. Put

$$\mathcal{H}^X = \ker d^X \cap \ker d^{X*}. \tag{1.2.3}$$

Then

$$\mathcal{H}^X = \ker D^X = \ker D^{X,2}. \tag{1.2.4}$$

Moreover, Hodge theory asserts that

$$\mathcal{H}^X \simeq H^{\cdot}(X, F). \tag{1.2.5}$$

By (1.2.5), $H^{\cdot}(X,F)$ inherits a Hermitian product $g^{H^{\cdot}(X,F)}$ from the restriction of $g^{\Omega^{\cdot}(X,F)}$ to $\mathcal{H}^X$.

Put

$$\omega\left(\nabla^F, g^F\right) = \left(g^F\right)^{-1} \nabla^F g^F. \tag{1.2.6}$$

Then $\omega\left(\nabla^F, g^F\right)$ is a smooth 1-form on X with values in self-adjoint elements in $\mathrm{End}(F)$. Set

$$\nabla^{F,u} = \nabla^F + \frac{1}{2}\omega\left(\nabla^F, g^F\right). \tag{1.2.7}$$

Then $\nabla^{F,u}$ is a unitary connection on F, and its curvature R^F is given by

$$R^F = -\frac{1}{4}\omega\left(\nabla^F, g^F\right)^2. \tag{1.2.8}$$

From (1.2.6), we get

$$\nabla^F \omega\left(\nabla^F, g^F\right) = -\omega\left(\nabla^F, g^F\right)^2. \tag{1.2.9}$$

From (1.2.7), (1.2.9), we obtain

$$\nabla^{F,u}\omega\left(\nabla^F, g^F\right) = 0. \tag{1.2.10}$$

This expresses the fact that $\nabla^{F,u}_A \omega\left(\nabla^F, g^F\right)(B)$ is a symmetric tensor in $A, B \in TX$.

Let ∇^{TX} be the Levi-Civita connection on TX, and let R^{TX} be its curvature. Let $\nabla^{\Lambda^{\cdot}(T^*X)\widehat{\otimes}F}, \nabla^{\Lambda^{\cdot}(T^*X)\widehat{\otimes}F,u}$ be the connections on $\Lambda^{\cdot}(T^*X)\widehat{\otimes}F$ induced by ∇^{TX} and $\nabla^F, \nabla^{F,u}$.

Let $e_1, \ldots, e_n$ be a locally defined smooth orthonormal basis of TX. By [BZ92, Proposition 4.12],

$$D^X = \sum_1^n c(e_i)\nabla^{\Lambda^{\cdot}(T^*X)\widehat{\otimes}F,u}_{e_i} - \frac{1}{2}\sum_{i=1}^n \widehat{c}(e_i)\,\omega\left(\nabla^F, g^F\right)(e_i). \tag{1.2.11}$$

Let Δ^H be the horizontal Laplacian acting on $\Omega^{\cdot}(X,F)$. Then when acting on $\Omega^{\cdot}(X,F)$,

$$\Delta^H = \sum_{i=1}^n \nabla^{\Lambda^{\cdot}(T^*X)\otimes F,2}_{e_i} - \nabla^{\Lambda^{\cdot}(T^*X)\otimes F}_{\sum_{i=1}^n \nabla^{TX}_{e_i} e_i}. \tag{1.2.12}$$

Observe that Δ^H is not self-adjoint, except when g^F is flat. Similarly, we can defined the self-adjoint Laplacian $\Delta^{H,u}$ by replacing $\nabla^{\Lambda^\cdot(T^*X)\widehat{\otimes}F}$ by $\nabla^{\Lambda^\cdot(T^*X)\widehat{\otimes}F,u}$.

In the sequel, we use Einstein's summation conventions. Let $e_1,\dots,e_n$ be an orthonormal basis of TX, let $e^1,\dots,e^n$ be the corresponding dual basis of T^*X. The Weitzenböck formula says that

$$\begin{aligned}\square^X &= -\Delta^H + \left\langle R^{TX}(e_i,e_j)e_k,e_l\right\rangle e^i i_{e_j} e^k i_{e_l}\\ &\quad - \omega\left(\nabla^F,g^F\right)(e_i)\nabla_{e_i}^{\Lambda^\cdot(T^*X)\otimes F} - e^i i_{e_j}\nabla^F_{e_i}\omega\left(\nabla^F,g^F\right)(e_j). \quad (1.2.13)\end{aligned}$$

Let S^X be the Ricci tensor of X. Using the circular symmetry of R^{TX} as in [B05, eq. (3.48)], we can rewrite (1.2.13) in the form

$$\begin{aligned}\square^X &= -\Delta^{H,u} + \left\langle S^X e_i,e_j\right\rangle e^i i_{e_j} - \frac{1}{2}\left\langle R^{TX}(e_i,e_j)e_k,e_l\right\rangle e^i e^j i_{e_k} i_{e_l}\\ &+\frac{1}{2}\nabla^{F,u}_{e_i}\omega\left(\nabla^F,g^F\right)(e_i) + \frac{1}{4}\omega\left(\nabla^F,g^F\right)^2(e_i) - \nabla^F_{e_i}\omega\left(\nabla^F,g^F\right)(e_j)e^i i_{e_j}.\end{aligned} \quad (1.2.14)$$

In (1.2.14), $\nabla^{F,u}_{e_i}\omega\left(\nabla^F,g^F\right)(e_i)$ can be replaced by $\nabla^F_{e_i}\omega\left(\nabla^F,g^F\right)(e_i)$.

1.3 THE LEVI-CIVITA SUPERCONNECTION

Now we summarize the main results of Bismut and Lott [BLo95] in the context of families. Our summary will necessarily be brief. We refer to [BLo95] for more details.

Let M,S be smooth manifolds. Let $p: M\to S$ be a smooth submersion with compact fiber X of dimension n. Let T^HM be a horizontal vector bundle on M, so that $TM = T^HM\oplus TX$. Let g^{TX} be a Euclidean metric on TX. Let $\left(F,\nabla^F\right)$ be a flat vector bundle on M, let g^F be a Hermitian metric on F. We still define $\omega\left(\nabla^F,g^F\right)$ on M as in (1.2.5). Let $P^{TX}: TM\to TX$ be the projection associated to the splitting $TM = T^HM\oplus TX$. If $U\in TS$, let $U^H\in T^HM$ be the horizontal lift of U.

In [B86, section 1], a Euclidean connection ∇^{TX} on TX was constructed, which is canonically attached to $\left(T^HM,g^{TX}\right)$. This connection restricts to the Levi-Civita connection along the fibers X.

Let R^{TX} be the curvature of ∇^{TX}. A tensor T was obtained in [B86, section 1], which is a 2-form on M with values in TX, which vanishes identically on $TX\times TX$. Let $U,V\in TS, A\in TX$. Then by [B97, Theorem 1.1],

$$T\left(U^H,V^H\right) = -P^{TX}\left[U^H,V^H\right],\quad T\left(U^H,A\right) = \frac{1}{2}\left(g^{TX}\right)^{-1}\left(L_{U^H}g^{TX}\right)A. \quad (1.3.1)$$

In particular, if $U\in TS, A,B\in TX$,

$$\left\langle T\left(U^H,A\right),B\right\rangle = \left\langle T\left(U^H,B\right),A\right\rangle. \quad (1.3.2)$$

Let $\left(\Omega^\cdot(X,F),d^X\right)$ be the fiberwise de Rham complex of forms with coefficients in F. Then $\Omega^\cdot(X,F)$ is a $\mathbf{Z}$-graded vector bundle on S. We equip

$\Omega^{\cdot}(X,F)$ with the L^2 Hermitian product $g^{\Omega^{\cdot}(X,F)}$ associated to g^{TX}, g^F, which was defined in (1.2.1). In [BLo95], Bismut and Lott constructed a superconnection A and an odd section B of $\Lambda^{\cdot}(T^*S)\widehat{\otimes}\operatorname{End}(\Omega^{\cdot}(X,F))$ on $\Omega^{\cdot}(X,F)$, which are canonically associated to $\left(T^HM, g^{TX}, g^F\right)$, and such that

$$A^2 = -B^2. \tag{1.3.3}$$

The superconnection A is a special case of the Levi-Civita superconnection of [B86] which is used in the proof of a local version of the Atiyah-Singer index theorem for families. The construction of B uses in particular the flat superconnection A' on $\Omega^{\cdot}(X,F)$, which is just the total de Rham operator on M.

Set $m = \dim S$. Let $f_1,\ldots,f_m$ be a basis of TS, let $f^1,\ldots,f^m$ be the corresponding dual basis of T^*S.

Let ${}^1\nabla^{\Lambda^{\cdot}(T^*X)\widehat{\otimes}F}$ be the connection on $\Lambda^{\cdot}(T^*S)\widehat{\otimes}\Lambda^{\cdot}(T^*X)$ along the fibers X:

$${}^1\nabla_{\cdot}^{\Lambda^{\cdot}(T^*X)\widehat{\otimes}F} = \nabla_{\cdot}^{\Lambda^{\cdot}(T^*X)\widehat{\otimes}F} + \left\langle T\left(f_\alpha^H, e_i\right), \cdot\right\rangle f^\alpha c\left(e_i\right) + \left\langle T^H, \cdot\right\rangle. \tag{1.3.4}$$

Let ${}^1\nabla^{\Lambda^{\cdot}(T^*X)\widehat{\otimes}F,u}$ be the connection taken as before, replacing ∇^F by $\nabla^{F,u}$.

We use the notation of section 1.1. Also we still assume that $e_1,\ldots,e_n$ is an orthonormal basis of TX. Bismut and Lott [BLo95, Theorem 3.11] gave a Weitzenböck formula for the curvature A^2 of the superconnection A. The following version of this formula was given in [B05, Theorem 4.55].

Theorem 1.3.1. *The following identity holds:*

$$\begin{aligned} A^2 = \frac{1}{4}\Big(-{}^1\nabla_{e_i}^{\Lambda^{\cdot}(T^*X)\widehat{\otimes}F,2} + \left\langle e_i, R^{TX}e_j\right\rangle \widehat{c}(e_i)\widehat{c}(e_j)\Big) \\ + \frac{1}{4}\left\langle R^{TX}\left(\cdot, e_i\right)e_i, e_j\right\rangle \widehat{c}(e_j) - \frac{1}{4}\nabla^F\omega\left(\nabla^F, g^F\right)(e_i)\widehat{c}(e_i) \\ - \frac{1}{4}\omega\left(\nabla^F, g^F\right)(e_i)\,{}^1\nabla_{e_i}^{\Lambda^{\cdot}(T^*X)\widehat{\otimes}F} - \frac{1}{4}\omega\left(\nabla^F, g^F\right)^2. \end{aligned} \tag{1.3.5}$$

1.4 SUPERCONNECTIONS AND POINCARÉ DUALITY

We briefly summarize the results obtained in [BLo95, subsection 2 (g)] on the behavior of A, B under Poincaré duality. We will write the objects we just considered with a superscript F, to emphasize their dependence on F.

Let $o(TX)$ be the orientation bundle of TX. Let $*^X$ be the Hodge operator associated to g^{TX}. Let $\nu : \Omega^{\cdot}(X,F) \to \Omega^{n-\cdot}\left(X, \overline{F}^* \otimes o(TX)\right)$ be such that if $s \in \Omega^i(X,F)$, then

$$\nu s = (-1)^{i(i+1)/2+ni} *^X s. \tag{1.4.1}$$

Then

$$\nu^2 = (-1)^{n(n-1)/2}. \tag{1.4.2}$$

By [BLo95, eq. (2.106)] and by (1.4.2),

$$A^F = (-1)^n \nu^{-1} A^{\overline{F}^*\otimes o(TX)}\nu, \quad B^F = -(-1)^n \nu^{-1} B^{\overline{F}^*\otimes o(TX)}\nu. \tag{1.4.3}$$

1.5 A GROUP ACTION

Let G be a compact Lie group. We assume that G acts on M and preserves the fibers X, the vector bundle T^HM, and also that the metric g^{TX} is G-invariant. Also we suppose that the action of G on M lifts to F, and preserves the flat connection ∇^F, and the metric g^F.

Clearly G acts on $\Omega^{\cdot}(X,F)$, so that if $s \in \Omega^{\cdot}(X,F)$,

$$(gs)(x) = g.s\left(g^{-1}x\right). \tag{1.5.1}$$

The action of G on $\Omega^{\cdot}(X,F)$ induces a corresponding action on $H^{\cdot}(X,F)$.

The constructions we described before are obviously G-invariant. So the operators which we described before commute with G.

Let M_g be the fixed point set of g in M. Then M_g is a smooth submanifold of M, which fibers on S, with compact fiber X_g, the fixed point set of g in X. Then X_g is a totally geodesic submanifold of X. Let g^{TX_g} be the restriction of g^{TX} to TX_g. Clearly,

$$T^HM|_{M_g} \subset TM_g, \tag{1.5.2}$$

i.e., the restriction of T^HM to M_g defines a horizontal subbundle T^HM_g on M_g.

1.6 THE LEFSCHETZ FORMULA

We make the same assumptions as in sections 1.2 and 1.5 and we use the corresponding notation. It is enough here to consider the case of a single fiber X.

Take $g \in G$. We define the Lefschetz number $\chi_g(F)$ by the formula

$$\chi_g(F) = \mathrm{Tr_s}^{H^{\cdot}(X,F)}[g]. \tag{1.6.1}$$

Let $N_{X_g/X}$ be the orthogonal bundle to TX_g in $TX|_{X_g}$. Set

$$\ell = \dim X_g. \tag{1.6.2}$$

Let $e(TX_g) \in H^{\cdot}(M_g, \mathbf{Q})$ be the Euler class of TX_g. Recall that g acts as a flat automorphism of $F|_{M_g}$. Then the Lefschetz fixed point formula asserts that

$$\chi_g(F) = \int_{X_g} e(TX_g)\,\mathrm{Tr}^F[g]. \tag{1.6.3}$$

Of course g acts on $o(TX)$. Moreover, on X_g, the action of g on $o(TX)$ is given by

$$g|_{o(TX)} = (-1)^{n-\ell}. \tag{1.6.4}$$

Set

$$L_+(g) = \chi_g(F), \qquad L_-(g) = (-1)^n \chi_g(F \otimes o(TX)). \tag{1.6.5}$$

By Poincaré duality,

$$L_+(g) = L_-(g). \tag{1.6.6}$$

Whenever necessary, we will write $L(g)$ instead of $L_\pm(g)$. Note that since $e(TX_g)$ is nonzero only if ℓ is even, (1.6.6) is compatible with (1.6.3), (1.6.4).

1.7 THE RIEMANN-ROCH-GROTHENDIECK THEOREM

We make the same assumptions as in sections 1.3 and 1.5 and we use the corresponding notation. We define the 1-form $\omega\left(\nabla^F, g^F\right)$ as in (1.2.5).

In the sequel, we set

$$h(x) = xe^{x^2}. \tag{1.7.1}$$

Let φ be the endomorphism of $\Lambda^{\cdot}(T^*M)$ given by $\alpha \to (2\pi)^{-\deg\alpha/2}\alpha$. Note here that with respect to the conventions of [BLo95] and in [BG01], the normalizing factor is now 2π instead of $2i\pi$. Our conventions fit instead with the conventions in [BG04]. Of course we extend the definition of φ to any manifold.

Take $g \in G$. By [BG01, Proposition 3.7], whose proof uses in particular (1.5.2), the connection ∇^{TX} preserves TX_g. The restriction of ∇^{TX} to TX_g is just the Euclidean connection ∇^{TX_g} on TX_g which is canonically attached to $\left(T^H M_g, g^{TX_g}\right)$. Let R^{TX_g} be the curvature of ∇^{TX_g}. Let $e\left(TX_g, \nabla^{TX_g}\right)$ be the closed Euler form in Chern-Weil theory, which represents the Euler class of TX_g associated to the Euclidean connection ∇^{TX_g}. Then

$$\begin{aligned} e\left(TX_g, \nabla^{TX_g}\right) &= \operatorname{Pf}\left[\frac{R^{TX_g}}{2\pi}\right] \text{ if } \dim X_g \text{ is even,} \\ &= 0 \text{ if } \dim X_g \text{ is odd.} \end{aligned} \tag{1.7.2}$$

Then $e(TX_g)$ is the cohomology class of $e\left(TX_g, \nabla^{TX_g}\right)$.

Let $h_g\left(\nabla^F, g^F\right)$ be the odd form on M_g,

$$h_g\left(\nabla^F, g^F\right) = (2\pi)^{1/2}\varphi\operatorname{Tr}^F\left[gh\left(\omega\left(\nabla^F, g^F\right)/2\right)\right]. \tag{1.7.3}$$

By [BLo95, Theorems 1.8 and 1.11] and [BG01, Theorem 1.8], the form $h_g\left(\nabla^F, g^F\right)$ is closed, and its cohomology class does not depend on the metric g^F. This class will be denoted $h_g\left(\nabla^F\right)$. Note that

$$h_g\left(\nabla^{\overline{F}^*}, g^{\overline{F}^*}\right) = -h_g\left(\nabla^F, g^F\right). \tag{1.7.4}$$

Recall that A is a superconnection on $\Omega^{\cdot}(X,F)$, and that B is an odd section of $\Lambda^{\cdot}(T^*S)\widehat{\otimes}\operatorname{End}(\Omega^{\cdot}(X,F))$. First we state a result established in [BLo95, Theorem 3.15] and in [BG01, Proposition 3.22]. Recall that $\chi_g(F)$ is a locally constant function on S. The heat kernel $\exp\left(-A^2\right)$ is fiberwise trace class. Now we use the formalism of section 1.1. The supertrace $\operatorname{Tr_s}\left[g\exp\left(-A^2\right)\right]$ is a smooth even form on S.

Proposition 1.7.1. *We have the identity*

$$\operatorname{Tr_s}\left[g\exp\left(-A^2\right)\right] = \chi_g(F). \tag{1.7.5}$$

Recall that A' is the flat superconnection on $\Omega^{\cdot}(X,F)$ which was used in [BLo95] to define A and B. As explained in section 1.3, A' is just the de Rham operator on the total space of M.

Definition 1.7.2. Put

$$h_g\left(A', g^{\Omega^\cdot(X,F)}\right) = (2\pi)^{1/2}\,\varphi\mathrm{Tr_s}\left[gh\left(B\right)\right]. \tag{1.7.6}$$

The forms in (1.7.6) are called elliptic odd Chern forms.

By (1.4.3), we get

$$h_g\left(A', g^{\Omega^\cdot(X,F)}\right) = (-1)^{n+1}\,h_g\left(A', g^{\Omega^\cdot(X,\overline{F}^*\otimes o(TX))}\right). \tag{1.7.7}$$

From (1.7.7), we deduce in particular that if the metric g^F is flat, X is oriented, n is even, and g preserves the orientation

$$h_g\left(A', g^{\Omega^\cdot(X,F)}\right) = 0. \tag{1.7.8}$$

For $t > 0$, we replace the metric g^{TX} by $g_t^{TX} = g^{TX}/t$. Here $g_t^{\Omega^\cdot(X,F)}$ denotes the Hermitian product on $\Omega^\cdot(X,F)$ in (1.2.1) which is associated to g_t^{TX}, g^F. We denote by A_t, B_t the objects we just considered, which are associated to g_t^{TX}.

For $a \geq 0$, let $\psi_a : \Lambda^\cdot(T^*S) \to \Lambda^\cdot(T^*S)$ be given by

$$\psi_a\kappa = a^{\deg\kappa/2}\kappa. \tag{1.7.9}$$

Let N be the number operator of $\Omega^\cdot(X,F)$, i.e., the operator acting by multiplication by k on $\Omega^k(X,F)$. For $t > 0$, set

$$C_t = t^{N/2}A_t t^{-N/2}, \qquad D_t = t^{N/2}B_t t^{-N/2}. \tag{1.7.10}$$

Then by the results in [BLo95], we get

$$C_t = \psi_t^{-1}\sqrt{t}A_t\psi_t, \qquad D_t = \psi_t^{-1}\sqrt{t}B_t\psi_t. \tag{1.7.11}$$

Recall that $H^\cdot(X,F)$ is a $\mathbf{Z}$-graded vector bundle on S, equipped with the flat Gauss-Manin connection $\nabla^{H^\cdot(X,F)}$, and with the metric $g^{H^\cdot(X,F)}$ defined after (1.2.5). We define the odd closed form $h_g\left(\nabla^{H^\cdot(X,F)}, g^{H^\cdot(X,F)}\right)$ as in (1.7.3), by simply replacing Tr by $\mathrm{Tr_s}$, so that this form is simply the alternate sum of the corresponding forms for $H^i(X,F)$.

For $t > 0$, let α_t be a smooth form on S. We will write that as $t \to 0$, $\alpha_t = \mathcal{O}\left(\sqrt{t}\right)$ if for any compact $K \subset S$, and $m \in \mathbf{N}$, the sup over K of the derivatives of order $\leq m$ is dominated by $C_{K,m}\sqrt{t}$. A similar notation will be used when $t \to +\infty$.

Now we state a result established in [BLo95, Theorems 3.16 and 3.17] and in [BG01, Theorems 3.24 and 3.25].

Theorem 1.7.3. *The forms $h_g\left(A', g_t^{\Omega^\cdot(X,F)}\right)$ are odd, closed, and their cohomology class does not depend on $t > 0$. Moreover, as $t \to 0$,*

$$h_g\left(A', g_t^{\Omega^\cdot(X,F)}\right) = \int_{X_g} e\left(TX_g, \nabla^{TX_g}\right) h_g\left(\nabla^F, g^F\right) + \mathcal{O}\left(\sqrt{t}\right). \tag{1.7.12}$$

As $t \to +\infty$,

$$h_g\left(A', g_t^{\Omega^\cdot(X,F)}\right) = h_g\left(\nabla^{H^\cdot(X,F)}, g^{H^\cdot(X,F)}\right) + \mathcal{O}\left(1/\sqrt{t}\right). \tag{1.7.13}$$

In particular,

$$h_g\left(\nabla^{H^\cdot(X,F)}\right) = \int_{X_g} e\left(TX_g\right) h_g\left(\nabla^F\right) \ in\ H^{\mathrm{odd}}\left(S, \mathbf{R}\right). \tag{1.7.14}$$

Definition 1.7.4. For $t > 0$, set

$$h_g^{\wedge}\left(A', g^{\Omega^{\cdot}(X,F)}\right) = \varphi \mathrm{Tr_s}\left[\frac{N}{2} g h'(B)\right]. \tag{1.7.15}$$

Using (1.7.5) and proceeding as in (1.7.7), we get

$$h_g^{\wedge}\left(A', g^{\Omega^{\cdot}(X,F)}\right) + (-1)^n h_g^{\wedge}\left(A', g^{\Omega^{\cdot}(X,\overline{F}^* \otimes o(TX))}\right) = \frac{n}{2}\chi_g(F). \tag{1.7.16}$$

Put

$$\chi'_g(F) = \sum_{j=0}^{m} (-1)^j j \mathrm{Tr}^{H^j(X,F|x)}[g]. \tag{1.7.17}$$

Then $\chi'_g(F)$ is also a locally constant function on S.

Now we recall the results established in [BLo95, Theorems 3.20 and 3.21] and in [BG01, Theorems 3.29 and 3.30].

Theorem 1.7.5. *The form $h_g^{\wedge}\left(A', g_t^{\Omega^{\cdot}(X,F)}\right)$ is even. Moreover,*

$$\frac{\partial}{\partial t} h_g\left(A', g_t^{\Omega^{\cdot}(X,F)}\right) = d\frac{h_g^{\wedge}\left(A', g_t^{\Omega^{\cdot}(X,F)}\right)}{t}. \tag{1.7.18}$$

As $t \to 0$,

$$h_g^{\wedge}\left(A', g_t^{\Omega^{\cdot}(X,F)}\right) = \frac{n}{4}\chi_g(F) + \mathcal{O}\left(\sqrt{t}\right). \tag{1.7.19}$$

As $t \to +\infty$,

$$h_g^{\wedge}\left(A', g_t^{\Omega^{\cdot}(X,F)}\right) = \frac{1}{2}\chi'_g(F) + \mathcal{O}\left(1/\sqrt{t}\right). \tag{1.7.20}$$

1.8 THE ELLIPTIC ANALYTIC TORSION FORMS

Now we follow [BLo95, subsection 3 (j)] and [BG01, subsection 3.12].

Definition 1.8.1. Set

$$\mathcal{T}_{h,g}\left(T^H M, g^{TX}, \nabla^F, g^F\right) = -\int_0^{+\infty}\left[h_g^{\wedge}\left(A', g_t^{\Omega^{\cdot}(X,F)}\right) - \frac{1}{2}\chi'_g(F) h'(0)\right.$$
$$\left. - \left(\frac{n}{4}\chi_g(F) - \frac{1}{2}\chi'_g(F)\right) h'\left(i\sqrt{t}/2\right)\right]\frac{dt}{t}. \tag{1.8.1}$$

By Theorem 1.7.5, we find that the integral in the right-hand side of (1.8.1) is well-defined. The following result was established in [BLo95, Theorem 3.23] and in [BG01, Theorem 3.32].

Theorem 1.8.2. *The form $\mathcal{T}_{h,g}\left(T^H M, g^{TX}, \nabla^F, g^F\right)$ is even. Moreover,*

$$d\mathcal{T}_{h,g}\left(T^H M, g^{TX}, \nabla^F, g^F\right) = \int_{X_g} e\left(TX_g, \nabla^{TX_g}\right) h_g\left(\nabla^F, g^F\right)$$
$$- h_g\left(\nabla^{H^{\cdot}(X,F)}, g^{H^{\cdot}(X,F)}\right). \tag{1.8.2}$$

The forms $\mathcal{T}_{h,g}\left(T^H M, g^{TX}, \nabla^F, g^F\right)$ are called analytic torsion forms.

Remark 1.8.3. Suppose that the connected components of X_g have odd dimension. This is true if X is orientable, and either X is odd dimensional and g preserves the orientation, or X is even dimensional and g reverses the orientation. If $H^{\cdot}(X,F)=0$, by Theorem 1.8.2, $\mathcal{T}_{h,g}\left(T^H M, g^{TX}, \nabla^F, g^F\right)$ is a closed form on S. Its cohomology class does not depend on $\left(T^H M, g^{TX}, g^F\right)$.

Remark 1.8.4. Let $\left(D^X\right)^{-2}$ be the inverse of $D^{X,2}$ acting on the orthogonal bundle to $\ker D^X$ in $\Omega^{\cdot}(X,F)$. For $s\in\mathbf{C}, \operatorname{Re}(s)>\dim(X)/2$, set

$$\vartheta_g(s) = -\operatorname{Tr_s}\left[N\left(D^{X,2}\right)^{-s}\right]. \tag{1.8.3}$$

Then $\vartheta_g(s)$ extends to a meromorphic function of $s\in\mathbf{C}$, which is holomorphic near $s=0$. By definition, the equivariant Ray-Singer analytic torsion [RS71], [BZ92, BZ94] of the de Rham complex $\left(\Omega^{\cdot}(X,F), d^X\right)$ is given by $\frac{\partial\vartheta_g}{\partial s}(0)$. It was shown in [BLo95, Theorem 3.29] that

$$\mathcal{T}_{h,g}\left(T^H M, g^{TX}, \nabla^F, g^F\right)^{(0)} = \frac{1}{2}\frac{\partial\vartheta_g}{\partial s}(0). \tag{1.8.4}$$

In the sequel, when $g=1$, we will use the notation $\vartheta(s)$ instead of $\vartheta_1(s)$.

For $t>0$, set

$$b_t = \varphi\operatorname{Tr_s}\left[g\left(\frac{N^X}{2}-\frac{n}{4}\right)h'(B_t)\right]. \tag{1.8.5}$$

By (1.7.5) and (1.7.15), we get

$$b_t = h_g^{\wedge}\left(A', g_t^{\Omega^{\cdot}(X,F)}\right) - \frac{n}{4}\chi_g(F). \tag{1.8.6}$$

By (1.7.19), as $t\to 0$,

$$b_t = \mathcal{O}\left(\sqrt{t}\right), \tag{1.8.7}$$

and by (1.7.20), as $t\to+\infty$,

$$b_t = \frac{1}{2}\chi_g'(F) - \frac{n}{4}\chi_g(F) + \mathcal{O}\left(1/\sqrt{t}\right). \tag{1.8.8}$$

Also, (1.8.1) is equivalent to

$$\mathcal{T}_{h,g}\left(T^H M, g^{TX}, \nabla^F, g^F\right) = -\int_0^{+\infty}\Bigg(b_t - \left(\frac{n}{4}\chi_g(F) - \frac{1}{2}\chi_g'(F)\right)\left(h'\left(i\sqrt{t}/2\right)-h'(0)\right)\Bigg)\frac{dt}{t}. \tag{1.8.9}$$

Observe that by [BG01, eq. (9.71)]

$$\begin{aligned}\int_0^1\left(h'\left(i\sqrt{t}/2\right)-h'(0)\right)\frac{dt}{t} + \int_1^{+\infty} h'\left(i\sqrt{t}/2\right)\frac{dt}{t} \\ = \int_0^1\left(e^{-t/4}-1\right)\frac{dt}{t} + \int_1^{+\infty} e^{-t/4}\frac{dt}{t} - \frac{1}{2}\int_0^{+\infty} e^{-t/4}dt \\ = \Gamma'(1) + 2\left(\log(2)-1\right).\end{aligned} \tag{1.8.10}$$

By (1.8.9), (1.8.10), we get

$$\mathcal{T}_{h,g}\left(T^H M, g^{TX}, \nabla^F, g^F\right) = -\int_0^1 b_t \frac{dt}{t} - \int_1^{+\infty} \left(b_t - b_\infty\right) \frac{dt}{t}$$
$$- \left(\Gamma'(1) + 2\left(\log(2) - 1\right)\right)\left(\frac{1}{2}\chi'_g(F) - \frac{n}{4}\chi_g(F)\right). \quad (1.8.11)$$

1.9 THE CHERN ANALYTIC TORSION FORMS

Now we follow [BG01, subsections 2.7 and 3.17], where the appropriate normalization of the analytic torsion forms was established.

If $f(x)$ is holomorphic, put

$$(Ff)(x) = x\int_0^1 f'\left(4s(1-s)x^2\right) ds, \quad Qf(x) = \int_0^1 f\left(4s(1-s)x\right) ds. \quad (1.9.1)$$

Then $Ff(x)$ is an odd function of x. An easy computation given in [BG01, eq. (2.99)] shows that

$$\left(Fe^{\cdot}\right)(x) = \sum_{p=1}^{+\infty} \frac{(p-1)!}{(2p-1)!} 2^{2p-2} x^{2p-1}. \quad (1.9.2)$$

Observe that

$$xe^{x^2} = \sum_{p=1}^{+\infty} \frac{x^{2p-1}}{(p-1)!}. \quad (1.9.3)$$

The coefficient of x^{2p-1} in $\left(Fe^{\cdot}\right)(x)$ is obtained from the corresponding coefficient in the expansion of xe^{x^2} by multiplication by the factor $2^{2p-2}\frac{[(p-1)!]^2}{(2p-1)!}$. We can then define $\left(Fe^x\right)_g\left(\nabla^F, g^F\right)$ as in (1.7.6), by simply replacing the function h by Fe^x. As in [BG01], we will use the notation

$$\mathrm{ch}_g^{\circ}\left(\nabla^F, g^F\right) = \left(Fe^x\right)_g\left(\nabla^F, g^F\right). \quad (1.9.4)$$

Recall that for $a \geq 0$, ψ_a was defined in (1.7.9). Let $Q \in \mathrm{End}\left(\Lambda^{\cdot}\left(T^*S\right)\right)$ be given by

$$Q\alpha = \int_0^1 \psi_{4s(1-s)}\alpha ds. \quad (1.9.5)$$

If $\alpha \in \Lambda^{2p}\left(T^*S\right)$, then

$$Q\alpha = \frac{(p!)^2}{(2p+1)!} 4^p \alpha. \quad (1.9.6)$$

As in [BG01, Definition 3.46], set

$$\mathcal{T}_{\mathrm{ch},g}\left(T^H M, g^{TX}, \nabla^F, g^F\right) = Q\mathcal{T}_{h,g}\left(T^H M, g^{TX}, \nabla^F, g^F\right). \quad (1.9.7)$$

The even forms $\mathcal{T}_{\mathrm{ch},g}\left(T^HM, g^{TX}, \nabla^F, g^F\right)$ are called the Chern analytic torsion forms. In [BG01, Theorem 3.47], it is shown as a consequence of (1.8.2) that

$$d\mathcal{T}_{\mathrm{ch},g}\left(T^HM, g^{TX}, \nabla^F, g^F\right) = \int_{X_g} e\left(TX_g, \nabla^{TX_g}\right)\mathrm{ch}_g^\circ\left(\nabla^F, g^F\right) - \mathrm{ch}_g^\circ\left(\nabla^{H^\cdot(X,F)}, g^{H^\cdot(X,F)}\right). \quad (1.9.8)$$

1.10 ANALYTIC TORSION FORMS AND POINCARÉ DUALITY

As explained in [BLo95, section 2 and Theorem 3.26], the above constructions are compatible with Poincaré duality. In particular by [BG01, eq. 7.24] or by (1.7.16),

$$\mathcal{T}_{h,g}\left(T^HM, g^{TX}, \nabla^{\overline{F}^*\otimes o(TX)}, g^{\overline{F}^*\otimes o(TX)}\right) = (-1)^{n+1}\mathcal{T}_{h,g}\left(T^HM, g^{TX}, \nabla^F, g^F\right). \quad (1.10.1)$$

Of course, a similar identity holds for the Chern analytic torsion forms.

1.11 THE SECONDARY CLASSES FOR TWO METRICS

Let g_0^F, g_1^F be two smooth g-invariant Hermitian metrics on F. In [BLo95, Definition 1.12], [BG01, Definition 1.10], a secondary class $\widetilde{h}_g\left(\nabla^F, g_0^F, g_1^F\right) \in \Omega^\cdot(M_g)/d\Omega^\cdot(M_g)$ was defined such that

$$d\widetilde{h}_g\left(\nabla^F, g_0^F, g_1^F\right) = h_g\left(\nabla^F, g_1^F\right) - h_g\left(\nabla^F, g_0^F\right). \quad (1.11.1)$$

Let $\ell \in [0,1] \to g_\ell^F$ be a smooth family of Hermitian metrics which interpolates between g_0^F and g_1^F. An explicit representative of $\widetilde{h}_g\left(\nabla^F, g_0^F, g_1^F\right)$ is given by

$$\widetilde{h}_g\left(\nabla^F, g_\ell^F\right) = \int_0^1 \varphi\mathrm{Tr}\left[g\frac{1}{2}\left(g_\ell^F\right)^{-1}\frac{\partial g_\ell^F}{\partial \ell}h'\left(\frac{1}{2}\omega\left(\nabla^F, g^F\right)\left(\nabla^F, g_\ell^F\right)\right)\right]d\ell. \quad (1.11.2)$$

The class of $\widetilde{h}_g\left(\nabla^F, g_\ell^F\right)$ in $\Omega^\cdot(M_g)/d\Omega^\cdot(M_g)$ does not depend on the interpolation.

In [BG01, Definition 2.38 and Theorem 2.39], the class $\widetilde{\mathrm{ch}}_g^\circ\left(\nabla^F, g_0^F, g_1^F\right) \in \Omega^\cdot(M_g)/d\Omega^\cdot(M_g)$ was defined by a formula similar to (1.11.2) such that

$$d\widetilde{\mathrm{ch}}_g^\circ\left(\nabla^F, g_0^F, g_1^F\right) = \mathrm{ch}_g^\circ\left(\nabla^F, g_1^F\right) - \mathrm{ch}_g^\circ\left(\nabla^F, g_0^F\right). \quad (1.11.3)$$

A representative of $\mathrm{ch}_g^\circ\left(\nabla^F, g_0^F, g_1^F\right)$ is obtained by a formula similar to (1.11.2), by replacing h by $Fe^\cdot$. Recall that $Q \in \mathrm{End}\left(\Lambda^\cdot\left(T^*M_g\right)\right)$ can be defined as in (1.9.5). By [BG01, Theorem 2.39],

$$\widetilde{\mathrm{ch}}_g^\circ\left(\nabla^F, g_0^F, g_1^F\right) = Q\widetilde{h}_g\left(\nabla^F, g_0^F, g_1^F\right). \quad (1.11.4)$$

Let us also observe that one can as well replace the metric g^F by $-g^F$ while still preserving the above results. Of course $\omega\left(\nabla^F, g^F\right)$ is unchanged when replacing g^F by $-g^F$. The above formulas are therefore unchanged when replacing g^F by $-g^F$.

In [BLo95, Theorem 3.24] and [BG01, Theorem 3.34], the dependence in $\Omega^{\cdot}(S)/d\Omega^{\cdot}(S)$ of $T_{h,g}\left(T^HM, g^{TX}, \nabla^F, g^F\right), T_{\mathrm{ch},g}\left(T^HM, g^{TX}, \nabla^F, g^F\right)$ on the given data is easily expressed in terms of the above classes, as a consequence of (1.8.2), (1.9.8).

1.12 DETERMINANT BUNDLE AND RAY-SINGER METRIC

If λ is a complex line, let λ^{-1} be the corresponding dual line. If E is a complex finite dimensional vector space, set

$$\det E = \Lambda^{\max}(E). \tag{1.12.1}$$

More generally, if $E^{\cdot} = \bigoplus_{i=1}^{m} E^i$ is a complex finite dimensional **Z**-graded complex vector space, put

$$\det E^{\cdot} = \bigotimes_{i=1}^{m} \left(\det E^i\right)^{(-1)^i}. \tag{1.12.2}$$

Put

$$\lambda(F) = \det H^{\cdot}(X, F). \tag{1.12.3}$$

Note that by Poincaré duality,

$$\lambda\left(F^* \otimes o(TX)\right) = \left(\lambda(F)\right)^{(-1)^{n+1}}. \tag{1.12.4}$$

Recall that in section 1.2, we defined the metric $g^{H^{\cdot}(X,F)}$ on $H^{\cdot}(X,F)$ via the identification $\mathcal{H}^X \simeq H^{\cdot}(X,F)$. Let $|\,|^2_{\lambda(F)}$ be the corresponding Hermitian metric on $\lambda(F)$. The Ray-Singer metric $\|\,\|^2_{\lambda(F)}$ on the complex line $\lambda(F)$ is then defined in [BZ92, Definition 2.2] by the formula

$$\|\,\|^2_{\lambda(F)} = \exp\left(\frac{\partial\vartheta}{\partial s}(0)\right) |\,|^2_{\lambda(F)}. \tag{1.12.5}$$

Now following [BZ94, sections 1 and 2] and [B95, sections 1 and 2], we extend the above formalism to the equivariant situation.

Let $\widehat{G}$ be the set of equivalence classes of complex irreducible representations of G. An element of $\widehat{G}$ is specified by a complex finite dimensional vector space W together with an irreducible representation $\rho_W : G \to \mathrm{End}(W)$. Let χ_W be the character of the representation χ_W.

Recall that G acts naturally on $H^{\cdot}(X,F)$ or $H^{\cdot}(X, F\otimes o(TX))$. We have the isotypical decomposition

$$H^{\cdot}(X,F) = \bigoplus_{W\in\widehat{G}} \mathrm{Hom}_G\left(W, H^{\cdot}(X,F)\right)\otimes W. \tag{1.12.6}$$

If $W \in \widehat{G}$, set

$$\lambda_W(F) = \det\left(\operatorname{Hom}_G\left(W, H^{\cdot}(X,F)\right) \otimes W\right). \tag{1.12.7}$$

Then $\lambda_W(F)$ is a complex line. Put

$$\lambda(F) = \bigoplus_{W \in \widehat{G}} \lambda_W(F). \tag{1.12.8}$$

The vector space $\lambda(F)$ is called an equivariant determinant.

Recall that $\mathcal{H}^X = \ker \square^X$. The identification $\mathcal{H}^X \simeq H^{\cdot}(X,F)$ is an identification of G-vector spaces. For $W \in \widehat{G}$, let $|\ |^2_{\lambda_W(F)}$ be the corresponding Hermitian metric on $\Lambda_W(F)$.

Set

$$\log\left(|\ |^2_{\lambda(F)}\right) = \sum_{W \in \widehat{G}} \log\left(|\ |^2_{\lambda_W(F)}\right) \otimes \frac{\chi_W}{\operatorname{rk} W}. \tag{1.12.9}$$

Recall that $\vartheta_g(s)$ was defined in (1.8.3).

Definition 1.12.1. Put

$$\log\left(\|\ \|^2_{\lambda(F)}\right) = \log\left(|\ |^2_{\lambda(F)}\right) + \vartheta'_{\cdot}(0). \tag{1.12.10}$$

Note that (1.12.9), (1.12.10) are formal symbols, which depend on the choice of a $g \in G$. In particular $\log\left(\|\ \|^2_{\lambda(F)}\right)$ is by definition the logarithm of the equivariant Ray-Singer metric.

By the methods of [BLo95, section 3], one deduces from (1.8.2), (1.8.4) the anomaly formulas for Ray-Singer metrics which were established in [BZ92, BZ94]. Since $H^{\cdot}(X,F)$ is equipped with the flat connection $\nabla^{H^{\cdot}(X,F)}$, we can define the variation $d\log\left(\|\ \|^2_{\lambda(F)}\right)$ with respect to this flat connection. As explained in [BLo95, eq. (3.138)], these formulas take the following form.

Theorem 1.12.2. *The following identity of* 1*-forms holds on* S*:*

$$\frac{1}{2} d\log\left(\|\ \|^2_{\lambda(F)}\right)(g) = \int_{X_g} e\left(TX_g, \nabla^{TX_g}\right) \operatorname{Tr}^F\left[g \frac{1}{2}\omega\left(\nabla^F, g^F\right)\right]. \tag{1.12.11}$$

Chapter Two

The hypoelliptic Laplacian on the cotangent bundle

The purpose of this chapter is to recall the main results in Bismut [B05] on the hypoelliptic Laplacian and also on the families version of this operator.

This chapter is organized as follows. In section 2.1, we recall the construction of the exotic Hodge theory given in [B05]. This construction depends in particular on the choice of a Hamiltonian $\mathcal{H}$.

In section 2.2, we give the Weitzenböck formula for the corresponding hypoelliptic Laplacian.

In section 2.3, we state the results in [B05] according to which this new Hodge theory is a deformation of classical Hodge theory.

In section 2.4, we describe the results in [B05] in the context of families. The relevant object is a superconnection.

In section 2.5, we give the Weitzenböck formula for the curvature of the relevant superconnection. This is still a hypoelliptic operator. We also give other remarkable identities established in [B05], which will be needed when using local index techniques.

In section 2.6, we relate this curvature to the curvature of the elliptic Levi-Civita superconnection considered in section 1.3.

In section 2.7, we briefly consider the issue of Poincaré duality.

In section 2.8, we give a 2-parameter version of the new superconnection, in which both the give metric and the Hamiltonian are rescaled. This 2-parameter deformation will play an essential role in the proof of our results on Ray-Singer metrics and on analytic torsion forms.

Finally, in section 2.9, we introduce a group action.

Throughout the chapter, we make the same assumptions as in chapter 1, and we use the corresponding notation.

2.1 A DEFORMATION OF HODGE THEORY

We make the same assumptions as in section 1.2 and we use the corresponding notation. In particular X denotes a compact Riemannian manifold of dimension n.

Here we follow [B05, section 2]. Let $\pi : T^*X \to X$ be the cotangent bundle of X. Let p be the generic element of the fiber T^*X. Let $\theta = \pi^* p$ be the canonical 1-form on T^*X, and let $\omega = d^{T^*X}\theta$ be the canonical symplectic 2-form on T^*X. Let dv_{T^*X} be the symplectic volume form on T^*X.

If $\mathcal{H} : T^*X \to \mathbf{R}$ is a smooth function, let $Y^{\mathcal{H}}$ be the corresponding

Hamiltonian vector field, so that

$$d^{T^*X}\mathcal{H} + i_{Y^{\mathcal{H}}}\omega = 0. \tag{2.1.1}$$

We will often identify TX and T^*X by the metric g^{TX}. The connection ∇^{TX} induces a connection ∇^{T^*X} on T^*X. We still use the notation T^*X for the total space of this vector bundle. The connection ∇^{T^*X} induces a horizontal subbundle $T^HT^*X \simeq \pi^*TX$ of TT^*X, so that we get the splittings

$$TT^*X = \pi^*\left(TX \oplus T^*X\right), \qquad T^*T^*X = \pi^*\left(T^*X \oplus TX\right). \tag{2.1.2}$$

If $U \in TX$, let $U^H \in T^HT^*X \simeq TX$ be the lift of U.

By (2.1.2), we have the isomorphism of $\mathbf{Z}$-graded bundles of algebras

$$\Lambda^{\cdot}\left(T^*T^*X\right) = \pi^*\left(\Lambda^{\cdot}\left(T^*X\right)\widehat{\otimes}\Lambda^{\cdot}\left(TX\right)\right). \tag{2.1.3}$$

Let $e_1,\dots,e_n$ be a basis of TX, let $e^1,\dots,e^n$ be the associated dual basis of T^*X. Let $\widehat{e}_1,\dots,\widehat{e}_n$ and $\widehat{e}^1,\dots,\widehat{e}^n$ be other copies of these two bases. By (2.1.2), $e_1,\dots,e_n,\widehat{e}^1,\dots,\widehat{e}^n$ is a basis of TT^*X, and $e^1,\dots,e^n,\widehat{e}_1,\dots,\widehat{e}_n$ is the corresponding dual basis of T^*T^*X. Set

$$\lambda_0 = e^i \wedge i_{\widehat{e}^i}, \qquad \mu_0 = \widehat{e}_i \wedge i_{e_i}. \tag{2.1.4}$$

Let $\left(\Omega^{\cdot}\left(T^*X,\pi^*F\right), d^{T^*X}\right)$ be the de Rham complex of smooth forms on T^*X with coefficients in π^*F which have compact support. The operator $i_{\widehat{R^{TX}p}}$ acts on $\Omega^{\cdot}\left(T^*X,\pi^*F\right)$. We have the classical identity [B05, Proposition 2.5]

$$d^{T^*X} = e^i \wedge \nabla_{e_i}^{\Lambda^{\cdot}(T^*T^*X)\widehat{\otimes}F} + \widehat{e}_i \wedge \nabla_{\widehat{e}^i} + i_{\widehat{R^{TX}p}}. \tag{2.1.5}$$

Put

$$f = \begin{pmatrix} 1 & 1 \\ 1 & 2 \end{pmatrix}, \qquad F = \begin{pmatrix} 1 & 2 \\ 0 & -1 \end{pmatrix}, \qquad \mathfrak{f} = \begin{pmatrix} 1 & 1 \\ 1 & 0 \end{pmatrix}. \tag{2.1.6}$$

Then f is a scalar product on $\mathbf{R}^2$, and F is an involution of $\mathbf{R}^2$, which is an isometry with respect to f. Its $+1$ eigenspace is spanned by $(1,0)$, and the -1 eigenspace is spanned by $(1,-1)$. Note here there should be no confusion between the flat bundle F and the morphism F. Also $\mathfrak{f}$ is a symmetric matrix, and, moreover,

$$\mathfrak{f} = fF. \tag{2.1.7}$$

Using the identifications in (2.1.2), we observe that f defines a metric $\mathfrak{g}^{TT^*X}$ on TT^*X given by

$$\mathfrak{g}^{TT^*X} = \begin{pmatrix} g^{TX} & 1|_{T^*X} \\ 1|_{TX} & 2g^{T^*X} \end{pmatrix}. \tag{2.1.8}$$

Then the volume form on T^*X which is attached to $\mathfrak{g}^{TT^*X}$ is just dv_{T^*X}. Let $\mathfrak{p} : TT^*X \to T^*X$ be the obvious projection with respect to the splitting (2.1.2) of TT^*X. Then if $U \in TT^*X$,

$$\langle U,U\rangle_{\mathfrak{g}^{TT^*X}} = \langle \pi_*U,\pi_*U\rangle_{g^{TX}} + 2\langle \pi_*U,\mathfrak{p}U\rangle + 2\langle \mathfrak{p}U,\mathfrak{p}U\rangle_{g^{T^*X}}. \tag{2.1.9}$$

Similarly, we will identify F to the $\mathfrak{g}^{TT^*X}$ isometric involution of TT^*X,

$$F = \begin{pmatrix} 1|_{TX} & 2\left(g^{TX}\right)^{-1} \\ 0 & -1|_{T^*X} \end{pmatrix}. \tag{2.1.10}$$

Then F acts as $\widetilde{F}^{-1} = \widetilde{F}$ on $\Lambda^{\cdot}(T^*T^*X)$.

Let $r : T^*X \to T^*X$ be the involution $(x,p) \to (x,-p)$.

Definition 2.1.1. We denote by $\mathfrak{g}^{\Omega^{\cdot}(T^*X,\pi^*F)}$ the Hermitian product on $\Omega^{\cdot}(T^*X,\pi^*F)$ which is naturally associated to the metrics $\mathfrak{g}^{TT^*X}$ and g^F. Let u be the isometric involution of $\Omega^{\cdot}(T^*X,\pi^*F)$ for $\langle\,\rangle_{\mathfrak{g}^{\Omega^{\cdot}(T^*X,\pi^*F)}}$, which is such that if $s \in \Omega^{\cdot}(T^*X,\pi^*F)$,

$$us(x,p) = Fs(x,-p). \tag{2.1.11}$$

Let $\mathfrak{h}^{\Omega^{\cdot}(T^*X,\pi^*F)}$ be the Hermitian form on $\Omega^{\cdot}(T^*X,\pi^*F)$,

$$\langle s,s'\rangle_{\mathfrak{h}^{\Omega^{\cdot}(T^*X,\pi^*F)}} = \langle us,s'\rangle_{\mathfrak{g}^{\Omega^{\cdot}(T^*X,\pi^*F)}}. \tag{2.1.12}$$

Let $\mathcal{H} : T^*X \to \mathbf{R}$ be a smooth function which is r-invariant. If $s,s' \in \Omega^{\cdot}(T^*X,\pi^*F)$, set

$$\langle s,s'\rangle_{\mathfrak{h}_{\mathcal{H}}^{\Omega^{\cdot}(T^*X,\pi^*F)}} = \langle ue^{-2\mathcal{H}}s,s'\rangle_{\mathfrak{g}^{\Omega^{\cdot}(T^*X,\pi^*F)}}. \tag{2.1.13}$$

Then $\mathfrak{h}_{\mathcal{H}}^{\Omega^{\cdot}(T^*X,\pi^*F)}$ is still a Hermitian form on $\Omega^{\cdot}(T^*X,\pi^*F)$.

Let $\mathfrak{f}^{TT^*X}$ be the symmetric bilinear form on TT^*X given by

$$\mathfrak{f}^{TT^*X} = \begin{pmatrix} g^{TX} & 1|_{T^*X} \\ 1|_{TX} & 0 \end{pmatrix}. \tag{2.1.14}$$

By (2.1.7), we get

$$\mathfrak{f}^{TT^*X} = \mathfrak{g}^{TT^*X}F. \tag{2.1.15}$$

Set

$$vs(s,p) = s(x,-p). \tag{2.1.16}$$

Let $\mathfrak{f}^{\Omega^{\cdot}(T^*X,\pi^*F)}$ be the Hermitian form on $\Omega^{\cdot}(T^*X,\pi^*F)$ which is naturally associated with $\mathfrak{f}^{TT^*X}$ and g^F. By (2.1.12), (2.1.13), (2.1.15), (2.1.16), we get

$$\begin{aligned} \langle s,s'\rangle_{\mathfrak{h}^{\Omega^{\cdot}(T^*X,\pi^*F)}} &= \langle vs,s'\rangle_{\mathfrak{f}^{\Omega^{\cdot}(T^*X,\pi^*F)}}, \\ \langle s,s'\rangle_{\mathfrak{h}_{\mathcal{H}}^{\Omega^{\cdot}(T^*X,\pi^*F)}} &= \langle ve^{-2\mathcal{H}}s,s'\rangle_{\mathfrak{f}^{\Omega^{\cdot}(T^*X,\pi^*F)}}. \end{aligned} \tag{2.1.17}$$

Set

$$d_{\mathcal{H}}^{T^*X} = e^{-\mathcal{H}}d^{T^*X}e^{\mathcal{H}}. \tag{2.1.18}$$

Definition 2.1.2. We denote by $\overline{d}_{\phi,\mathcal{H}}^{T^*X}$ the formal adjoint of $d_{\mathcal{H}}^{T^*X}$ with respect to $\mathfrak{h}^{\Omega^{\cdot}(T^*X,\pi^*F)}$. If $\mathcal{H}=0$, we will write $\overline{d}_{\phi}^{T^*X}$ instead of $\overline{d}_{\phi,\mathcal{H}}^{T^*X}$.

Then one has the obvious,

$$\overline{d}_{\phi,\mathcal{H}}^{T^*X} = e^{\mathcal{H}}\overline{d}_{\phi}^{T^*X}e^{-\mathcal{H}}. \tag{2.1.19}$$

Moreover, $\overline{d}_{\phi,2\mathcal{H}}^{T^*X}$ is the formal adjoint of d^{T^*X} with respect to $\mathfrak{h}_{\mathcal{H}}^{\Omega^{\cdot}(T^*X,\pi^*F)}$.

Definition 2.1.3. Set

$$A_{\phi,\mathcal{H}} = \frac{1}{2}\left(\overline{d}^{T^*X}_{\phi,2\mathcal{H}} + d^{T^*X}\right), \qquad B_{\phi,\mathcal{H}} = \frac{1}{2}\left(\overline{d}^{T^*X}_{\phi,2\mathcal{H}} - d^{T^*X}\right), \tag{2.1.20}$$
$$\mathfrak{A}_{\phi,\mathcal{H}} = \frac{1}{2}\left(\overline{d}^{T^*X}_{\phi,\mathcal{H}} + d^{T^*X}_{\mathcal{H}}\right), \qquad \mathfrak{B}_{\phi,\mathcal{H}} = \frac{1}{2}\left(\overline{d}^{T^*X}_{\phi,\mathcal{H}} - d^{T^*X}_{\mathcal{H}}\right).$$

Clearly,

$$\mathfrak{A}_{\phi,\mathcal{H}} = e^{-\mathcal{H}} A_{\phi,\mathcal{H}} e^{\mathcal{H}}, \qquad \mathfrak{B}_{\phi,\mathcal{H}} = e^{-\mathcal{H}} B_{\phi,\mathcal{H}} e^{\mathcal{H}}. \tag{2.1.21}$$

By [B05, Theorem 2.21], $A_{\phi,\mathcal{H}}$ (resp. $B_{\phi,\mathcal{H}}$) is $\mathfrak{h}^{\Omega^{\cdot}(T^*X,\pi^*F)}_{\mathcal{H}}$ self-adjoint (resp. skew-adjoint), and $\mathfrak{A}_{\phi,\mathcal{H}}$ (resp. $\mathfrak{B}_{\phi,\mathcal{H}}$) is $\mathfrak{h}^{\Omega^{\cdot}(T^*X,\pi^*F)}$ self-adjoint (resp. skew-adjoint).

We denote by $\mathfrak{A}'_{\phi,\mathcal{H}}, \mathfrak{B}'_{\phi,\mathcal{H}}$ the operators obtained from $\mathfrak{A}_{\phi,\mathcal{H}}, \mathfrak{B}_{\phi,\mathcal{H}}$ by replacing e^i by $e^i - \widehat{e}_i$, $i_{\widehat{e}^i}$ by $i_{e_i+\widehat{e}^i}$ for $1 \le i \le n$, while leaving unchanged the other annihilation and creation variables. Equivalently,

$$\mathfrak{A}'_{\phi,\mathcal{H}} = e^{-\mu_0}\mathfrak{A}_{\phi,\mathcal{H}} e^{\mu_0}, \qquad \mathfrak{B}'_{\phi,\mathcal{H}} = e^{-\mu_0}\mathfrak{B}_{\phi,\mathcal{H}} e^{\mu_0}. \tag{2.1.22}$$

We will not give the explicit expressions for these operators. They can be found in [B05, section 2]. The operators $\mathfrak{A}'_{\phi,\mathcal{H}}, \mathfrak{B}'_{\phi,\mathcal{H}}$ are associated as before to the operators

$$d^{T^*X\prime}_{\phi,\mathcal{H}} = e^{-\mu_0} d^{T^*X}_{\mathcal{H}} e^{\mu_0}, \qquad \overline{d}^{T^*X\prime}_{\phi,\mathcal{H}} = e^{-\mu_0} \overline{d}^{T^*X}_{\phi,\mathcal{H}} e^{\mu_0}. \tag{2.1.23}$$

Let $g^{TT^*X} = g^{TX} \oplus g^{T^*X}$ be the obvious natural metric on $TT^*X = TX \oplus T^*X$. Let $g^{\Omega^{\cdot}(T^*X,\pi^*F)}$ be the corresponding Hermitian product on $\Omega^{\cdot}(T^*X,\pi^*F)$. Let $h^{\Omega^{\cdot}(T^*X,\pi^*F)}$ be the Hermitian form on $\Omega^{\cdot}(T^*X,\pi^*F)$ such that if $s, s' \in \Omega^{\cdot}(T^*X,\pi^*F)$, then

$$\langle s, s'\rangle_{h^{\Omega^{\cdot}(T^*X,\pi^*F)}} = \langle r^* s, s'\rangle_{g^{\Omega^{\cdot}(T^*X,\pi^*F)}}. \tag{2.1.24}$$

By [B05, eq. (2.120)],

$$\langle s, s'\rangle_{\mathfrak{h}^{\Omega^{\cdot}(T^*X,\pi^*F)}} = \left\langle e^{-\mu_0} s, e^{-\mu_0} s'\right\rangle_{h^{\Omega^{\cdot}(T^*X,\pi^*F)}}. \tag{2.1.25}$$

By [B05, Theorem 2.30], the operator $\mathfrak{A}'_{\phi,\mathcal{H}^c}$ (resp. $\mathfrak{B}'_{\phi,\mathcal{H}^c}$) is $h^{\Omega^{\cdot}(T^*X,\pi^*F)}$ self-adjoint (resp. skew-adjoint).

Remark 2.1.4. In [B05, subsection 2.12], for $b \in \mathbf{R}^*$, an extension of the above constructions is given, these constructions themselves corresponding to the case $b = 1$. Indeed, in (2.1.6), we replace $f, F, \mathfrak{f}$ by the more general $f_b, F_b, \mathfrak{f}_b$ given by

$$f_b = \begin{pmatrix} 1 & b \\ b & 2b^2 \end{pmatrix}, \qquad F_b = \begin{pmatrix} 1 & 2b \\ 0 & -1 \end{pmatrix}, \qquad \mathfrak{f}_b = \begin{pmatrix} 1 & b \\ b & 0 \end{pmatrix}. \tag{2.1.26}$$

The objects corresponding to $A_{\phi,\mathcal{H}}, B_{\phi,\mathcal{H}}, \mathfrak{A}_{\phi,\mathcal{H}}, \mathfrak{B}_{\phi,\mathcal{H}}$ will now be denoted with the subscript ϕ_b instead of ϕ. The definition of $\mathfrak{A}'_{\phi_b,\mathcal{H}}, \mathfrak{B}'_{\phi_b,\mathcal{H}}$ is slightly more involved and is given in [B05].

For $a \in \mathbf{R}$, let $r_a : T^*X \to T^*X$ be the map $(x,p) \to (x,ap)$. Let $K_a : \Omega^{\cdot}(T^*X,\pi^*F) \to \Omega^{\cdot}(T^*X,\pi^*F)$ be the map $s(x,p) \to s(x,ap)$. The difference between K_a and r_a^* is that K_a has no action on the exterior algebra. Put

$$\mathcal{H}_a = r_a^* \mathcal{H}. \tag{2.1.27}$$

Then by [B05, Proposition 2.32],

$$\begin{aligned} d^{T^*X} &= r_b^* d^{T^*X} r_b^{*-1}, & \overline{d}_{\phi_b,\mathcal{H}}^{T^*X} &= r_b^* \overline{d}_{\phi,\mathcal{H}_{1/b}}^{T^*X\prime} r_b^{*-1}, \\ A_{\phi_b,\mathcal{H}} &= r_b^* A_{\phi,\mathcal{H}_{1/b}} r_b^{*-1}, & B_{\phi_b,\mathcal{H}} &= r_b^* B_{\phi,\mathcal{H}_{1/b}} r_b^{*-1}, \\ \mathfrak{A}'_{\phi_b,\mathcal{H}} &= K_b \mathfrak{A}'_{\phi,\mathcal{H}_{1/b}} K_b^{-1}, & \mathfrak{B}'_{\phi_b,\mathcal{H}} &= K_b \mathfrak{B}'_{\phi,\mathcal{H}_{1/b}} K_b^{-1}. \end{aligned} \tag{2.1.28}$$

2.2 THE HYPOELLIPTIC WEITZENBÖCK FORMULAS

From now on, we use the notation

$$\mathcal{H} = \frac{1}{2} |p|^2 . \tag{2.2.1}$$

For $c \in \mathbf{R}^*$, set

$$\mathcal{H}^c = \frac{c}{2} |p|^2 . \tag{2.2.2}$$

We will often distinguish the + case, with $c > 0$, and the − case, with $c < 0$.

Recall that $Y^{\mathcal{H}}$ is the Hamiltonian vector field on T^*X associated to $\mathcal{H}$. Let $L_{Y^{\mathcal{H}}}$ be the associated Lie derivative operator acting on $\Omega^{\cdot}(T^*X, \pi^*F)$. Then we have the easy formula

$$L_{Y^{\mathcal{H}}} = \nabla_{Y^{\mathcal{H}}}^{\Lambda^{\cdot}(T^*X)\widehat{\otimes}\Lambda^{\cdot}(TX)\widehat{\otimes}F} + \widehat{e}_i i_{e_i} + \left\langle R^{T^*X}(p, e_i)\, p, e_j \right\rangle e^i i_{\widehat{e}^j}. \tag{2.2.3}$$

If $p \in T^*X$, we will write instead $\widehat{p}$ when we want to emphasize that p is considered as a vertical 1-form, or as the corresponding radial vector field. For example, $L_{\widehat{p}}$ denotes the Lie derivative operator which is associated to the fiberwise radial vector field $\widehat{p}$. We have the easy formula

$$L_{\widehat{p}} = \nabla_{\widehat{p}} + \widehat{e}_i i_{\widehat{e}^i}. \tag{2.2.4}$$

Let Δ^V be the standard Laplacian along the fibers of T^*X. Now we state the Weitzenböck formulas of [B05, Theorem 3.4].

Theorem 2.2.1. *The following identities hold:*

$$A^2_{\phi,\mathcal{H}^c} = \frac{1}{4}\left(-\Delta^V + 2cL_{\widehat{p}} - \frac{1}{2}\left\langle R^{TX}(e_i,e_j)e_k,e_l\right\rangle e^ie^j i_{\widehat{e}^k} i_{\widehat{e}^l}\right)$$
$$-\frac{1}{2}\left(L_{Y^{\mathcal{H}^c}} + \frac{1}{2}e^i i_{\widehat{e}^j}\nabla^F_{e_i}\omega\left(\nabla^F,g^F\right)(e_j) + \frac{1}{2}\omega\left(\nabla^F,g^F\right)(e_i)\nabla_{\widehat{e}^i}\right),$$
$$\mathfrak{A}^2_{\phi,\mathcal{H}^c} = \frac{1}{4}\Bigg(-\Delta^V + c^2|p|^2 + c\left(2\widehat{e}_i i_{\widehat{e}^i} - n\right)$$
$$-\frac{1}{2}\left\langle R^{TX}(e_i,e_j)e_k,e_l\right\rangle e^ie^j i_{\widehat{e}^k} i_{\widehat{e}^l}\Bigg)$$
$$-\frac{1}{2}\Bigg(L_{Y^{\mathcal{H}^c}} + \frac{1}{2}\omega\left(\nabla^F,g^F\right)\left(Y^{\mathcal{H}^c}\right) + \frac{1}{2}e^i i_{\widehat{e}^j}\nabla^F_{e_i}\omega\left(\nabla^F,g^F\right)(e_j)$$
$$+\frac{1}{2}\omega\left(\nabla^F,g^F\right)(e_i)\nabla_{\widehat{e}^i}\Bigg), \tag{2.2.5}$$
$$\mathfrak{A}'^2_{\phi,\mathcal{H}^c} = \frac{1}{4}\Bigg(-\Delta^V + c^2|p|^2 + c\left(2\widehat{e}_i i_{\widehat{e}^i} - n\right)$$
$$-\frac{1}{2}\left\langle R^{TX}(e_i,e_j)e_k,e_l\right\rangle \left(e^i-\widehat{e}_i\right)\left(e^j-\widehat{e}_j\right) i_{e_k+\widehat{e}^k} i_{e_\ell+\widehat{e}^l}\Bigg)$$
$$-\frac{1}{2}\Bigg(\nabla^{\Lambda^{\cdot}(T^*T^*X)\widehat{\otimes}F,u}_{Y^{\mathcal{H}^c}} + \Bigg(c\left\langle R^{TX}(p,e_i)p,e_j\right\rangle$$
$$+\frac{1}{2}\nabla^F_{e_i}\omega\left(\nabla^F,g^F\right)(e_j)\Bigg)\left(e^i-\widehat{e}_i\right) i_{e_j+\widehat{e}^j} + \frac{1}{2}\omega\left(\nabla^F,g^F\right)(e_i)\nabla_{\widehat{e}^i}\Bigg).$$

By using Hörmander's theorem [Hör67], it is shown in [B05, Theorem 3.6] that if $c \neq 0$, if $u \in \mathbf{R}$ is an extra variable, the operator $\frac{\partial}{\partial u} - A^2_{\phi,\mathcal{H}}$ is hypoelliptic.

2.3 HYPOELLIPTIC LAPLACIAN AND STANDARD LAPLACIAN

Put

$$\mathfrak{a}_{\pm} = \frac{1}{2}\left(-\Delta^V \pm 2L_{\widehat{p}} - \frac{1}{2}\left\langle R^{TX}(e_i,e_j)e_k,e_l\right\rangle e^ie^j i_{\widehat{e}^k} i_{\widehat{e}^l}\right), \tag{2.3.1}$$
$$\mathfrak{b}_{\pm} = -\left(\pm L_{Y^{\mathcal{H}}} + \frac{1}{2}e^i i_{\widehat{e}^j}\nabla^F_{e_i}\omega\left(\nabla^F,g^F\right)(e_j) + \frac{1}{2}\omega\left(\nabla^F,g^F\right)(e_i)\nabla_{\widehat{e}^i}\right).$$

Then by [B05, Theorem 3.8], which itself is deduced from Theorem 2.2.1, we find that for $b \in \mathbf{R}^*_+$,

$$2A^2_{\phi_b,\pm\mathcal{H}} = \frac{\mathfrak{a}_{\pm}}{b^2} + \frac{\mathfrak{b}_{\pm}}{b}. \tag{2.3.2}$$

Recall that $o(TX)$ is the orientation bundle of TX. This is a $\mathbf{Z}_2$ line bundle, which we identity to the corresponding obvious complex flat Euclidean line bundle.

Let Φ^{T^*X} be the Thom form of Mathai-Quillen [MatQ86] on the total space of T^*X which is associated to the connection ∇^{T^*X}. The n-form Φ^{T^*X} is a form on T^*X with values in $\pi^*o(TX)$, which is closed and Gaussian shaped along the fibers of T^*X. To fix the normalization of Φ^{T^*X} unambiguously, let us just say that

$$\Phi^{T^*X} = \frac{1}{\pi^{n/2}} \exp\left(-|p|^2 + \dots\right). \tag{2.3.3}$$

In (2.3.3), $\dots$ denotes explicit differential forms.

Let $j : T^*_xX \to T^*X$ be the embedding of one given fiber into the total space of T^*X. Let η be a n-form of norm 1 along T^*_xX. Then by [MatQ86],

$$j^*\Phi^{T^*X} = \frac{1}{\pi^{n/2}} \exp\left(-|p|^2\right)\eta. \tag{2.3.4}$$

By construction,

$$\pi_*\Phi^{T^*X} = 1. \tag{2.3.5}$$

Note that (2.3.5) follows from (2.3.4).

In the $+$ case, $\Omega^{\cdot}(T^*X, \pi^*F)$ now denotes the vector space of smooth sections s of $\Lambda^{\cdot}(T^*T^*X)\widehat{\otimes}\pi^*F$ on T^*X such that $s\exp\left(-|p|^2/2\right)$ lies in the Schwartz space, i.e., for any $k, m \in \mathbf{N}$, $|p|^k \nabla^m s \exp\left(-|p|^2/2\right)$ is uniformly bounded on T^*X. We identify $\Omega^{\cdot}(X, F)$ to its image in $\Omega^{\cdot}(T^*X, \pi^*F)$ by the map $s \to \pi^*s$. Then $\Omega^{\cdot}(X, F)$ is the image of the projector $Q_+^{T^*X} : \Omega^{\cdot}(T^*X, \pi^*F) \to \Omega^{\cdot}(T^*X, \pi^*F)$ given by

$$Q_+^{T^*X}\beta = \pi_*\left(\beta \wedge \Phi^{T^*X}\right). \tag{2.3.6}$$

In the $-$ case, $\Omega^{\cdot}(T^*X, \pi^*F)$ denotes the vector space of smooth sections s of $\Lambda^{\cdot}(T^*T^*X)\widehat{\otimes}\pi^*F$ on T^*X, such that $\exp\left(|p|^2/2\right)s$ lies in the corresponding Schwartz space. We identify $\Omega^{\cdot}(X, F \otimes o(TX))$ to its image in $\Omega^{\cdot+n}(T^*X, \pi^*F)$ by the map $s \to \pi^*s \wedge \Phi^{T^*X}$. Then $\Omega^{\cdot}(X, F \otimes o(TX))$ is the image of the projector $Q_-^{T^*X} : \Omega^{\cdot}(T^*X, \pi^*F) \to \Omega^{\cdot}(T^*X, \pi^*F)$,

$$Q_-^{T^*X}\beta = (\pi_*\beta) \wedge \Phi^{T^*X}. \tag{2.3.7}$$

By [B05, Theorem 3.11], the operators $\mathfrak{a}_+, \mathfrak{a}_-$ are semisimple. Moreover, the kernel of $\mathfrak{a}_+$ is spanned by the 0-form 1, and the kernel of $\mathfrak{a}_-$ by the n-form Φ^{T^*X}. Finally, the operators $Q_+^{T^*X}, Q_-^{T^*X}$ are simply the projectors over the kernels of $\mathfrak{a}_+, \mathfrak{a}_-$ with respect to the splitting

$$\Omega^{\cdot}(T^*X, \pi^*F) = \ker \mathfrak{a}_\pm \oplus \operatorname{Im} \mathfrak{a}_\pm. \tag{2.3.8}$$

Let $\mathfrak{a}_\pm^{-1}$ denote the inverse of $\mathfrak{a}_\pm$ acting on $\operatorname{Im}\mathfrak{a}_\pm$. As observed in [B05, subsection 3.7], $\mathfrak{b}_\pm$ maps $\ker\,\mathfrak{a}_\pm$ into $\operatorname{Im}\mathfrak{a}_\pm$.

In the sequel $\square^X$ will be the standard elliptic Laplacian acting on $\Omega^{\cdot}(X, F)$ for $c > 0$, on $\Omega^{\cdot}(X, F \otimes o(TX))$ for $c < 0$. Now we state an important result established in [B05, Theorem 3.13].

Theorem 2.3.1. *The following identity holds:*

$$-Q_{\pm}^{T^*X}\mathfrak{b}_{\pm}\mathfrak{a}_{\pm}^{-1}\mathfrak{b}_{\pm}Q_{\pm}^{T^*X}=\frac{1}{2}\square^{X}. \tag{2.3.9}$$

Now we denote by $\Omega^{\cdot}(T^*X,\pi^*F)$ the vector space of smooth sections of $\Lambda^{\cdot}(T^*T^*X)\widehat{\otimes}\pi^*F$ which are square integrable with respect to dv_{T^*X}. By [B05, Propositions 2.36 and 2.39],

$$\begin{aligned}\mathfrak{B}'_{\phi_b,\pm\mathcal{H}}=&-\frac{1}{2b}\left(\widehat{c}\left(\widehat{e}_i\right)\nabla_{\widehat{e}^i}\pm c\left(\widehat{p}\right)\right)-\frac{1}{2}\left(\widehat{c}\left(e_i\right)-c\left(\widehat{e}^i\right)\right)\nabla_{e_i}^{\Lambda^{\cdot}(T^*T^*X)\widehat{\otimes}F,u}\\&+\frac{1}{4}\left(c\left(e_i\right)-\widehat{c}\left(\widehat{e}^i\right)\right)\omega\left(\nabla^F,g^F\right)(e_i)-\frac{b}{4}\Bigg(\left(e^i-\widehat{e}_i\right)\left(e^j-\widehat{e}_j\right)i_{e_k+\widehat{e}^k}\\&-i_{e_i+\widehat{e}^i}i_{e_j+\widehat{e}^j}\left(e^k-\widehat{e}_k\right)\Bigg)\left\langle R^{TX}\left(e_i,e_j\right)p,e_k\right\rangle.\end{aligned} \tag{2.3.10}$$

We rewrite (2.3.10) in the form

$$\mathfrak{B}'_{\phi_b,\pm\mathcal{H}}=-\frac{1}{2b}\left(\widehat{c}\left(\widehat{e}_i\right)\nabla_{\widehat{e}^i}\pm c\left(\widehat{p}\right)\right)+H+bJ. \tag{2.3.11}$$

Put

$$\begin{aligned}\alpha_{\pm}&=\frac{1}{2}\left(-\Delta^V+|p|^2\pm\left(2\widehat{e}_i i_{\widehat{e}^i}-n\right)\right),\\\beta_{\pm}&=-\left(\pm\nabla_{Y^{\mathcal{H}}}^{\Lambda^{\cdot}(T^*T^*X)\widehat{\otimes}F,u}+\frac{1}{2}\omega\left(\nabla^F,g^F\right)(e_i)\nabla_{\widehat{e}^i}\right),\\\gamma_{\pm}&=-\frac{1}{4}\left\langle R^{TX}\left(e_i,e_j\right)e_k,e_\ell\right\rangle\left(e^i-\widehat{e}_i\right)\left(e^j-\widehat{e}_j\right)i_{e_k+\widehat{e}^k}i_{e_\ell+\widehat{e}^\ell}\\&\quad-\left(\pm\left\langle R^{TX}\left(p,e_i\right)p,e_j\right\rangle+\frac{1}{2}\nabla_{e_i}^F\omega\left(\nabla^F,g^F\right)(e_j)\right)\left(e^i-\widehat{e}_i\right)i_{e_j+\widehat{e}^j}.\end{aligned} \tag{2.3.12}$$

By [B05, Theorem 3.8],

$$2\mathfrak{A}'^{2}_{\phi_b,\pm\mathcal{H}}=\frac{\alpha_{\pm}}{b^2}+\frac{\beta_{\pm}}{b}+\gamma_{\pm}. \tag{2.3.13}$$

The operators $\alpha_{\pm}$ are self-adjoint. The fiberwise kernel of the restriction of α_+ to fiberwise forms is 1-dimensional and spanned by $\exp\left(-|p|^2/2\right)$. If η is a fiberwise volume form with norm 1, the fiberwise kernel of α_- restricted to fiberwise forms is also 1-dimensional and spanned by $\exp\left(-|p|^2/2\right)\eta$. Note that $\ker\alpha_{\pm}$ is also $\ker\left(\widehat{c}\left(\widehat{e}_i\right)\nabla_{\widehat{e}^i}\pm c\left(\widehat{p}\right)\right)$. Let $P_{\pm}$ be the fiberwise orthogonal projection on $\ker\alpha_{\pm}$ with respect to the standard Hermitian L^2 product. Of course we have the orthogonal splitting

$$\Omega^{\cdot}(T^*X,\pi^*F)=\ker\alpha_{\pm}\oplus\operatorname{Im}\alpha_{\pm}. \tag{2.3.14}$$

We denote by $\alpha_{\pm}^{-1}$ the inverse of $\alpha_{\pm}$ acting on $\operatorname{Im}\alpha_{\pm}$. Observe that $\beta_{\pm}$ maps $\ker\alpha_{\pm}$ into $\operatorname{Im}\alpha_{\pm}$.

We identify $\Omega^{\cdot}(X,F)$ to its image in $\Omega^{\cdot}(T^*X,\pi^*F)$ by the embedding $i_+:\alpha\to\pi^*s\exp\left(-|p|^2/2\right)/\pi^{n/4}$, and $\Omega^{\cdot}(X,F\otimes o(TX))$ to its image in

$\Omega^{\cdot}(T^*X, \pi^*F)$ by the embedding $i_- : s \to \pi^* s \exp\left(-|p|^2/2\right) \wedge \frac{\eta}{\pi^{n/4}}$. In the sequel $d^{X*} - d^X$ acts on $\Omega^{\cdot}(X, F)$ in the $+$ case, on $\Omega^{\cdot}(X, F \otimes o(TX))$ in the $-$ case. The same convention will apply to $\square^X$.

The following result, closely related to Theorem 2.3.1, is established in [B05, Proposition 2.41 and Theorem 3.14].

Theorem 2.3.2. *The following identities hold:*

$$P_{\pm} H P_{\pm} = \frac{1}{2}\left(d^{X*} - d^X\right), \tag{2.3.15}$$

$$P_{\pm}\left(\gamma_{\pm} - \beta_{\pm}\alpha_{\pm}^{-1}\beta_{\pm}\right) P_{\pm} = \frac{\square^X}{2}.$$

Remark 2.3.3. In [B05, Propositions 2.39-2.41], corresponding results are established on the asymptotics of the operator $K_b \mathfrak{A}'_{\phi, \mathcal{H}^c} K_{1/b}$ as $b \to 0$. These results will not be used here.

2.4 A DEFORMATION OF HODGE THEORY IN FAMILIES

Here we make the same assumptions as in section 1.3. Namely, $p : M \to S$ denotes a submersion with compact fiber X of dimension n. We use otherwise the same notation as in this section, and also in sections 2.1-2.3.

Let $\mathcal{M}$ be the total space of the cotangent bundle T^*X. Let $\pi : \mathcal{M} \to M, q : \mathcal{M} \to S$ be the obvious projections. Take $(x,p) \in T^*X$. Then p is a 1-form on T_xX. We extend it into a 1-form on M which vanishes on T^HM. This way we obtain a 1-form on T_xM, which lifts to a 1-form on $\mathcal{M}$. Equivalently, we get a canonical 1-form θ on $\mathcal{M}$. Set

$$\omega = d^{\mathcal{M}}\theta. \tag{2.4.1}$$

Then ω is a 2-form on $\mathcal{M}$, which restricts to the canonical symplectic form on the fibers of T^*X.

In [B05, subsection 4.5], a horizontal vector bundle $T^H\mathcal{M}$ is defined, which is just the orthogonal vector bundle to TT^*X in $T\mathcal{M}$ with respect to ω. Then ω splits naturally into

$$\omega = \omega^V + \omega^H, \tag{2.4.2}$$

where ω^V, ω^H are the restrictions of ω to $TT^*X, T^H\mathcal{M}$. By [B05, Proposition 4.6], if T^H is the restriction of T to $T^HM \times T^HM$,

$$\omega^H = \left\langle p, T^H \right\rangle. \tag{2.4.3}$$

Let $\mathcal{T}^H$ be the fiberwise Hamiltonian vector field whose associated Hamiltonian is just ω^H. Then $\mathcal{T}^H$ is a 2-form on S with values in vector fields along the fibers T^*X.

Note that $\left(\Omega^{\cdot}(T^*X, \pi^*F), d^{T^*X}\right)$ is a $\mathbf{Z}$-graded complex of vector bundles over S.

Let $\mathcal{H} : \mathcal{M} \to \mathbf{R}$ be a smooth function. A superconnection $A^{\mathcal{M}}_{\phi,\mathcal{H}-\omega^H}$ and an odd section $B^{\mathcal{M}}_{\phi,\mathcal{H}-\omega^H}$ of $\Lambda^{\cdot}(T^*S)\widehat{\otimes}\mathrm{End}(\Omega^{\cdot}(T^*X,\pi^*F))$ are constructed in [B05, section 4], which are such that

$$A^{\mathcal{M},2}_{\phi,\mathcal{H}-\omega^H} = -B^{\mathcal{M},2}_{\phi,\mathcal{H}-\omega^H}, \qquad \left[A^{\mathcal{M}}_{\phi,\mathcal{H}-\omega^H}, B^{\mathcal{M}}_{\phi,\mathcal{H}-\omega^H}\right] = 0. \tag{2.4.4}$$

Put

$$\begin{aligned}\mathfrak{C}^{\mathcal{M}}_{\phi,\mathcal{H}-\omega^H} &= e^{-(\mathcal{H}-\omega^H)} A^{\mathcal{M}}_{\phi,\mathcal{H}-\omega^H} e^{\mathcal{H}-\omega^H},\\ \mathfrak{D}^{\mathcal{M}}_{\phi,\mathcal{H}-\omega^H} &= e^{-(\mathcal{H}-\omega^H)} B^{\mathcal{M}}_{\phi,\mathcal{H}-\omega^H} e^{\mathcal{H}-\omega^H}.\end{aligned} \tag{2.4.5}$$

Identities similar to (2.4.4) are valid for $\mathfrak{C}^{\mathcal{M}}_{\phi,\mathcal{H}-\omega^H}, \mathfrak{D}^{\mathcal{M}}_{\phi,\mathcal{H}-\omega^H}$.

If S is a point, then

$$\begin{aligned}A^{\mathcal{M}}_{\phi,\mathcal{H}-\omega^H} &= A_{\phi,\mathcal{H}}, & B^{\mathcal{M}}_{\phi,\mathcal{H}-\omega^H} &= B_{\phi,\mathcal{H}},\\ \mathfrak{C}^{\mathcal{M}}_{\phi,\mathcal{H}-\omega^H} &= \mathfrak{A}_{\phi,\mathcal{H}}, & \mathfrak{D}^{\mathcal{M}}_{\phi,\mathcal{H}-\omega^H} &= \mathfrak{B}_{\phi,\mathcal{H}}.\end{aligned} \tag{2.4.6}$$

More generally, the component of degree 0 in $\Lambda^{\cdot}(T^*S)$ of the objects considered above are just the operators we described in section 2.1. In particular the components of degree 0 of $A'^{\mathcal{M}}, \mathfrak{C}'^{\mathcal{M}}_{\mathcal{H}-\omega^H}$ coincide with $d^{T^*X}, d^{T^*X}_{\mathcal{H}}$.

By using the splitting $T\mathcal{M} = T^H\mathcal{M} \oplus TT^*X$, 1-forms along the fibers T^*X can be considered as forms on $\mathcal{M}$ which vanish on $T^H\mathcal{M}$.

Now we use the same notation as after (2.1.3). We still define μ_0 as in (2.1.4). Set

$$\mathfrak{E}^{\mathcal{M}}_{\phi,\mathcal{H}-\omega^H} = e^{-\mu_0}\mathfrak{C}^{\mathcal{M}}_{\phi,\mathcal{H}-\omega^H} e^{\mu_0}, \quad \mathfrak{F}^{\mathcal{M}}_{\phi,\mathcal{H}-\omega^H} = e^{-\mu_0}\mathfrak{D}^{\mathcal{M}}_{\phi,\mathcal{H}-\omega^H} e^{\mu_0}. \tag{2.4.7}$$

Equivalently, $\mathfrak{E}^{\mathcal{M}}_{\phi,\mathcal{H}-\omega^H}, \mathfrak{F}^{\mathcal{M}}_{\phi,\mathcal{H}-\omega^H}$ are obtained from $\mathfrak{C}^{\mathcal{M}}_{\phi,\mathcal{H}-\omega^H}, \mathfrak{D}^{\mathcal{M}}_{\phi,\mathcal{H}-\omega^H}$ by making the replacements indicated after (2.1.21). If S is a point, then

$$\mathfrak{E}^{\mathcal{M}}_{\phi,\mathcal{H}-\omega^H} = \mathfrak{A}'_{\phi,\mathcal{H}}, \qquad \mathfrak{F}^{\mathcal{M}}_{\phi,\mathcal{H}-\omega^H} = \mathfrak{B}'_{\phi,\mathcal{H}}. \tag{2.4.8}$$

Now we give some more details on $A^{\mathcal{M}}_{\phi,\mathcal{H}-\omega^H}, \mathfrak{C}^{\mathcal{M}}_{\phi,\mathcal{H}-\omega^H}$. The construction of $A^{\mathcal{M}}_{\phi,\mathcal{H}-\omega^H}$ is made using the superconnection $A'^{\mathcal{M}}$, which is the total de Rham operator acting on $\Omega^{\cdot}(\mathcal{M},\pi^*F)$, considered as a flat superconnection on $\Omega^{\cdot}(T^*X,\pi^*F)$ as in [BLo95]. The superconnection $A'^{\mathcal{M}}$ is such that

$$\left[A'^{\mathcal{M}}, A^{\mathcal{M},2}_{\phi,\mathcal{H}-\omega^H}\right] = 0. \tag{2.4.9}$$

In [B05], $A^{\mathcal{M}}_{\phi,\mathcal{H}-\omega^H}, B^{\mathcal{M}}_{\phi,\mathcal{H}-\omega^H}$ are written in the form

$$A^{\mathcal{M}}_{\phi,\mathcal{H}-\omega^H} = \frac{1}{2}\left(\overline{\mathfrak{C}}'^{\mathcal{M}}_{\phi,2(\mathcal{H}-\omega^H)} + A'^{\mathcal{M}}\right), \quad B^{\mathcal{M}}_{\phi,\mathcal{H}-\omega^H} = \frac{1}{2}\left(\overline{\mathfrak{C}}'^{\mathcal{M}}_{\phi,2(\mathcal{H}-\omega^H)} - A'^{\mathcal{M}}\right). \tag{2.4.10}$$

Note that $\mathfrak{h}^{\Omega^{\cdot}(T^*X,\pi^*F)}_{\mathcal{H}}$ was defined in (2.1.13). We define $\mathfrak{h}^{\Omega^{\cdot}(T^*X,\pi^*F)}_{\mathcal{H}-\omega^H}$ by replacing $\mathcal{H}$ by $\mathcal{H}-\omega^H$ in the right-hand side of (2.1.13). By [B05, Proposition 4.24], $\overline{\mathfrak{C}}'^{\mathcal{M}}_{\phi,2(\mathcal{H}-\omega^H)}$ is the $\mathfrak{h}^{\Omega^{\cdot}(T^*X,\pi^*F)}_{\mathcal{H}-\omega^H}$-adjoint of $A'^{\mathcal{M}}$.

Let $\overline{A}'^{\mathcal{M}}$ be the symplectic adjoint of $A'^{\mathcal{M}}$ with respect to the fiberwise symplectic form ω^V in the sense of [B05]. Let $\overline{A}'^{\mathcal{M}}_{\phi}$ be the adjoint flat superconnection to $A'^{\mathcal{M}}$ in the sense of [BLo95] with respect to the Hermitian form $\mathfrak{h}^{\Omega^{\cdot}(T^*X,\pi^*F)}$ (which is just the Hermitian form in (2.1.13) with $\mathcal{H}=0$) on $\Omega^{\cdot}(T^*X,\pi^*F)$. By [B05, Proposition 4.18],

$$\overline{A}_{\phi}^{\prime\mathcal{M}} = e^{\lambda_0}\overline{A}^{\prime\mathcal{M}}e^{-\lambda_0}. \tag{2.4.11}$$

By [B05, Definition 4.20],

$$\overline{\mathfrak{C}}_{\phi,2(\mathcal{H}-\omega^H)}^{\prime\mathcal{M}} = e^{2(\mathcal{H}-\omega^H)}\overline{A}_{\phi}^{\prime\mathcal{M}}e^{-2(\mathcal{H}-\omega^H)}. \tag{2.4.12}$$

Set

$$\mathfrak{C}_{\mathcal{H}-\omega^H}^{\prime\mathcal{M}} = e^{-(\mathcal{H}-\omega^H)}A^{\prime\mathcal{M}}e^{(\mathcal{H}-\omega^H)}. \tag{2.4.13}$$

Then

$$\begin{aligned}\mathfrak{C}_{\phi,\mathcal{H}-\omega^H}^{\mathcal{M}} &= \frac{1}{2}\left(\overline{\mathfrak{C}}_{\phi,\mathcal{H}-\omega^H}^{\prime\mathcal{M}} + \mathfrak{C}_{\mathcal{H}-\omega^H}^{\prime\mathcal{M}}\right),\\ \mathfrak{D}_{\phi,\mathcal{H}-\omega^H}^{\mathcal{M}} &= \frac{1}{2}\left(\overline{\mathfrak{C}}_{\phi,\mathcal{H}-\omega^H}^{\prime\mathcal{M}} - \mathfrak{C}_{\mathcal{H}-\omega^H}^{\prime\mathcal{M}}\right).\end{aligned} \tag{2.4.14}$$

Note in particular that (2.4.5) follows from (2.4.10) and (2.4.12)-(2.4.14).

Remark 2.4.1. We use the same notation as in Remark 2.1.4. The corresponding objects are still denoted with a subscript ϕ_b, and ω^H is replaced by $b\omega^H$. By [B05, Propositions 2.32 and 4.33], we get

$$\begin{aligned}\mathfrak{C}_{\phi_b,\mathcal{H}-b\omega^H}^{\mathcal{M}} &= r_b^*\mathfrak{C}_{\phi,\mathcal{H}_{1/b}-\omega^H}^{\mathcal{M}}r_b^{*-1}, \qquad \mathfrak{D}_{\phi_b,\mathcal{H}-b\omega^H}^{\mathcal{M}} = r_b^*\mathfrak{D}_{\phi,\mathcal{H}_{1/b}-\omega^H}^{\mathcal{M}}r_b^{*-1},\\ \mathfrak{E}_{\phi_b,\mathcal{H}-b\omega^H}^{\mathcal{M}} &= K_b\mathfrak{E}_{\phi,\mathcal{H}_{1/b}-\omega^H}^{\mathcal{M}}K_b^{-1}, \qquad \mathfrak{F}_{\phi_b,\mathcal{H}-b\omega^H}^{\mathcal{M}} = K_b\mathfrak{F}_{\phi,\mathcal{H}_{1/b}-\omega^H}^{\mathcal{M}}K_b^{-1}.\end{aligned} \tag{2.4.15}$$

2.5 WEITZENBÖCK FORMULAS FOR THE CURVATURE

From now on, we still take $\mathcal{H}$ as in (2.2.1), and for $c \in \mathbf{R}$, we define $\mathcal{H}^c$ as in (2.2.2).

Let $e_1,\ldots,e_n$ be an orthonormal basis of TX as in section 2.1. We use otherwise the same notation as in that section on the $e^i, \widehat{e}_i, \widehat{e}^i$.

Let $f_1,\ldots,f_m$ be a basis of TS, let $f^1,\ldots,f^m$ be the corresponding dual basis of T^*S.

The following result was established in [B05, Propositions 2.39, 4.31, and 4.35]. Take $b \in \mathbf{R}_+^*, c = \pm 1/b^2$.

Proposition 2.5.1. *The following identity holds:*

$$\begin{aligned}\mathfrak{F}^{\mathcal{M}}_{\phi_b,\pm\mathcal{H}-b\omega^H} = &-\frac{1}{2b}\left(\widehat{c}\left(\widehat{e}_i\right)\nabla_{\widehat{e}^i} \pm c\left(\widehat{p}\right)\right) - \frac{1}{2}\left(\widehat{c}\left(e_i\right) - c\left(\widehat{e}^i\right)\right)\nabla^{\Lambda^{\cdot}(T^*T^*X)\widehat{\otimes}F,u}_{e_i} \\ &+\frac{1}{4}\left(c\left(e_i\right) - \widehat{c}\left(\widehat{e}^i\right)\right)\omega\left(\nabla^F, g^F\right)\left(e_i\right) - \frac{b}{4}\Bigg(\left(e^i - \widehat{e}_i\right)\left(e^j - \widehat{e}_j\right) i_{e_k+\widehat{e}^k} \\ &\qquad - i_{e_i+\widehat{e}^i} i_{e_j+\widehat{e}^j}\left(e^k - \widehat{e}_k\right)\Bigg)\left\langle R^{TX}\left(e_i, e_j\right)p, e_k\right\rangle \\ &\qquad - \left\langle T\left(f^H_\alpha, e_i\right), e_j\right\rangle f^\alpha\left(e^i - \widehat{e}_i\right) i_{e_j+\widehat{e}^j} \\ &\qquad + f^\alpha\left(\frac{1}{2}\omega\left(\nabla^F, g^F\right)\left(f^H_\alpha\right) \pm \left\langle T\left(f^H_\alpha, p\right), p\right\rangle\right) \\ &\qquad - \frac{1}{2}\left\langle T^H, e_i\right\rangle\left(e^i - \widehat{e}_i + i_{e_i+\widehat{e}^i}\right). \qquad (2.5.1)\end{aligned}$$

Now we replace S by $S \times \mathbf{R}^*$, where $c \in \mathbf{R}^*$, so that c is now allowed to vary. By [B05, Theorem 4.40], for $c \neq 0$, the operator $\frac{\partial}{\partial u} - \mathfrak{C}^{\mathcal{M},2}_{\phi,\mathcal{H}^c-\omega^H}$ is fiberwise hypoelliptic.

Set

$$\begin{aligned}\nu_c &= \frac{1}{2}\left(e^i + i_{\widehat{e}^i}\right)\left(e^i + \widehat{e}_i + i_{e_i}\right) - \nabla_{\widehat{p}} + \frac{3c}{2}|p|^2, \qquad (2.5.2)\\ \mathfrak{G}^{\mathcal{M}}_{\phi,\mathcal{H}^c-\omega^H} &= \mathfrak{D}^{\mathcal{M}}_{\phi,\mathcal{H}^c-\omega^H} + \left[\mathfrak{D}^{\mathcal{M}}_{\phi,\mathcal{H}^c-\omega^H}, \nu_c\right].\end{aligned}$$

Now we give the result established in [B05, Theorems 4.38 and 4.42].

Theorem 2.5.2. *The following identity holds:*

$$\begin{aligned}
\mathfrak{C}^{\mathcal{M},2}_{\phi,\mathcal{H}^c-\omega^H} = \frac{1}{4}\Bigg(&-\Delta^V + c^2|p|^2 + c\left(2\widehat{e}_i i_{\widehat{e}^i - e_i} - n\right) \\
&- \frac{1}{2}\left\langle R^{TX}(e_i,e_j)e_k, e_l\right\rangle e^i e^j i_{\widehat{e}^k} i_{\widehat{e}^l}\Bigg) \\
- \frac{c}{2}\Bigg(&\nabla_p^{\Lambda^\cdot(T^*T^*X)\widehat{\otimes}F,u} + \left\langle T\left(f_\alpha^H, p\right), e_i\right\rangle f^\alpha\left(e^i + 2\widehat{e}_i + i_{\widehat{e}^i - 2e_i}\right) \\
+ \left\langle T^H, p\right\rangle &- \left(e^i + i_{\widehat{e}^i}\right) f^\alpha \left\langle \nabla_{e_i}^{TX} T\left(f_\alpha^H, p\right), p\right\rangle - e^i i_{\widehat{e}^j}\left\langle R^{TX}(p,e_i)e_j, p\right\rangle\Bigg) \\
&- \left(\frac{1}{4}\omega\left(\nabla^F, g^F\right)(e_i) + \frac{1}{2}\left\langle T\left(f_\alpha^H, e_i\right), e_j\right\rangle f^\alpha\left(e^j + i_{\widehat{e}^j}\right)\right)\nabla_{\widehat{e}^i} \\
&- \frac{1}{2}\left(e^k + i_{\widehat{e}^k}\right) f^\alpha e^i i_{\widehat{e}^j}\left\langle \nabla_{e_k}^{TX} T\left(f_\alpha^H, e_i\right), e_j\right\rangle \\
&- \frac{1}{4}\left(e^i + i_{\widehat{e}^i}\right)\left(e^j + i_{\widehat{e}^j}\right)\left\langle \nabla_{e_i}^{TX} T^H, e_j\right\rangle \\
&- \frac{1}{4} e^i i_{\widehat{e}^j}\nabla_{e_i}^F \omega\left(\nabla^F, g^F\right)(e_j) - \frac{1}{8}\left(e^i - i_{\widehat{e}^i}\right) f^\alpha \omega^2\left(\nabla^F, g^F\right)\left(e_i, f_\alpha^H\right) \\
&+ \frac{1}{4}\left(e^i + i_{\widehat{e}^i}\right) f^\alpha \nabla_{e_i}^{F,u}\omega\left(\nabla^F, g^F\right)\left(f_\alpha^H\right) \\
&- \frac{1}{8} f^\alpha f^\beta \omega^2\left(\nabla^F, g^F\right)\left(f_\alpha^H, f_\beta^H\right) + \frac{1}{2} dc\left(\widehat{e}_i + i_{\widehat{e}^i - e_i}\right)\langle p, e_i\rangle . \qquad (2.5.3)
\end{aligned}$$

Moreover,

$$\begin{aligned}
\mathfrak{G}^{\mathcal{M}}_{\phi,\mathcal{H}^c-\omega^H} &= \frac{1}{2}\omega\left(\nabla^F, g^F\right) - \frac{1}{4}\left(e^i + i_{\widehat{e}^i}\right)\omega\left(\nabla^F, g^F\right)(e_i) \\
&- \frac{c}{2}\left(p + i_{\widehat{p}} + 6\widehat{p} - 6f^\alpha\left\langle T\left(f_\alpha^H, p\right), p\right\rangle\right) - \frac{3}{2}dc|p|^2 . \qquad (2.5.4)
\end{aligned}$$

Finally,

$$\begin{aligned}
&\mathfrak{G}^{\mathcal{M}}_{\phi,\mathcal{H}^c-\omega^H} + \left[\mathfrak{C}'^{\mathcal{M}}_{\mathcal{H}^c-\omega^H}, \frac{3c}{2}|p|^2\right] \\
&= \frac{1}{2}\omega\left(\nabla^F, g^F\right) - \frac{1}{4}\left(e^i + i_{\widehat{e}^i}\right)\omega\left(\nabla^F, g^F\right)(e_i) - \frac{c}{2}\left(p + i_{\widehat{p}}\right) . \qquad (2.5.5)
\end{aligned}$$

Remark 2.5.3. By (2.4.7), we obtain the formula for $\mathfrak{C}^{\mathcal{M},2}_{\phi,\mathcal{H}-\omega^H}$ by replacing e^i by $e^i - \widehat{e}_i$ and $i_{\widehat{e}^i}$ by $i_{e_i + \widehat{e}^i}$, while leaving the other creation and annihilation operators unchanged.

In [B05, subsection 4.21], various conjugations are made on the operators considered above to put them in a more geometric form. We briefly describe the main steps of these constructions. First we replace $e^i, i_{e_i}, \widehat{e}_i, i_{\widehat{e}^i}$ by $e^i, i_{e_i + \widehat{e}_i}, i_{\widehat{e}_i}, \widehat{e}^i - e^i$. This transformation does not change the obvious commutation relations.

The above transformation can also be described as follows. First we use the canonical isomorphism

$$\Lambda^{\cdot}(TX) \simeq \Lambda^{n-\cdot}(T^*X)\widehat{\otimes}\Lambda^n(TX). \tag{2.5.6}$$

Using (2.1.3), (2.5.6), we get the isomorphism

$$\Lambda^{\cdot}(T^*T^*X) \simeq \pi^*\left(\Lambda^{\cdot}(T^*X)\widehat{\otimes}\Lambda^{n-\cdot}(T^*X)\widehat{\otimes}\Lambda^n(TX)\right). \tag{2.5.7}$$

The second copy of $\Lambda^{\cdot}(T^*X)$ in (2.5.7) will now be generated by $\widehat{e}^1,\dots,\widehat{e}^n$.

Set

$$\underline{\lambda}_0 = e^i i_{\widehat{e}_i}. \tag{2.5.8}$$

Then we conjugate the operator obtained by the first transformation by $e^{-\underline{\lambda}_0}$.

Let $\langle T^0, p\rangle$ be given by

$$\langle T^0, p\rangle = \langle T(f_\alpha^H, e_i), p\rangle f^\alpha \widehat{e}^i. \tag{2.5.9}$$

A final conjugation is done by conjugating the operator we obtained before by $\exp\left(\langle T^0, p\rangle\right)$. Starting from $\mathfrak{C}^{\mathcal{M}}_{\phi,\mathcal{H}-\omega^H}, \mathfrak{D}^{\mathcal{M}}_{\phi,\mathcal{H}-\omega^H}$, we obtain the operators $\underline{\widehat{\mathfrak{C}}}^{\mathcal{M}}_{\phi,\mathcal{H}^c-\omega^H}, \underline{\widehat{\mathfrak{D}}}^{\mathcal{M}}_{\phi,\mathcal{H}^c-\omega^H}$. The transform of other operators which were previously considered will be denoted using a similar notation.

The Lie derivative operator $L_{Y^{\mathcal{H}}}$ acts naturally on $\Omega^{\cdot}(T^*X, \pi^*F)$. Also the vector field $Y^{\mathcal{H}}$ can also be considered as a vector field on $\mathcal{M}$. We denote by $\mathcal{L}_{Y^{\mathcal{H}}}$ the corresponding Lie derivative operator acting on smooth sections of $\Lambda^{\cdot}(T^*\mathcal{M})\widehat{\otimes}F$. We still have the Cartan formula,

$$L_{Y^{\mathcal{H}}} = \left[d^{T^*X}, i_{Y^{\mathcal{H}}}\right], \qquad \mathcal{L}_{Y^{\mathcal{H}}} = \left[d^{\mathcal{M}}, i_{Y^{\mathcal{H}}}\right]. \tag{2.5.10}$$

Let ∇^{T^*X} be the connection corresponding to the connection ∇^{TX} by the metric g^{TX}. Then the connection ∇^{T^*X} induces the splitting

$$T\mathcal{M} = T^{H'}\mathcal{M} \oplus TT^*X. \tag{2.5.11}$$

As explained in [B05, subsection 4.22], in general $T^{H'}\mathcal{M}$ does not coincide with $T^H\mathcal{M}$. From (2.5.11), we get the isomorphism

$$\Lambda^{\cdot}(T^*\mathcal{M}) = \Lambda^{\cdot}(T^*S)\widehat{\otimes}\Lambda^{\cdot}(TT^*X). \tag{2.5.12}$$

In particular the $e^i, \widehat{e}_i$ can be considered as forms on $\mathcal{M}$.

Then by [B05, eq. (4.174)] and by (2.2.3), we have the identity

$$\begin{aligned} L_{Y^{\mathcal{H}}} &= \nabla_{Y^{\mathcal{H}}}^{\Lambda^{\cdot}(T^*X)\widehat{\otimes}\Lambda^{\cdot}(TX)\widehat{\otimes}F} + \widehat{e}_i i_{e_i} + \left\langle R^{T^*X}(p, e_i)p, e_j\right\rangle e^i i_{\widehat{e}^j}, \\ \mathcal{L}_{Y^{\mathcal{H}}} &= L_{Y^{\mathcal{H}}} - \left\langle T(f_\alpha^H, p), e^i\right\rangle f^\alpha i_{e_i} + \left\langle R^{T^*X}(p, f_\alpha^H)p, e_i\right\rangle f^\alpha i_{\widehat{e}^i}. \end{aligned} \tag{2.5.13}$$

We still use the same notation $L_{Y^{\mathcal{H}}}, \mathcal{L}_{Y^{\mathcal{H}}}$ for the above operators in which $\widehat{e}_i, i_{\widehat{e}^i}$ have been replaced by $i_{\widehat{e}_i}, \widehat{e}^i$. By (2.5.13), we get

$$\begin{aligned} L_{Y^{\mathcal{H}}} &= \nabla_{Y^{\mathcal{H}}}^{\Lambda^{\cdot}(T^*X)\widehat{\otimes}\Lambda^{\cdot}(T^*X)\widehat{\otimes}F} + i_{\widehat{e}_i} i_{e_i} + \left\langle R^{T^*X}(p, e_i)p, e_j\right\rangle e^i \widehat{e}^j, \\ \mathcal{L}_{Y^{\mathcal{H}}} &= L_{Y^{\mathcal{H}}} - \left\langle T(f_\alpha^H, p), e^i\right\rangle f^\alpha i_{e_i} + \left\langle R^{T^*X}(p, f_\alpha^H)p, e_i\right\rangle f^\alpha \widehat{e}^i. \end{aligned} \tag{2.5.14}$$

Moreover, recall that θ has been extended into a canonical 1-form on $\mathcal{M}$. More precisely,

$$\theta = \langle p, e_i \rangle e^i. \tag{2.5.15}$$

Then by [B05, eq. (4.175)], we get

$$d^{\mathcal{M}}\theta = \widehat{e}_i e^i + \left\langle T\left(f_\alpha^H, e_i\right), p\right\rangle f^\alpha e^i + \left\langle T^H, p\right\rangle. \tag{2.5.16}$$

We denote by $\widehat{A}_{\dagger}^{\prime\mathcal{M}}\theta$ the expression obtained from $d^{\mathcal{M}}\theta$ by replacing $\widehat{e}_i, i_{\widehat{e}^i}$ by $i_{\widehat{e}_i}, \widehat{e}^i$. By (2.5.15), we get

$$\widehat{A}_{\dagger}^{\prime\mathcal{M}}\theta = i_{\widehat{e}_i} e^i + \left\langle T\left(f_\alpha^H, e_i\right), p\right\rangle f^\alpha e^i + \left\langle T^H, p\right\rangle. \tag{2.5.17}$$

In the sequel, we will use the notation

$$\widehat{\omega}\left(\nabla^F, g^F\right) = \widehat{e}^i \omega\left(\nabla^F, g^F\right)(e_i). \tag{2.5.18}$$

Also forms like $\omega\left(\nabla^F, g^F\right)^2$ will be viewed as 2-forms on M, and so they can be expanded as a sum of monomials of degree 2 in the e^i, f^α. The same is true for objects like $\frac{1}{4}\left\langle e_i, R^{TX} e_j\right\rangle \widehat{e}^i\widehat{e}^j$, where R^{TX} should itself be expanded as a sum of monomials in the e^i, f^α.

Now we state the formula given in [B05, Theorems 4.45 and 4.52].

Theorem 2.5.4. *The following identity holds:*

$$\begin{aligned}\widehat{\underline{\mathfrak{C}}}_{\phi,\mathcal{H}^c-\omega^H}^{\mathcal{M},2} &= \frac{1}{4}\Big(-\Delta^V + c^2|p|^2 + c\left(2 i_{\widehat{e}_i}\widehat{e}^i - n\right)\Big) + \frac{1}{4}\left\langle e_i, R^{TX} e_j\right\rangle \widehat{e}^i\widehat{e}^j \\ &- \frac{1}{4}\omega\left(\nabla^F, g^F\right)(e_i)\nabla_{\widehat{e}^i} - \frac{1}{4}\nabla^{\Lambda^{\cdot}(T^*T^*X)\widehat{\otimes}F}\widehat{\omega}\left(\nabla^F, g^F\right) - \frac{1}{4}\omega\left(\nabla^F, g^F\right)^2 \\ &- \frac{c}{2}\left(\mathcal{L}_{Y^{\mathcal{H}}} + \frac{1}{2}\omega\left(\nabla^F, g^F\right)\left(Y^{\mathcal{H}}\right) + \widehat{A}_{\dagger}^{\prime\mathcal{M}}\theta\right) + \frac{1}{2}dc\left(\widehat{e}^i - e^i - i_{e_i}\right)\langle p, e_i\rangle.\end{aligned} \tag{2.5.19}$$

Moreover,

$$\begin{aligned}\widehat{\underline{\nu}}_c &= \frac{1}{2}\widehat{e}^i\left(e^i + i_{e_i + 2\widehat{e}_i}\right) - \nabla_{\widehat{p}} + \frac{3c}{2}|p|^2, \\ \widehat{\underline{\mathfrak{B}}}_{\phi,\mathcal{H}^c-\omega^H}^{\mathcal{M}} &= \frac{1}{2}\omega\left(\nabla^F, g^F\right) - \frac{1}{4}\widehat{\omega}\left(\nabla^F, g^F\right) - \frac{c}{2}\left(\widehat{p} + 6 i_{\widehat{p}}\right) - \frac{3}{2}dc|p|^2,\end{aligned} \tag{2.5.20}$$

$$\widehat{\underline{\mathfrak{B}}}_{\phi,\mathcal{H}^c-\omega^H}^{\mathcal{M}} + \left[\widehat{\underline{\mathfrak{C}}}_{\mathcal{H}^c-\omega^H}^{\prime\mathcal{M}}, \frac{3c}{2}|p|^2\right] = \frac{1}{2}\omega\left(\nabla^F, g^F\right) - \frac{1}{4}\widehat{\omega}\left(\nabla^F, g^F\right) - \frac{c}{2}\widehat{p}.$$

Remark 2.5.5. It should be observed that the considerations we made in (2.5.11)-(2.5.12) to interpret the $e^i, \widehat{e}_i$ as forms on $\mathcal{M}$ will be partially irrelevant in the sequel. The reader may as well take formulas (2.5.13)-(2.5.17) as definitions, whenever necessary.

2.6 $\mathfrak{F}^{\mathcal{M}}_{\phi_b,\pm\mathcal{H}-b\omega^H}, \mathfrak{E}^{\mathcal{M},2}_{\phi_b,\pm\mathcal{H}-b\omega^H}$ AND LEVI-CIVITA SUPERCONNECTION

Now we recall the results of [B05, subsection 4.22], which extend to families the results already described in section 2.3. We will give the simplest version as possible of the results of [B05].

By equation (2.5.1) in Proposition 2.5.1, we can write $\mathfrak{F}^{\mathcal{M}}_{\phi_b,\pm\mathcal{H}-b\omega^H}$ in the form

$$\mathfrak{F}^{\mathcal{M}}_{\phi_b,\pm\mathcal{H}-b\omega^H} = -\frac{1}{2b}\left(\widehat{c}\left(\widehat{e}_i\right)\nabla_{\widehat{e}^i} \pm c\left(\widehat{p}\right)\right) + \mathfrak{H}_{\pm} + b\mathfrak{J}_{\pm}. \tag{2.6.1}$$

Recall that $\alpha_{\pm}$ was defined in (2.3.12). Inspection of the formulas in [B05, Theorem 4.38] or use of equation (2.5.3) together with Remark 2.5.3 shows that that there are operators $\beta'_{\pm}, \gamma'_{\pm}$ such that for $b \in \mathbf{R}^*_+, c = \pm 1/b^2$,

$$2\mathfrak{E}^{\mathcal{M},2}_{\phi_b,\pm\mathcal{H}-b\omega^H} = \frac{\alpha_{\pm}}{b^2} + \frac{\beta'_{\pm}}{b} + \gamma'_{\pm}. \tag{2.6.2}$$

We make the same fiberwise identifications as after equation (2.3.13). Also we define the operators $P_{\pm}$ as in the corresponding section. These operators now act fiberwise. We denote by A_+, A_- the Levi-Civita superconnections on $\Omega^{\cdot}(X,F), \Omega^{\cdot}(X, F \otimes o(TX))$. Let B_+, B_- be the odd sections of $\Lambda^{\cdot}(T^*S)\widehat{\otimes}\mathrm{End}(\Omega^{\cdot}(X,F)), \Lambda^{\cdot}(T^*S)\widehat{\otimes}\mathrm{End}(\Omega^{\cdot}(X,F\otimes o(TX)))$ as in section 1.3.

The following related results were established in [B05, Proposition 4.36 and Theorem 4.57], which extends Theorem 2.3.2 to the case of families.

Theorem 2.6.1. *The following identities hold:*

$$\begin{aligned} P_{\pm}\mathfrak{H}_{\pm}P_{\pm} &= B_{\pm}, \\ P_{\pm}\left(\gamma'_{\pm} - \beta'_{\pm}\alpha_{\pm}^{-1}\beta'_{\pm}\right)P_{\pm} &= 2A^2_{\pm}. \end{aligned} \tag{2.6.3}$$

2.7 THE SUPERCONNECTION $A^{\mathcal{M}}_{\phi,\mathcal{H}-\omega^H}$ AND POINCARÉ DUALITY

Here we assume again that $\mathcal{H} : \mathcal{M} \to \mathbf{R}$ is arbitrary. Let $*^{T^*X}$ be the Hodge operator associated to the metric $\mathfrak{g}^{TT^*X}$ and to the orientation of T^*X by the fiberwise symplectic form ω^V. Let $\kappa^F : \Omega^{\cdot}(T^*X, \pi^*F) \to \Omega^{2n-\cdot}\left(T^*X, \pi^*\overline{F}^*\right)$ be the linear map such that if $s \in \Omega^i(T^*X, \pi^*F)$, then

$$\kappa^F s = (-1)^{i(i+1)/2}\, u *^{T^*X} g^F s. \tag{2.7.1}$$

Set

$$\kappa^F_{\mathcal{H}-\omega^H} = \kappa^F e^{-2\left(\mathcal{H}-\omega^H\right)}. \tag{2.7.2}$$

Temporarily, we will use the notation $A^{\mathcal{M},F}_{\phi,\mathcal{H}-\omega^H}$ instead of $A^{\mathcal{M}}_{\phi,\mathcal{H}-\omega^H}$, the notation for other objects being modified in the same way. In [B05, eq.

(4.91)], it is shown that if $\mathcal{H}$ is r-invariant,

$$\begin{aligned} A^{\mathcal{M},F}_{\phi,\mathcal{H}-\omega^H} &= \kappa^{F,-1}_{\mathcal{H}-\omega^H} A^{\mathcal{M},\overline{F}^*}_{\phi,-\mathcal{H}-\omega^H} \kappa^F_{\mathcal{H}-\omega^H}, \\ B^{\mathcal{M},F}_{\phi,\mathcal{H}-\omega^H} &= -\kappa^{F,-1}_{\mathcal{H}-\omega^H} B^{\mathcal{M},\overline{F}^*}_{\phi,-\mathcal{H}-\omega^H} \kappa^F_{\mathcal{H}-\omega^H}, \\ \mathfrak{C}^{\mathcal{M},F}_{\phi,\mathcal{H}-\omega^H} &= \kappa^{F,-1} \mathfrak{C}^{\mathcal{M},\overline{F}^*}_{\phi,-\mathcal{H}-\omega^H} \kappa^F, \\ \mathfrak{D}^{\mathcal{M},F}_{\phi,\mathcal{H}-\omega^H} &= -\kappa^{F,-1} \mathfrak{D}^{\mathcal{M},\overline{F}^*}_{\phi,-\mathcal{H}-\omega^H} \kappa^F. \end{aligned} \tag{2.7.3}$$

It is interesting to relate (1.4.3) and (2.7.3) in the light of Theorem 2.6.1. Of course here we take $\mathcal{H}$ as in (2.2.1). Although our statement may look cryptic, let us just mention that the extra factor $(-1)^{ni}$ in (1.4.1) with respect to to (2.7.1) can be explained by commutation with $dp_1 \wedge \ldots \wedge dp_n$. The extra factor $(-1)^n$ in (1.4.3) with respect to (2.7.3) appears for the same reason.

2.8 A 2-PARAMETER RESCALING

We make the same assumptions as in section 2.4. We still consider the case where $\mathcal{H} : \mathcal{M} \to \mathbf{R}$ is an arbitrary smooth function.

In this section, M, S will be replaced by $M \times \mathbf{R}^{*2}_+, S \times \mathbf{R}^{*2}_+$. For $t \in \mathbf{R}^*_+$, set

$$g^{TX}_t = \frac{g^{TX}}{t}. \tag{2.8.1}$$

Along the fiber X_s over (s, b, t), we equip TX with the metric g^{TX}_t. Let $\overline{\mathcal{H}}$ be the function defined on $\mathcal{M} \times \mathbf{R}^{*2}_+$, which restricts to $r^*_{t/b}\mathcal{H}$ on $\mathcal{M} \times (b, t)$.

We will denote with the subscript $\phi, \overline{\mathcal{H}}$ the objects which were considered in the previous sections. Still, when restricting these objects to $\mathcal{M} \times (b, t)$, so that $dt = 0, db = 0$, we will denote them with the subscript b, t. For instance $A^{\mathcal{M}}_{b,t}, B^{\mathcal{M}}_{b,t}$ denote the restriction of $A^{\mathcal{M}\times\mathbf{R}^{*2}_+}_{\overline{\mathcal{H}}-\omega^H}, B^{\mathcal{M}\times\mathbf{R}^2}_{\overline{\mathcal{H}}-\omega^H}$ to given values of b, t.

Let N^{T^*X} be the number operator of $\Lambda^\cdot (T^*T^*X)$. Using (2.1.3), we find that N^{T^*X} splits as

$$N^{T^*X} = N^H + N^V, \tag{2.8.2}$$

where N^H, N^V are the number operators of $\Lambda^\cdot (T^*X)$ and $\Lambda^\cdot (TX)$.

Set

$$U_{b,t} = t^{N^{T^*X}/2} r^*_{b/t}. \tag{2.8.3}$$

It will often be convenient to conjugate the objects which were considered above by the operator $U_{b,t}$.

In this section, $\overline{A}'^{\mathcal{M}}_{\phi_b}$ denotes the object constructed in Remark 2.4.1, which is associated to the fixed metric g^{TX}, and b is kept fixed. Similarly, $\mathfrak{C}^{\mathcal{M}}_{\phi_b,\mathcal{H}-b\omega^H}, \mathfrak{D}^{\mathcal{M}}_{\phi_b,\mathcal{H}-b\omega^H}$ denote the corresponding objects associated to the metric g^{TX}, with b fixed. Set

$$E = \left\langle L_{f^H_\alpha} g^{TX} e_i, e_j \right\rangle f^\alpha \wedge e^i \wedge i_{\widehat{e}^j}. \tag{2.8.4}$$

Using (1.3.1), we can rewrite (2.8.4) in the form

$$E = 2\left\langle T\left(f_\alpha^H, e_i\right), e_j\right\rangle f^\alpha \wedge e^i \wedge i_{\widehat{e}^j}. \tag{2.8.5}$$

As we explained in Remark 2.4.1, the operators $\mathfrak{A}_{\phi_b,\mathcal{H}}, \mathfrak{B}_{\phi_b,\mathcal{H}}$ are well defined, as are $\overline{A}^{\prime\mathcal{M}}_{\phi_b}, \mathfrak{E}^{\mathcal{M}}_{\phi_b,\mathcal{H}-b\omega^H}, \mathfrak{F}^{\mathcal{M}}_{\phi_b,\mathcal{H}-b\omega^H}$.

Let τ_b be the morphism of $\Lambda^\cdot(T^*X)\widehat{\otimes}\widehat{\Lambda}(TX)$ which acts trivially on $\Lambda^\cdot(T^*X)$ and which maps $\widehat{e}_i$ into $b\widehat{e}_i$ and $i_{\widehat{e}^i}$ into $i_{\widehat{e}^i}/b$. Then τ_b extends to $\Lambda^\cdot(T^*X)\widehat{\otimes}\widehat{\Lambda}^\cdot(TX)\widehat{\otimes}F$.

Finally, by analogy with (1.7.11), set

$$\mathfrak{C}^{\mathcal{M}}_{\phi_b,\mathcal{H}-b\omega^H,t} = \psi_t^{-1}\sqrt{t}\mathfrak{C}^{\mathcal{M}}_{\phi_b,\mathcal{H}-b\omega^H}\psi_t, \quad \mathfrak{D}^{\mathcal{M}}_{\phi_b,\mathcal{H}-b\omega^H,t} = \psi_t^{-1}\sqrt{t}\mathfrak{D}^{\mathcal{M}}_{\phi_b,\mathcal{H}-b\omega^H}\psi_t, \tag{2.8.6}$$

$$\mathfrak{E}^{\mathcal{M}}_{\phi_b,\mathcal{H}-b\omega^H,t} = \psi_t^{-1}\sqrt{t}\mathfrak{E}^{\mathcal{M}}_{\phi_b,\mathcal{H}-b\omega^H}\psi_t, \quad \mathfrak{F}^{\mathcal{M}}_{\phi_b,\mathcal{H}-b\omega^H,t} = \psi_t^{-1}\sqrt{t}\mathfrak{F}^{\mathcal{M}}_{\phi_b,\mathcal{H}-b\omega^H}\psi_t.$$

Theorem 2.8.1. *The following identities hold:*

$$\begin{aligned}
U_{b,t}\mathfrak{C}^{\prime\mathcal{M}\times\mathbf{R}_+^{*2}}_{\mathcal{H}-\omega^H}U_{b,t}^{-1} &= e^{-(\mathcal{H}-b\omega^H/t)}t^{N^{T^*X}/2}A^{\prime\mathcal{M}}t^{-N^{T^*X}/2}e^{(\mathcal{H}-b\omega^H/t)}\\
&+ dt\left(\frac{\partial}{\partial t} + \frac{b}{t^2}L_{\widehat{p}} + \frac{\nabla^V_{\widehat{p}}\mathcal{H}}{t} - \frac{N^{T^*X}}{2t}\right) + db\left(\frac{\partial}{\partial b} - \frac{L_{\widehat{p}}}{t} - \frac{\nabla_{\widehat{p}}\mathcal{H}}{b}\right),
\end{aligned} \tag{2.8.7}$$

$$\begin{aligned}
U_{b,t}\overline{\mathfrak{C}}^{\prime\mathcal{M}\times\mathbf{R}_+^{*2}}_{\phi,\overline{\mathcal{H}}-\omega^H}U_{b,t}^{-1} &= e^{(\mathcal{H}-b\omega^H/t)}t^{-N^{T^*X}/2}\overline{A}^{\prime\mathcal{M}}_{\phi_b}t^{N^{T^*X}/2}e^{-(\mathcal{H}-b\omega^H/t)}\\
&+ dt\left(\frac{\partial}{\partial t} + \frac{b}{t^2}L_{\widehat{p}} - \frac{\nabla^V_{\widehat{p}}\mathcal{H}}{t} - \frac{N^{T^*X}}{2t} + \frac{\lambda_0}{tb}\right) + db\left(\frac{\partial}{\partial b} - \frac{L_{\widehat{p}}}{t} + \frac{\nabla^V_{\widehat{p}}\mathcal{H}}{b}\right).
\end{aligned}$$

Moreover,

$$\begin{aligned}
U_{b,t}\mathfrak{C}^{\mathcal{M}\times\mathbf{R}_+^{*2}}_{\phi,\overline{\mathcal{H}}-\omega^H}U_{b,t}^{-1} &= \sqrt{t}\mathfrak{A}_{\phi_b,\mathcal{H}} + \nabla^{\Omega^\cdot(T^*X,\pi^*F)} - \frac{1}{2b}E\\
&+ \frac{1}{2}f^\alpha\omega\left(\nabla^F, g^F\right)\left(f_\alpha^H\right) + i_{T^H/\sqrt{t}} - bd^{T^*X}\omega^H/\sqrt{t}\\
&- \frac{1}{2\sqrt{t}}\left\langle T^H, e^i\right\rangle\left(e^i + i_{\widehat{e}^i/b}\right) + dt\left(\frac{\partial}{\partial t} + \frac{b}{t^2}L_{\widehat{p}} - \frac{N^{T^*X}}{2t} + \frac{\lambda_0}{2tb}\right)\\
&+ db\left(\frac{\partial}{\partial b} - \frac{L_{\widehat{p}}}{t}\right),
\end{aligned} \tag{2.8.8}$$

$$\begin{aligned}
U_{b,t}\mathfrak{D}^{\mathcal{M}\times\mathbf{R}_+^{*2}}_{\phi,\overline{\mathcal{H}}-\omega^H}U_{b,t}^{-1} &= \sqrt{t}\mathfrak{B}_{\phi_b,\mathcal{H}} - \frac{1}{2b}E + f^\alpha\left(\frac{1}{2}\omega\left(\nabla^F, g^F\right)\left(f_\alpha^H\right) - \nabla_{f_\alpha^H}\mathcal{H}\right)\\
&- \frac{1}{2\sqrt{t}}\left\langle T^H, e^i\right\rangle\left(e^i + i_{\widehat{e}^i/b}\right) - \left(\frac{dt}{t} - \frac{db}{b}\right)\nabla^V_{\widehat{p}}\mathcal{H} + \frac{dt}{2tb}\lambda_0.
\end{aligned}$$

Also,

$$U_{b,t}\mathfrak{C}^{\mathcal{M}\times\mathbf{R}_+^{*2}}_{b,t}U_{b,t}^{-1} = \mathfrak{C}^{\mathcal{M}}_{\phi_b,\mathcal{H}-b\omega^H,t}, \quad U_{b,t}\mathfrak{D}^{\mathcal{M}\times\mathbf{R}_+^{*2}}_{b,t}U_{b,t}^{-1} = \mathfrak{D}^{\mathcal{M}}_{\phi_b,\mathcal{H}-b\omega^H,t}. \tag{2.8.9}$$

Finally,

$$\begin{aligned}
\mathfrak{E}^{\mathcal{M}}_{\phi_b,\mathcal{H}-b\omega^H} &= e^{-\mu_0}\tau_b^{-1}\mathfrak{C}^{\mathcal{M}}_{\phi_b,\mathcal{H}-b\omega^H}\tau_b e^{\mu_0},\\
\mathfrak{F}^{\mathcal{M}}_{\phi_b,\mathcal{H}-b\omega^H} &= e^{-\mu_0}\tau_b^{-1}B^{\mathcal{M}}_{\phi_b,\mathcal{H}-b\omega^H}\tau_b e^{\mu_0}.
\end{aligned} \tag{2.8.10}$$

Proof. To establish the first identity in (2.8.7), we first make $\mathcal{H} = 0$, we replace ω^H by 0, and we make $dt = 0$. Then the first identity in (2.8.7) follows easily from the first identity in [B05, Proposition 4.33], which expresses the fact that the de Rham operator $d^{\mathcal{M}}$ on $\mathcal{M}$ commutes with the operators r_b^*. The equality of the dt parts follows from a trivial calculation which is left to the reader. Also note that

$$L_{\widehat{p}}\omega^H = \omega^H. \tag{2.8.11}$$

Using (2.8.11), we obtain the first identity in (2.8.7) by conjugation.

By [B05, Proposition 4.33], we get

$$r_b^* \overline{A}_\phi^{\prime\mathcal{M}} r_b^{*-1} = \overline{A}_{\phi_b}^{\prime\mathcal{M}}, \qquad \overline{A}_{\phi_b}^{\prime\mathcal{M}} = e^{\lambda_0/b} b^{N^{T^*X}} \overline{A}^{\prime\mathcal{M}} b^{-N^{T^*X}} e^{-\lambda_0/b}. \tag{2.8.12}$$

Observe that here, over $S \times (b,t)$, the metric is given by g^{TX}/t, so that λ_0 is replaced by λ_0/t. Then the second identity in (2.8.7) follows easily from (2.8.12).

To establish (2.8.8), we note first that the identity of the terms containing dt or db follows from (2.8.7). We can now make $dt = 0, db = 0$. Then (2.8.8) follows from [B05, Proposition 2.18, Theorem 4.23, and Proposition 4.33].

For $t = 1$, the identities in (2.8.9) follow from [B05, Proposition 4.33]. The case of a general t is now obvious by (2.8.8).

Finally, equation (2.8.10) follows from (2.4.7) and (2.4.15). The proof of our theorem is completed. □

Remark 2.8.2. Identity (2.8.9) already indicates that the above scaling in the t variable exhibits the same naturality properties as the scaling which is done in [B86], [BLo95], and [BG01]. Also note that when squaring any of the identities (2.8.7), we should get 0.

2.9 A GROUP ACTION

We make the same assumptions as in section 1.5. The action of G on M lifts to $\mathcal{M}$. Also G acts on $\Omega^{\cdot}(T^*X, \pi^*F)$ by a formula similar to (1.5.1). Moreover, by construction, $T^H\mathcal{M}$ is preserved by G. The geometric data we started with being G-invariant, the operators in [B05] which we just described commute with G. For example, $T^H\mathcal{M}$ is also preserved by G. The action of G on $\Omega^{\cdot}(X, F)$ lifts to $\Omega^{\cdot}(T^*X, \pi^*F)$. Moreover, the operators which we considered before commute with G.

Chapter Three

Hodge theory, the hypoelliptic Laplacian and its heat kernel

The purpose of this chapter is to give a short summary of the results on the analysis of the operator $A^2_{\phi,\mathcal{H}^c}$, these results being established in detail in chapters 15 and 17. In particular we state in detail convergence results on the resolvent of $\mathfrak{A}'^2_{\phi_b,\pm\mathcal{H}}$ as $b \to 0$ which are established in chapter 17. Also we derive various results on the spectral theory of $A^2_{\phi,\mathcal{H}^c}$. In particular we show that for $b > 0$ small enough, the standard consequences of Hodge theory hold, and we prove that the set of $b > 0$ where the Hodge theorem does hold is discrete. Finally, we prove that at finite distance, for $b > 0$ small enough, the spectrum of $A^2_{\phi,\mathcal{H}^c}$ is real.

This chapter is organized as follows. In section 3.1, we briefly describe the relations between the cohomology of X and the cohomology of T^*X.

In section 3.2, we state general results relating the finite dimensional kernel of $A^2_{\phi,\mathcal{H}^c}$ to the cohomology of X.

In section 3.3, we state the main properties of the heat kernel for $\mathfrak{A}^2_{\phi,\mathcal{H}^c}$.

In section 3.4, we state results on the convergence of the heat kernel for $\mathfrak{A}'^2_{\phi_b,\pm\mathcal{H}}$ to the heat kernel for $\square^X/4$ as $b \to 0$.

In section 3.5, we describe more precisely the spectrum of $\mathfrak{A}'^2_{\phi_b,\pm\mathcal{H}}$ as $b \to 0$.

In section 3.6, we show that the set of $b > 0$ where the Hodge theorem does not hold is discrete.

Finally, in section 3.7, we show that the above results extend to the case of families.

3.1 THE COHOMOLOGY OF T*X AND THE THOM ISOMORPHISM

As before, $o(TX)$ is the orientation bundle of TX. Note that T^*X is oriented by its symplectic form ω.

Let $k : X \to T^*X$ be the embedding of X as the zero section of T^*X. Let $H^{\cdot}(T^*X, \pi^*F)$ (resp. $H^{\mathrm{c},\cdot}(T^*X, \pi^*F)$) be the cohomology of T^*X (resp. with compact support) with coefficients in π^*F. Then, classically,

$$H^{\cdot}(T^*X, \pi^*F) = H^{\cdot}(X, F), \quad H^{\mathrm{c},\cdot}(T^*X, \pi^*F) = H^{\cdot-n}(X, F \otimes o(TX)). \tag{3.1.1}$$

The first isomorphism comes from the maps $k^* : \Omega^{\cdot}(T^*X, \pi^*F) \to \Omega^{\cdot}(X, F)$

and $\pi^* : \Omega^{\cdot}(X,F) \to \Omega^{\cdot}(T^*X, \pi^*F)$. The second is the Thom isomorphism. Namely, if $\left[\Phi^{T^*X}\right] \in H^{c,n}(T^*X, \pi^* \otimes o(TX))$ is the Thom class, then

$$s \in H^{c,\cdot}(T^*X, \pi^*F) \to \pi_* s \in H^{\cdot-n}(X, F \otimes o(TX)) \tag{3.1.2}$$

is the inverse to

$$s' \in H^{\cdot}(X, F \otimes o(TX)) \to \pi^* s' \wedge \left[\Phi^{T^*X}\right] \in H^{c,\cdot+n}(T^*X, \pi^*F). \tag{3.1.3}$$

It will be convenient to use the notation

$$\begin{aligned} \mathfrak{H}^{\cdot}(X,F) &= H^{\cdot}(X,F) \text{ if } c>0, \\ &= H^{\cdot-n}(X, F \otimes o(TX)) \text{ if } c<0. \end{aligned} \tag{3.1.4}$$

Equivalently, $\mathfrak{H}^{\cdot}(X,F)$ is just $H^{\cdot}(T^*X, \pi^*F)$ for $c>0$ and $H^{c,\cdot}(T^*X, \pi^*F)$ for $c<0$.

3.2 THE HODGE THEORY OF THE HYPOELLIPTIC LAPLACIAN

We make the same assumptions as in sections 2.1-2.3, and we use the corresponding notation. In particular g^{TX} denotes a Riemannian metric on X, and S is now reduced to a point. Moreover, we fix $c = \pm 1/b^2 \in \mathbf{R}^*$, with $b>0$. The constants in the estimations which follow will depend in general on b.

Let $\Omega^{\cdot}(T^*X, \pi^*F)$ now denote the vector space of smooth sections of $\Lambda^{\cdot}(T^*T^*X)\widehat{\otimes}F$ over T^*X. Let $\mathcal{S}^{\cdot}(T^*X, \pi^*F)$ be the vector space of the $s \in \Omega^{\cdot}(T^*X, \pi^*F)$ which decay, together with their derivatives of any order, faster that any $|p|^{-m}$, with $m \in \mathbf{N}$. For obvious reasons, we will use the notation $\mathfrak{h}^{\mathcal{S}^{\cdot}(T^*X,\pi^*F)}$ instead of $\mathfrak{h}^{\Omega^{\cdot}(T^*X,\pi^*F)}$, and a similar notation for the other Hermitian forms which were considered in chapter 2.

As in (2.2.1), (2.2.2), we use the notation

$$\mathcal{H} = \frac{|p|^2}{2}, \qquad \mathcal{H}^c = \frac{c}{2}|p|^2. \tag{3.2.1}$$

Recall that by (2.1.21), (2.1.22), (2.4.15),

$$\begin{aligned} &\mathfrak{A}_{\phi,\mathcal{H}^c} = e^{-\mathcal{H}^c} A_{\phi,\mathcal{H}^c} e^{\mathcal{H}^c}, \qquad \mathfrak{A}'_{\phi,\mathcal{H}^c} = e^{-\mathcal{H}^c-\mu_0} A_{\phi,\mathcal{H}^c} e^{\mathcal{H}^c+\mu_0}, \\ &\mathfrak{A}'_{\phi_b,\pm\mathcal{H}} = K_b \mathfrak{A}'_{\phi,\mathcal{H}^c} K_b^{-1}. \end{aligned} \tag{3.2.2}$$

By (3.2.2), we find that any statement we make for one of the operators is valid for the other one, when making the obvious changes. We will use this procedure constantly, without further mention, in particular when referring to spectral theory.

As was explained in section 2.1, $A_{\phi,\mathcal{H}^c}$ is self-adjoint with respect to $\mathfrak{h}_{\mathcal{H}^c}^{\mathcal{S}^{\cdot}(T^*X,\pi^*F)}$, $\mathfrak{A}_{\phi,\mathcal{H}^c}$ is self-adjoint with respect to $\mathfrak{h}^{\mathcal{S}^{\cdot}(T^*X,\pi^*F)}$, and $\mathfrak{A}'_{\phi,\mathcal{H}^c}$ is self-adjoint with respect to $h^{\mathcal{S}^{\cdot}(T^*X,\pi^*F)}$.

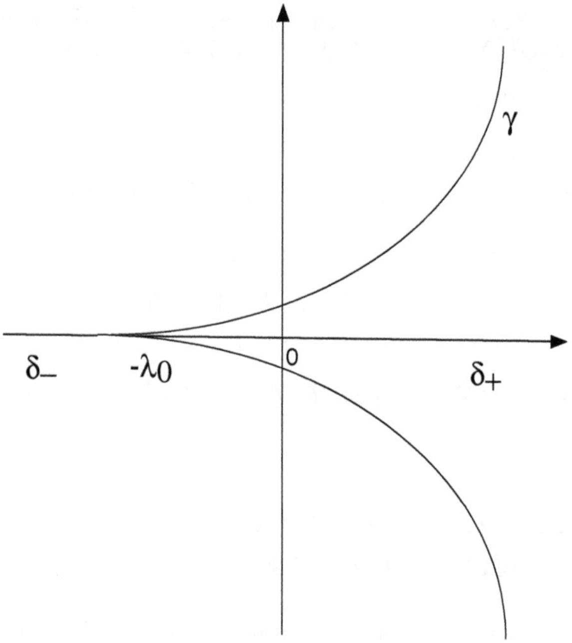

Figure 3.1

Let $\Omega^{\cdot}(T^*X, \pi^*F)^0$ be the space of sections of $\Lambda^{\cdot}(T^*T^*X)\widehat{\otimes}\pi^*F$ which are square integrable over T^*X. By Theorem 15.7.1, the operator $\mathfrak{A}^2_{\phi,\mathcal{H}^c}$ has discrete spectrum and compact resolvent in $\operatorname{End}\left(\Omega^{\cdot}(T^*X, \pi^*F)^0\right)$.

To make our terminology unambiguous, let us just say that elements of the spectrum will be called eigenvalues.

Moreover, the corresponding characteristic spaces are finite dimensional and included in $\mathcal{S}^{\cdot}(T^*X, \pi^*F)$. Let us just mention that given an eigenvalue λ, the associated characteristic space is the analogue of a Jordan block in finite dimensions. It is the image of the spectral projector associated with the eigenvalue λ.

Take $\lambda \in \mathbf{C}\backslash \operatorname{Sp}\mathfrak{A}^2_{\phi,\mathcal{H}^c}$. Then the resolvent $\left(\mathfrak{A}^2_{\phi,\mathcal{H}^c} - \lambda\right)^{-1}$ maps bijectively $\mathcal{S}^{\cdot}(T^*X, \pi^*F)$ into itself.

Let γ be the contour in $\mathbf{C}$ in Figure 3.1, which separates the closed domains $\delta_\pm$, which contain $\pm\infty$. The precise description of γ is as follows. Given $b > 0, c = \pm 1/b^2$, constants $\lambda_0 > 0, c_0 > 0$ depending on $b > 0$ are defined in Theorem 15.7.1, so that

$$\gamma = \left\{\lambda = -\lambda_0 + \sigma + i\tau, \sigma, \tau \in \mathbf{R}, \sigma = c_0 \left|\tau\right|^{1/6}\right\}. \tag{3.2.3}$$

Note that γ depends explicitly on b.

By Theorem 15.7.1,

$$\operatorname{Sp}\mathfrak{A}^2_{\phi,\mathcal{H}^c} \subset \overset{\circ}{\delta}_+. \tag{3.2.4}$$

First we prove the obvious extension of [B05, Proposition 1.1].

Proposition 3.2.1. *The spectrum* $\operatorname{Sp}\mathfrak{A}^2_{\phi,\mathcal{H}^c}$ *is invariant by conjugation. Moreover, given* $b>0$, *if* $\lambda \in \operatorname{Sp}\mathfrak{A}^2_{\phi,\mathcal{H}^c}$, *then*

$$\operatorname{Re}\lambda \geq -\lambda_0. \tag{3.2.5}$$

Proof. Let $\mathfrak{A}^*_{\phi,\mathcal{H}^c}$ be the formal adjoint of $\mathfrak{A}_{\phi,\mathcal{H}^c}$ with respect to the Hermitian product $\mathfrak{g}^{\Omega^\cdot(T^*X,\pi^*F)}$ considered in Definition 2.1.1. If u is the isometric involution of $\Omega^\cdot(T^*X,\pi^*F)$ defined in (2.1.11), we deduce from (2.1.12) and from the above that

$$\mathfrak{A}^*_{\phi,\mathcal{H}^c} = u\mathfrak{A}_{\phi,\mathcal{H}^c}u^{-1}. \tag{3.2.6}$$

By (3.2.6), we find that $\mathfrak{A}^2_{\phi,\mathcal{H}^c}$ and $\mathfrak{A}^{2*}_{\phi,\mathcal{H}^c}$ have the same spectrum. Therefore $\operatorname{Sp}\mathfrak{A}^2_{\phi,\mathcal{H}^c}$ is conjugation-invariant. Since δ_+ contains the spectrum of $\mathfrak{A}^2_{\phi,\mathcal{H}^c}$, (3.2.5) is obvious. □

If $\lambda \in \operatorname{Sp}\mathfrak{A}^2_{\phi,\mathcal{H}^c}$, let $\mathcal{S}^\cdot(T^*X,\pi^*F)_\lambda \in \mathcal{S}^\cdot(T^*X,\pi^*F)$ be the corresponding characteristic subspace. If $n_\lambda = \dim \mathcal{S}^\cdot(T^*X,\pi^*F)_\lambda$, then

$$\mathcal{S}^\cdot(T^*X,\pi^*F)_\lambda = \ker\left(\mathfrak{A}^2_{\phi,\mathcal{H}^c} - \lambda\right)^{n_\lambda}. \tag{3.2.7}$$

Recall that $d^{T^*X}_{\mathcal{H}^c}$ commutes with $\mathfrak{A}^2_{\phi,\mathcal{H}^c}$, and so $d^{T^*X}_{\mathcal{H}^c}$ acts on $\mathcal{S}^\cdot(T^*X,\pi^*F)_\lambda$. Then $\left(\mathcal{S}^\cdot(T^*X,\pi^*F)_\lambda, d^{T^*X}_{\mathcal{H}^c}\right)$ is a subcomplex of $\left(\mathcal{S}^\cdot(T^*X,\pi^*F), d^{T^*X}_{\mathcal{H}^c}\right)$.

Consider the subcomplex $\left(\mathcal{S}^\cdot(T^*X,\pi^*F)_0, d^{T^*X}_{\mathcal{H}^c}\right)$. By (3.2.7),

$$\mathcal{S}^\cdot(T^*X,\pi^*F)_0 = \ker\mathfrak{A}^{2n_0}_{\phi,\mathcal{H}^c}. \tag{3.2.8}$$

Take $\epsilon>0$ small enough so that the disk of center 0 and radius ϵ intersects $\operatorname{Sp}\mathfrak{A}^2_{\phi,\mathcal{H}^c}$ only possibly at 0. Let δ be the circle of center 0 and radius ϵ. Set

$$\mathfrak{P} = \frac{1}{2i\pi}\int_\delta \frac{d\lambda}{\lambda - \mathfrak{A}^2_{\phi,\mathcal{H}^c}}, \qquad \mathfrak{Q} = 1 - \mathfrak{P}. \tag{3.2.9}$$

Then $\mathfrak{P}$ is a projector on $\mathcal{S}^\cdot(T^*X,\pi^*F)_0$, and $\mathfrak{Q}$ is a complementary projector. Moreover, $\mathfrak{P}$ does not depend on ϵ. In particular $\mathfrak{P}$ and $\mathfrak{Q}$ map $\mathcal{S}^\cdot(T^*X,\pi^*F)$ into itself. Set

$$\mathcal{S}^\cdot(T^*X,\pi^*F)_* = \operatorname{Im}\mathfrak{Q}|_{\mathcal{S}^\cdot(T^*X,\pi^*F)}. \tag{3.2.10}$$

Then

$$\mathcal{S}^\cdot(T^*X,\pi^*F) = \mathcal{S}^\cdot(T^*X,\pi^*F)_0 \oplus \mathcal{S}^\cdot(T^*X,\pi^*F)_*. \tag{3.2.11}$$

Also the operators $d^{T^*X}_{\mathcal{H}^c}$ and $\overline{d}^{T^*X}_{\phi,\mathcal{H}^c}$ preserve the above splitting.

Moreover, $\mathfrak{A}^2_{\phi,\mathcal{H}^c}$ acts as an invertible operator on $\mathcal{S}^\cdot(T^*X,\pi^*F)_*$. Indeed let $\Omega^\cdot(T^*X,\pi^*F)^0_*$ be the image of $\Omega^\cdot(T^*X,\pi^*F)^0$ by the projector $\mathfrak{Q}$. Then 0 does not lie in the spectrum of the restriction of $\mathfrak{A}^2_{\phi,\mathcal{H}^c}$ to $\mathcal{S}^\cdot(T^*X,\pi^*F)^0_*$, so that $\left(\mathfrak{A}^2_{\phi,\mathcal{H}^c}\right)^{-1}$ acts as a bounded operator on this vector space. Using equation (15.5.3) in Theorem 15.5.1, we find that $\left(\mathfrak{A}^2_{\phi,\mathcal{H}^c}\right)^{-1}$ maps $\mathcal{S}^\cdot(T^*X,\pi^*F)_*$ into itself.

By (3.2.8), we find that the splitting (3.2.11) is just the splitting

$$\mathcal{S}^{\cdot}(T^*X, \pi^*F) = \ker \mathfrak{A}^{2n_0}_{\phi,\mathcal{H}^c} \oplus \mathrm{Im}\mathfrak{A}^{2n_0}_{\phi,\mathcal{H}^c}. \tag{3.2.12}$$

Of course, in the above, we can as well replace n_0 by $k \in \mathbf{N}, k \geq n_0$.

Now we establish the obvious analogue of [B05, Theorem 1.2].

Theorem 3.2.2. *The complex* $\left(\mathcal{S}^{\cdot}(T^*X, \pi^*F)_*, d^{T^*X}_{\mathcal{H}^c}\right)$ *is exact. In particular*

$$H^{\cdot}\left(\mathcal{S}^{\cdot}(T^*X, \pi^*F), d^{T^*X}_{\mathcal{H}^c}\right) = H^{\cdot}\left(\mathcal{S}^{\cdot}(T^*X, \pi^*F)_0, d^{T^*X}_{\mathcal{H}^c}\right). \tag{3.2.13}$$

The vector spaces $\mathcal{S}^{\cdot}(T^*X, \pi^*F)^{\cdot}_0$ *and* $\mathcal{S}^{\cdot}(T^*X, \pi^*F)^{\cdot}_*$ *are orthogonal with respect to* $\mathfrak{h}^{\mathcal{S}^{\cdot}(T^*X,\pi^*F)}$. *The restrictions of* $\mathfrak{h}^{\mathcal{S}^{\cdot}(T^*X,\pi^*F)}$ *to* $\mathcal{S}^{\cdot}(T^*X, \pi^*F)^{\cdot}_0$ *and* $\mathcal{S}^{\cdot}(T^*X, \pi^*F)^{\cdot}_*$ *are nondegenerate. Moreover,*

$$\mathcal{S}^{\cdot}(T^*X, \pi^*F)^{\cdot}_* = \mathrm{Im}\, d^{T^*X}_{\mathcal{H}^c}|_{\mathcal{S}^{\cdot}(T^*X,\pi^*F)_*} \oplus \mathrm{Im}\, \overline{d}^{T^*X}_{\phi,\mathcal{H}^c}|_{\mathcal{S}^{\cdot}(T^*X,\pi^*F)^{\cdot}_*}, \tag{3.2.14}$$

and the decomposition (3.2.14) is $\mathfrak{h}^{\mathcal{S}^{\cdot}(T^*X,\pi^*F)}$ *orthogonal. Finally, the map* $s \in \mathcal{S}^{\cdot}(T^*X, \pi^*F) \to e^{\mathcal{H}^c}s \in \Omega^{\cdot}(T^*X, \pi^*F)$ *induces the canonical isomorphism*

$$H^{\cdot}\left(\mathcal{S}^{\cdot}(T^*X, \pi^*F), d^{T^*X}_{\mathcal{H}^c}\right) \simeq \mathfrak{H}^{\cdot}(X, F). \tag{3.2.15}$$

Proof. We proceed as in the proof of [B05, Theorem 1.2]. Since $\left(\mathfrak{A}^2_{\phi,\mathcal{H}^c}\right)^{-1}$ acts on $\mathcal{S}^{\cdot}(T^*X, \pi^*F)_*$, we get the identity in $\mathrm{End}(\mathcal{S}^{\cdot}(T^*X, \pi^*F)_*)$,

$$1 = \left[d^{T^*X}_{\mathcal{H}^c}, \overline{d}^{T^*X}_{\phi,\mathcal{H}^c}\left(\mathfrak{A}^2_{\phi,\mathcal{H}^c}\right)^{-1}\right]. \tag{3.2.16}$$

From (3.2.16), we deduce that $\left(\mathcal{S}^{\cdot}(T^*X, \pi^*F)_*, d^{T^*X}_{\mathcal{H}^c}\right)$ is exact. By (3.2.11), we get (3.2.13). Using (3.2.8) and the fact that $\mathfrak{A}^2_{\phi,\mathcal{H}^c}$ is $\mathfrak{h}^{\Omega^{\cdot}(T^*X,\pi^*F)}$ self-adjoint, if $a \in \mathcal{S}^{\cdot}(T^*X, \pi^*F)_0, a' \in \mathcal{S}^{\cdot}(T^*X, \pi^*F)$, we get

$$\left\langle a, \mathfrak{A}^{2n_0}_{\phi,\mathcal{H}^c} a'\right\rangle_{\mathfrak{h}^{\mathcal{S}^{\cdot}(T^*X,\pi^*F)}} = 0. \tag{3.2.17}$$

Since $\mathfrak{A}^{2n_0}_{\phi,\mathcal{H}^c}$ acts as an invertible operator on $\mathcal{S}^{\cdot}(T^*X, \pi^*F)_*$, we deduce from (3.2.17) that $\mathcal{S}^{\cdot}(T^*X, \pi^*F)_0$ and $\mathcal{S}^{\cdot}(T^*X, \pi^*F)_*$ are mutually $\mathfrak{h}^{\mathcal{S}^{\cdot}(T^*X,\pi^*F)}$ orthogonal. Since $\mathfrak{h}^{\mathcal{S}^{\cdot}(T^*X,\pi^*F)}$ is nondegenerate, we get the second part of our theorem. By (3.2.16) the images of $d^{T^*X}_{\mathcal{H}^c}$ and of $\overline{d}^{T^*X}_{\phi,\mathcal{H}^c}$ in $\mathcal{S}^{\cdot}(T^*X, \pi^*F)_*$ span $\mathcal{S}^{\cdot}(T^*X, \pi^*F)_*$. Moreover, since $\overline{d}^{T^*X}_{\phi,\mathcal{H}^c}$ is the $\mathfrak{h}^{\mathcal{S}^{\cdot}(T^*X,\pi^*F)}$ adjoint of $d^{T^*X}_{\mathcal{H}^c}$, these images are orthogonal with respect to $\mathfrak{h}^{\mathcal{S}^{\cdot}(T^*X,\pi^*F)}$. Since the restriction of $\mathfrak{h}^{\mathcal{S}^{\cdot}(T^*X,\pi^*F)}$ to $\mathcal{S}^{\cdot}(T^*X, \pi^*F)_*$ is nondegenerate, we get (3.2.14).

Now we establish (3.2.15). Let $\widehat{d}^{T^*X}$ be the de Rham operator along the fiber T^*X. Set

$$\widehat{d}^{T^*X}_{\mathcal{H}^c} = e^{-\mathcal{H}^c}\widehat{d}^{T^*X}e^{\mathcal{H}^c}. \tag{3.2.18}$$

Let $\widehat{d}^{T^*X*}, \widehat{d}^{T^*X}_{\mathcal{H}^c}$ be the obvious formal adjoints of $\widehat{d}^{T^*X}, \widehat{d}^{T^*X}_{\mathcal{H}^c}$ with respect to the standard Hermitian product on smooth forms along the fibers of T^*X. Clearly,

$$\widehat{d}^{T^*X}_{\mathcal{H}^c} = \widehat{d}^{T^*X} + c\widehat{p}\wedge, \qquad \widehat{d}^{T^*X*}_{\mathcal{H}^c} = \widehat{d}^{T^*X*} + ci_{\widehat{p}}. \tag{3.2.19}$$

Let $\widehat{\square}^{T^*X}_{\mathcal{H}^c} = \left[\widehat{d}^{T^*X}_{\mathcal{H}^c}, \widehat{d}^{T^*X*}_{\mathcal{H}^c}\right]$ be the corresponding Laplacian. Recall that N^V is the number operator of $\Lambda^{\cdot}(TX)$, the exterior algebra along the fiber. An easy computation shows that if Δ^V is the scalar Laplacian along the fibers T^*X, then

$$\widehat{\square}^{T^*X}_{\mathcal{H}^c} = -\Delta^V + c^2 |p|^2 + 2c\left(N^V - n\right). \tag{3.2.20}$$

It is now a basic observation of Witten [Wi82] that for $c > 0$, the kernel of $\widehat{\square}^{T^*X}_{\mathcal{H}^c}$ is concentrated in degree 0 and generated by $\exp\left(-c|p|^2/2\right)$, while for $c < 0$, it is concentrated in degree n and generated by $\exp\left(c|p|^2/2\right)\eta$, where η is a n form of norm 1 along the fibers of T^*X, which is defined up to sign. Using the basic properties of the harmonic oscillator (and in particular the fact that its resolvent acts on $\mathcal{S}(T^*X, \pi^*F)$), we deduce from the above that the cohomology of the fiberwise complex $\left(\mathcal{S}(T^*X, \pi^*F), \widehat{d}^{T^*X}_{\mathcal{H}^c}\right)$ is concentrated in degree 0 or n, and generated by the forms we just considered.

For $c > 0$, the function $\exp\left(-c|p|^2/2\right)$ is $d^{T^*X}_{\mathcal{H}^c}$ closed. Using the Leray-Hirsch theorem, we get (3.2.15) for $c > 0$. Similarly, if Φ^{T^*X} is the Mathai-Quillen Thom form of T^*X [MatQ86], as we saw in (2.3.4), the restriction of Φ^{T^*X} to one given fiber is given by $\exp\left(-|p|^2\right)\frac{\eta}{\pi^{n/2}}$. Using the above and the Leray-Hirsch theorem, we get (3.2.15) for $c < 0$. The proof of our theorem is completed. □

Theorem 3.2.3. *Take $\lambda, \mu \in \mathrm{Sp}\,\mathfrak{A}^2_{\phi,\mathcal{H}^c}, \overline{\lambda} \neq \mu$. Then $\mathcal{S}^{\cdot}(T^*X, \pi^*F)_\lambda$ and $\mathcal{S}^{\cdot}(T^*X, \pi^*F)_\mu$ are $\mathfrak{h}^{\mathcal{S}^{\cdot}(T^*X,\pi^*F)}$ orthogonal. If $\lambda \in \mathrm{Sp}\,\mathfrak{A}^2_{\phi,\mathcal{H}^c}$, the restriction of $\mathfrak{h}^{\mathcal{S}^{\cdot}(T^*X,\pi^*F)}$ to $\mathcal{S}^{\cdot}(T^*X, \pi^*F)_\lambda + \mathcal{S}^{\cdot}(T^*X, \pi^*F)_{\overline{\lambda}}$ is nondegenerate.*

Proof. The proof of our theorem is similar to the proof of [B05, Theorem 1.11]. Take λ, μ as above. Take $k \in \mathbf{N}$ large enough so that

$$\mathcal{S}^{\cdot}(T^*X, \pi^*F)_\lambda = \ker\left(\mathfrak{A}^2_{\phi,\mathcal{H}^c} - \lambda\right)^k. \tag{3.2.21}$$

If $a \in \mathcal{S}^{\cdot}(T^*X, \pi^*F)_\lambda, b \in \mathcal{S}^{\cdot}(T^*X, \pi^*F)_\mu$, since $\mathfrak{A}^2_{\phi,\mathcal{H}^c}$ is $\mathfrak{h}^{\mathcal{S}^{\cdot}(T^*X,\pi^*F)}$ self-adjoint, from (3.2.21), we get

$$\left\langle a, \left(\mathfrak{A}^2_{\phi,\mathcal{H}^c} - \overline{\lambda}\right)^k b\right\rangle_{\mathfrak{h}^{\mathcal{S}^{\cdot}(T^*X,\pi^*F)}} = 0. \tag{3.2.22}$$

Since $\mu \neq \overline{\lambda}$, the restriction of $\left(\mathfrak{A}^2_{\phi,\mathcal{H}^c} - \overline{\lambda}\right)^k$ to $\mathcal{S}^{\cdot}(T^*X, \pi^*F)_\mu$ is invertible. By (3.2.22), $\mathcal{S}^{\cdot}(T^*X, \pi^*F)_\lambda$ and $\mathcal{S}^{\cdot}(T^*X, \pi^*F)_\mu$ are $\mathfrak{h}^{\mathcal{S}^{\cdot}(T^*X,\pi^*F)}$ orthogonal.

Let δ be a small circle centered at λ. We define the projectors $\mathfrak{P}_\lambda, \mathfrak{Q}_\lambda$ by a formula similar to (3.2.9), so that $\mathfrak{P}_\lambda, \mathfrak{Q}_\lambda$ are projectors on supplementary subspaces $\mathcal{S}^{\cdot}(T^*X, \pi^*F)_\lambda, \mathcal{S}^{\cdot}(T^*X, \pi^*F)_{\lambda,*}$.

If $\lambda \in \mathbf{R}$, the same argument as in (3.2.22) shows that $\mathcal{S}^{\cdot}(T^*X, \pi^*F)_\lambda$ and $\mathcal{S}^{\cdot}(T^*X, \pi^*F)_{\lambda,*}$ are $\mathfrak{h}^{\mathcal{S}^{\cdot}(T^*X,\pi^*F)}$ orthogonal. Since $\mathfrak{h}^{\mathcal{S}^{\cdot}(T^*X,\pi^*F)}$ is nondegenerate, the restriction of $\mathfrak{h}^{\mathcal{S}^{\cdot}(T^*X,\pi^*F)}$ to $\mathcal{S}^{\cdot}(T^*X, \pi^*F)_\lambda$ is nondegenerate.

If $\lambda \notin \mathbf{R}$, then $\mathfrak{P}_\lambda, \mathfrak{P}_{\overline{\lambda}}$ are commuting projectors such that

$$\mathfrak{P}_\lambda \mathfrak{P}_{\overline{\lambda}} = \mathfrak{P}_{\overline{\lambda}} \mathfrak{P}_\lambda = 0. \tag{3.2.23}$$

Put

$$\mathcal{S}^{\cdot}(T^*X, \pi^*F)_{\lambda,\overline{\lambda}} = \mathcal{S}^{\cdot}(T^*X, \pi^*F)_\lambda \oplus \mathcal{S}^{\cdot}(T^*X, \pi^*F)_{\overline{\lambda}}. \tag{3.2.24}$$

Set

$$\mathfrak{P}_{\lambda,\overline{\lambda}} = \mathfrak{P}_\lambda + \mathfrak{P}_{\overline{\lambda}}, \qquad \mathfrak{Q}_{\lambda,\overline{\lambda}} = 1 - \mathfrak{P}_{\lambda,\overline{\lambda}}. \tag{3.2.25}$$

Then $\mathfrak{P}_{\lambda,\overline{\lambda}}$ is a projector on $\mathcal{S}^{\cdot}(T^*X, \pi^*F)_{\lambda,\overline{\lambda}}$, and $\mathfrak{Q}_{\lambda,\overline{\lambda}}$ is a projector on a vector space $\mathcal{S}^{\cdot}(T^*X, \pi^*F)_{\lambda,\overline{\lambda}*}$, so that

$$\mathcal{S}^{\cdot}(T^*X, \pi^*F) = \mathcal{S}^{\cdot}(T^*X, \pi^*F)_{\lambda,\overline{\lambda}} \oplus \mathcal{S}^{\cdot}(T^*X, \pi^*F)_{\lambda,\overline{\lambda}*}. \tag{3.2.26}$$

By proceeding as in (3.2.22), $\mathcal{S}^{\cdot}(T^*X, \pi^*F)_{\lambda,\overline{\lambda}}$ and $\mathcal{S}^{\cdot}(T^*X, \pi^*F)_{\lambda,\overline{\lambda}*}$ are $\mathfrak{h}^{\mathcal{S}^{\cdot}(T^*X,\pi^*F)}$ orthogonal. This shows that the restriction of $\mathfrak{h}^{\mathcal{S}^{\cdot}(T^*X,\pi^*F)}$ is nondegenerate. The proof of our theorem is completed. □

3.3 THE HEAT KERNEL FOR $\mathfrak{A}^2_{\phi,\mathcal{H}^c}$

Again we fix $c = \pm 1/b^2$.

The domain of the operator $\mathfrak{A}^2_{\phi,\mathcal{H}^c}$ is dense in $\Omega^{\cdot}(T^*X, \pi^*F)^0$ since it contains $\mathcal{S}^{\cdot}(T^*X, \pi^*F)$. Moreover, by equation (15.7.4) in Theorem 15.7.1, there exists $\lambda_0 > 0$ such that if $\lambda \in \mathbf{R}, \lambda \leq -\lambda_0$,

$$\left\| \left(\mathfrak{A}^2_{\phi,\mathcal{H}^c} - \lambda\right)^{-1} \right\| \leq C \left(1 + |\lambda|\right)^{-1}. \tag{3.3.1}$$

By the theorem of Hille-Yosida [Y68, section IX-7, p. 266], we find that there is a unique well-defined semigroup $\exp\left(-t\mathfrak{A}^2_{\phi,\mathcal{H}^c}\right)$.

Moreover, by (3.2.4) or by (15.7.3), if $\lambda \in \delta_-$, the resolvent $\left(\mathfrak{A}^2_{\phi,\mathcal{H}^c} - \lambda\right)^{-1}$ exists. Also by equation (15.7.3) in Theorem 15.7.1, if $\lambda \in \delta_-$, the operator $\left(\mathfrak{A}^2_{\phi,\mathcal{H}^c} - \lambda\right)^{-1}$ is compact, and

$$\left\| \left(\mathfrak{A}^2_{\phi,\mathcal{H}^c} - \lambda\right)^{-1} \right\| \leq \frac{C}{\left(1 + |\lambda|\right)^{1/6}}. \tag{3.3.2}$$

Proposition 3.3.1. *For $t > 0$, the heat operator $\exp\left(-t\mathfrak{A}^2_{\phi,\mathcal{H}^c}\right)$ is given by the contour integral*

$$\exp\left(-t\mathfrak{A}^2_{\phi,\mathcal{H}^c}\right) = \frac{1}{2i\pi} \int_\gamma e^{-t\lambda} \left(\lambda - \mathfrak{A}^2_{\phi,\mathcal{H}^c}\right)^{-1} d\lambda. \tag{3.3.3}$$

Proof. Let ϵ be the downward oriented straight line $\{\lambda \in \mathbf{C}, \operatorname{Re}\lambda = -\lambda_0\}$. We claim that we have the equality operators distributions in the variable $t > 0$,

$$\exp\left(-t\mathfrak{A}^2_{\phi,\mathcal{H}^c}\right) = \frac{1}{2i\pi} \int_\epsilon e^{-t\lambda} \left(\lambda - \mathfrak{A}^2_{\phi,\mathcal{H}^c}\right)^{-1} d\lambda. \tag{3.3.4}$$

Equation (3.3.4) is obvious by taking the Fourier transform in the variable $t > 0$ of the parabolic equation associated to the operator $\exp\left(-t\mathfrak{A}^2_{\phi,\mathcal{H}^c}\right)$ and then using the inverse Fourier transform.

Using (3.3.4) and integration by parts, we find that for any $N \in \mathbf{N}$,

$$\exp\left(-t\mathfrak{A}^2_{\phi,\mathcal{H}^c}\right) = \frac{(-1)^N N!}{2i\pi t^N} \int_\epsilon e^{-t\lambda} \left(\lambda - \mathfrak{A}^2_{\phi,\mathcal{H}^c}\right)^{-(N+1)} d\lambda. \tag{3.3.5}$$

Now by (3.3.2), we can choose N large enough so that the integral in (3.3.5) is converging absolutely in the space of bounded operators. Since the integrand in the right-hand side of (3.3.5) is holomorphic in $\lambda \in \delta_-$, by using the theorem of residues, we get

$$\frac{(-1)^N N!}{2i\pi t^N} \int_\epsilon e^{-t\lambda} \left(\lambda - \mathfrak{A}^2_{\phi,\mathcal{H}^c}\right)^{-(N+1)} d\lambda = \frac{(-1)^N N!}{2i\pi t^N} \int_\gamma e^{-t\lambda} \left(\lambda - \mathfrak{A}^2_{\phi,\mathcal{H}^c}\right)^{-(N+1)} d\lambda. \tag{3.3.6}$$

Finally, using the bound (3.3.2) and integration by parts, we get

$$\frac{1}{2i\pi} \int_\gamma e^{-t\lambda} \left(\lambda - \mathfrak{A}^2_{\phi,\mathcal{H}^c}\right)^{-1} d\lambda = \frac{(-1)^N N!}{2i\pi t^N} \int_\gamma e^{-t\lambda} \left(\lambda - \mathfrak{A}^2_{\phi,\mathcal{H}^c}\right)^{-(N+1)} d\lambda. \tag{3.3.7}$$

By (3.3.6), (3.3.7), we get (3.3.3). □

Using (3.3.2), (3.3.3) and integration by parts, we find that for any $N \in \mathbf{N}$,

$$\exp\left(-t\mathfrak{A}^2_{\phi,\mathcal{H}^c}\right) = \frac{(-1)^N N!}{2i\pi t^N} \int_\gamma e^{-t\lambda} \left(\lambda - \mathfrak{A}^2_{\phi,\mathcal{H}^c}\right)^{-(N+1)} d\lambda. \tag{3.3.8}$$

If A is an operator acting on $\Omega^{\cdot}\left(T^*X, \pi^*F\right)^0$ which is trace class, let $\|A\|_1$ be its norm as a trace class operator. By Theorem 15.7.1, if $\lambda \in \delta_-$, for $N \in \mathbf{N}, N > 12n$, the operator $\left(\mathfrak{A}^2_{\phi,\mathcal{H}^c} - \lambda\right)^{-N}$ is trace class, and

$$\left\| \left(\mathfrak{A}^2_{\phi,\mathcal{H}^c} - \lambda\right)^{-N} \right\|_1 \leq C\left(1 + |\lambda|\right)^N. \tag{3.3.9}$$

By (3.3.8), (3.3.9), we conclude that given $t > 0$, there exists $C_t > 0$

$$\left\| \exp\left(-t\mathfrak{A}^2_{\phi,\mathcal{H}^c}\right) \right\|_1 \leq C_t. \tag{3.3.10}$$

In section 15.2, a standard chain of Sobolev spaces H^s on T^*X is defined using the $s/2$ powers of the positive self-adjoint operator

$$S = -\Delta^H - \Delta^V + |p|^2. \tag{3.3.11}$$

In section 15.3, another chain of Sobolev spaces $\mathcal{H}^s$ on T^*X is also defined. Note that

$$H^0 = \mathcal{H}^0 = \Omega^{\cdot}\left(T^*X, \pi^*F\right)^0. \tag{3.3.12}$$

Moreover, in Remark 15.3.2, it is shown that for $s \in \mathbf{R}$, we have the continuous embedding $\mathcal{H}^s \subset H^s$, and also that for $s' \geq s$ large enough, we have the continuous embedding $H^{s'} \subset \mathcal{H}^s$. In particular,

$$H^\infty = \mathcal{H}^\infty = \mathcal{S}^{\cdot}\left(T^*X, \pi^*F\right). \tag{3.3.13}$$

By equation (15.7.6) in Theorem 15.7.1, if $\lambda \in \delta_-$, $\left(\mathfrak{A}^2_{\phi,\mathcal{H}^c} - \lambda\right)^{-1}$ maps $\mathcal{H}^s$ into $\mathcal{H}^{s+1/4}$ with a norm dominated by $C_s\left(1+|\lambda|\right)^{4|s|+1}$. Therefore, under the same conditions, $\left(\mathfrak{A}^2_{\phi,\mathcal{H}^c} - \lambda\right)^{-N}$ maps $\mathcal{H}^s$ into $\mathcal{H}^{s+N/4}$ with a norm dominated by $C_{s,N}\left(1+|\lambda|\right)^{(4|s|+1)N}$.

By (3.3.8), it follows that the operator $\exp\left(-t\mathfrak{A}^2_{\phi,\mathcal{H}^c}\right)$ is regularizing, i.e., it maps any $\mathcal{H}^s$ into $\mathcal{H}^\infty = \mathcal{S}^{\cdot}\left(T^*X, \pi^*F\right)$. More precisely given $s, s' \in \mathbf{R}$, this operator is continuous from $\mathcal{H}^s$ into $\mathcal{H}^{s'}$. From the above, it follows that $\exp\left(-t\mathfrak{A}^2_{\phi,\mathcal{H}^c}\right)$ maps H^s into $H^{s'}$ with a continuous norm. By standard arguments, $\exp\left(-t\mathfrak{A}^2_{\phi,\mathcal{H}^c}\right)$ has a smooth kernel $\exp\left(-t\mathfrak{A}^2_{\phi,\mathcal{H}^c}\right)\left((x,p),(x',p')\right)$ which is rapidly decreasing in the variables p, p' together with all the derivatives.

From the above it follows that

$$\operatorname{Tr}\left[\exp\left(-t\mathfrak{A}^2_{\phi,\mathcal{H}^c}\right)\right] = \int_{T^*X} \operatorname{Tr}\left[\exp\left(-t\mathfrak{A}^2_{\phi,\mathcal{H}^c}\right)(z,z)\right] dv_{T^*X}. \tag{3.3.14}$$

Of course in (3.3.14), instead of taking the trace, we may as well take the supertrace.

As we already saw, for $\lambda \in \delta_-$, the resolvent $\left(\mathfrak{A}^2_{\phi,\mathcal{H}^c} - \lambda\right)^{-1}$ is compact, the spectrum $\operatorname{Sp}\mathfrak{A}^2_{\phi,\mathcal{H}^c}$ is discrete. However, contrary to what is known for self-adjoint operators, we do not know if the closure of the direct sum of the characteristic subspaces of $\mathfrak{A}^2_{\phi,\mathcal{H}^c}$ spans $\Omega^{\cdot}\left(T^*X, \pi^*F\right)^0$.

Let $\lambda_n \in \mathbf{C}, n \in \mathbf{N}$ denote the elements of $\operatorname{Sp}\mathfrak{A}^2_{\phi,\mathcal{H}^c}$, counted with the multiplicity of the corresponding characteristic space. Then the $e^{-t\lambda_n}, n \in \mathbf{N}$ are the nonzero eigenvalues of the operator $\exp\left(-t\mathfrak{A}^2_{\phi,\mathcal{H}^c}\right)$. Indeed using equation (3.3.3), it is clear that the $e^{-t\lambda_n}$ are characteristic values of $\exp\left(-t\mathfrak{A}^2_{\phi,\mathcal{H}^c}\right)$. Moreover, since $\exp\left(-t\mathfrak{A}^2_{\phi,\mathcal{H}^c}\right)$ is a compact operator, it has a discrete spectrum, which can accumulate only at 0. Let $\mu \in \mathbf{C}$ be a nonzero eigenvalue. Let c be a small circle with center μ, such that the corresponding disk does not contain 0 or any eigenvalue other than μ. The corresponding spectral projector $\mathcal{P}_\mu$ can be written in the form

$$\mathcal{P}_\mu = \frac{1}{2i\pi}\int_c \frac{d\mu'}{\mu' - \exp\left(-t\mathfrak{A}^2_{\phi,\mathcal{H}^c}\right)}. \tag{3.3.15}$$

Since c does not contain 0, we can rewrite (3.3.15) in the form

$$\mathcal{P}_\mu = \exp\left(-t\mathfrak{A}^2_{\phi,\mathcal{H}^c}\right)\frac{1}{2i\pi}\int_c \frac{d\mu'}{\mu'\left(\mu' - \exp\left(-t\mathfrak{A}^2_{\phi,\mathcal{H}^c}\right)\right)}. \tag{3.3.16}$$

By (3.3.16), the spectral projector $\mathcal{P}_\mu$ is compact, and so it has finite range $\mathcal{E}_\mu$, which is the characteristic subspace for the operator $\exp\left(-t\mathfrak{A}^2_{\phi,\mathcal{H}^c}\right)$ associated to the eigenvalue μ. Then $\mathfrak{A}^2_{\phi,\mathcal{H}^c}$ commutes with $\exp\left(-t\mathfrak{A}^2_{\phi,\mathcal{H}^c}\right)$, and so it acts on $\mathcal{E}_\mu$. Note here that the fact that $\mathfrak{A}^2_{\phi,\mathcal{H}^c}$ is unbounded is irrelevant. Let $M \subset \mathbf{C}$ be the associated family of eigenvalues of $\mathfrak{A}^2_{\phi,\mathcal{H}^c}$ on $\mathcal{E}_\mu$. By (3.3.3), it is now clear that if $\lambda \in M$, then $e^{-t\lambda} = \mu$. This concludes the proof of our statement on the relation between the eigenvalues of $\mathfrak{A}^2_{\phi,\mathcal{H}^c}$ and the eigenvalues of $\exp\left(-t\mathfrak{A}^2_{\phi,\mathcal{H}^c}\right)$.

By Weyl's inequality [ReSi78, Theorem XIII.10.3, p. 318], we know that

$$\sum_{n\in\mathbf{N}} \left|e^{-t\lambda_n}\right| \leq \left\|\exp\left(-t\mathfrak{A}^2_{\phi,\mathcal{H}^c}\right)\right\|_1 . \tag{3.3.17}$$

Moreover, by Lidskii's theorem [ReSi78, Corollary, p. 328], we get

$$\sum_{n\in\mathbf{N}} e^{-t\lambda_n} = \operatorname{Tr}\left[\exp\left(-t\mathfrak{A}^2_{\phi,\mathcal{H}^c}\right)\right] . \tag{3.3.18}$$

Because of (3.2.4) and using the fact that $\mathfrak{A}^2_{\phi,\mathcal{H}^c}$ has compact resolvent, we find that given $a > 0$, the set $\{n \in \mathbf{N}, \operatorname{Re}\lambda_n \leq a\}$ is finite. Note that this also follows from (3.3.17).

3.4 UNIFORM CONVERGENCE OF THE HEAT KERNEL AS $\mathbf{b} \to 0$

In this section, we give uniform estimates on the resolvent of $\mathfrak{A}'^2_{\phi_b,\pm\mathcal{H}}$ when $b \to 0$.

Let $\delta = (\delta_0, \delta_1, \delta_2)$ with $\delta_0 \in \mathbf{R}, \delta_1 > 0, \delta_2 > 0$. Put

$$\mathcal{W}_\delta = \left\{\lambda \in \mathbf{C}, \operatorname{Re}\lambda \leq \delta_0 + \delta_1 \left|\operatorname{Im}\lambda\right|^{\delta_2}\right\}. \tag{3.4.1}$$

For $r > 0, b > 0$, set

$$\mathcal{W}_{\delta',b,r} = \left\{\lambda \in \mathcal{W}_{\delta'}/b^2, r\operatorname{Re}\lambda + 1 \leq \left|\operatorname{Im}\lambda\right|\right\}. \tag{3.4.2}$$

By Theorem 17.21.3, for any $r > 0$, there exists $b_0 > 0, \delta' = (\delta'_0, \delta'_1, \delta'_2)$ with $\delta'_0 \in]0,1], \delta'_1 > 0, \delta'_2 = 1/6$ such that if $b \in]0, b_0], \lambda \in \mathcal{W}_{\delta',b,r}$, the resolvent $\left(\mathfrak{A}'^2_{\phi_b,\pm\mathcal{H}} - \lambda\right)^{-1}$ exists. Moreover, it verifies the obvious uniform bounds with respect to the norms considered above.

Figure 3.2 should make this quite clear. Indeed we have denoted by γ_b the boundary of $W_{\delta'}/b^2$. Then the spectrum of $\mathfrak{A}'^2_{\phi_b,\pm\mathcal{H}}$ is located in a domain which lies either to the right of γ_b or inside the cone domain limited by the two lines indicated on Figure 3.2.

Now we use the same conventions as in section 2.3. In particular $\square^X$ is the Hodge Laplacian acting on $\Omega^{\cdot}(X, F)$ for $c > 0$, on $\Omega^{\cdot}(X, F \otimes o(TX))$ for $c < 0$.

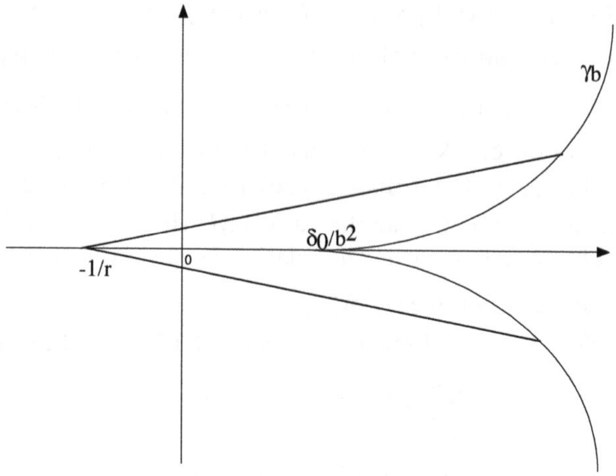

Figure 3.2

We fix $r > 0$. With the notation of section 3.2 and in Figure 3.1, by equation (17.21.58) in Theorem 17.21.5, there exists $b_0 \in]0,1]$ such that given $v \in]0,1[, \ell \in \mathbf{N}$, for $N \in \mathbf{N}^*$ large enough, if $b \in]0, b_0], \lambda \in \mathcal{W}_{\delta',b,r}$,

$$\left|\left|\left| \left(\mathfrak{A}'^2_{\phi_b,\pm\mathcal{H}} - \lambda\right)^{-(N+1)} - i_\pm \left(\Box^X/4 - \lambda\right)^{-(N+1)} P_\pm \right|\right|\right|_\ell \leq C_N b^v. \quad (3.4.3)$$

The precise definition of the norm in (3.4.3) is given just after equation (17.21.55).

More generally, by Remark 17.21.6, if $K \subset \mathbf{C} \setminus 0$ is a compact set such that $\mathrm{Sp}\,\Box^X/4 \cap K = \emptyset$, the estimates in (3.4.3) remain valid for $\lambda \in K$.

By the above, it follows that for any $\lambda_0 > 0$, if γ is a contour taken as in Figure 3.1, for r large enough and $b > 0$ small enough, γ lies entirely to the left of $\mathcal{W}_{\delta',b,r}$.

Using the uniform estimates (17.21.23), (17.21.24) in Theorem 17.21.3, for $b > 0$ small enough, we get

$$\exp\left(-t\mathfrak{A}'^2_{\phi,\mathcal{H}^c}\right) = \frac{1}{2i\pi}\int_\gamma e^{-t\lambda}\left(\lambda - \mathfrak{A}'^2_{\phi,\mathcal{H}^c}\right)^{-1} d\lambda. \quad (3.4.4)$$

By proceeding as in (3.3.8), for $N \in \mathbf{N}$, we also get

$$\exp\left(-t\mathfrak{A}'^2_{\phi_b,\pm\mathcal{H}}\right) = \frac{(-1)^N N!}{2i\pi t^N}\int_\gamma e^{-t\lambda}\left(\lambda - \mathfrak{A}'^2_{\phi_b,\pm\mathcal{H}}\right)^{-(N+1)} d\lambda. \quad (3.4.5)$$

Also we have the trivial

$$\exp\left(-t\Box^X/4\right) = \frac{1}{2i\pi}\int_\gamma e^{-t\lambda}\left(\lambda - \Box^X/4\right) d\lambda, \quad (3.4.6)$$

from which we also get

$$\exp\left(-t\Box^X/4\right) = \frac{(-1)^N N!}{2i\pi t^N}\int_\gamma e^{-t\lambda}\left(\lambda - \Box^X/4\right)^{-(N+1)} d\lambda. \quad (3.4.7)$$

By (3.4.3) and (3.4.5)-(3.4.7), we get

$$\left|\left|\left|\exp\left(-t\mathfrak{A}'^2_{\phi_b,\pm\mathcal{H}}\right) - i_\pm \exp\left(-t\square^X/4\right) P_\pm\right|\right|\right|_\ell \leq C_{t,L} b^v. \tag{3.4.8}$$

Let $\exp\left(-t\square^X\right)(x,x')$ be the smooth kernel for the operator $\exp\left(-t\square^X\right)$ with respect to the volume $dv_X(x')$. Then

$$\begin{aligned} i_+ \exp\left(-t\square^X\right) P_+ \left((x,p),(x',p')\right) \\ = \frac{1}{\pi^{n/2}} \exp\left(-|p|^2/2\right) \exp\left(-t\square^X\right)(x,x') \exp\left(-|p'|^2/2\right). \end{aligned} \tag{3.4.9}$$

There is a corresponding formula in the $-$ case, which is left to the reader.

Note that by the definition of the norms $|||\ |||_L$, the estimate (3.4.8) implies a corresponding estimate for the smooth kernels. From (3.4.8), (3.4.9), for any $m, m' \in \mathbf{N}$, and any multiindex α, $|\alpha| \leq m'$,

$$\left|(1+|p|)^m \partial^\alpha_{x,p}\left(\exp\left(-t\mathfrak{A}'^2_{\phi_b,\pm\mathcal{H}}\right) - i_\pm \exp\left(-t\square^X/4\right) P_\pm\right)\right.$$
$$\left.\left((x,p),(x',p')\right)\right| \leq C_{t,m,m'} b^v. \tag{3.4.10}$$

3.5 THE SPECTRUM OF $\mathfrak{A}'^2_{\phi_b,\pm\mathcal{H}}$ AS $\mathbf{b} \to 0$

Recall that by Proposition 3.2.1, the spectrum of $\mathfrak{A}'^2_{\phi_b,\pm\mathcal{H}}$ is conjugation-invariant. If $\lambda \in \mathrm{Sp}\,\mathfrak{A}'^2_{\phi_b,\pm\mathcal{H}}$, we still denote by $\mathcal{S}^\cdot\left(T^*X, \pi^*F\right)_\lambda$ the corresponding characteristic space.

Recall that $\mathrm{Sp}\,\square^X$ is real. Let $K \subset \mathbf{C} \setminus 0$ be a compact subset of $\mathbf{C}$ such that

$$\mathrm{Sp}\,\square^X/4 \cap K = \emptyset. \tag{3.5.1}$$

By Theorem 17.21.3 and by Remark 17.21.6, there exists $b_0 > 0$ such that for $b \in]0, b_0]$,

$$\mathrm{Sp}\,\mathfrak{A}'^2_{\phi_b,\pm\mathcal{H}} \cap K = \emptyset. \tag{3.5.2}$$

As we shall see later, the condition that $0 \notin K$ can easily be dropped.

By Figure 3.2, for $M > 0, b \in]0, b_0]$, $\{\lambda \in \mathrm{Sp}\,\mathfrak{A}'^2_{\phi_b,\pm\mathcal{H}}, \mathrm{Re}\,\lambda \leq M\}$ is uniformly bounded.

Let $M > 0$, and let $\epsilon \in]0, 1/2]$ be small enough so that the eigenvalues $\lambda \in \mathrm{Sp}\,\square^X/4, \lambda \leq M+1$ are spaced by more that 3ϵ. If $\lambda \in \mathrm{Sp}\,\square^X, \lambda \leq M$, let $c_\lambda \in \mathbf{C}$ be the circle of center λ and small radius ϵ. Let d_λ, d'_λ be the disks of center λ and radius $\epsilon, \epsilon/2$.

By the above, when 0 is an eigenvalue of $\square^X$, there exists $b_0 > 0$ such that for $b \in]0, b_0]$,

$$\left\{\mu \in \mathrm{Sp}\,\mathfrak{A}'^2_{\phi_b,\pm\mathcal{H}}, \mathrm{Re}\mu \leq M\right\} \subset \cup_{\substack{\lambda\in\mathrm{Sp}\,\square^X/4 \\ \mathrm{Re}\,\lambda \leq M}} d'_\lambda. \tag{3.5.3}$$

If 0 is not an eigenvalue of $\square^X$, one should in principle also add to the right-hand side the disk d'_0, which means that $\mathrm{Sp}\,\square^X/4$ should be replaced by $\mathrm{Sp}\,\square^X/4 \cup 0$. However, we will see shortly that this addition is unnecessary.

By Proposition 17.21.4 and by Remark 17.21.6, for $\ell \in \mathbf{N}$, for $N \in \mathbf{N}^*$ large enough, if μ lies in the finite union of the circles $c_\lambda, \mathrm{Re}\,\lambda \leq M$, we have uniform bounds on $\left|\left|\left|\left(\mathfrak{A}'^2_{\phi_b,\pm\mathcal{H}} - \mu\right)^{-N}\right|\right|\right|_\ell$.

Set

$$\mathfrak{P}^\lambda = \frac{1}{2i\pi}\int_{c_\lambda}\left(\mu - \mathfrak{A}'^2_{\phi_b,\pm\mathcal{H}}\right)^{-1} d\mu. \tag{3.5.4}$$

As explained in the proof of Theorem 15.7.1, the operator $\mathfrak{P}^\lambda$ is a compact projector which projects on a finite dimensional vector space $\mathcal{S}^{\cdot}\left(T^*X, \pi^*F\right)^\lambda$ included in $\mathcal{S}^{\cdot}\left(T^*X, \pi^*F\right)$. This vector space is the direct sum of the characteristic subspaces $\mathcal{S}^{\cdot}\left(T^*X, \pi^*F\right)_\mu$ for the $\mu \in \mathrm{Sp}\,\mathfrak{A}'^2_{\phi_b,\pm\mathcal{H}}$ which are included in the disk d_λ. We have the trivial inclusion

$$\mathcal{S}^{\cdot}\left(T^*X, \pi^*F\right)_\lambda \subset \mathcal{S}^{\cdot}\left(T^*X, \pi^*F\right)^\lambda. \tag{3.5.5}$$

For any $N \in \mathbf{N}$, we can reexpress the projector $\mathfrak{P}^\lambda$ in the form

$$\mathfrak{P}^\lambda = \frac{1}{2i\pi}\int_{c_\lambda}\mu^N\left(\mu - \mathfrak{A}'^2_{\phi_b,\pm\mathcal{H}}\right)^{-(N+1)} d\mu. \tag{3.5.6}$$

Set

$$P_\lambda = \frac{1}{2i\pi}\int_{c_\lambda}\left(\mu - \square^X/4\right)^{-1} d\mu. \tag{3.5.7}$$

Then P_λ is the spectral projection on the eigenspace E_λ of $\square^X/4$ associated to the eigenvalue λ. As in (3.5.6), we can write

$$P_\lambda = \frac{1}{2i\pi}\int_{c_\lambda}\mu^N\left(\mu - \square^X/4\right)^{-(N+1)} d\mu. \tag{3.5.8}$$

Using the bounds in (3.4.3) and the observations which follow, and proceeding as in (3.4.10), we easily deduce that for $b \in]0, b_0]$,

$$\left\|\mathfrak{P}^\lambda - i_\pm P_\lambda P_\pm\right\|_1 \leq C b^v. \tag{3.5.9}$$

By (3.5.9), we find that for $b > 0$ small enough,

$$\begin{aligned}\dim \mathcal{S}^{\cdot}\left(T^*X, \pi^*F\right)^\lambda &= \dim E^{\cdot}_\lambda \text{ if } c > 0,\\ &= \dim E^{\cdot-n}_\lambda \text{ if } c < 0.\end{aligned} \tag{3.5.10}$$

Equation (3.5.10) shows that if $0 \notin \mathrm{Sp}\square^X$, then the dimensions in (3.5.10) vanish for $b > 0$ small enough. This vindicates our claim that equation (3.5.3) is valid in full generality.

Since the radius ϵ of d_λ is arbitrary small, we find that as $b \to 0$,

$$\mathrm{Sp}\,\mathfrak{A}'^2_{\phi_b,\pm\mathcal{H}} \cap d_\lambda \to \{\lambda\}. \tag{3.5.11}$$

Theorem 3.5.1. *Take $M > 0$. There exists $b_0 > 0$ such that for $b \in]0, b_0]$,*

$$\begin{aligned} &\operatorname{Sp} \mathfrak{A}'^2_{\phi_b, \pm\mathcal{H}} \cap \{\lambda \in \mathbf{C}, \operatorname{Re}\lambda \le M\} \subset \mathbf{R}_+, \\ &\dim \mathcal{S}^\cdot (T^*X, \pi^*F)_0 = \dim \mathfrak{h}^\cdot (X, F), \\ &\mathcal{S}^\cdot (T^*X, \pi^*F)_0 = \ker d^{T^*X\prime}_{\phi_b, \pm\mathcal{H}} \cap \ker \overline{d}^{T^*X\prime}_{\phi_b, \pm\mathcal{H}}. \end{aligned} \tag{3.5.12}$$

Finally, for $b \in]0, b_0]$, there is a canonical isomorphism

$$\mathcal{S}^\cdot (T^*X, \pi^*F)_0 \simeq \mathfrak{H}^\cdot (X, F). \tag{3.5.13}$$

Proof. Assume that $\mu \in \operatorname{Sp} \mathfrak{A}'^2_{\phi_b, \pm\mathcal{H}} \cap d'_\lambda$, and that $\mu \notin \mathbf{R}$. By Theorem 3.2.3, the bilinear form $h^{\mathcal{S}^\cdot (T^*X, \pi^*F)}$ vanishes identically on $\mathcal{S}^\cdot (T^*X, \pi^*F)_\mu$. Recall that in the $+$ case, i_+ is an embedding of $\Omega^\cdot (X, F)$ into $\mathcal{S}^\cdot (T^*X, \pi^*F)$, and that in the $-$ case, i_- is an embedding of $\Omega^\cdot (X, F \otimes o(TX))$ into $\mathcal{S}^\cdot (T^*X, \pi^*F)$. The restriction of $h^{\mathcal{S}^\cdot (T^*X, \pi^*F)}$ to $i_+\Omega^\cdot (X, F)$ is a positive metric, and the restriction of $h^{\mathcal{S}^\cdot (T^*X, \pi^*F)}$ to $i_-\Omega^\cdot (X, F \otimes o(TX))$ is the product of $(-1)^n$ by a positive metric.

We will now establish the first part of (3.5.12) in the $+$ case, the proof in the $-$ case being identical. Indeed by (3.5.9), for $b > 0$ small enough and $\lambda \in \operatorname{Sp} \square^X/4, \lambda \le M$, the map $e \in E_\lambda \to \mathfrak{P}^\lambda i_+ e \in \mathcal{S}^\cdot (T^*X, \pi^*F)_\lambda$ is one to one. In particular for $b > 0$ small enough and λ, e taken as before,

$$\left\| \mathfrak{P}^\lambda i_+ e \right\|_{L^2} \ge \frac{1}{2} \left\| e \right\|_{L^2}. \tag{3.5.14}$$

Let R_- be the orthogonal projector from $\mathcal{S}^\cdot (T^*X, \pi^*F)$ on vector space of r-antiinvariant forms. Since R_- vanishes on $i_+\Omega^\cdot (X, F)$, we find that for $b > 0$ small enough, $\operatorname{Re}\lambda \le M, e \in E_\lambda$,

$$\left\| R_- \mathfrak{P}^\lambda i_+ e \right\|_{L^2} \le \frac{1}{8} \left\| e \right\|_{L^2}. \tag{3.5.15}$$

If $e \in E_\lambda$ is such that $\mathfrak{P}^\lambda i_+ e$ lies in a characteristic subspace associated to an eigenvalue $\mu \in \mathbf{C} \setminus \mathbf{R}$, by Theorem 3.2.3, $h^{\mathcal{S}^\cdot (T^*X, \pi^*F)}$ vanishes on $\mathcal{S}^\cdot (T^*X, \pi^*F)_\mu$, and so

$$\left\| R_- \mathfrak{P}^\lambda i_+ e \right\|_{L^2} = \frac{1}{2} \left\| \mathfrak{P}^\lambda i_+ e \right\|_{L^2}. \tag{3.5.16}$$

From (3.5.14)-(3.5.16), we obtain a contradiction, i.e., we have shown that the considered μ lie in $\mathbf{R}$.

Now we proceed as in [B05, proof of Theorem 1.5]. By (3.1.4) and (3.5.10), we get

$$\dim \mathcal{S}^\cdot (T^*X, \pi^*F)^0 = \dim \mathfrak{H}^\cdot (X, F). \tag{3.5.17}$$

Moreover, by (3.2.13) and (3.2.15) in Theorem 3.2.2,

$$\dim \mathcal{S}^\cdot (T^*X, \pi^*F)_0 \ge \dim \mathfrak{H}^\cdot (X, F). \tag{3.5.18}$$

By combining (3.5.5), (3.5.17), and (3.5.18), we see that there is equality in (3.5.5) for $\lambda = 0$, and also there is equality in (3.5.18). Equality in (3.5.5) means that 0 is the only eigenvalue of $\mathfrak{A}'^2_{\phi_b, \pm\mathcal{H}}$ contained in the disk d_0.

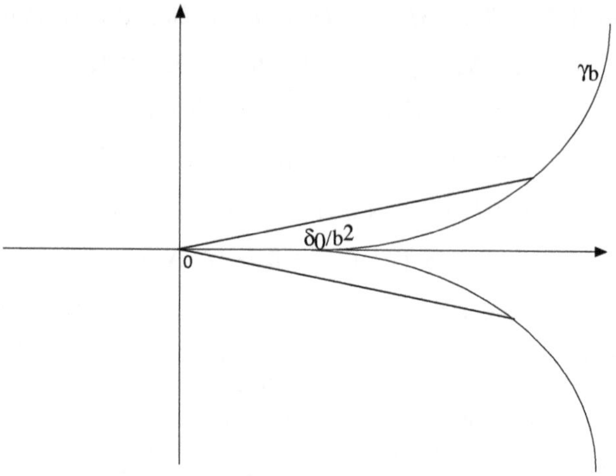

Figure 3.3

Using the above results, we have therefore established the first two identities in (3.5.11).

The fact that there is equality in (3.5.18) forces the map $d^{T^*X}_{\phi_b,\pm\mathcal{H}}$ to vanish on $\mathcal{S}^{\cdot}(T^*X,\pi^*F)_0$. By Theorem 3.2.2, the restriction of $h^{\mathcal{S}^{\cdot}(T^*X,\pi^*F)}$ to $\mathcal{S}^{\cdot}(T^*X,\pi^*F)_0$ is nondegenerate. Since $\mathcal{S}^{\cdot}(T^*X,\pi^*F)_0$ is stable by $\overline{d}^{T^*X}_{\phi_b,\pm\mathcal{H}}$, $\overline{d}^{T^*X}_{\phi_b,\pm\mathcal{H}}$ also vanishes on $\mathcal{S}^{\cdot}(T^*X,\pi^*F)_0$. Therefore,

$$\mathcal{S}^{\cdot}(T^*X,\pi^*F)_0 \subset \ker d^{T^*X}_{\phi_b,\pm\mathcal{H}} \cap \ker \overline{d}^{T^*X}_{\phi_b,\pm\mathcal{H}}. \tag{3.5.19}$$

The opposite inclusion is trivial. So we obtain the last identity in (3.5.12).

By (3.2.13), (3.2.15) in Theorem 3.2.2, and using the fact that $d^{T^*X}_{\phi_b,\pm\mathcal{H}}$ vanishes on $\mathcal{S}^{\cdot}(T^*X,\pi^*F)_0$, we get (3.5.13). This concludes the proof of our theorem. □

Remark 3.5.2. An interesting corollary of Theorem 3.5.1 is that Figure 3.2 can be replaced by Figure 3.3. Namely, for $b>0$ small enough, the spectrum of $\mathfrak{A}'^2_{\phi_b,\pm\mathcal{H}}$ is contained in the union of the domain to the right of γ_b and of a small cone based at 0.

3.6 THE HODGE CONDITION

Now we follow [B05, Definition 1.4].

Definition 3.6.1. We will say that $b>0$ is of Hodge type if

$$\mathcal{S}^{\cdot}(T^*X,\pi^*F)_0 = \ker d'^{T^*X}_{\phi_b,\pm\mathcal{H}} \cap \ker \overline{d}^{T^*X}_{\phi_b,\pm\mathcal{H}}. \tag{3.6.1}$$

Using Theorem 3.2.2 and proceeding as in the proof of [B05, Theorem 1.5], we find that $b>0$ is of Hodge type if and only if

$$\dim \mathcal{S}^{\cdot}(T^*X,\pi^*F)_0 = \dim \mathfrak{H}^{\cdot}(X,F). \tag{3.6.2}$$

As we saw in Theorem 3.5.1, there exists b_0 such that if $b \in]0, b_0]$, then b is of Hodge type.

Theorem 3.6.2. *The set of $b \in \mathbf{R}_+^*$ such that b is not of Hodge type is discrete.*

Proof. If the set of b which are not of Hodge type is not discrete, there is a $\underline{b} \in \mathbf{R}_+$ where such non-Hodge b accumulate. By Theorem 3.5.1, we have $\underline{b} > 0$.

Our proof will now proceed along the lines of the proof of a corresponding result [B05, Proposition 1.23], established in a finite dimensional context. Set

$$M_c = b^{4/3} K_{b^{2/3}} \mathfrak{A}'^2_{\phi_b, \pm\mathcal{H}} K_{b^{-2/3}}. \tag{3.6.3}$$

Using (2.2.5), we can write M_c in the form

$$\begin{aligned} M_c = \frac{1}{4}\Bigg(& -\Delta^V + b^{-4/3} |p|^2 \pm b^{-2/3} \left(2\widehat{e}_i i_{\widehat{e}^i} - n\right) \\ & - \frac{b^{4/3}}{2} \left\langle R^{TX}(e_i, e_j) e_k, e_l \right\rangle \left(e^i - \widehat{e}_i\right) \left(e^j - \widehat{e}_j\right) i_{e_k + \widehat{e}^k} i_{e_\ell + \widehat{e}^l} \Bigg) \\ & - \frac{1}{2} \Bigg(\pm \nabla^{\Lambda^\cdot(T^*T^*X)\widehat{\otimes}F, u}_{Y^{\mathcal{H}}} + \Bigg(\pm b^{2/3} \left\langle R^{TX}(p, e_i) p, e_j \right\rangle \\ + \frac{1}{2} \nabla^F_{e_i} \omega\left(\nabla^F, g^F\right)(e_j) \Bigg) & \left(e^i - \widehat{e}_i\right) i_{e_j + \widehat{e}^j} + \frac{b^{2/3}}{2} \omega\left(\nabla^F, g^F\right)(e_i) \nabla_{\widehat{e}^i} \Bigg). \end{aligned} \tag{3.6.4}$$

Take $r > 0$. Let $V_r \subset \mathbf{R}_+^*$ be the open set of the b such that the circle $c_r \subset \mathbf{C}$ of center 0 and radius r does not intersect $\operatorname{Sp} M_c$. Then the V_r form an open covering of $\mathbf{R}_+^*$.

For $r > 0, b \in V_r$, set

$$\mathfrak{P}'_r = \frac{1}{2i\pi} \int_{c_r} \left(\mu - M_c\right)^{-1} d\mu. \tag{3.6.5}$$

Then $\mathfrak{P}'_r$ projects on a finite dimensional vector space $E^\cdot_r \subset \mathcal{S}^\cdot(T^*X, \pi^*F)$.

For $0 \le i \le 2n$, let $\mathfrak{M}_{r,c,i}$ be the restriction of M_c to E^i_r. Let $P_{r,b,i}(z)$ be the characteristic polynomial of $\mathfrak{M}_{r,c,i}$, i.e.,

$$P_{r,b,i}(z) = \det\left(\mathfrak{M}_{r,c,i} - z\right). \tag{3.6.6}$$

By Theorem 3.2.2, the multiplicity of 0 as a zero of $P_{r,b,i}(z)$ is at least equal to $\dim \mathfrak{H}^i(X, F)$. For $b \in V_r$, set

$$Q_{r,b,i} = P^{(\dim \mathfrak{H}^i(X,F)+1)}_{r,b,i}(0). \tag{3.6.7}$$

By (3.6.2), $b \in V_r$ is of Hodge type if and only if $Q_{r,b,i} \neq 0$. Note here the fundamental but obvious fact that this condition depends only on b and not on r as long as $b \in V_r$.

To complete the proof, let us accept for the moment the fact that $Q_{r,b,i}$ is an analytic function of $b \in V_r$.

Consider the set $A \subset \mathbf{R}_+^*$ such that if $b \in A$, and $b \in V_r$, function $Q_{r,b,i}$ vanishes on an open neighborhood of b. Then A is open. We claim that A is closed. Indeed if $b > 0$ lies in the closure of A and $b \in V_r$, the function $Q_{r,\cdot,i}$ vanishes infinitely many times near b. Since this function is analytic on V_r, it vanishes identically near b, and so $b \in A$. Therefore A is either empty or is equal to $\mathbf{R}_+^*$. However, by Theorem 3.5.1, for $b > 0$ small enough, b is of Hodge type, and so $b \in^c A$. Therefore A is empty.

The above implies that the non-Hodge $b > 0$ cannot accumulate on $\underline{b} > 0$, since otherwise, by analyticity, $\underline{b}$ would lie in A.

Now we concentrate on the proof of the analyticity of $Q_{r,b,i}$. It is enough to show that $\mathfrak{P}_{r,b}$ depends analytically on b or, equivalently, that $\mathfrak{P}_{r,b}$ extends to a holomorphic function on a small open $\mathcal{V} \in \mathbf{C}$ of a given $b > 0$. It is convenient to take $b' = b^{2/3}$ as a new variable, so that the right-hand side of (3.6.4) is a finite sum containing $b'^{-2}, b'^{-1}, b', b'^2$ as well as a constant term.

We claim that the resolvent $(M_c - \lambda)^{-1}$ is still well-defined when $b' \in \mathbf{C}$ lies in a small open neighborhood of $b_0' > 0$. Indeed take $b' = x + iy, x > 0, y \in \mathbf{R}$. We claim that for $|y|$ small enough, the arguments of chapter 15 can be used. We study the extra terms which appear only because b' is now complex. First there is the term $|p|^2$ which appears with an extra small purely imaginary factor. This term is easily dealt with, since it has no impact on the estimates (15.4.7), (15.4.8). Otherwise it can be dealt with like the customary real factor containing $|p|^2$. The term in the second line of (3.6.4) is irrelevant. The term in the third line of (3.6.4) is already a source of concern in chapter 15, but the fact it contains now an extra imaginary factor is irrelevant.

The only serious difficulty comes from the last term in the fourth line of (3.6.4). Indeed this term is naturally skew-adjoint, and now it acquires a small self-adjoint component. However, again because this component is small, it is easily absorbed by the "big" nonnegative term $-\frac{1}{4}\Delta^V$, and so does not affect the estimate (15.4.7).

Now M_c is a holomorphic function of b', and so is the resolvent $(M_c - \lambda)^{-1}$. This completes the proof of our theorem. □

Remark 3.6.3. It is important to observe here that the analyticity property which is used in the proof of Theorem 3.6.2 does not extend to $b = 0$.

3.7 THE HYPOELLIPTIC CURVATURE

Observe that by (2.4.6), (2.4.8), the component of degree 0 in $\Lambda^\cdot(T^*S)$ in the operators $\mathfrak{C}^{\mathcal{M},2}_{\phi,\mathcal{H}^c-\omega^H}, \mathfrak{C}'^{\mathcal{M},2}_{\phi,\mathcal{H}^c-\omega^H}$ is equal to $\mathfrak{A}^2_{\phi,\mathcal{H}^c}, \mathfrak{A}'^2_{\phi,\mathcal{H}^c}$. The same considerations apply to $\mathfrak{C}^{\mathcal{M},2}_{\phi_b,\pm\mathcal{H}-b\omega^H}$ and $\mathfrak{A}^{\mathcal{M},\prime 2}_{\phi_b,\pm\mathcal{H}}$.

Moreover, inspection of equation (2.5.3) shows that all the terms of positive degree in these curvatures can be handled by the methods of chapters 15-17. This is made very easy because of the fact that the f^α act as nilpotent operators which supercommute with the other operators. In particular the spectrum of the above curvature operators is the same as the spectrum of their component of degree 0, which we studied before.

All the results which were stated before in the case of one single fiber extend to the case of families. In particular as $b \to 0$, the resolvent of $\mathfrak{E}^{\mathcal{M},2}_{\phi_b,\pm\mathcal{H}-b\omega^H}$ converges to the resolvent of $A^{\mathcal{M},2}_{\pm}$ in exactly the same sense as before, Theorem 2.6.1 being used instead of Theorem 2.3.2 in the proof of the convergence.

Details are easy to fill and are left to the reader.

Chapter Four

Hypoelliptic Laplacians and odd Chern forms

In this chapter, given $b > 0$, we construct the odd Chern forms associated to a family of hypoelliptic Laplacians. The idea is to adapt the construction of the forms of [BLo95, BG01] which was explained in chapter 1.

Our Chern forms depend on the parameters $b > 0, t > 0$. We will study their asymptotics as $t \to 0$. The asymptotics of the forms rely on local index theoretic techniques which we adapt to the hypoelliptic context.

The proofs of some of the probabilistic results which are needed in the proof of the localization properties of the heat kernels for $t > 0$ small are deferred to chapter 14.

This chapter is organized as follows. In section 4.1, we introduce the formalism of Berezin integration.

In section 4.2, we show that as in the elliptic case, the even Chern forms associated with the hypoelliptic Laplacian are trivial.

In section 4.3, we construct the odd Chern forms, and also a fundamental closed 1-form in the parameters b, t. This 1-form plays a key role in the proof of our main results on the hypoelliptic torsion forms and on Ray-Singer metrics.

In section 4.4, we give the limit as $t \to 0$ of the odd Chern forms. The proofs are delayed to sections 4.5-4.13.

In section 4.5, we use a commutator identity established in Theorem 2.5.2 to give another expression for the odd Chern forms. This identity plays a key role in our local index computations.

In section 4.6, we make a rescaling on the coordinate p. The limit as $t \to 0$ of our odd Chern forms will be studied in this scale.

In section 4.7, we show that given $g \in G$, the evaluation of the above limit can be localized near $\pi^{-1}X_g$. This is done using probabilistic techniques, and in particular arguments obtained via the Malliavin calculus [M78, B81b] which are in part given in chapter 14. Alternative localization techniques can be found in [L05].

In section 4.8, we show that given $x \in X_g$, locally near x, we can replace the total space of T^*X by $T_xX \oplus T_x^*X$. Also a Getzler rescaling [G86] adapted to this new situation is introduced to compensate for the singularities of the corresponding heat kernel as $t \to 0$.

In section 4.9, we show that the rescaled operator has a limit as $t \to 0$. This last result is the exact analogue of the corresponding result by Getzler [G86] for the square of the classical elliptic Dirac operator.

In section 4.10, we establish the convergence of the associated heat kernels, and we establish the appropriate uniform bounds on these kernels.

In section 4.11, we give an explicit formula for the model hypoelliptic heat kernel on a flat space.

In section 4.12, we give an explicit formula for the supertraces involving the matrix part of the limit operator.

In section 4.13, we obtain the asymptotics as $t \to 0$ of our Chern forms.

We make the same assumptions and we use the same notation as in chapter 2.

4.1 THE BEREZIN INTEGRAL

Let E and V be real finite dimensional vector spaces of dimension n and m. Let g^E be a Euclidean metric on E. We will often identify E and E^* by the metric g^E. Let $e_1, \dots, e_n$ be an orthonormal basis of E, and let $e^1, \dots, e^n$ be the corresponding dual basis of E^*.

Let $\Lambda^{\cdot}(E^*)$ be the exterior algebra of E^*. It will be convenient to introduce another copy $\widehat{\Lambda}^{\cdot}(E^*)$ of this exterior algebra. If $e \in E^*$, we will denote by $\widehat{e}$ the corresponding element in $\widehat{\Lambda}^{\cdot}(E^*)$.

Suppose temporarily that E is oriented and that $e_1, \dots, e_n$ is an oriented basis of E. Let $\int^{\widehat{B}}$ be the linear map from $\Lambda^{\cdot}(V^*) \widehat{\otimes} \widehat{\Lambda}^{\cdot}(E^*)$ into $\Lambda(V^*)$, such that if $\alpha \in \Lambda(V^*), \beta \in \widehat{\Lambda}(E^*)$,

$$\int^{\widehat{B}} \alpha\beta = 0 \text{ if } \deg\beta < \dim E, \tag{4.1.1}$$

$$\int^{\widehat{B}} \alpha \widehat{e}^1 \wedge \dots \wedge \widehat{e}^n = (-1)^{n(n+1)/2} \alpha.$$

More generally, let $o(E)$ be the orientation line of E. Then $\int^{\widehat{B}}$ defines a linear map from $\Lambda^{\cdot}(V^*) \widehat{\otimes} \widehat{\Lambda}^{\cdot}(E^*)$ into $\Lambda^{\cdot}(V^*) \widehat{\otimes} o(E)$, which is called a Berezin integral.

Let A be an antisymmetric endomorphism of E. We identify A with the element of $\Lambda(E^*)$,

$$A = \frac{1}{2} \sum_{1 \le i,j \le n} \langle e_i, Ae_j \rangle \widehat{e}^i \wedge \widehat{e}^j. \tag{4.1.2}$$

By definition, the Pfaffian $\operatorname{Pf}[A]$ of A is given by

$$\int^{\widehat{B}} \exp(-A) = \operatorname{Pf}[A]. \tag{4.1.3}$$

Then $\operatorname{Pf}[A]$ lies in $o(E)$. Moreover, $\operatorname{Pf}[A]$ vanishes if n is odd.

Let S be a manifold, and let E be a real vector bundle on S of dimension n, equipped with a Euclidean metric g^E and a metric preserving connection ∇^E. Let $o(E)$ be the orientation bundle of E. Let R^E be the curvature of ∇^E. Let $e(E) \in H^{\cdot}(S, o(E) \otimes \mathbf{Q})$ be the rational Euler class of E. Then $e(E)$ vanishes if n is odd. Moreover, if n is even, the class $e(E)$ is represented in Chern-Weil theory by the closed form $e\left(E, \nabla^E\right)$ given by

$$e\left(E, \nabla^E\right) = \mathrm{Pf}\left[\frac{R^E}{2\pi}\right]. \tag{4.1.4}$$

Let $e_1, \dots, e_n$ be an orthonormal basis of E. We will use the Berezin integration formalism, with $V = TS$. By (4.1.3) and (4.1.4), we get

$$e\left(E, \nabla^E\right) = \frac{1}{\pi^{n/2}} \int^{\widehat{B}} \exp\left(-\frac{1}{4}\left\langle e_i, R^E e_j\right\rangle \widehat{e}^i \widehat{e}^j\right). \tag{4.1.5}$$

4.2 THE EVEN CHERN FORMS

We make the same assumptions as in sections 1.3, 1.5, 2.4, 2.8, and 2.9.

By (3.1.4), $\mathfrak{H}^{\cdot}(X, F)$ is naturally $\mathbf{Z}$-graded, and the operator defining the grading is just N^{T^*X}. Set

$$\overline{\chi}_g(F) = \mathrm{Tr_s}^{\mathfrak{H}^{\cdot}(X,F)}[g], \qquad \overline{\chi}'_g(F) = \mathrm{Tr_s}^{\mathfrak{H}^{\cdot}(X,F)}\left[gN^{T^*X}\right]. \tag{4.2.1}$$

By (1.6.1), (1.6.5), (1.6.6), we get

$$\overline{\chi}_g(F) = \chi_g(F) = L(g). \tag{4.2.2}$$

Moreover, using the Thom isomorphism in (3.1.4) and Poincaré duality, we also get

$$\begin{aligned} \overline{\chi}'_g(F) &= \chi'_g(F) \ \text{ if } c > 0, \\ &= \left(2n\chi_g(F) - \chi'_g(F)\right) \ \text{ if } c < 0. \end{aligned} \tag{4.2.3}$$

In the sequel, we use the notation of section 2.8 and of chapter 3. In particular we take

$$\overline{\mathcal{H}} = \pm\frac{t^2}{2b^2}|p|^2. \tag{4.2.4}$$

Note that if we used the conventions of section 2.8, then we would have $\mathcal{H} = \pm|p|^2/2$. However, for notational convenience we stick to the notation in equation (3.2.1), so that $\mathcal{H} = |p|^2/2$.

Let $g_t^{T^*X}$ be the metric on T^*X which is associated to the metric g^{TX}/t on TX. Then if $g^{T^*X} = g_1^{T^*X}$, we get $g_t^{T^*X} = tg^{T^*X}$. We can write (4.2.4) in the form

$$\overline{\mathcal{H}} = \pm\frac{t}{2b^2}|p|_t^2. \tag{4.2.5}$$

By (4.2.5), we can then use the results of [B05] and of chapter 2, with $c = \pm\frac{t}{b^2}$ and $\mathcal{H} = \frac{1}{2}|p|_t^2$.

Also, all the objects we considered in section 2.8 should have an extra index $\pm$. This index will not be explicitly written. As explained in section 1.6, $L(g)$ is the common value of $L_+(g)$ and $L_-(g)$.

By the results established in chapter 3, the operator $\exp\left(-A_{b,t}^{2}\right)$ is fiberwise trace class. We claim that the even form $\mathrm{Tr_s}\left[\exp\left(-A_{b,t}^{\mathcal{M},2}\right)\right]$ is a smooth form on S. Indeed smoothness is a consequence of the fact that the fiberwise hypoelliptic operators $A_{b,t}^{\mathcal{M},2}$ depend smoothly on the parameter $s \in S$, and also on the uniformity of the hypoelliptic estimates in chapters 3 and 15, as long as b remains in a compact set in $\mathbf{R}_{+}^{*}$.

First we state an analogue of Proposition 1.7.1.

Theorem 4.2.1. *The following identity holds:*

$$\mathrm{Tr_s}\left[g\exp\left(-A_{b,t}^{\mathcal{M},2}\right)\right] = L(g). \tag{4.2.6}$$

Proof. Clearly,

$$\mathrm{Tr_s}\left[g\exp\left(-A_{b,t}^{\mathcal{M},2}\right)\right] = \mathrm{Tr_s}\left[g\exp\left(B_{b,t}^{\mathcal{M},2}\right)\right]. \tag{4.2.7}$$

Using (4.2.7) and the fact that supertraces vanish on supercommutators, we get

$$\frac{\partial}{\partial t}\mathrm{Tr_s}\left[g\exp\left(B_{b,t}^{\mathcal{M},2}\right)\right] = \mathrm{Tr_s}\left[g\left[B_{b,t}^{\mathcal{M}},\frac{\partial B_{b,t}^{\mathcal{M}}}{\partial t}\exp\left(B_{b,t}^{\mathcal{M},2}\right)\right]\right] = 0. \tag{4.2.8}$$

Therefore, the left-hand side of (4.2.6) does not depend on t. In sections 4.5-4.13, we will obtain our theorem by taking the limit as $t = 0$ of the left-hand side of (4.2.6). □

4.3 THE ODD CHERN FORMS AND A 1-FORM ON $\mathbf{R}^{*2}$

We use the notation of section 2.8. As in (1.7.1), set

$$h(x) = xe^{x^2}. \tag{4.3.1}$$

By the second identity in [B05, Theorem 4.23], we get

$$B_{\phi,\overline{\mathcal{H}}-\omega^H}^{\mathcal{M}\times\mathbf{R}_{+}^{*2}} = B_{b,t}^{\mathcal{M}} + \frac{dt}{2t^2}\lambda_0 \mp d\left(t/b\right)t/b\left|p\right|^2. \tag{4.3.2}$$

From the results on the heat kernel of $\mathfrak{A}_{\phi,\mathcal{H}^c}^{2}$ which were explained in section 3.3, the operator $h\left(B_{\phi,\overline{\mathcal{H}}-\omega^H}^{\mathcal{M}\times\mathbf{R}_{+}^{*2}}\right)$ is fiberwise trace class.

Definition 4.3.1. Set

$$\begin{aligned} a &= (2\pi)^{1/2}\varphi\mathrm{Tr_s}\left[gh\left(B_{\phi,\overline{\mathcal{H}}-\omega^H}^{\mathcal{M}\times\mathbf{R}_{+}^{*2}}\right)\right], \\ u_{b,t} &= (2\pi)^{1/2}\varphi\mathrm{Tr_s}\left[gh\left(B_{b,t}^{\mathcal{M}}\right)\right], \\ v_{b,t} &= \pm\varphi\mathrm{Tr_s}\left[g\frac{t^2}{b^2}\left|p\right|^2 h'\left(B_{b,t}^{\mathcal{M}}\right)\right], \\ w_{b,t} &= \varphi\mathrm{Tr_s}\left[g\left(\frac{\lambda_0}{2t}\mp\frac{t^2}{b^2}\left|p\right|^2\right)h'\left(B_{b,t}^{\mathcal{M}}\right)\right]. \end{aligned} \tag{4.3.3}$$

Then a is a smooth form on $S\times\mathbf{R}_{+}^{*2}$, and the other objects are smooth forms on S which depend on the parameters (b,t).

By the considerations we made after (2.4.10), $\mathfrak{h}^{\Omega^{\cdot}(T^*X,\pi^*F)}_{\mathcal{H}-\omega^H}$ is well-defined. To fit with the conventions of Definition 1.7.2, we will also write a in the form

$$a = h_g\left(A'^{\mathcal{M}\times\mathbf{R}^{*2}_+}, \mathfrak{h}^{\Omega^{\cdot}(T^*X,\pi^*F)}_{\mathcal{H}-\omega^H}\right). \tag{4.3.4}$$

Similarly, let $\mathfrak{h}^{\Omega^{\cdot}(T^*X,\pi^*F)}_{t,\mathcal{H}t^2/b^2-\omega^H}$ be the bilinear form which was considered after (2.4.10), which is associated to the metric $g^{TX}_t = g^{TX}/t$ and to the function $\pm\frac{t^2}{b^2}\mathcal{H}$. Then we can write the form $u_{b,t}$ as

$$u_{b,t} = h_g\left(A'^{\mathcal{M}}, \mathfrak{h}^{\Omega^{\cdot}(T^*X,\pi^*F)}_{t,\mathcal{H}t^2/b^2-\omega^H}\right). \tag{4.3.5}$$

The forms in (4.3.5) will be called the hypoelliptic odd Chern forms.

Theorem 4.3.2. *The forms a and $u_{b,t}$ are odd, and the forms $v_{b,t}, w_{b,t}$ are even. There is a smooth odd form $\mathfrak{r}_{b,t}$ on S such that*

$$a = u_{b,t} + \frac{db}{b}v_{b,t} + \frac{dt}{t}w_{b,t} + dbdt\mathfrak{r}_{b,t}. \tag{4.3.6}$$

The form a is closed on $S\times\mathbf{R}^{*2}_+$. In particular the odd forms $u_{b,t}$ are closed on S, and their cohomology class does not depend on (b,t).

Proof. Recall that $B^{\mathcal{M}\times\mathbf{R}^{*2}_+}_{\phi,\overline{\mathcal{H}}-\omega^H}$ is odd in $\Lambda^{\cdot}(T^*S)\widehat{\otimes}\mathrm{End}(\Omega^{\cdot}(T^*X,\pi^*F))$, so that a is an odd form. By (4.3.2), we get (4.3.6).

Using the second identity in (2.4.4), we get

$$\begin{aligned} d\frac{a}{2\pi} =& \varphi d\mathrm{Tr_s}\left[gh\left(B^{\mathcal{M}\times\mathbf{R}^{+2}_*}_{\phi,\overline{\mathcal{H}}-\omega^H}\right)\right] \\ =&\varphi\mathrm{Tr_s}\left[g\left[A^{\mathcal{M}\times\mathbf{R}^{*2}_+}_{\phi,\overline{\mathcal{H}}-\omega^H}, B^{\mathcal{M}\times\mathbf{R}^{*2}_+}_{\phi,\overline{\mathcal{H}}-\omega^H}\right]h'\left(B^{\mathcal{M}}_{\phi,\overline{\mathcal{H}}-\omega^H}\right)\right] = 0. \end{aligned} \tag{4.3.7}$$

So we have shown that a is closed. The remaining part of the theorem is now trivial. □

Definition 4.3.3. Put

$$\underline{w}_{b,t} = \varphi\mathrm{Tr_s}\left[g\left(\frac{N^{T^*X}-n}{2} - \omega^H\right)h'\left(B^{\mathcal{M}}_{b,t}\right)\right]. \tag{4.3.8}$$

Proposition 4.3.4. *The form $\underline{w}_{b,t}$ is even. Moreover,*

$$w_{b,t} - \underline{w}_{b,t} = (2\pi)^{-1/2}d\varphi\mathrm{Tr_s}\left[g\frac{1}{2}(p - i_{\widehat{p}})h'\left(B^{\mathcal{M}}_{b,t}\right)\right]. \tag{4.3.9}$$

In particular,

$$w^{(0)}_{b,t} = \underline{w}^{(0)}_{b,t}. \tag{4.3.10}$$

Proof. By [B05, Proposition 4.34], where we replace g^{TX} by g^{TX}/t and $\mathcal{H}$ by $\pm r^*_{t/b}\mathcal{H}$, we get

$$\frac{1}{2}\left(\left[\overline{\mathfrak{C}}'^{\mathcal{M}}_{b,t}, p\right] - \left[A'^{\mathcal{M}}, i_{\widehat{p}}\right]\right) = \frac{\lambda_0}{2t} \mp \frac{t^2}{b^2}|p|^2 - \left(\frac{N^{T^*X}-n}{2} - \omega^H\right). \tag{4.3.11}$$

Moreover,

$$h'(x) = \left(1 + 2x^2\right) \exp\left(x^2\right). \tag{4.3.12}$$

Therefore,

$$h'\left(B_{b,t}^{\mathcal{M}}\right) = \left(1 - 2A_{b,t}^{\mathcal{M},2}\right) \exp\left(-A_{b,t}^{\mathcal{M},2}\right). \tag{4.3.13}$$

Finally, by (2.4.9),

$$\left[A'^{\mathcal{M}}, A_{b,t}^{\mathcal{M},2}\right] = 0, \qquad \left[\overline{\mathfrak{C}}_{b,t}'^{\mathcal{M}}, A_{b,t}^{\mathcal{M},2}\right] = 0. \tag{4.3.14}$$

Using (4.3.3), (4.3.8), (4.3.11)-(4.3.14), we get (4.3.9). □

Now we define $\nu, \mathfrak{G}_{\phi,\overline{\mathcal{H}}-\omega^H}^{\mathcal{M}\times\mathbf{R}_*^2}$ as in (2.5.2).

Definition 4.3.5. Put

$$\overline{a} = (2\pi)^{1/2} \varphi \mathrm{Tr_s}\left[g\left(\mathfrak{G}_{\phi,\overline{\mathcal{H}}-\omega^H}^{\mathcal{M}\times\mathbf{R}_*^2} + \left[\mathfrak{C}_{\overline{\mathcal{H}}-\omega^H}'^{\mathcal{M}\times\mathbf{R}_*^{\in}}, \pm\frac{3t^2}{2b^2}|p|^2\right]\right)\exp\left(\mathfrak{D}_{\phi,\overline{\mathcal{H}}-\omega^H}^{\mathcal{M}\times\mathbf{R}_*^2,2}\right)\right]. \tag{4.3.15}$$

Then $\overline{a}$ is also an odd form on $S \times \mathbf{R}_*^2$.

Theorem 4.3.6. *The following identity holds:*

$$a = (2\pi)^{1/2} \varphi \mathrm{Tr_s}\left[g\mathfrak{G}_{\phi,\overline{\mathcal{H}}-\omega^H}^{\mathcal{M}\times\mathbf{R}_*^2} \exp\left(\mathfrak{D}_{\phi,\overline{\mathcal{H}}-\omega^H}^{\mathcal{M}\times\mathbf{R}_*^2,2}\right)\right]. \tag{4.3.16}$$

Moreover, $\overline{a}$ is a closed form on $S \times \mathbf{R}_*^2$, which is cohomologous to a. More precisely,

$$\overline{a} = a \pm d\varphi \mathrm{Tr_s}\left[g\frac{3t^2}{2b^2}|p|^2 \exp\left(\mathfrak{D}_{\phi,\overline{\mathcal{H}}-\omega^H}^{\mathcal{M}\times\mathbf{R}_*^2,2}\right)\right]. \tag{4.3.17}$$

Proof. Since supertraces vanish on supercommutators,

$$\mathrm{Tr_s}\left[g\left[\mathfrak{D}_{\phi,\overline{\mathcal{H}}-\omega^H}^{\mathcal{M}\times\mathbf{R}_*^2}, \nu\right] \exp\left(\mathfrak{D}_{\phi,\overline{\mathcal{H}}-\omega^H}^{\mathcal{M}\times\mathbf{R}_*^2,2}\right)\right] = 0. \tag{4.3.18}$$

Using (2.5.2) and (4.3.18), we get (4.3.16). Moreover, by (2.4.9), $\mathfrak{C}_{\overline{\mathcal{H}}-\omega^H}'^{\mathcal{M}\times\mathbf{R}_*^2}$ commutes with $\mathfrak{D}_{\phi,\overline{\mathcal{H}}-\omega^H}^{\mathcal{M}\times\mathbf{R}_*^2,2}$, and so

$$\mathrm{Tr_s}\left[g\left[\mathfrak{C}_{\overline{\mathcal{H}}-\omega^H}'^{\mathcal{M}\times\mathbf{R}_*^2}, \frac{3t^2}{2b^2}|p|^2\right] \exp\left(\mathfrak{D}_{\phi,\overline{\mathcal{H}}-\omega^H}^{\mathcal{M}\times\mathbf{R}_*^2,2}\right)\right] \\ = d\mathrm{Tr_s}\left[g\frac{3t^2}{2b^2}|p|^2 \exp\left(\mathfrak{D}_{\phi,\overline{\mathcal{H}}-\omega^H}^{\mathcal{M}\times\mathbf{R}_*^2,2}\right)\right]. \tag{4.3.19}$$

Identity (4.3.17) is now a consequence of (4.3.15), (4.3.16), and (4.3.19). □

4.4 THE LIMIT AS t → 0 OF THE FORMS $\mathbf{u}_{\mathbf{b},\mathbf{t}}, \mathbf{v}_{\mathbf{b},\mathbf{t}}, \mathbf{w}_{\mathbf{b},\mathbf{t}}$

Now we state the main result of this chapter.

Theorem 4.4.1. *As $t \to 0$,*

$$u_{b,t} = \int_{X_g} e\left(TX_g, \nabla^{TX_g}\right) h_g\left(\nabla^F, g^F\right) + \mathcal{O}\left(\sqrt{t}\right), \tag{4.4.1}$$
$$v_{b,t} = \mathcal{O}\left(\sqrt{t}\right), \qquad w_{b,t} = \mathcal{O}\left(\sqrt{t}\right).$$

Proof. Our theorem will be proved in the next sections. Still we give a direct proof that given the first equation, the second and the third equations can be deduced from each other. Indeed, we introduce a third copy of $\mathbf{R}_+^*$, and over $h \in \mathbf{R}_+^*$, the metric g^{TX} is replaced by g^{TX}/h. Let $\overline{u}_{b,t}$ be the form corresponding to $u_{b,t}$ over $S \times \mathbf{R}_+^*$. Using the analogue of (4.3.2), we find easily that when evaluated at $h = 1$, we have the equality

$$\overline{u}_{b,t} = u_{b,t} + dh\left(\frac{1}{2}v_{b,t} + w_{b,t}\right). \tag{4.4.2}$$

The connection $\overline{\nabla}^{TX}$ over $M \times \mathbf{R}_+^*$ which corresponds to ∇^{TX} is given by

$$\overline{\nabla}^{TX} = \nabla^{TX} + dh\left(\frac{\partial}{\partial h} - \frac{1}{2h}\right). \tag{4.4.3}$$

By (4.4.3), we get

$$e\left(TX_g, \overline{\nabla}^{TX_g}\right) = e\left(TX_g, \nabla^{TX_g}\right), \tag{4.4.4}$$

i.e., the form $e\left(TX_g, \overline{\nabla}^{TX_g}\right)$ does not contain dh. By the first equation in (4.4.1) and by (4.4.2), we find that as $t \to 0$,

$$\frac{1}{2}v_{b,t} + w_{b,t} = \mathcal{O}\left(\sqrt{t}\right). \tag{4.4.5}$$

By (4.4.5), we find that the second and third equations in (4.4.1) are equivalent. □

Remark 4.4.2. It is remarkable that the asymptotics as $t \to 0$ of the hypoelliptic odd Chern forms $u_{b,t} = h_g\left(A'^{\mathcal{M}}, \mathfrak{h}^{\Omega^{\cdot}(T^*X,\pi^*F)}_{t,\pm\frac{t^2}{b^2}\mathcal{H}-\omega^H}\right)$ is the same as the asymptotics of the elliptic odd Chern forms $h_g\left(A', g_t^{\Omega^{\cdot}(X,F)}\right)$, which was given in (1.7.12).

4.5 A FUNDAMENTAL IDENTITY

We use the notation in section 2.8. Let z be an odd Grassmann variable, which anticommutes with all the other odd variables considered before.

Definition 4.5.1. Set

$$\mathfrak{L}_{b,t} = \mathfrak{C}_{b,t}^{\mathcal{M},2} - z\mathfrak{G}_{b,t}^{\mathcal{M}}. \tag{4.5.1}$$

Proposition 4.5.2. *The following identities hold:*

$$(2\pi)^{1/2}\varphi\mathrm{Tr_s}\left[g\exp\left(-\mathfrak{L}_{b,t}\right)\right] = (2\pi)^{1/2}\varphi\mathrm{Tr_s}\left[g\exp\left(-A_{b,t}^{\mathcal{M},2}\right)\right] + zu_{b,t}. \tag{4.5.2}$$

Proof. The identity (4.5.2) with $z=0$ is trivial. In the definition of $u_{b,t}$, we may as well replace $B_{b,t}^{\mathcal{M}}$ by $\mathfrak{D}_{b,t}^{\mathcal{M}}$. By definition,

$$\mathrm{Tr_s}\left[gh\left(\mathfrak{D}_{b,t}^{\mathcal{M}}\right)\right] = \mathrm{Tr_s}\left[g\mathfrak{D}_{b,t}^{\mathcal{M}}\exp\left(\mathfrak{D}_{b,t}^{\mathcal{M},2}\right)\right]. \tag{4.5.3}$$

Using (2.5.2), (4.5.3), and the fact that supertraces vanish on supercommutators, we get

$$\mathrm{Tr_s}\left[gh\left(\mathfrak{D}_{b,t}^{\mathcal{M}}\right)\right] = \mathrm{Tr_s}\left[g\mathfrak{G}_{b,t}^{\mathcal{M}}\exp\left(\mathfrak{D}_{b,t}^{\mathcal{M},2}\right)\right]. \tag{4.5.4}$$

Our proposition follows from (4.5.4). □

Remark 4.5.3. The reader can ask what is the point of replacing $\mathfrak{D}_{b,t}^{\mathcal{M}}$ by $\mathfrak{G}_{b,t}^{\mathcal{M}}$ in (4.5.4). This is because methods of local index theory can be used in the second expression, which would fail when applied to the first one.

We denote by $\underline{\widehat{\mathfrak{L}}}_{b,t}$ the operator obtained from $\mathfrak{L}_{b,t}$ by making the transformations indicated in [B05, subsections 4.21 and 4.22] and also in section 2.5 in (2.5.6)-(2.5.9). Observe that this transformation is g-equivariant. The operator $\underline{\widehat{\mathfrak{L}}}_{b,t}$ now acts on smooth sections of

$$\Lambda^{\cdot}\left(T^*S\right)\widehat{\otimes}\Lambda^{\cdot}\left(T^*X\right)\widehat{\otimes}\Lambda^{\cdot}\left(T^*X\right)\widehat{\otimes}\Lambda^{n}\left(TX\right)\widehat{\otimes}F\widehat{\otimes}\mathbf{C}\left[z\right] \tag{4.5.5}$$

over the fibers of T^*X over S.

Proposition 4.5.4. *The following identity holds:*

$$\mathrm{Tr_s}\left[g\exp\left(-\mathfrak{L}_{b,t}\right)\right] = (-1)^n\,\mathrm{Tr_s}\left[g\exp\left(-\underline{\widehat{\mathfrak{L}}}_{b,t}\right)\right]. \tag{4.5.6}$$

Proof. The identification (2.5.7) is g-equivariant. Taking into account the shift by n in the grading, equation (4.5.6) follows. □

4.6 A RESCALING ALONG THE FIBERS OF T*X

Recall that for $a\in\mathbf{R}^*$, r_a is the dilation $p\to ap$. Then r_a^* acts naturally on the smooth sections of the vector bundle in (4.5.5). Also the operator $\underline{\widehat{\mathfrak{L}}}_{b,t}$ acts on the smooth sections of (4.5.5). Clearly, conjugation by r_a^* leaves the operators e_i, i_{e_i} unchanged, and, moreover,

$$r_a^*\widehat{e}^i r_a^{*-1} = \widehat{e}^i/a, \qquad r_a^* i_{\widehat{e}_i} r_a^{*-1} = a i_{\widehat{e}_i}. \tag{4.6.1}$$

Set

$$\widehat{\mathfrak{M}}_{b,t} = r^*_{1/\sqrt{t}} \widehat{\mathfrak{L}}_{b,t} r^*_{\sqrt{t}}. \tag{4.6.2}$$

By (4.5.6), (4.6.2),

$$\operatorname{Tr_s}\left[g \exp\left(-\mathfrak{L}_{b,t}\right)\right] = (-1)^n \operatorname{Tr_s}\left[g \exp\left(-\widehat{\mathfrak{M}}_{b,t}\right)\right]. \tag{4.6.3}$$

Let $\nabla^{\Lambda^{\cdot}(T^*X)\widehat{\otimes}\Lambda^{\cdot}(T^*X)\widehat{\otimes}\Lambda^n(TX)\widehat{\otimes}F,u}$ be the connection on

$$\Lambda^{\cdot}(T^*X)\widehat{\otimes}\Lambda^{\cdot}(T^*X)\widehat{\otimes}\Lambda^n(TX)\widehat{\otimes}F$$

which is induced by ∇^{TX} and $\nabla^{F,u}$.

Proposition 4.6.1. *The following identity holds*

$$\begin{aligned}
\widehat{\mathfrak{M}}_{b,t} = \frac{1}{4}\Bigg(-\Delta^V + \frac{t^2}{b^4}|p|^2 \pm \frac{t}{b^2}\left(2 i_{\widehat{e}_i}\left(\widehat{e}^i - e^i/\sqrt{t} - \sqrt{t} i_{e_i}\right) - n\right)\Bigg) \\
+ \frac{1}{4}\left\langle e_i, R^{TX} e_j\right\rangle \widehat{e}^i \widehat{e}^j - \frac{\sqrt{t}}{4}\omega\left(\nabla^F, g^F\right)(e_i)\nabla_{\widehat{e}^i} \\
- \frac{\sqrt{t}}{4}\nabla^{\Lambda^{\cdot}(T^*T^*X)\widehat{\otimes}F}\widehat{\omega}\left(\nabla^F, g^F\right) - \frac{1}{4}\omega\left(\nabla^F, g^F\right)^2 \\
\mp \frac{t}{2b^2}\Bigg(\nabla^{\Lambda^{\cdot}(T^*X)\widehat{\otimes}\Lambda^{\cdot}(T^*X)\widehat{\otimes}\Lambda^n(TX)\widehat{\otimes}F,u}_{\sqrt{t}Y^{\mathcal{H}}} \\
+ \left\langle T(f_\alpha, e_i) f^\alpha\left(e^i - t i_{e_i}\right) + T^H, p/\sqrt{t}\right\rangle + \left\langle R^{TX}(\cdot, p) e_i, p\right\rangle \sqrt{t}\widehat{e}^i\Bigg) \\
- z\left(\frac{1}{2}\omega\left(\nabla^F, g^F\right) - \frac{\sqrt{t}}{4}\widehat{\omega}\left(\nabla^F, g^F\right) \mp \frac{t}{2b^2}\left(\widehat{p} + 6 i_{\widehat{p}}\right)\right).
\end{aligned} \tag{4.6.4}$$

Proof. This is an easy consequence of (1.3.2), (2.5.14), (2.5.17), and of Theorem 2.5.4. □

Remark 4.6.2. Recall that dv_{T^*X} is the symplectic volume form on T^*X. Let $\exp\left(-\widehat{\mathfrak{M}}_{b,t}\right)(z, z')$ be the smooth kernel with respect to $dv_{T^*X}(z')$ which is associated to the operator $\exp\left(-\widehat{\mathfrak{M}}_{b,t}\right)$.

Clearly,

$$\operatorname{Tr_s}\left[g \exp\left(-\widehat{\mathfrak{M}}_{b,t}\right)\right] = \int_{T^*X} \operatorname{Tr_s}\left[g \exp\left(-\widehat{\mathfrak{M}}_{b,t}\right)(z, gz)\right] dv_{T^*X}(z). \tag{4.6.5}$$

4.7 LOCALIZATION OF THE PROBLEM

In the sequel, for simplicity we will assume that S is compact. If this is not the case, we can as well restrict ourselves to compact subsets of S.

Let d_X be the Riemannian distance along the fibers X with respect to g^{TX}. Let a_X be a lower bound for the injectivity radius of the fibers X.

Let $N_{X_g/X}$ be the orthogonal bundle to TX_g in $TX|_{X_g}$. We identify X_g to the zero section of $N_{X_g/X}$.

Given $\eta > 0$, let $\mathcal{V}_\eta$ be the η-neighborhood of X_g in $N_{X_g/X}$. Then there exists $\eta_0 \in]0, a_X/32]$ such that if $\eta \in]0, 8\eta_0]$, the map $(x, Z) \in N_{X_g/X} \to \exp_x^X(Z) \in X$ is a diffeomorphism from $\mathcal{V}_\eta$ on the tubular neighborhood $\mathcal{U}_\eta$ of X_g in X. In the sequel, we identify $\mathcal{V}_\eta$ and $\mathcal{U}_\eta$. This identification is g-equivariant. Let $\alpha \in]0, \eta_0]$ be small enough so that if $d_X\left(g^{-1}x, x\right) \le \alpha$, then $x \in \mathcal{U}_{\eta_0}$.

Let dv_{X_g} be the volume element on X_g, and let $dv_{N_{X_g/X}}$ be the volume element along the fibers of $N_{X_g/X}$. Let $k(x,y), x \in X_g, y \in N_{X_g/X,x}, |y| \le \eta_0$ be the smooth function with values in $\mathbf{R}_+$ such that on $\mathcal{U}_{\eta_0}$,

$$dv_X(x,y) = k(x,y)\, dv_{N_{X_g/X}}(y)\, dv_{X_g}(x). \tag{4.7.1}$$

Note that

$$k(x,0) = 1. \tag{4.7.2}$$

Set

$$\widehat{\underline{\mathfrak{M}}}'_{b,t} = r^*_{b^2/t} \widehat{\underline{\mathfrak{M}}}_{b,t} r^*_{t/b^2}. \tag{4.7.3}$$

Then by (4.6.1), (4.6.4), we get

$$\begin{aligned}\widehat{\underline{\mathfrak{M}}}'_{b,t} = \frac{1}{4}\Bigg(&-\frac{t^2}{b^4}\Delta^V + |p|^2 \pm \frac{t}{b^2}\left(2i_{\widehat{e}_i}\left(\widehat{e}^i - b^2 e^i/t^{3/2} - b^2 i_{e_i}/\sqrt{t}\right) - n\right)\Bigg) \\ &+ \frac{t^2}{4b^4}\left\langle e_i, R^{TX} e_j\right\rangle \widehat{e}^i \widehat{e}^j - \frac{t^{3/2}}{4b^2}\omega\left(\nabla^F, g^F\right)(e_i)\nabla_{\widehat{e}^i} \\ &- \frac{t^{3/2}}{4b^2}\nabla^{\Lambda^\cdot(T^*T^*X)\widehat{\otimes}F}\widehat{\omega}\left(\nabla^F, g^F\right) - \frac{1}{4}\omega\left(\nabla^F, g^F\right)^2 \\ \mp\frac{1}{2}\Bigg(&\nabla^{\Lambda^\cdot(T^*X)\widehat{\otimes}\Lambda^\cdot(T^*X)\widehat{\otimes}\Lambda^n(TX)\widehat{\otimes}F,u}_{\sqrt{t}Y^{\mathcal{H}}} + \left\langle T(f_\alpha, e_i) f^\alpha \left(e^i - t i_{e_i}\right) + T^H, p/\sqrt{t}\right\rangle \\ &+ \left\langle R^{TX}(\cdot, p) e_i, p\right\rangle \sqrt{t}\widehat{e}^i\Bigg) \\ &- z\left(\frac{1}{2}\omega\left(\nabla^F, g^F\right) - \frac{t^{3/2}}{4b^2}\widehat{\omega}\left(\nabla^F, g^F\right) \mp \frac{t}{2b^2}\left(\widehat{p} + 6b^4 i_{\widehat{p}}/t^2\right)\right).\end{aligned} \tag{4.7.4}$$

Clearly,

$$\mathrm{Tr_s}\left[g\exp\left(-\widehat{\underline{\mathfrak{L}}}_{b,t}\right)\right] = \mathrm{Tr_s}\left[g\exp\left(-\widehat{\underline{\mathfrak{M}}}'_{b,t}\right)\right]. \tag{4.7.5}$$

Let $\exp\left(-\widehat{\underline{\mathfrak{M}}}'_{b,t}\right)(\cdot,\cdot)(z,z')$ be the smooth kernel for $\exp\left(-\widehat{\underline{\mathfrak{M}}}'_{b,t}\right)$ with respect to $dv_{T^*X}(z')$. Then

$$\mathrm{Tr_s}\left[g\exp\left(-\widehat{\underline{\mathfrak{M}}}'_{b,t}\right)\right] = \int_{T^*X}\mathrm{Tr_s}\left[g\exp\left(-\widehat{\underline{\mathfrak{M}}}'_{b,t}\right)(z, gz)\right] dv_{T^*X}(z). \tag{4.7.6}$$

Now we will show why the asymptotics of (4.7.6) as $t \to 0$ can be localized near $\pi^{-1}X_g \subset T^*X$.

Throughout the chapter, we fix $b_0 \geq 1$.

Proposition 4.7.1. *There exist $m \in \mathbf{N}, c > 0, C > 0$ such that if $a \in \left[\frac{1}{2}, 1\right], t \in]0,1], b \in \left[\sqrt{t}, b_0\right], z = (x,p), z' = (x',p') \in T^*X$, then*

$$\left|\exp\left(-a\widehat{\mathfrak{M}}'_{b,t}\right)(z,z')\right| \leq \frac{C}{t^m}\exp\left(-c\left(|p|^2 + |p'|^2 + d^{X,2}(x,x')/t\right)\right). \tag{4.7.7}$$

Proof. Consider the scalar operator

$$S_{b,t} = -\frac{t^2}{2b^4}\Delta^V + \frac{1}{2}|p|^2 \mp \sqrt{t}\nabla_p. \tag{4.7.8}$$

First we will establish (4.7.7) when replacing $\widehat{\mathfrak{M}}'_{b,t}$ by $S_{b,t}$. More precisely we take $a > 0$. We will show that for a given $b > 0$, the heat kernel $\exp(-aS_{b,t})(\cdot,\cdot)$ verifies an estimate similar to (4.7.7). Take $(x,p) \in T^*X$. To prove such an estimate we will use the Malliavin calculus [M78] as in [B81a, B81b] and in chapter 14.

Let P be the probability law of the Brownian motion $w_\cdot \in T_xX$, and let E^P denote the corresponding expectation. Consider the stochastic differential equation for $z_s = (x_s, p_s) \in T^*X$,

$$\dot{x} = \pm\sqrt{t}p, \qquad \dot{p} = \frac{t}{b^2}\tau_s^0\dot{w}, \tag{4.7.9}$$
$$x_0 = x, \qquad p_0 = p.$$

In (4.7.9), $\dot{p}$ is the covariant derivative of p with respect to the Levi-Civita connection, and τ_s^0 denotes parallel transport from T_xX into $T_{x_s}X$ with respect to the Levi-Civita connection. If $g : T^*X \to \mathbf{R}$ is a bounded smooth function, by the Feynman-Kac formula, we get

$$\exp(-aS_{b,t})\,g(z) = E^P\left[\exp\left(-\frac{1}{2}\int_0^a |p_s|^2\,ds\right)g(z_a)\right]. \tag{4.7.10}$$

Note that to establish (4.7.10), we use the Itô calculus on the process (x,p) together with the existence of the smooth kernel $\exp(-aS_{b,t})(z,z')$.

Let $h : \mathbf{R}_+ \to T_xX$ be a bounded adapted process. Consider the differential equation

$$\ddot{J} + R^{TX}(J,\dot{x})\,\dot{x} = \pm\frac{t^{3/2}}{b^2}h, \tag{4.7.11}$$
$$J_0 = 0, \qquad \dot{J}_0 = 0.$$

Recall that $TT^*X \simeq TX \oplus T^*X$. Then by proceeding as in the proof of (14.2.7), we get

$$E^P\left[\exp\left(-\frac{1}{2}\int_0^a |p_s|^2\,ds\right)\left\langle g'(z_a), \left(J_a, \pm\frac{1}{\sqrt{t}}\dot{J}_a\right)\right\rangle\right]$$
$$= E^P\left[\exp\left(-\frac{1}{2}\int_0^a |p_s|^2\,ds\right)g(z_a)\int_0^a\left(\langle h, \delta w\rangle \pm \left\langle \dot{J}, \frac{p}{\sqrt{t}}\right\rangle ds\right)\right]. \tag{4.7.12}$$

In (4.7.12), $\int_0^a \langle h, \delta w\rangle$ is a notation for the corresponding Itô integral. Of course,

$$\int_0^a \left\langle \dot{J}, \frac{p}{\sqrt{t}} \right\rangle ds = \left\langle J_a, \frac{p_a}{\sqrt{t}} \right\rangle - \int_0^a \left\langle J, \frac{\sqrt{t}}{b^2} \delta w \right\rangle. \tag{4.7.13}$$

Consider now the functions

$$\phi_s = (s/a)^2 (3 - 2s/a), \qquad \psi_s = -(s/a)^2 (a - s). \tag{4.7.14}$$

Then the functions ϕ, ψ vanish at 0 together with their first derivatives, and, moreover,

$$\phi_a = 1, \phi'_a = 0, \qquad \psi_a = 0, \psi'_a = 1. \tag{4.7.15}$$

In the sequel, we take $U = (Y, Z) \in T_x X \oplus T_x X$. Put

$$J_s^U = \tau_s^0 (\phi_s Y + \psi_s Z). \tag{4.7.16}$$

The linear map $(Y, Z) \to \left(J_a^U, \dot{J}_a^U \right)$ is just the parallel transport operator τ_a^0. With this choice of J_s, we can write equation (4.7.12) in the form

$$\begin{aligned} E^P \left[\exp\left(-\frac{1}{2} \int_0^a |p_s|^2 ds \right) \left\langle g'(z_a), \left(\tau_a^0 Y, \pm \frac{1}{\sqrt{t}} \tau_a^0 Z \right) \right\rangle \right] \\ = E^P \Bigg[\exp\left(-\frac{1}{2} \int_0^a |p_s|^2 ds \right) g(z_a) \\ \int_0^a \left(\left\langle \pm \frac{b^2}{t^{3/2}} \left(\ddot{J}^U + R^{TX} (J^U, \dot{x}) \dot{x} \right), \delta w \right\rangle \pm \left\langle \dot{J}^U, \frac{p}{\sqrt{t}} \right\rangle ds \right) \Bigg]. \end{aligned} \tag{4.7.17}$$

Now we use the basic technique of the Malliavin calculus. Let Y, Z be smooth sections of TX over X. Set $U = (Y, Z) \in TT^*X$. We wish to write an integration by parts for

$$E^P \left[\exp\left(-\frac{1}{2} \int_0^a |p_s|^2 ds \right) \langle g'(z_a), U_{x_a} \rangle \right]. \tag{4.7.18}$$

Clearly,

$$(Y_{x_a}, Z_{x_a}) = \tau_a^0 (\tau_0^a Y_{x_a}, \tau_0^a Z_{x_a}). \tag{4.7.19}$$

Clearly $\tau_0^a Y_{x_a}, \tau_0^a Z_z$ can be written as linear combinations of the $e_i, \widehat{e}^i$. By (4.7.11)-(4.7.14), we can express

$$E^P \left[\exp\left(-\frac{1}{2} \int_0^a |p_s|^2 ds \right) \langle g'(z_a), (\tau_a^0 e_i, \tau_a^0 \widehat{e}^j) \rangle \right] \tag{4.7.20}$$

as the expectation of a quantity where only g appears, and none of its derivatives. Still, to obtain the required integration by parts formula, $\tau_0^a U_{x_a} = (\tau_0^a Y_{x_a}, \tau_0^a Z_{x_a})$ also has to be differentiated. By taking into account the contribution of ϖ in (14.2.8), this computation can be easily done. Therefore we get an integration by parts formula very similar to (4.7.11) for (4.7.18).

The above procedure can be iterated as many times as necessary. We get a formula of integration by parts for

$$E^P\left[\exp\left(-\frac{1}{2}\int_0^a |p_s|^2\,ds\right)U_1\dots U_m g(z_a)\right]$$

of the same kind as above. By taking m large enough, this leads in principle to a uniform bound for $\exp(-aS_{b,t})(z,z')$ and its derivatives, when $a\in\left[\frac{1}{4},\frac{1}{2}\right]$, and when the other parameters vary as indicated in our proposition.

Still we have to be careful in proving that the right-hand side of the analogue of (4.7.17) is indeed integrable with respect to P, and also that we can obtain the uniform bound in (4.7.7). First we consider equation (4.7.17) itself. We will show that if $a\in\left[\frac{1}{4},\frac{1}{2}\right]$, if $z=(x,p)$, and if the other parameters are taken as before, the right-hand sides of (4.7.10) and (4.7.17) can be uniformly bounded by

$$\frac{C}{t^m}\exp\left(-c|p|^2\right)(|Y|+|Z|)\|g\|_\infty. \tag{4.7.21}$$

Indeed recall that by (4.7.9), $\dot{x}=\pm\sqrt{t}p$. Classically [IM74, p. 27], there exist $c>0, C>0$ such that for any $M>0, a>0$,

$$P\left[\sup_{0\le s\le a}|w_s|\ge M\right]\le C\exp\left(-M^2/2a\right). \tag{4.7.22}$$

By (4.7.22), we get a bound like (4.7.21) easily, with $c=0$ and C still depending on $|p|$. However, we now take into account the exponential factor in the right-hand side of (4.7.12). Take p with $|p|\ge 1$, and choose $M=b^2|p|/2t$ in (4.7.22). Then on $\left(\sup_{0\le s\le a}|w_s|\le b^2|p|/2t\right)$,

$$\int_0^a|p_s|^2\,ds\ge\frac{a}{4}|p|^2. \tag{4.7.23}$$

Moreover, by (4.7.22),

$$P\left[\sup_{0\le s\le a}|w_s|\ge b^2|p|/2t\right]\le C\exp\left(-b^4|p|^2/8at^2\right). \tag{4.7.24}$$

The above then leads easily to the uniform bound in (4.7.21).

We claim that the same method can be used to control

$$E^P\left[\exp\left(-\frac{1}{2}\int_0^a|p_s|^2\,ds\right)\langle g'(z_a),U_{z_a}\rangle\right].$$

Indeed the extra term which appears in the integration by parts formula does not raise any new difficulty. The same idea can be used to control instead

$$E^P\left[\exp\left(-\frac{1}{2}\int_0^a|p_s|^2\,ds\right)U_1\dots U_m g(z_a)\right]$$

for arbitrary m. Therefore, we get the analogue of the bound (4.7.7) for $\exp(-aS_{b,t})$. We leave to the reader to verify that similar bounds hold for integration by parts formulas involving derivatives of arbitrary order of g. This way, we find that in the range of parameters which was specified before,

$$\exp(-aS_{b,t})(z,z')\le\frac{C}{t^m}\exp\left(-\left(|p|^2+|p'|^2\right)\right). \tag{4.7.25}$$

Note that the appearance of $|p'|^2$ in the right-hand side of (4.7.25) can be obtained by the same arguments as before, or by using instead the adjoint $S^*_{b,t}$ of $S_{b,t}$.

Assume now that $x, x' \in X$ are such that $d_X(x, x') \geq \beta$, and that $z = (x, p), z' = (x', p')$. If in (4.7.9) $z_0 = z, z_a = z'$, we find that

$$\sup_{0 \leq s \leq a} |p_s| \geq \frac{\beta}{a\sqrt{t}}. \tag{4.7.26}$$

If p is such that $|p| \leq \beta/2a\sqrt{t}$, by (4.7.9), (4.7.22), we get

$$P\left[\sup_{0 \leq s \leq a} |p_s| \geq \frac{\beta}{a\sqrt{t}}\right] \leq C \exp\left(-cb^4\beta^2/8a^3t^3\right). \tag{4.7.27}$$

By proceeding as above, and using (4.7.27) and the Cauchy-Schwarz inequality, we find that there exist $c > 0, C > 0, m \in \mathbf{N}$ such that if a, b, t vary in the above range of parameters, if $\beta > 0$ and $d_X(x, x') \geq \beta$,

$$\exp\left(-aS_{b,t}\right)(z, gz) \leq \frac{C}{t^m} \exp\left(-c\left(|p|^2 + |p'|^2 + \beta^2/t\right)\right). \tag{4.7.28}$$

This establishes in particular the obvious analogue of (4.7.7) when $\underline{\widehat{\mathfrak{M}}}'_{b,t}$ is replaced by $S_{b,t}/2$.

Now we briefly explain how to obtain the exact form of (4.7.7). We should inspect the precise form of equation (4.7.4) for $\underline{\widehat{\mathfrak{M}}}'_{b,t}$. We claim that we can construct the heat kernel $\exp\left(-2a\underline{\widehat{\mathfrak{M}}}'_{b,t}\right)$ by using the same probability space as above, by combining this with the Itô calculus and the Feynman-Kac formula. This is indeed very classical [M78], [B84], except maybe for the term $-\frac{t^{3/2}}{4}\omega\left(\nabla^F, g^F\right)\left(e_i \nabla_{\widehat{e}^i}\right)$. We trivialize F along $x.$ using parallel transport with respect to the unitary connection $\nabla^{F,u}$. When constructing the kernel $\exp\left(-2a\underline{\widehat{\mathfrak{M}}}'_{b,t}\right)$, the noncommutative Feynman-Kac formula which incorporates this term is of the type

$$\begin{aligned} dV = V\Bigg[&\frac{\sqrt{t}}{2}\omega\left(\nabla^F, g^F\right)(\delta w) + \Bigg(\frac{t^{3/2}}{2b^2}\nabla^{\Lambda^\cdot(T^*T^*X)\widehat{\otimes}F}\widehat{\omega}\left(\nabla^F, g^F\right) \\ &+ \frac{1}{2}\omega\left(\nabla^F, g^F\right)^2 + z\left(\omega\left(\nabla^F, g^F\right) - \frac{t^{3/2}}{2b^2}\widehat{\omega}\left(\nabla^F, g^F\right)\right)\Bigg)ds\Bigg], \\ V_0 = 1.& \end{aligned} \tag{4.7.29}$$

Equation (4.7.29) incorporates only the terms acting on F.

Another equation incorporates all the other terms in the right-hand side of (4.7.4) acting on $\Lambda^\cdot(T^*X)\widehat{\otimes}\Lambda^\cdot(T^*X)$. Observe that some of these terms are diverging as $t \to 0$, like the term $\mp\frac{1}{2\sqrt{t}} i_{\widehat{e}^i} e^i$. To overcome this divergence, we conjugate the operator $\underline{\widehat{\mathfrak{M}}}'_{b,t}$ by $t^{N^H/2}$. This conjugation has no effect on the required estimate. The only diverging unbounded term for $b \in \left[\sqrt{t}, b_0\right]$ and $t \to 0$ is given by $\pm\frac{3b^2z}{t} i_{\widehat{p}}$. Still since this term contains z and $z^2 = 0$, the divergence can be absorbed in a diverging term of the type $1/t^m$. The techniques of estimation of the kernel $\exp\left(-\widehat{\underline{\mathfrak{M}}}'_{b,t}\right)(\cdot,\cdot)$ remain the same as above.

The proof of our proposition is completed. □

Remark 4.7.2. For $\beta > 0$, recall that $\mathcal{U}_\beta$ is the β-neighborhood of X_g in X. By (4.7.6) and (4.7.7), there is $C > 0, c > 0$ such that for $a \in \left[\frac{1}{2}, 1\right], t \in]0,1], b \in \left[\sqrt{t}, b_0\right], \beta \in]0,1]$,

$$\left| \int_{\pi^{-1}X \setminus \mathcal{U}_\beta} \mathrm{Tr_s}\left[g \exp\left(-a\widehat{\underline{\mathfrak{M}}}'_{b,t}\right)(z, gz)\right] dv_{T^*X}(z) \right| \le C \exp\left(-c\beta^2/t\right). \tag{4.7.30}$$

Of course, in (4.7.30), we can as well replace $\widehat{\underline{\mathfrak{M}}}'_{b,t}$ by $\widehat{\underline{\mathfrak{M}}}_{b,t}$. The above shows that the integral in the right-hand sides of (4.6.5) or of (4.7.6) localize near $\pi^{-1}X_g$.

4.8 REPLACING $\mathbf{T^*X}$ BY $\mathbf{T_xX \oplus T_x^*X}$ AND THE RESCALING OF CLIFFORD VARIABLES ON $\mathbf{T^*X}$

Let $\gamma(s) : \mathbf{R} \to [0,1]$ be a smooth even function such that

$$\begin{aligned} \gamma(s) &= 1 \text{ if } |s| \le 1/2, \\ &= 0 \text{ if } |s| \ge 1. \end{aligned} \tag{4.8.1}$$

If $y \in TX$, set

$$\rho(y) = \gamma\left(\frac{|y|}{4\eta_0}\right). \tag{4.8.2}$$

Then

$$\begin{aligned} \rho(y) &= 1 \text{ if } |y| \le 2\eta_0, \\ &= 0 \text{ if } |y| \ge 4\eta_0. \end{aligned} \tag{4.8.3}$$

First we describe the case where S is reduced to one point, i.e., the case of a single fiber X.

Take $\epsilon \in]0, a_X/2]$. If $s \in S, x \in X_g$, let $B^X(x,\epsilon)$ be the geodesic ball of center x and radius ϵ in X, and let $B^{T_xX}(0,\epsilon)$ be the open ball of center 0 and radius ϵ in T_xX. The exponential map $\exp_x$ identifies $B^{T_xX}(0,\epsilon)$ to $B^X(x,\epsilon)$.

Along radial lines centered at x along the fiber X_s, we identify

$$\Lambda^{\cdot}(T^*X)\widehat{\otimes}\Lambda^{\cdot}(T^*X)\widehat{\otimes}\Lambda^n(TX)\widehat{\otimes}F$$

to

$$\left(\Lambda^{\cdot}(T^*X)\widehat{\otimes}\Lambda^{\cdot}(T^*X)\widehat{\otimes}\Lambda^n(TX)\widehat{\otimes}F\right)_x$$

by parallel transport with respect to the connection

$$\nabla^{\Lambda^{\cdot}(T^*X)\widehat{\otimes}\Lambda^{\cdot}(T^*X)\widehat{\otimes}\Lambda^n(TX)\widehat{\otimes}F}.$$

In particular, the fibers $T_y^*X|_{y\in B^X(x,\epsilon)}$ are identified to T_x^*X by parallel transport with respect to the connection ∇^{T^*X} along the radial geodesic connecting y to x. The total space of T^*X over $B^X(x,\epsilon)$ is then identified with $B^{T_xX}(0,\epsilon)\times T_x^*X$. Note that all the above identifications are g-equivariant.

The flat vector bundle F has been trivialized on $B^X(x,\epsilon)$ using the flat connection ∇^F. Therefore we can consider F as the trivial flat vector bundle on T_xX.

Definition 4.8.1. Let $g_x^{T_xX}$ be the metric on T_xX given by

$$g_x^{T_xX}=\rho^2(y)\,g^{T_yX}+\left(1-\rho^2(y)\right)g^{T_xX}. \tag{4.8.4}$$

In particular the metric g_x^{TX} is just the given metric g^{TX} on $B^{T_xX}(0,2\eta_0)\simeq B^X(x,2\eta_0)$, and coincides with the flat metric g^{T_xX} outside of $B^{T_xX}(0,4\eta_0)$. Note that the above constructions are g-invariant.

Similarly let $g_x^{F_x}$ be the metric on F_x over T_xX which is given by

$$g_x^{F_x}=\rho^2(y)\,g^{F_y}+\left(1-\rho^2(y)\right)g^{F_x}. \tag{4.8.5}$$

Let $\widehat{\mathfrak{N}}_{b,t}$ be the operator of the type $\widehat{\mathfrak{M}}_{b,t}$ which is associated to the metrics $g_x^{T_xX}, g_x^{F_x}$.

The operator $\widehat{\mathfrak{N}}_{b,t}$ acts on smooth sections of

$$\Lambda^{\cdot}(T^*S)_s\,\widehat{\otimes}\left(\Lambda^{\cdot}(T^*X)\,\widehat{\otimes}\Lambda^{\cdot}(T^*X)\,\widehat{\otimes}\Lambda^n(TX)\,\widehat{\otimes}F\right)_x$$

on $(TX\oplus T^*X)_x$. By (4.8.3), if $|y|\le 2\eta_0$, the operators $\widehat{\mathfrak{M}}_{b,t}$ and $\widehat{\mathfrak{N}}_{b,t}$ coincide. Also we define the operator $\widehat{\mathfrak{N}}'_{b,t}$ from $\widehat{\mathfrak{N}}_{b,t}$ as in (4.7.3).

Now we consider the case where S is not necessarily reduced to one point. We will be especially careful here, although this precise construction will be needed in chapters 11 and 13 only. Indeed near $x\in M_g$, there is a coordinate system identifying a neighborhood $\mathcal{V}$ of x in M_g to an open ball centered at 0 in $\mathbf{R}^m\times\mathbf{R}^\ell$, so that the projection $\pi_g : M_g\to S$ is just the obvious projection $\mathbf{R}^m\times\mathbf{R}^\ell\to\mathbf{R}^m$. The vector bundle TX can be trivialized as a Euclidean vector bundle over $\mathcal{V}$, so that the action of g on $TX|_{M_g}$ is constant. In particular TX is trivialized near 0 on $\mathbf{R}^m\times\{0\}$.

If $x'\in\mathcal{V}$, we still use the exponential map $\exp_{x'}^X$ to identify the ball $B^{T_{x'}X}(0,4\eta_0)$ to an open ball along the fiber containing x'. The map $(s,Y)\in\mathbf{R}^m\times T_xX\to\exp_s^X Y\in M$ provides us with a chart for M near $x\in M_g$, such that the projection $\pi : M\to S$ is just $(s,Y)\to s$, and moreover $g(s,Y)=(s,gY)$. Using this chart, we find that the metric g^{TX} pulls back to a metric on T_xX. More precisely, the metric $g^{T_{s,y}X}$ pulls back to a metric on T_xX, which we still denote $g^{T_{s,y}X}$.

We still define the metric $g_x^{T_xX}$ on T_xX as in (4.8.4). In particular for $|y|\ge 4\eta_0$, this new metric is a "constant metric", which does not depend on (s,y). Note that the metric $g_x^{T_xX}$ is g-invariant.

Similarly, we can choose a new g-invariant horizontal vector bundle T^HM_x which coincides with the given T^HM for $|y|\le 2\eta_0$, and is given by a "constant" horizontal vector space for $|y|\ge 4\eta_0$.

Let dy, dp be the volume forms on the fibers of TX, T^*X. We denote by $\exp\left(-\underline{\widehat{\mathfrak{N}}}_{b,t}\right)((y,p),(y',p'))$ the smooth kernel on $T_xX \oplus T_x^*X$ associated with $\exp\left(-\underline{\widehat{\mathfrak{N}}}_{b,t}\right)$ with respect to the symplectic volume $dy'dp'$. We use a similar notation when replacing $\underline{\widehat{\mathfrak{N}}}_{b,t}$ by $\underline{\widehat{\mathfrak{N}}}'_{b,t}$.

Proposition 4.8.2. *There exist* $c > 0, C > 0$ *such that for* $a \in \left[\frac{1}{2}, 1\right], t \in]0,1], b \in \left[\sqrt{t}, b_0\right], x \in X_g$,

$$\left|\int_{\pi^{-1}\left\{y \in N_{X_g/X,x}, |y| \leq \eta_0\right\}} \left(\operatorname{Tr_s}\left[g \exp\left(-a\underline{\widehat{\mathfrak{M}}}'_{b,t}\right)((y,p), g(y,p))\right] k(x,y)\right.\right.$$
$$\left.\left. - \operatorname{Tr_s}\left[g \exp\left(-a\underline{\widehat{\mathfrak{N}}}'_{b,t}\right)((y,p), g(y,p))\right]\right) dydp\right| \leq C \exp(-c/t). \quad (4.8.6)$$

Proof. We will give a probabilistic proof of (4.8.6). Indeed by the same procedure as in the proof of Proposition 4.7.1, we can give a probabilistic representation of the kernels which appear in the left-hand side of (4.8.6) as a path integral involving paths connecting z to gz in time $a/2$, which project on X into paths connecting $y = \pi z$ to $y' = g\pi z$. Let T be the first time before a where the path $y_s = \pi z_s$ exits the ball $B^X(x, 2\eta_0)$, with the convention that $T = +\infty$ if this event does not occur. By (4.8.3), (4.8.4), and (4.8.5), the contribution of the paths which are such that $T = +\infty$ to the path integrals are the same. To evaluate the difference of the heat kernels, we have to consider only those paths such that $T < +\infty$. Now using (4.7.9), we find that on such paths

$$\sup_{0 \leq s \leq a} |p_s| \geq 2\frac{\eta_0}{\sqrt{ta}}. \quad (4.8.7)$$

We can now use the uniform bounds in (4.7.7) and proceed as in (4.7.26)-(4.7.27) and after (4.7.29) to obtain (4.8.6). In particular the diverging terms one gets by the rescaling indicated after (4.7.29) are ultimately killed by the term $\exp(-c/t)$. The proof of our proposition is completed. □

Of course, in (4.8.6), we can replace $\underline{\widehat{\mathfrak{M}}}'_{b,t}, \underline{\widehat{\mathfrak{N}}}'_{b,t}$ by $\underline{\widehat{\mathfrak{M}}}_{b,t}, \underline{\widehat{\mathfrak{N}}}_{b,t}$.

Definition 4.8.3. For $a > 0$, if f is a smooth section of

$$\Lambda^{\cdot}(T^*S)_s \widehat{\otimes} \left(\Lambda^{\cdot}(T^*X) \widehat{\otimes} \Lambda^{\cdot}(T^*X) \widehat{\otimes} \Lambda^n(TX) \widehat{\otimes} F\right)_x$$

on $T_xX \oplus T_x^*X$, if $(y,p) \in (TX \oplus T^*X)_x$, set

$$I_a f(y,p) = f(ay, p). \quad (4.8.8)$$

Put

$$\underline{\widehat{\mathfrak{O}}}_{b,t} = I_{t^{3/2}/b^2} \underline{\widehat{\mathfrak{N}}}_{b,t} I_{b^2/t^{3/2}}. \quad (4.8.9)$$

Note that in (4.8.9), the operator $I_{t^{3/2}/b^2}$ just acts as a scalar operator. It does not act on the Grassmann variables e^i.

Recall that by (1.6.2), $\ell = \dim X_g$. In the sequel, we may and we will assume that $e_1, \ldots, e_\ell$ is an orthonormal basis of $T_x X_g$, and $e_{\ell+1}, \ldots, e_n$ is an orthonormal basis of $N_{X_g/X}$.

Let $\mathfrak{e}_i, \widehat{\mathfrak{e}}_i, 1 \leq i \leq \ell$ be other orthonormal bases of TX_g. These variables will be considered as generating other copies of $\Lambda^{\cdot}(T_x X_g)$. Let $\mathfrak{e}^i, 1 \leq i \leq \ell$ be the basis of $T^* X_g$ dual to the basis $\mathfrak{e}_i, 1 \leq i \leq \ell$.

Definition 4.8.4. Let $\widehat{\mathfrak{P}}_{b,t}$ be the operator obtained from $\underline{\widehat{\mathfrak{O}}}_{b,t}$ by making the following replacements for $1 \leq i \leq \ell$:

- e^i is unchanged.
- i_{e_i} is changed into $-e^i/t + i_{e_i} + \frac{b}{t}\mathfrak{e}_i + i_{\mathfrak{e}^i}/b$.
- $\widehat{e}^i$ is unchanged.
- $i_{\widehat{e}_i}$ is changed into $i_{\widehat{e}_i} + \frac{b}{\sqrt{t}}\widehat{\mathfrak{e}}_i$,

and for $\ell + 1 \leq i \leq n$:

- e^i is changed into $\sqrt{t}e^i$.
- i_{e_i} is changed into $i_{e_i}/\sqrt{t}$.
- $\widehat{e}^i$ is changed into $\frac{\sqrt{t}}{b}\widehat{e}^i$.
- $i_{\widehat{e}_i}$ is changed into $\frac{b}{\sqrt{t}} i_{\widehat{e}_i}$.

Needless to say, the replacements made in Definition 4.8.4 are compatible with the commutation relations verified by the given operators.

The kernels for $\exp\left(-\underline{\widehat{\mathfrak{O}}}_{b,t}\right), \exp\left(-\underline{\widehat{\mathfrak{P}}}_{b,t}\right)$ will be denoted as before, and will be calculated with respect to $dydp$.

Let $\widehat{\mathrm{Tr_s}}$ be the linear map defined on the algebra $\mathcal{A}$ spanned by the $e^i, \widehat{e}^i, \mathfrak{e}_i, \widehat{\mathfrak{e}}_i$ for $1 \leq i \leq \ell$ with values in $\mathbf{R}$, which, up to permutation, vanishes on all the monomials except on the monomial of maximal length, with

$$\widehat{\mathrm{Tr_s}}\left[\prod_{i=1}^{\ell} e^i \widehat{e}^i \mathfrak{e}_i \widehat{\mathfrak{e}}_i\right] = 1. \tag{4.8.10}$$

We extend the functional $\widehat{\mathrm{Tr_s}}$ to a functional mapping

$$\Lambda^{\cdot}(T^*S)\widehat{\otimes}\mathcal{A}\widehat{\otimes}\mathrm{End}\left(\Lambda^{\cdot}\left(N^*_{X_g/X} \oplus N^*_{X_g/X}\right)\widehat{\otimes}F\right)_x$$

into $\Lambda^{\cdot}(T^*S)$, by taking the classical supertrace on the last factor above.

Clearly g acts as the identity on $TX_g \oplus T^*X_g$. In the sequel if $I \subset \{1, \ldots, \ell\}$, set

$$e^I = \prod_{i \in I} e^i, \tag{4.8.11}$$

$$\left(-e^{\cdot}/t + i_{e_{\cdot}} + \frac{b}{t}\mathfrak{e}_{\cdot}/t + i_{\mathfrak{e}^{\cdot}}/b\right)^I = \prod_{i \in I}\left(-e^i/t + i_{e_i} + \frac{b}{t}\mathfrak{e}_i + i_{\mathfrak{e}^i}/b\right).$$

Other products will be denoted in the same way.

Note that we have the expansion

$$g\exp\left(-\underline{\widehat{\mathfrak{P}}}_{b,t}\right)\left(g^{-1}(y,p),(y,p)\right)=\sum e^{I}\left(-e^{\cdot}/t+i_{e_{\cdot}}+\frac{b}{t}\mathfrak{e}_{\cdot}+i_{\mathfrak{e}^{\cdot}}/b\right)^{J}$$

$$\widehat{e}^{K}\left(i_{\widehat{e}_{\cdot}}+\frac{b}{\sqrt{t}}\widehat{\mathfrak{e}}_{\cdot}\right)^{L}H_{IJKL}(y,p), \quad (4.8.12)$$

where the $H_{IJKL}(y,p)$ are smooth sections of

$$\Lambda^{\cdot}(T^{*}S)\widehat{\otimes}\mathrm{End}\left(\Lambda^{\cdot}\left(N^{*}_{X_g/X}\oplus N_{X_g/X}\right)\widehat{\otimes}F\right)_{x}.$$

We define

$$\widehat{\mathrm{Tr}_{\mathrm{s}}}\left[g\exp\left(-\underline{\widehat{\mathfrak{P}}}_{b,t}\right)\left(g^{-1}(y,p),(y,p)\right)\right]$$

by writing the expansion (4.8.12) in normal form, i.e., by putting the annihilation operators $i_{e_i}, i_{\widehat{e}_i}, i_{\mathfrak{e}^i}, 1\leq i\leq \ell$ to the very right of the expansion, by ignoring any of the terms containing any of these annihilation operators, and by applying otherwise the above rule on the definition of $\widehat{\mathrm{Tr}_{\mathrm{s}}}$.

Proposition 4.8.5. *The following identity holds:*

$$\left(\frac{t^{3/2}}{b^{2}}\right)^{n-\ell}\mathrm{Tr}_{\mathrm{s}}\left[g\exp\left(-\underline{\widehat{\mathfrak{N}}}_{b,t}\right)\left(g^{-1}\left(\frac{t^{3/2}}{b^{2}}y,p\right),\left(\frac{t^{3/2}}{b^{2}}y,p\right)\right)\right]=$$

$$(-1)^{\ell}\widehat{\mathrm{Tr}_{\mathrm{s}}}\left[g\exp\left(-\underline{\widehat{\mathfrak{P}}}_{b,t}\right)\left(g^{-1}(y,p),(y,p)\right)\right]. \quad (4.8.13)$$

Proof. Consider the vector space $\mathbf{R}$ with its canonical basis e. Let e^* be the dual basis. The operators e^*, i_e act on $\Lambda^{\cdot}(\mathbf{R}^*)$. Then $e^* i_e$ is the only monomial in the algebra spanned by $1, e^*, i_e$ whose supertrace is nonzero, moreover,

$$\mathrm{Tr}_{\mathrm{s}}\left[e^{*}i_{e}\right]=-1. \quad (4.8.14)$$

By applying (4.8.14) to $TX_g\oplus TX_g$, we get (4.8.13) easily. □

4.9 THE LIMIT AS $t\to 0$ OF THE RESCALED OPERATOR

Let $i: M_g\to M$ be the obvious embedding. Let $\underline{\widehat{\mathfrak{P}}}$ be the operator given by

$$\underline{\widehat{\mathfrak{P}}}=\frac{1}{4}\left(-\Delta^{V}\mp 2\sum_{1\leq i\leq\ell}\widehat{\mathfrak{e}}_{i}\mathfrak{e}_{i}\right)\mp\frac{1}{2}\nabla_{p}$$

$$+\frac{1}{4}\sum_{1\leq i,j\leq\ell}\left\langle e_{i},R^{TX_g}e_{j}\right\rangle\widehat{e}^{i}\widehat{e}^{j}-i^{*}\frac{\omega\left(\nabla^{F},g^{F}\right)^{2}}{4}-zi^{*}\frac{\omega\left(\nabla^{F},g^{F}\right)}{2}. \quad (4.9.1)$$

Note that $\underline{\widehat{\mathfrak{P}}}$ depends on $x\in M_g$, but this dependence will not be written explicitly.

In the sequel, we will write $\mathcal{O}(|y|)$ for an expression which, in the given range of parameters, is uniformly bounded by $C|y|$. In the given trivialization, we denote by $\nabla, \widehat{\nabla}$ first order differentiations in the directions y or p. Also $\mathcal{O}(p)$ will denote an expression which depends linearly on p with uniformly bounded coefficients. Other notation is self-explanatory.

The fundamental algebraic fact of this chapter is as follows.

Theorem 4.9.1. *As* $t \to 0$,

$$\widehat{\underline{\mathfrak{P}}}_{b,t} \to \widehat{\underline{\mathfrak{P}}}. \tag{4.9.2}$$

More precisely, for $t \in]0,1], b \in \left[\sqrt{t}, b_0\right], |y| \leq 2b^2\eta_0/t^{3/2}$,

$$\begin{aligned}\widehat{\underline{\mathfrak{P}}}_{b,t} = \widehat{\underline{\mathfrak{P}}} + \frac{t^2}{4b^4}|p|^2 \pm \frac{\sqrt{t}}{2b}\left(\sum_{1\leq i\leq \ell}\left(\widehat{\mathfrak{e}}_i\widehat{e}^i + \mathfrak{e}_i i_{\widehat{e}_i}\right) - \sum_{\ell+1\leq i\leq n} i_{\widehat{e}_i}\left(e^i + i_{e_i}\right)\right)\\ + \mathcal{O}\left(\frac{t}{b^2}\right) + \mathcal{O}\left(\frac{t^{3/2}}{b^2}|y|\right) + \mathcal{O}\Bigg(\sqrt{t}\widehat{\nabla} + \frac{t^{3/2}}{b^2}|y||\nabla_p + \frac{t^{3/2}}{b^2}|p|^2|\widehat{\nabla} + \frac{\sqrt{t}}{b^2}p\\ + \frac{t^{3/2}}{b^2}|p|^2 + \frac{t^2}{b^4}p|y|\Bigg).\end{aligned} \tag{4.9.3}$$

Proof. Our theorem is an easy consequence of Proposition 4.6.1. We use implicitly the fact that in our trivialization of $\Lambda^{\cdot}(T^*X)\widehat{\otimes}\Lambda^{\cdot}(T^*X)\widehat{\otimes}\Lambda^n(TX)$, the connection form is a combination of operators of the type $e^i i_{e_j}$ and $\widehat{e}^i i_{\widehat{e}_j}$, which ultimately disappear in the given rescaling. Also we use the fact that i^*R^{TX} restricts to R^{TX_g} on TX_g.

Now we explain in more detail the various terms which appear in the right-hand side of (4.9.3). The first two terms are obvious. The last series of terms in the second line in the right-hand side of (4.6.4) contribute to the two next terms. The difference between the term evaluated at $\frac{t^{3/2}}{b^2}y$ and the corresponding term evaluated at 0 produces the contribution $\mathcal{O}\left(\frac{t^{3/2}}{b^2}y\right)$ in the first line in the right-hand side of (4.9.3).

Let us consider the term $\mp\frac{t^{3/2}}{2b^2}\nabla_p^{\Lambda^{\cdot}(T^*X)\widehat{\otimes}\Lambda^{\cdot}(T^*X)\widehat{\otimes}\Lambda^n(TX)\widehat{\otimes}F,u}$ in the right-hand side of (4.6.4). We split this term into its scalar part and its matrix part.

The scalar part is the operator $\mp\frac{t^{3/2}}{2b^2}\nabla_p$, which we split in its horizontal part and vertical part with respect to the coordinates (y,p). After the rescaling in the variable y, the difference of the horizontal part with the standard operator ∇_p is of the form $\mathcal{O}\left(\frac{t^{3/2}}{b^2}|y|\nabla_p\right)$. As to the vertical differentiation, it can be dominated by $\mathcal{O}\left(\frac{t^{3/2}}{b^2}|p|^2\widehat{\nabla}\right)$.

Now we estimate the matrix part of $\mp\frac{t^{3/2}}{2b^2}\nabla_p^{\Lambda^{\cdot}(T^*X)\widehat{\otimes}\Lambda^{\cdot}(T^*X)\widehat{\otimes}\Lambda^n(TX)\widehat{\otimes}F,u}$. In the given trivialization, the corresponding connection form vanishes at $y=0$. Using the replacements in Definition 4.8.4 and the considerations we made at the beginning of the proof, we find that this term contributes by $\mathcal{O}\left(\frac{t^2}{b^4}|y|\,p\right)$.

The remaining terms in the fourth line in the right-hand side of (4.6.4) are of the form $\mathcal{O}\left(\frac{\sqrt{t}}{b^2}p\right)$. The term in the fifth line in (4.6.4) can be dominated by $\mathcal{O}\left(\frac{t^{3/2}}{b^2}|p|^2\right)$. The sixth line can also be easily controlled.

By summing up all the terms, we get (4.9.2). The proof of our theorem is completed. □

Remark 4.9.2. In the right-hand side of (4.9.3), for fixed b, as $t \to 0$, the remainder tends to 0. However, when $b \simeq \sqrt{t}$, this is not the case for a number of terms, including a few of them which do diverge. These divergences will be dealt with in chapters 11 and 13. They are indeed spurious divergences.

Clearly the operators which appear in the right-hand side of (4.9.1) commute. Therefore,

$$\exp\left(-\underline{\widehat{\mathfrak{P}}}\right) = \exp\left(\frac{\Delta^V}{4} \pm \frac{1}{2}\nabla_p\right)\exp\left(\mp\frac{1}{2}\sum_{1\le i\le \ell}\mathfrak{e}_i\widehat{\mathfrak{e}}_i\right)$$
$$\exp\left(-\frac{1}{4}\sum_{1\le i,j\le \ell}\left\langle e_i, R^{TX_g}e_j\right\rangle \widehat{e}^i\widehat{e}^j\right)$$
$$i^*\exp\left(\frac{\omega\left(\nabla^F,g^F\right)^2}{4} + z\frac{\omega\left(\nabla^F,g^F\right)}{2}\right). \quad (4.9.4)$$

4.10 THE LIMIT OF THE RESCALED HEAT KERNEL

Theorem 4.10.1. *For $\eta_0 > 0$ small enough, there exist $c > 0, C > 0, m \in \mathbf{N}$ such that for $a \in \left[\frac{1}{2}, 1\right], t \in]0,1], b \in \left[\sqrt{t}, b_0\right], x \in X_g, y \in N_{X_g/X,x}, |y| \le b^2\eta_0/t^{3/2}, p \in T^*_xX$,*

$$\left|\exp\left(-a\underline{\widehat{\mathfrak{P}}}_{b,t}\right)\left(g^{-1}(y,p),(y,p)\right)\right|$$
$$\le C\left(1+\left(\frac{\sqrt{t}}{b^2}\right)^m\right)\exp\left(-c\left(|y|^{1/3}+|p|^{2/3}\right)\right). \quad (4.10.1)$$

Moreover, as $t \to 0$, for $x \in X_g, y \in N_{X_g/X,x}, p \in T^*X$,

$$\widehat{\mathrm{Tr}}_{\mathrm{s}}\left[g\exp\left(-a\underline{\widehat{\mathfrak{P}}}_{b,t}\right)\left(g^{-1}(y,p),(y,p)\right)\right]$$
$$\to \widehat{\mathrm{Tr}}_{\mathrm{s}}\left[g\exp\left(-a\underline{\widehat{\mathfrak{P}}}\right)\left(g^{-1}(y,p),(y,p)\right)\right]. \quad (4.10.2)$$

Proof. To simplify the arguments, we will first assume that $g = 1$, so that $X_g = X$. In this case we should take $y = 0$ in our proposition. As a first step, by imitating the strategy in the proof of Proposition 4.7.1, we consider only the scalar part $\mathfrak{S}_{b,t}$ of the operator $2\widehat{\underline{\mathfrak{P}}}_{b,t}$. Note that in the whole proof, we will not change our notation when dealing with T_xX instead of considering the whole manifold X. In particular ∇^{TX} denotes the Levi-Civita connection with respect to the metric $g_x^{T_xX}$.

As we said before, the manifold X is now replaced by T_xX equipped with the metric g_x^{TX}. The analogue of equation (4.7.9) is now

$$\begin{aligned} \dot{y}_s &= \pm p_s, & \dot{p}_s &= \tau_s^0 \dot{w}_s, \\ y_0 &= 0, & p_0 &= g^{-1}p. \end{aligned} \tag{4.10.3}$$

In (4.10.3), τ_s^0 represents the parallel transport with respect to the Levi-Civita connection associated to the metric $g_x^{T_xX}$.

For $a \in]\frac{1}{4}, \frac{1}{2}]$, we want to estimate $\exp\left(-a\mathfrak{S}_{b,t}\right)\left(\left(0,p\right),\left(0,p\right)\right)$. Inspection of the right-hand side of (4.6.4) shows that salvation will not come only from the factor $\frac{t^2}{b^4}\left|p\right|^2$, which tends to 0 as $t \to 0$. However, in (4.10.3), we want to have $y_a = 0$.

For the paths which are such that $|p| \geq b^2/2at^{3/2}$, we can use the dampening factor $t^2\left|p\right|^2/b^4$ in the right-hand side of (4.6.4). By proceeding as before, these contributions can be uniformly bounded by $C\exp\left(-cat^2\left|p\right|^2/b^4\right)$, with $c > 0, C > 0$. Now $t \geq \left(b^2\eta_0/2a\left|p\right|\right)^{2/3}$, and so there is $c' > 0$ such that

$$\exp\left(-cat^2\left|p\right|^2/b^4\right) \leq \exp\left(-\frac{c'}{b_0^{4/3}}\left|p\right|^{2/3}\right). \tag{4.10.4}$$

Now we consider those paths which are such that $|p| \leq b^2/2at^{3/2}$. Since $y_0 = y_a = 0$, when considering $\dot{y}$ as a section of T_xX, we must have

$$\int_0^a \dot{y}ds = 0. \tag{4.10.5}$$

By (4.10.3), we get

$$\dot{y}_s = \pm\tau_s^0\left(p + w_s\right). \tag{4.10.6}$$

In (4.10.6), τ_s^0 denotes parallel transport with respect to the Levi-Civita connection along the curve $y_\cdot$ from T_xX into $T_{y_s}X$. By (4.10.5), (4.10.6), we get

$$\int_0^a \tau_s^0 w_s ds = -\int_0^a \tau_s^0 p ds. \tag{4.10.7}$$

The operators τ_s^0 are preserving the metric $g_x^{T_xX}$. Moreover, since the metric $g_x^{T_xX}$ differs from the constant metric g^{T_xX} only on a compact set, when viewed as acting on T_xX, the operators τ_s^0 are uniformly bounded. Therefore,

$$\left|\int_0^a \tau_s^0 w_s ds\right| \leq Ca \sup_{0\leq s\leq a}\left|w_s\right|. \tag{4.10.8}$$

We denote temporarily by ∇^{TX} the Levi-Civita connection on T_xX with respect to the metric g_x^{TX} on T_xX. Let Γ^{TX} be the Christoffel symbol of the connection ∇^{TX}. Equivalently, Γ^{TX} is the connection form for ∇^{TX} in the trivialization of TX associated to the given chart near x. If $U \in T_xX$, if $U_s = \tau_s^0 U$, then

$$\frac{d}{ds}U_s + \frac{t^{3/2}}{b^2}\Gamma^{TX}_{\frac{t^{3/2}}{b^2}y_s}\left(\dot{y}_s\right)U_s = 0. \tag{4.10.9}$$

By (4.10.9), we get

$$U_s - U = -\frac{t^{3/2}}{b^2}\int_0^s \Gamma^{TX}_{\frac{t^{3/2}}{b^2}y_u}\left(\dot{y}_u\right)U_u du. \tag{4.10.10}$$

Using (4.10.6), (4.10.10) and the fact that the τ_s^0 are uniformly bounded, we get

$$\left|\tau_s^0 - 1\right| \leq C\frac{t^{3/2}}{b^2}s \sup_{0\leq s\leq a}\left|p_s\right|. \tag{4.10.11}$$

By (4.10.11), we obtain

$$\left|\int_0^a \tau_s^0 p ds\right| \geq a\left|p\right|\left(1 - C'\frac{t^{3/2}}{b^2}\sup_{0\leq s\leq a}\left|p_s\right|\right). \tag{4.10.12}$$

By (4.10.7), (4.10.8), (4.10.12), we get

$$\sup_{0\leq s\leq a}\left|w_s\right| \geq \left|p\right|\left(1 - C'\frac{t^{3/2}}{b^2}\sup_{0\leq s\leq a}\left|p_s\right|\right). \tag{4.10.13}$$

Now by (4.10.3), we obtain

$$\sup_{0\leq s\leq a}\left|p_s\right| \leq \left|p\right| + \sup_{0\leq s\leq a}\left|w_s\right|. \tag{4.10.14}$$

We deduce from (4.10.13), (4.10.14) that

$$\left(1 - C\frac{t^{3/2}}{b^2}\left|p\right|\right)\left|p\right| \leq \left(1 + C'\frac{t^{3/2}}{b^2}\left|p\right|\right)\sup_{0\leq s\leq a}\left|w_s\right|. \tag{4.10.15}$$

Recall that now $\left|p\right| \leq \frac{b^2}{2Ct^{3/2}}$. By (4.10.15),

$$\sup_{0\leq s\leq a}\left|w_s\right| \geq C''\left|p\right|/2. \tag{4.10.16}$$

We can control these paths by the same argument we already used to dominate the contribution of such p by $\exp\left(-c\left|p\right|^2\right)$.

The proof of our proposition is completed for the scalar part of the operator. Controlling the full operator does not raise any substantially new difficulty. Still two kinds of terms have to be taken care of:

- In the fourth line in the right-hand side of (4.6.4), there is a term $p/\sqrt{t}$. For fixed b, this singularity is compensated by the factor $\frac{t}{2b^2}$. Still the resulting expression diverges when $b = \sqrt{t}$. However, this term can be controlled by an adequate rescaling of the Grassmann variables f^α, also using the fact that by (4.7.22), for any $\gamma > 0$,

$$E\left[\exp\left(\gamma \sup_{0\leq s\leq 1}\left|w_s\right|\right)\right] < +\infty. \tag{4.10.17}$$

The rescaling of the f^α by the factor $\frac{b^2}{\sqrt{t}(1+|p|)}$ combined with (4.10.17) ultimately introduces a correcting factor $1+\left(\frac{\sqrt{t}|p|}{b^2}\right)^m$, the factor $|p|^m$ ultimately disappearing because of $\exp\left(-c\,|p|^{2/3}\right)$.

- The only serious point is to control $\mp\frac{t}{2b^2}\left\langle R^{TX}\left(\cdot,p\right)e_i,p\right\rangle\sqrt{t}\widehat{e}^i$ in the right-hand side of (4.6.4). This term is made indeed smaller by the rescalings of Definition 4.8.4. The main difficulty is that it appears in a first order differential equation which defines the associated noncommutative Feynman-Kac formula, and the weight $t^{3/2}$ is bigger than t^2 which appears as a factor of $|p|^2$. The difficulty in the estimation comes exactly at the stage leading to equation (4.10.4), where this quadratic term becomes too big to control.

We briefly explain how to take care of this second term. Take $\kappa\geq 1$. In Definition 4.8.4, we make the rescaling also depend on the extra parameter κ. For $1\leq i\leq \ell$, $\widehat{e}^i$ is changed into $\widehat{e}^i/\kappa$, and $i_{\widehat{e}_i}$ is changed into $\kappa i_{\widehat{e}_i}+\frac{b}{\sqrt{t}}\widehat{\mathfrak{e}}_i$. For $\ell+1\leq i\leq n$, $\widehat{e}^i$ is changed into $\frac{\sqrt{t}}{b\kappa}\widehat{e}^i$, and $i_{\widehat{e}_i}$ into $\frac{\kappa b}{\sqrt{t}}i_{\widehat{e}_i}$. The other rescalings are kept unchanged. Let $\underline{\widehat{\mathfrak{P}}}^{\kappa}_{b,t}$ be the obvious analogue of $\underline{\widehat{\mathfrak{P}}}_{b,t}$. Clearly

$$\widehat{\mathrm{Tr_s}}\left[g\exp\left(-\underline{\widehat{\mathfrak{P}}}_{b,t}\right)\left(g^{-1}\left(y,p\right),\left(y,p\right)\right)\right]$$
$$=\kappa^n\widehat{\mathrm{Tr_s}}\left[g\exp\left(-\underline{\widehat{\mathfrak{P}}}^{\kappa}_{b,t}\right)\left(g^{-1}\left(y,p\right),\left(y,p\right)\right)\right]. \quad (4.10.18)$$

Recall that for the moment, we assume that $g=1$. The quadratic term in the right-hand side of (4.6.4) is now $\mp\frac{t^{3/2}}{2b^2\kappa}\left\langle R^{TX}\left(\cdot,p\right)e_i,p\right\rangle\widehat{e}^i$. Moreover, the terms containing $i_{\widehat{e}_i}$ appear in the first line in the right-hand side of (4.6.4). The contribution of the annihilation creation operators to the first line in the right-hand side of (4.6.4) is now

$$\pm\frac{t}{2b^2}\left(i_{\widehat{e}_i}+\frac{b}{\kappa\sqrt{t}}\widehat{\mathfrak{e}}_i\right)\left(\widehat{e}^i-\kappa\sqrt{t}i_{e_i}-\frac{\kappa b}{\sqrt{t}}\mathfrak{e}_i-\frac{\kappa\sqrt{t}}{b}i_{\mathfrak{e}^i}\right). \quad (4.10.19)$$

Taking into account the fact that $t\in]0,1]$, $b\in\left[\sqrt{t},b_0\right]$, we find that (4.10.19) is dominated by

$$C\left(1+\frac{\sqrt{t}\kappa}{b}\right). \quad (4.10.20)$$

Now consider the noncommutative Feynman-Kac formula which produces the heat kernel $\exp\left(-2a\underline{\widehat{\mathfrak{P}}}^{\kappa}_{b,t}\right)\left(\left(0,p\right),\left(0,p\right)\right)$. Using the above considerations, and in particular (4.10.20), we find that at least if $\omega\left(\nabla^F,g^F\right)=0$, it can be controlled by the expectation of

$$C\exp\left(-\frac{t^2}{2b^4}\int_0^a\left|p_s\right|^2ds+\frac{ct^{3/2}}{b^2\kappa}\int_0^a\left|p_s\right|^2ds+c\frac{\sqrt{t}a\kappa}{b}\right). \quad (4.10.21)$$

Now by (4.10.3), we get

$$p_s = \tau_s^0 \left(p + w_s\right). \tag{4.10.22}$$

Using (4.10.17)-(4.10.21), given $b \in \mathbf{R}^*$, for $t \in]0,1]$ small enough

$$\begin{aligned}\left|\exp\left(-2a\widehat{\underline{\mathfrak{P}}}^{\kappa}_{b,t}\right)\left(\left(0,p\right),\left(0,p\right)\right)\right| \\ \leq C\exp\left(-\frac{cat^2}{b^4}\left|p\right|^2 + \frac{c'at^{3/2}}{b^2}\frac{\left|p\right|^2}{\kappa} + \frac{c''a\kappa\sqrt{t}}{b}\right) \\ = C\exp\left(-\frac{cat^2}{b^4}\left|p\right|^2\left(1-\frac{c'b^2}{c\kappa\sqrt{t}}\right) + c''a\kappa\right).\end{aligned} \tag{4.10.23}$$

Put

$$\kappa = 1 + \left|p\right|^{1/2}. \tag{4.10.24}$$

Under the conditions given before (4.10.4), we get

$$\frac{\kappa\sqrt{t}}{b^2} \geq \frac{\sqrt{t}\left|p\right|^{1/2}}{b^2} \geq \frac{c}{bt^{1/4}} \geq \frac{c'}{t^{1/4}}. \tag{4.10.25}$$

By proceeding as in (4.10.4) and using (4.10.23)-(4.10.25), we find that for $t \in]0,1]$ small enough,

$$\left|\exp\left(-2a\widehat{\underline{\mathfrak{P}}}^{\kappa}_{b,t}\right)\left(\left(0,p\right),\left(0,p\right)\right)\right| \leq C\exp\left(-c\left|p\right|^{2/3}\right). \tag{4.10.26}$$

Taking into account (4.10.18), (4.10.24), and (4.10.26), the remainder of the proof of the estimate (4.10.1) continues as before, at least for t small enough. Note that if t is bounded away from 0, it would be enough to take κ to be a large constant, and get an estimate better than (4.10.26).

Now we consider the case of a general $g \in G$. Equation (4.10.3) now becomes

$$\begin{aligned} \dot{y}_s &= \pm p_s, & \dot{p}_s &= \tau_s^0 \dot{w}_s, \\ y_a &= g y_0, & p_a &= g p_0. \end{aligned} \tag{4.10.27}$$

In (4.10.27), we take $y_0 \in N_{X_g/X,x}$.

By (4.10.27), we get

$$\left|\left(g-1\right)p_0\right| \leq \sup_{0\leq s\leq a}\left|w_s\right|, \qquad \left|\left(g-1\right)y_0 - \int_0^a \tau_s^0 p_0 ds\right| \leq a \sup_{0\leq s\leq a}\left|w_s\right|. \tag{4.10.28}$$

Let $p_0^{\|}, p_0^{\perp}$ be the components of p in $T_xX_g, N_{X_g/X,x}$. By (4.10.28), we get

$$\left|p_0^{\perp}\right| \leq C \sup_{0\leq a\leq s}\left|w_s\right|. \tag{4.10.29}$$

By (4.10.11), (4.10.28), (4.10.29), we obtain

$$\left|y_0\right| + \left|p_0^{\|}\right| \leq C\left(\sup_{0\leq a\leq s}\left|w_s\right| + \left|p_0\right|\frac{t^{3/2}}{b^2}\sup_{0\leq s\leq a}\left|p_s\right|\right). \tag{4.10.30}$$

By (4.10.14), (4.10.29), (4.10.30), we get

$$|y_0| + \left(1 - C\frac{t^{3/2}}{b^2}|p_0|\right)|p_0| \leq \left(1 + C\frac{t^{3/2}}{b^2}|p_0|\right) \sup_{0\leq a\leq s} |w_s|. \tag{4.10.31}$$

Equation (4.10.31) is the obvious extension of equation (4.10.15) to this more general situation.

From (4.10.31), we get (4.10.15). By proceeding as before, we obtain the bound

$$\left|\exp\left(-a\widehat{\underline{\mathfrak{P}}}_{b,t}\right)\left(g^{-1}(y_0,p_0),(y_0,p_0)\right)\right| \leq C\left(1+\left(\frac{\sqrt{t}}{b^2}\right)^m\right)\exp\left(-c|p_0|^{2/3}\right). \tag{4.10.32}$$

We consider first the case where $|y_0| \leq 2C\frac{t^{3/2}}{b^2}|p_0|^2$. Since $b \geq \sqrt{t}$, $|p_0|^{2/3} \geq \left(\frac{1}{2Ct^{1/2}}\right)^{1/3}|y_0|^{1/3}$ and so (4.10.1) holds in this case.

Now consider the case where $|y_0| \geq 2C\frac{t^{3/2}}{b^2}|p_0|^2$. By (4.10.31),

$$\frac{|y_0|}{1+\left(C\frac{t^{3/2}}{2b^2}\right)^{1/2}|y_0|^{1/2}} \leq 2\sup_{0\leq a\leq s}|w_s|. \tag{4.10.33}$$

In particular, since $b \geq \sqrt{t}$,

$$\frac{|y_0|}{1+\left(\frac{C}{2}t^{1/2}\right)^{1/2}|y_0|^{1/2}} \leq 2\sup_{0\leq a\leq s}|w_s|. \tag{4.10.34}$$

Using (4.7.22) and also (4.10.34), we obtain the bound

$$\left|\exp\left(-a\widehat{\underline{\mathfrak{P}}}_{b,t}\right)\left(g^{-1}(y_0,p_0),(y_0,p_0)\right)\right| \leq C\left(1+\left(\frac{\sqrt{t}}{b^2}\right)^m\right)\exp\left(-C'|y_0|\right). \tag{4.10.35}$$

By combining the bounds in (4.10.32) and (4.10.35), we still get (4.10.1).

Now we establish (4.10.2). We use Theorem 4.9.1, in combination with uniform estimates on the heat kernel $\exp\left(-\widehat{\underline{\mathfrak{P}}}_{b,t}\right)$ and its derivatives. These uniform estimates can be established using the Malliavin calculus as in chapter 14 . Combining these uniform estimates with Duhamel's formula, we get (4.10.2). The proof of our theorem is completed. □

4.11 EVALUATION OF THE HEAT KERNEL FOR $\frac{\Delta^V}{4} + a\nabla_p$

Let V be a finite dimensional vector space of dimension n equipped with a scalar product g^V. Let V^* be the dual of V equipped with the dual metric. Let Δ^H, Δ^V be the Laplacians of V, V^*. Set

$$\Delta^{H,V} = \frac{\partial^2}{\partial y^i \partial p_i}. \tag{4.11.1}$$

Note that $\Delta^{H,V}$ does not depend on the metric g^V, but just on the symplectic form ω of $V \oplus V^*$.

In [Kol34], Kolmogorov computed the heat kernel for the operator $\frac{\Delta^V}{4} + a\nabla_p$. The fact that it has a smooth heat kernel was one motivation for Hörmander [Hör67] to prove his theorem on hypoelliptic second order differential operators.

Proposition 4.11.1. *For any $a \in \mathbf{R}, s > 0$, the following identity holds:*

$$\exp\left(s\left(\frac{\Delta^V}{4} + a\nabla_p\right)\right) = \exp\left(\frac{s}{4}\Delta^V - \frac{as^2}{4}\Delta^{H,V} + \frac{a^2s^3}{12}\Delta^H\right)\exp\left(as\nabla_p\right). \tag{4.11.2}$$

Equivalently,

$$\exp\left(s\left(\frac{\Delta^V}{4} + a\nabla_p\right)\right) = \exp\left(\frac{as}{2}\nabla_p\right) \exp\left(\frac{s}{4}\Delta^V + \frac{a^2s^3}{48}\Delta^H\right)\exp\left(\frac{as}{2}\nabla_p\right). \tag{4.11.3}$$

Proof. We give an algebraic proof of (4.11.2). The arguments which will be used can of course be justified analytically. Set

$$U_s = \exp\left(s\left(\frac{\Delta^V}{4} + a\nabla_p\right)\right), \qquad V_s = U_s \exp\left(-as\nabla_p\right). \tag{4.11.4}$$

Clearly,

$$[\nabla_p, \nabla_{\widehat{e}^i}] = -\nabla_{e_i}, \tag{4.11.5}$$

and the commutator of ∇_p with ∇_{e_i} vanishes. From (4.11.5), we get

$$\exp\left(s\nabla_p\right)\Delta^V \exp\left(-s\nabla_p\right) = \left(\nabla_{\widehat{e}^i} - s\nabla_{e_i}\right)^2. \tag{4.11.6}$$

By (4.11.4), (4.11.6), we obtain

$$\frac{dV_s}{ds} = V_s\frac{1}{4}\left(\nabla_{\widehat{e}^i} - as\nabla_{e_i}\right)^2 = V_s\frac{1}{4}\left(\Delta^V - 2as\Delta^{H,V} + a^2s^2\Delta^H\right). \tag{4.11.7}$$

By (4.11.4), (4.11.7), we get (4.11.2). Moreover,

$$\frac{s}{4}\Delta^V - \frac{as^2}{4}\Delta^{H,V} + \frac{a^2s^3}{12}\Delta^H = \frac{s}{4}\left(\nabla_{\widehat{e}^i} - \frac{as}{2}\nabla_{e_i}\right)^2 + \frac{a^2s^3}{48}\Delta^H. \tag{4.11.8}$$

Equivalently,

$$\frac{s}{4}\Delta^V - \frac{as^2}{4}\Delta^{H,V} + \frac{a^2s^3}{12}\Delta^H = \exp\left(\frac{as}{2}\nabla_p\right)\left(\frac{s}{4}\Delta^V + \frac{a^2s^3}{48}\Delta^H\right)\exp\left(-\frac{as}{2}\nabla_p\right). \tag{4.11.9}$$

By (4.11.2) and (4.11.9), we get (4.11.3). □

Let g now be a linear isometry of V. Then g induces the isometry $\widetilde{g}^{-1}$ of V^*. If V^* is identified to V by the metric, this is just g itself. In the sequel, we assume that 1 is not an eigenvalue of g on V.

Proposition 4.11.2. *If $a \in \mathbf{R}^*$, the following identities hold:*

$$\int_{V^*} \exp\left(\frac{\Delta^V}{4} + a\nabla_p\right)((0,q),(0,q))\,dq = \left(\frac{1}{|a|\sqrt{\pi}}\right)^n, \tag{4.11.10}$$
$$\int_{V\oplus V^*} \exp\left(\frac{\Delta^V}{4} + a\nabla_p\right)\left(g^{-1}(y,p),(y,p)\right)dydq = \frac{1}{\det(1-g)^2}.$$

Proof. We will give two different proofs of our proposition, one based on Proposition 4.11.1, the other using Fourier analysis and not relying on explicit computations.

- A first proof

  To prove the first identity, we begin by using (4.11.3) to get

$$\exp\left(\frac{\Delta^V}{4} + a\nabla_p\right)((0,q),(0,q))$$
$$= \exp\left(\frac{\Delta^V}{4}\right)(q,q)\exp\left(\frac{a^2}{48}\Delta^H\right)\left(\frac{aq}{2},-\frac{aq}{2}\right). \tag{4.11.11}$$

  By (4.11.11), we get

$$\int_{V^*} \exp\left(\frac{\Delta^V}{4} + a\nabla_p\right)((0,q),(0,q))\,dq$$
$$= \left(\frac{12}{a^2\pi^2}\right)^{n/2}\int_{V^*}\exp\left(-12|q|^2\right)dq = \left(\frac{1}{|a|\sqrt{\pi}}\right)^n, \tag{4.11.12}$$

  which is the first identity in (4.11.10).

  Similarly, using again (4.11.3), we get

$$\exp\left(\frac{\Delta^V}{4} + a\nabla_p\right)\left(g^{-1}(y,q),(y,q)\right) = \exp\left(\frac{\Delta^V}{4}\right)(\widetilde{g}q,q)$$
$$\exp\left(\frac{a^2}{48}\Delta^H\right)\left(g^{-1}y + \frac{a}{2}\widetilde{g}q, y - \frac{a}{2}q\right) = \left(\frac{12}{a^2\pi^2}\right)^{n/2}$$
$$\exp\left(-\left|\widetilde{g}q - q\right|^2\right)\exp\left(-\frac{12}{a^2}\left|g^{-1}y - y + \frac{a}{2}(\widetilde{g}q+q)\right|^2\right). \tag{4.11.13}$$

  Note that (4.11.13) is precisely the formula obtained by Kolmogorov [Kol34]. Using (4.11.13) and the fact that no eigenvalue of g is equal to 1, we get the second identity in (4.11.10). The first proof of our proposition is completed.

- A second proof

  We use first the Fourier transform in the variable $y \in V$. Also, to avoid notational confusion, we will consider p as an operator, and use the notation q to denote the variable which is integrated in V^*. We get

$$\int_{V^*} \exp\left(\frac{\Delta^V}{4} + a\nabla_p\right)((0,q),(0,q))\,dq$$
$$= \int_{V^*\oplus V^*} \exp\left(\frac{\Delta^V}{4} + 2i\pi a\langle p,\xi\rangle\right)(q,q)\,dqd\xi. \tag{4.11.14}$$

Note that the operator which appears in the right-hand side of (4.11.14) acts on V^* and depends on the parameter $\xi \in V^*$. Moreover, given $\xi \in V^*$, we have the obvious

$$\exp\left(\frac{\Delta^V}{4} + 2i\pi a \langle p, \xi \rangle\right)(q, q) = \exp\left(\frac{\Delta^V}{4} + 2i\pi a \langle p + q, \xi \rangle\right)(0, 0). \tag{4.11.15}$$

By (4.11.14), (4.11.15), we get

$$\begin{aligned}\int_{V^*} &\exp\left(\frac{\Delta^V}{4} + a\nabla_p\right)((0, q), (0, q))\, dq \\ &= \int_{V^* \oplus V^*} \exp\left(\frac{\Delta^V}{4} + 2i\pi a \langle p, \xi \rangle\right)(0, 0) \exp\left(2i\pi a \langle q, \xi \rangle\right) dq d\xi.\end{aligned} \tag{4.11.16}$$

Also

$$\int_V \exp\left(2i\pi a \langle q, \xi \rangle\right) dq = \frac{1}{|a|^n} \delta_{\xi=0}. \tag{4.11.17}$$

By (4.11.16), (4.11.17), we obtain

$$\begin{aligned}\int_{V^*} &\exp\left(\frac{\Delta^V}{4} + a\nabla_p\right)((0, q), (0, q))\, dq \\ &= \frac{1}{|a|^n} \exp\left(\frac{\Delta^V}{4}\right)(0, 0) = \left(\frac{1}{|a|\sqrt{\pi}}\right)^n.\end{aligned} \tag{4.11.18}$$

A similar computation shows that

$$\begin{aligned}\int_{V \oplus V^*} &\exp\left(\frac{\Delta^V}{4} + a\nabla_p\right)\left(g^{-1}(y, q), (y, q)\right) dy dq \\ &= \int_{V \oplus V^* \oplus V^*} \exp\left(\frac{\Delta^V}{4} + 2i\pi a \langle p, \xi \rangle\right)\left(g^{-1} q, q\right) \\ &\qquad \exp\left(2i\pi \left\langle \left(1 - g^{-1}\right) y, \xi \right\rangle\right) dy dq d\xi.\end{aligned} \tag{4.11.19}$$

Now

$$\int_V \exp\left(2i\pi \left\langle \left(1 - g^{-1}\right) y, \xi \right\rangle\right) dy = \frac{1}{\det(1 - g)} \delta_{\xi=0}. \tag{4.11.20}$$

By (4.11.19), (4.11.20), we obtain

$$\begin{aligned}\int_{V \oplus V^*} &\exp\left(\frac{\Delta^V}{4} + a\nabla_p\right)\left(g^{-1}(y, q), (y, q)\right) dy dq \\ &= \frac{1}{\det(1 - g)} \int_{V^*} \exp\left(\frac{\Delta^V}{4}\right)\left(g^{-1} q, q\right) dq = \frac{1}{\det(1 - g)^2}.\end{aligned} \tag{4.11.21}$$

The second proof of our proposition is completed.

□

4.12 AN EVALUATION OF CERTAIN SUPERTRACES

We will use the formalism of Berezin integration of section 4.1, as in equation (4.1.5), with S replaced by M_g, and E by TX_g. We will evaluate the supertrace of the expression which appears in the right-hand side of (4.9.4).

If $\alpha \in \Lambda^{\cdot}(T^*S)\widehat{\otimes}\Lambda^{\cdot}(T^*X_g)$, we denote by $\alpha^{\max} \in \Lambda^{\cdot}(T^*S)\widehat{\otimes}o(TX_g)$ the form such that $\alpha^{\max}e^1 \wedge \ldots \wedge e^\ell$ is the form of top vertical degree appearing in the decomposition of α.

Proposition 4.12.1. *The following identity holds:*

$$
\begin{aligned}
(-1)^{n-\ell}\,\widehat{\mathrm{Tr_s}}\Bigg[g\exp\Bigg(\mp\frac{1}{2}\sum_{1\le i\le \ell}\mathfrak{e}_i\widehat{\mathfrak{e}}_i\Bigg)\exp\Bigg(-\frac{1}{4}\sum_{1\le i,j\le \ell}\left\langle e_i,R^{TX_g}e_j\right\rangle\widehat{e}^i\widehat{e}^j\Bigg)\\
i^*\exp\left(\omega\left(\nabla^F,g^F\right)^2/4+z\omega\left(\nabla^F,g^F\right)/2\right)\Bigg]\\
=\det\left(\left(1-g\right)|_{N_{X_g/X}}\right)^2\Bigg[\left(\frac{1}{2}\right)^\ell\int^{\widehat{B}}\exp\Bigg(-\frac{1}{4}\sum_{1\le i,j\le \ell}\left\langle e_i,R^{TX_g}e_j\right\rangle\widehat{e}^i\widehat{e}^j\Bigg)\\
\mathrm{Tr}^F\left[gi^*\exp\left(\omega\left(\nabla^F,g^F\right)^2/4+z\omega\left(\nabla^F,g^F\right)/2\right)\right]\Bigg]^{\max}. \qquad (4.12.1)
\end{aligned}
$$

Proof. We have the identity

$$
\begin{aligned}
(-1)^{n-\ell}\,\widehat{\mathrm{Tr_s}}\Bigg[g\exp\Bigg(\mp\frac{1}{2}\sum_{1\le i\le \ell}\mathfrak{e}_i\widehat{\mathfrak{e}}_i\Bigg)\\
\exp\Bigg(-\frac{1}{4}\sum_{1\le i,j\le \ell}\left\langle e_i,R^{TX_g}e_j\right\rangle\widehat{e}^i\widehat{e}^j\Bigg)\\
i^*\exp\left(\omega\left(\nabla^F,g^F\right)^2/4+z\omega\left(\nabla^F,g^F\right)/2\right)\Bigg]\\
=\mathrm{Tr}^{\Lambda^{\cdot}\left(N^*_{X_g/X}\oplus N_{X_g/X}\right)}[g]\,\widehat{\mathrm{Tr_s}}\Bigg[\exp\Bigg(\mp\frac{1}{2}\sum_{1\le i\le \ell}\mathfrak{e}_i\widehat{\mathfrak{e}}_i\Bigg)\\
\exp\Bigg(-\frac{1}{4}\sum_{1\le i,j\le \ell}\left\langle e_i,R^{TX_g}e_j\right\rangle\widehat{e}^i\widehat{e}^j\Bigg)\\
i^*\exp\left(\omega\left(\nabla^F,g^F\right)^2/4+z\omega\left(\nabla^F,g^F\right)/2\right)\Bigg]. \qquad (4.12.2)
\end{aligned}
$$

Note that the factor $(-1)^{n-\ell}$ in the right-hand side of (4.12.2) has disappeared so as to lead to $\mathrm{Tr}^{\Lambda^{\cdot}\left(N^*_{X_g/X}\oplus N_{X_g/X}\right)}[g]$.

Clearly,

$$\mathrm{Tr_s}^{\Lambda^{\cdot}\left(N^*_{X_g/X}\oplus N_{X_g/X}\right)}[g] = \det\left(1 - g|_{N_{X_g/X}}\right)^2. \tag{4.12.3}$$

Moreover,

$$\exp\left(\mp\frac{1}{2}\sum_{1\le i\le \ell} \mathfrak{e}_i\widehat{\mathfrak{e}}_i\right) = \prod_{1\le i\le \ell}\left(1\mp\frac{1}{2}\mathfrak{e}_i\widehat{\mathfrak{e}}_i\right). \tag{4.12.4}$$

When calculating $\widehat{\mathrm{Tr_s}}$ in the right-hand side of (4.12.2), (4.12.4) introduces a sign $(\mp 1)^{\ell}$. When comparing with the sign conventions in (4.1.1) and (4.8.10), we see that there is an extra sign $(-1)^{\ell}$ which appears when ultimately replacing $\widehat{\mathrm{Tr_s}}$ by the Berezin integral $\int^{\widehat{B}}$ in the right-hand side of (4.12.2). Ultimately the sign correction is $(\pm 1)^{\ell}$. However, in the present case, the right-hand side is nonzero only if an even number of $\widehat{e}^i, 1\le i\le \ell$ appears, so that ℓ has to be even. This ensures that the sign correction ultimately disappears. The proof of our proposition is completed. □

The final step in the computation of the local supertrace of $\exp\left(-\underline{\mathfrak{P}}\right)$ is as follows.

Proposition 4.12.2. *The following identity holds*

$$\begin{aligned}\int_{N_{X_g/X}\times T^*X} &(-1)^{n-\ell}\,\widehat{\mathrm{Tr_s}}\left[g\exp\left(-\underline{\mathfrak{P}}\right)\left(g^{-1}(y,q),(y,q)\right)\right]\\ &= \left[\frac{1}{\pi^{\ell/2}}\int^{\widehat{B}}\exp\left(-\frac{1}{4}\sum_{1\le i,j\le \ell}\left\langle e_i, R^{TX_g}e_j\right\rangle \widehat{e}^i\widehat{e}^j\right)\right.\\ &\left.\mathrm{Tr}^F\left[gi^*\exp\left(\omega\left(\nabla^F,g^F\right)^2/4 + z\omega\left(\nabla^F,g^F\right)/2\right)\right]\right]^{\max}.\end{aligned} \tag{4.12.5}$$

Proof. Our Proposition follows from (4.9.4) and from Propositions 4.11.2 and 4.12.1. □

4.13 A PROOF OF THEOREMS 4.2.1 AND 4.4.1

We use (4.6.3), (4.6.5), (4.7.6), (4.7.30) to conclude that to evaluate the limit as $t\to 0$ of $\mathrm{Tr_s}\left[g\exp\left(-\mathfrak{L}_{b,t}\right)\right]$, we should only evaluate the limit of

$$(-1)^n\int_{\pi^{-1}U_{\eta_0}}\mathrm{Tr_s}\left[g\exp\left(-\underline{\mathfrak{M}}_{b,t}\right)(z,gz)\right]dv_{T^*X}(z).$$

By (4.7.1), we get

$$\int_{\pi^{-1}U_{\eta_0}} \mathrm{Tr_s}\left[g\exp\left(-\widehat{\mathfrak{M}}_{b,t}\right)(z,gz)\right]dv_{T^*X}(z)$$
$$= \int_{X_g} dv_{X_g}(x)\int_{\left\{y\in N_{X_g/X},|y|\le\eta_0\right\}\times T^*X}$$
$$\mathrm{Tr_s}\left[g\exp\left(-\widehat{\mathfrak{M}}_{b,t}\right)\left(g^{-1}(y,p),(y,p)\right)\right]k(x,y)\,dv_{N_{X_g/X}}(y)\,dp. \quad (4.13.1)$$

By (4.8.6), in the right-hand side of (4.13.1), we can as well replace $\widehat{\mathfrak{M}}_{b,t}$ by $\widehat{\mathfrak{N}}_{b,t}$, while making $k(x,y)=1$. Moreover, using (4.8.13), given $x\in X_g$, we get

$$\int_{\left\{y\in N_{X_g/X},|y|\le\eta_0\right\}\times T^*_xX} \mathrm{Tr_s}\left[g\exp\left(-\widehat{\mathfrak{N}}_{b,t}\right)\left(g^{-1}(y,p),(y,p)\right)\right]$$
$$dv_{N_{X_g/X}}(y)\,dp = (-1)^{\ell}\int_{\left\{y\in N_{X_g/X},|y|\le b^2\eta_0/t^{3/2}\right\}\times T^*_xX}$$
$$\widehat{\mathrm{Tr}}_{\mathrm{s}}\left[g\exp\left(-\widehat{\mathfrak{P}}_{b,t}\right)\left(g^{-1}(y,p),(y,p)\right)\right]dv_{N_{X_g/X}}(y)\,dp. \quad (4.13.2)$$

Using Theorem 4.10.1, Proposition 4.12.2, and the above considerations, we find that as $t\to 0$,

$$\mathrm{Tr_s}\left[g\exp\left(-\mathfrak{L}_{b,t}\right)\right]\to\int_{X_g}\frac{1}{\pi^{\ell/2}}\int^{\widehat{B}}\exp\left(-\frac{1}{4}\sum_{1\le i,j\le\ell}\left\langle e_i,R^{TX_g}e_j\right\rangle\widehat{e}^i\widehat{e}^j\right)$$
$$\mathrm{Tr}^F\left[gi^*\exp\left(\omega\left(\nabla^F,g^F\right)^2/4+z\omega\left(\nabla^F,g^F\right)/2\right)\right]. \quad (4.13.3)$$

By [BG01, Proposition 1.6], we get

$$\mathrm{Tr}^F\left[gi^*\exp\left(\omega\left(\nabla^F,g^F\right)^2/4+z\omega\left(\nabla^F,g^F\right)/2\right)\right]$$
$$=\mathrm{Tr}^F[g]+z\mathrm{Tr}^F\left[gh\left(\omega\left(\nabla^F,g^F\right)/2\right)\right]. \quad (4.13.4)$$

By (1.6.3), (1.6.5), (1.7.3), (4.1.5), (4.13.3), (4.13.4), we obtain

$$\mathrm{Tr_s}\left[g\exp\left(-\mathfrak{L}_{b,t}\right)\right]\to L(g)+z\frac{1}{(2\pi)^{1/2}}\varphi^{-1}\int_{X_g}e\left(TX_g,\nabla^{TX_g}\right)h_g\left(\nabla^F,g^F\right). \quad (4.13.5)$$

Moreover, the difference $\widehat{\mathfrak{P}}_{b,t}-\widehat{\mathfrak{P}}_b$ has been estimated in equation (4.9.3) of Theorem 4.9.1. Using Duhamel's formula, one concludes that for given $b\in\mathbf{R}^*_+$, the speed of convergence in (4.13.5) is $\mathcal{O}\left(\sqrt{t}\right)$. By (4.2.8), by Proposition 4.5.2 and by (4.13.5), we get Theorem 4.2.1 and the first identity in (4.4.1) in Theorem 4.4.1.

Now we establish the second identity in (4.4.1). Instead of using the formula in (4.3.3) for $v_{b,t}$, we can observe that varying $c=\pm\frac{t}{b^2}$ in $\mathcal{H}^c=c\,|p|^2_t$

can also be obtained via Theorem 2.5.4. Ultimately, we find that to obtain $\frac{db}{b} v_{b,t}$, instead of the operator $\widehat{\underline{\mathfrak{L}}}_{b,t}$, we should now consider the operator

$$\widehat{\underline{\mathfrak{L}}}_{b,t} \mp t \frac{db}{b^3} \left(\widehat{p} - p - t i_p \right) \mp 3z \frac{db}{b^3} t^2 \left| p \right|^2 . \tag{4.13.6}$$

It is then easy to see that when making the rescalings in (4.6.2), in (4.8.8), and in Definition 4.8.4, when $t \to 0$, in the analogue of (4.9.2), (4.9.3), no term containing db appears in the right-hand side, so that the second identity in (4.4.1) holds.

An alternative method is to consider directly the expression in (4.3.3) for $v_{b,t}$. By (4.3.12), we get

$$h' (x) = \left(1 + 2 \frac{\partial}{\partial a} \right) \exp \left(a x^2 \right) |_{a=1}. \tag{4.13.7}$$

Let $\widehat{\underline{\mathfrak{M}}}^0_{b,t}$ be the operator obtained from $\widehat{\underline{\mathfrak{M}}}_{b,t}$ by making $z = 0$. By proceeding as in Proposition 4.5.4 and using (4.3.3), (4.13.7), we get

$$v_{b,t} = \pm (-1)^n \varphi \left(1 + 2 \frac{\partial}{\partial a} \right) \mathrm{Tr_s} \left[g \frac{t}{b^2} \left| p \right|^2 \exp \left(-a \widehat{\underline{\mathfrak{M}}}^0_{b,t} \right) \right] |_{a=1}. \tag{4.13.8}$$

By (4.13.8), we can then proceed as before and we get the second identity in (4.4.1).

As we saw in (4.4.5), we already know from the above that as $t \to 0$, $w_{b,t} = \mathcal{O} \left(\sqrt{t} \right)$. Still, it is interesting to use the same sort of arguments as above for $w_{b,t}$. Indeed, we have the identity

$$w_{b,t} = \left(1 + 2 \frac{\partial}{\partial a} \right) \varphi \mathrm{Tr_s} \left[g \left(\frac{\lambda_0}{2t} \mp \frac{t^2}{b^2} \left| p \right|^2 \right) \exp \left(-a A^{\mathcal{M},2}_{b,t} \right) \right] |_{a=1}. \tag{4.13.9}$$

Set

$$\mathbb{L} = \frac{1}{2} \sum_{1 \le i \le n} e^i \wedge \widehat{e}^i . \tag{4.13.10}$$

By proceeding as before, we get

$$\begin{aligned} \mathrm{Tr_s} & \left[g \left(\frac{\lambda_0}{2t} \mp \frac{t^2}{b^2} \left| p \right|^2 \right) \exp \left(-a A^{\mathcal{M},2}_{b,t} \right) \right] \\ & = (-1)^n \mathrm{Tr_s} \left[g \left(\frac{\mathbb{L}}{\sqrt{t}} \mp \frac{t}{b^2} \left| p \right|^2 \right) \exp \left(-a \widehat{\underline{\mathfrak{M}}}^0_{b,t} \right) \right] |_{a=1}. \end{aligned} \tag{4.13.11}$$

Set

$$\mathbb{M} = \frac{1}{2} \sum_{1 \le i \le \ell} e^i \wedge \widehat{e}^i . \tag{4.13.12}$$

Using (4.13.9) and proceeding as in the proof of Proposition 4.12.1, we find easily that as $t \to 0$,

$$\begin{aligned} & (-1)^n \sqrt{t} \mathrm{Tr_s} \left[g \left(\frac{\mathbb{L}}{\sqrt{t}} \mp \frac{t}{b^2} \left| p \right|^2 \right) \exp \left(- \widehat{\underline{\mathfrak{M}}}^0_{b,t} \right) \right] \\ & \to \int_{X_g} (\pm 1)^l \frac{1}{\pi^{\ell/2}} \int^{\widehat{B}} \mathbb{M} \exp \left(-\frac{1}{4} \left\langle e_i, R^{TX_g} e_j \right\rangle \widehat{e}^i \widehat{e}^j \right) \mathrm{Tr}^F [g] . \end{aligned} \tag{4.13.13}$$

The sign $(-1)^{\ell}$ in the right-hand side of (4.13.13) appears by the discussion at the end of the proof of Proposition 4.12.1. Note that the right-hand side of (4.13.13) is a form of degree 0 on S. Also only odd dimensional components of X_g give a nonzero contribution, so that $(\pm 1)^{\ell}$ is just ± 1. By (4.13.11), (4.13.13), we find that as $t \to 0$,

$$\operatorname{Tr_s}\left[g\left(\frac{\lambda_0}{2t} \mp \frac{t^2}{b^2}|p|^2\right)\exp\left(-A_{b,t}^{\mathcal{M},2}\right)\right] \\ = \frac{\pm 1}{\sqrt{t}}\int_{X_g}\frac{1}{\pi^{\ell/2}}\int^{\widehat{B}}\mathbb{M}\exp\left(-\frac{1}{4}\left\langle e_i, R^{TX_g}e_j\right\rangle\widehat{e}^i\widehat{e}^j\right)\operatorname{Tr}^F[g] + \mathcal{O}(1). \tag{4.13.14}$$

We claim that $\mathcal{O}(1)$ can be replaced by $\mathcal{O}\left(\sqrt{t}\right)$ in the right-hand side of (4.13.14). First we consider the case where $g = 1$. In this case, the term of weight $\sqrt{t}$ which appears in the first line in the right-hand side of (4.9.3) is given by $\pm\frac{\sqrt{t}}{2b}\left(\widehat{\mathfrak{e}}_i\widehat{e}^i + \mathfrak{e}_i i_{\widehat{e}_i}\right)$. The same computations as before show that the contribution of this term vanishes identically. In the general case where g is not necessarily equal to 1, the contribution of the remaining terms in the first line of (4.9.3) are also irrelevant. Note that the operator $\underline{\widehat{\mathfrak{P}}}$ is invariant by the map $(y,p) \to (-y,-p)$. In the second line in (4.9.3), the terms $O\left(\sqrt{t}\widehat{\nabla}\right)$ and $\mathcal{O}\left(\frac{\sqrt{t}}{b^2}p\right)$ do not contribute for the same reason. Therefore we have proved that instead of (4.13.14), we have

$$\operatorname{Tr_s}\left[g\left(\frac{\lambda_0}{2t} \mp \frac{t^2}{b^2}|p|^2\right)\exp\left(-A_{b,t}^{\mathcal{M},2}\right)\right] \\ = \frac{\pm 1}{\sqrt{t}}\int_{X_g}\frac{1}{\pi^{\ell/2}}\int^{\widehat{B}}\mathbb{M}\exp\left(-\frac{1}{4}\left\langle e_i, R^{TX_g}e_j\right\rangle\widehat{e}^i\widehat{e}^j\right)\operatorname{Tr}^F[g] + \mathcal{O}\left(\sqrt{t}\right). \tag{4.13.15}$$

We claim that for $a > 0$, as $t \to 0$,

$$\operatorname{Tr_s}\left[\left(\frac{\lambda_0}{2t} \mp \frac{t^2}{b^2}|p|^2\right)\exp\left(-aA_{b,t}^{\mathcal{M},2}\right)\right] \\ = \frac{\pm 1}{\sqrt{at}}\int_{X_g}\frac{1}{\pi^{\ell/2}}\int^{\widehat{B}}\mathbb{M}\exp\left(-\frac{1}{4}\left\langle e_i, R^{TX_g}e_j\right\rangle\widehat{e}^i\widehat{e}^j\right)\operatorname{Tr}^F[g] + \mathcal{O}\left(\sqrt{t}\right). \tag{4.13.16}$$

Indeed the only difference with respect to (4.13.15) is that we should instead evaluate the contribution of $\exp\left(-a\underline{\widehat{\mathfrak{P}}}\right)$ instead of $\exp\left(-\underline{\widehat{\mathfrak{P}}}\right)$ as was done before. However, when replacing $\frac{\Delta^V}{4} \pm \frac{1}{2}\nabla_p$ by $a\left(\frac{\Delta^V}{4} \pm \frac{\nabla_p}{2}\right)$, we get instead in the right-hand side of the first equation of (4.11.10) $\left(2/a^{3/2}\sqrt{\pi}\right)^n$. Ultimately we obtain (4.13.16) easily.

By (4.13.9), (4.13.16), we see that as $t \to 0$,

$$w_{b,t} = \mathcal{O}\left(\sqrt{t}\right), \tag{4.13.17}$$

which is just the third equation in (4.4.1).

The proof of Theorem 4.4.1 is completed. □

As an aside, we state the following result, part of which was already proved.

Proposition 4.13.1. *For $a > 0$, as $t \to 0$,*

$$\operatorname{Tr_s}\left[\left(\frac{\lambda_0}{2t} \mp \frac{t^2}{b^2}|p|^2\right)\exp\left(-aA_{b,t}^{\mathcal{M},2}\right)\right] = \frac{\pm 1}{\sqrt{at}}\int_{X_g}\frac{1}{\pi^{\ell/2}}\int^{\widehat{B}}\mathbb{M}\exp\left(-\frac{1}{4}\left\langle e_i, R^{TX_g}e_j\right\rangle \widehat{e}^i\widehat{e}^j\right)\operatorname{Tr}^F[g] + \mathcal{O}\left(\sqrt{t}\right), \tag{4.13.18}$$

$$\operatorname{Tr_s}\left[g\left(\frac{N-n}{2} - \omega^H\right)\exp\left(-aA_{b,t}^{\mathcal{M},2}\right)\right] = \frac{\pm 1}{\sqrt{at}}\int_{X_g}\frac{1}{\pi^{\ell/2}}\int^{\widehat{B}}\mathbb{M}\exp\left(-\frac{1}{4}\left\langle e_i, R^{TX_g}e_j\right\rangle \widehat{e}^i\widehat{e}^j\right)\operatorname{Tr}^F[g] + \mathcal{O}(1).$$

In particular as $t \to 0$,

$$\underline{w}_{b,t} = \mathcal{O}(1). \tag{4.13.19}$$

Proof. The first identity in (4.13.18) was already established in (4.13.16). So, we concentrate on the proof of the second one.

By using the transformations which were described after Remark 2.5.3 and also the conjugation in (4.6.2), we find that the analogue of (4.13.11) is the identity

$$\begin{aligned}\operatorname{Tr_s}&\left[\left(\frac{N^{T^*X}-n}{2} - \omega^H\right)\exp\left(-aA_{b,t}^{\mathcal{M},2}\right)\right] \\ &= \frac{(-1)^n}{2}\operatorname{Tr_s}\Bigg[\Bigg(e^i i_{e_i} - \widehat{e}^i i_{\widehat{e}_i} + 2e^i i_{\widehat{e}_i}/\sqrt{t} + \left\langle T\left(f_\alpha^H, e_i\right), p/\sqrt{t}\right\rangle f^\alpha\left(\sqrt{t}\widehat{e}^i - 2e^i\right) \\ &\qquad - 2\left\langle T^H, p/\sqrt{t}\right\rangle\Bigg)\exp\left(-a\widehat{\mathfrak{M}}_{b,t}^0\right)\Bigg].\end{aligned} \tag{4.13.20}$$

To make the argument simpler, we first assume that $g = 1$. When making the changes indicated in Definition 4.8.4, the first term in the right-hand side of (4.13.20) is changed into

$$e^i\left(i_{e_i} + \frac{b}{t}\mathfrak{e}_i + i_{\mathfrak{e}^i}/b\right) + \left(2e^i/\sqrt{t} - \widehat{e}^i\right)\left(i_{\widehat{e}_i} + b\widehat{\mathfrak{e}}_i/\sqrt{t}\right) \tag{4.13.21}$$

and the expression appearing in the second line is unchanged.

In (4.13.21), the leading singular term as $t \to 0$ is given by $\frac{b}{t}e^i\left(\mathfrak{e}_i + 2\widehat{\mathfrak{e}}_i\right)$. Inspection of equation (4.9.4) shows that as $t \to 0$,

$$t\operatorname{Tr_s}\left[g\left(\frac{N-n}{2} - \omega^H\right)\exp\left(-A_{b,t}^{\mathcal{M},2}\right)\right] \to 0. \tag{4.13.22}$$

In the asymptotic expansion of (4.13.21), to compute the term in $1/\sqrt{t}$, we first observe that the coefficient of $1/\sqrt{t}$ in the second line of (4.13.20) or in

(4.13.21) does not contribute to the evaluation of the corresponding coefficient, either because of the considerations we just made, or by the argument we gave after (4.13.14). So what remains to understand is the contribution of the term which is $\sqrt{t}$ in the right-hand side of (4.9.3) combined with the $1/t$ term in (4.13.21). We find that the relevant expression to be considered is

$$\mp \frac{1}{2} \left(\widehat{\mathfrak{e}}_i \widehat{e}^i + \mathfrak{e}_i i_{\widehat{e}_i} \right) e^i \left(\mathfrak{e}_i + 2\widehat{\mathfrak{e}}_i \right). \tag{4.13.23}$$

In (4.13.23) only the component containing $\mathfrak{e}_i \widehat{\mathfrak{e}}_i$ is relevant. This component is exactly given by

$$\mp \frac{1}{2} \widehat{\mathfrak{e}}_i \mathfrak{e}_i \left(\widehat{e}^i + 2 i_{\widehat{e}_i} \right) e^i. \tag{4.13.24}$$

The term containing $i_{\widehat{e}_i}$ in (4.13.24) is also irrelevant, since it disappears under $\widehat{\mathrm{Tr}}_{\mathrm{s}}$. Ultimately, we get the second identity in (4.13.18).

In the case where g is not equal to 1, a similar computation still using (4.9.3) leads to the second identity in (4.13.18).

Equation (4.13.19) follows from (4.13.7) and (4.13.18). The proof of our proposition is completed. □

Remark 4.13.2. It should be pointed out that the right-hand side of (4.13.18) already appeared in [BZ92, Theorem 7.10] in a study of the small time asymptotics of the supertraces which appear in the definition of the Ray-Singer analytic torsion [RS71]. This is not an accident.

The remainder in the first equation in (4.13.18) is $\mathcal{O}\left(\sqrt{t}\right)$ and in the second equation it is $\mathcal{O}(1)$. Indeed the first leading term in the second equation is obtained by considering the expansion in (4.9.3) to order $\sqrt{t}$, while in the first equation, it is just computed using the constant term in this expansion. Using equation (4.3.9) does not compensate for the discrepancy, except in degree 0, because of (4.3.11), which makes the left-hand sides of the two identities in (4.13.18) coincide. The purpose of identities like (2.5.2), (4.5.2) and (4.5.4) is to go back to the situation where only the constant part of the asymptotic expansion in (4.9.3) is used.

As we showed in (4.4.2)-(4.4.5), the expansion of $w_{b,t}$ in (4.4.1) can be directly obtained from the expansion of $u_{b,t}$ and $v_{b,t}$, which only requires the consideration of the leading term in the expansion (4.9.3).

Finally, as should be clear from the methods used above, all the quantities which were considered above have an asymptotic expansion in $\sqrt{t}$, and this to arbitrary order, the coefficients of the expansion being given by integrals of functions which can be computed locally over X.

Chapter Five

The limit as $t \to +\infty$ and $b \to 0$ of the superconnection forms

The purpose of this chapter is to establish the asymptotics as $t \to +\infty$ or $b \to 0$ of the hypoelliptic superconnection forms which were constructed in chapter 4. We show that for $b > 0$ small enough, convergence as $t \to +\infty$ is uniform, and also that as $b \to 0$, the hypoelliptic superconnection forms converge to the elliptic superconnection forms, and this occurs uniformly when $t > 0$ stays away from 0.

This chapter is organized as follows. In section 5.1, we define what will eventually be limit superconnection forms, for $t = +\infty$ or for $b = 0$.

In section 5.2, we state the convergence results.

In section 5.3, we split our even superconnection forms as the sum of two pieces defined via contour integrals, the first integral excluding the 0 eigenvalue, and the other one referring only to the 0 eigenvalue. Also we state two results on the asymptotics as $t \to +\infty$ or $b \to 0$ of these two pieces, from which the asymptotics as $t \to +\infty$ of our three superconnection follows.

Sections 5.4 and 5.5 are devoted to the proofs of the above two results.

Finally, in section 5.6 we establish the asymptotics of the odd superconnection forms.

We make the same assumptions and we use the same notation as in chapters 2 and 4. Also throughout this chapter, we assume that S is compact. if S is noncompact, the uniform convergence results are still valid over compact subsets of S.

5.1 THE DEFINITION OF THE LIMIT FORMS

Recall that $c = \pm 1/b^2$, with $b > 0$. In the present section, we will apply the results of Theorems 3.2.2 and 3.2.3 to the operator $A^2_{\phi,\mathcal{H}^c}$, by simply using the conjugation in (3.2.2). In particular $\exp\left(c\left|p\right|^2/2\right) \mathcal{S}^{\cdot}\left(T^*X, \pi^*F\right)$ now replaces $\mathcal{S}^{\cdot}(T^*X, \pi^*F)$. For $\lambda \in \mathrm{Sp}\mathfrak{A}^2_{\phi,\mathcal{H}^c}$, set

$$\begin{aligned} \Omega^{\cdot}(T^*X, \pi^*F) &= \exp\left(c\left|p\right|^2/2\right)\mathcal{S}^{\cdot}(T^*X, \pi^*F), \\ \Omega^{\cdot}(T^*X, \pi^*F)_\lambda &= \exp\left(c\left|p\right|^2/2\right)\mathcal{S}^{\cdot}(T^*X, \pi^*F)_\lambda, \\ \Omega^{\cdot}(T^*X, \pi^*F)_* &= \exp\left(c\left|p\right|^2/2\right)\mathcal{S}^{\cdot}(T^*X, \pi^*F)_*, \end{aligned} \tag{5.1.1}$$

so that

$$\Omega^{\cdot}(T^*X, \pi^*F) = \Omega^{\cdot}(T^*X, \pi^*F)_0 \oplus \Omega^{\cdot}(T^*X, \pi^*F)_*. \tag{5.1.2}$$

By Theorem 3.2.2, the splitting (5.1.2) is $\mathfrak{h}^{\Omega^{\cdot}(T^*X,\pi^*F)}_{\mathcal{H}^c}$ orthogonal, and the restriction of $\mathfrak{h}^{\Omega^{\cdot}(T^*X,\pi^*F)}_{\mathcal{H}^c}$ to each of the vector spaces in the right-hand side of (5.1.1) is nondegenerate.

For $b > 0$, let P_b be the projector from $\Omega^{\cdot}(T^*X, \pi^*F)$ on $\Omega^{\cdot}(T^*X, \pi^*F)_0$ with respect to the splitting (5.1.2). Then d^{T^*X} acts on both terms of the splitting (5.1.2). By Theorem 3.2.2, the complex $\left(\Omega^{\cdot}(T^*X, \pi^*F)_*, d^{T^*X}\right)$ is exact, so that

$$H^{\cdot}\left(\Omega^{\cdot}(T^*X, \pi^*F), d^{T^*X}\right) = H^{\cdot}\left(\Omega^{\cdot}(T^*X, \pi^*F)_0, d^{T^*X}\right). \tag{5.1.3}$$

Finally, by (3.2.15),

$$H^{\cdot}\left(\Omega^{\cdot}(T^*X, \pi^*F), d^{T^*X}\right) = \mathfrak{H}^{\cdot}(X, F). \tag{5.1.4}$$

By Theorem 3.5.1, there exists $b_0 > 0$ such that for $b \in]0, b_0]$, over S, the generalized metric $h^{\Omega^{\cdot}(T^*X,\pi^*F)}_{\mathcal{H}^c}$ is of Hodge type, i.e.,

$$\Omega^{\cdot}(T^*X, \pi^*F)_0 = \ker d^{T^*X} \cap \ker \overline{d}^{T^*X}_{\phi,2\mathcal{H}^c}. \tag{5.1.5}$$

Besides by equation (3.5.13) in Theorem 3.5.1, for $b \in]0, b_0]$, we have the canonical isomorphism

$$\Omega^{\cdot}(T^*X, \pi^*F)_0 \simeq \mathfrak{H}^{\cdot}(X, F). \tag{5.1.6}$$

Set

$$\mathbb{H}^{\cdot}_b(X, F) = \ker A^2_{\phi,\mathcal{H}^c}. \tag{5.1.7}$$

From (5.1.5), we find that for $b \in]0, b_0]$,

$$\Omega^{\cdot}(T^*X, \pi^*F)_0 = \mathbb{H}^{\cdot}_b(X, F). \tag{5.1.8}$$

From (5.1.6), (5.1.8), we deduce that for $b \in]0, b_0]$,

$$\mathbb{H}^{\cdot}_b(X, F) \simeq \mathfrak{H}^{\cdot}(X, F). \tag{5.1.9}$$

By Theorem 3.2.2, the restriction of the Hermitian form $\mathfrak{h}^{\Omega^{\cdot}(T^*X,\pi^*F)}_{\mathcal{H}^c}$ to $\mathbb{H}^{\cdot}_b(X, F)$ is nondegenerate. Let $\mathfrak{h}^{\mathfrak{H}^{\cdot}(X,F)}_b$ be the Hermitian form on $\mathfrak{H}^{\cdot}(X, F)$ which is induced by the restriction of $\mathfrak{h}^{\Omega^{\cdot}(T^*X,\pi^*F)}_{\mathcal{H}^b}$ to $\mathbb{H}^{\cdot}_b(X, F)$ via the canonical isomorphism (5.1.6).

Clearly the $\mathfrak{H}^{\cdot}(X,F)$ are the fibers of a smooth vector bundle on S, which is equipped with the flat Gauss-Manin connection $\nabla^{\mathfrak{H}^{\cdot}(X,F)}$. Observe that under the canonical isomorphism in (3.1.4), the Gauss-Manin connections $\nabla^{H^{\cdot}(X,F)}$ and $\nabla^{\mathfrak{H}^{\cdot}(X,F)}$ coincide.

Moreover, the Hermitian form $\mathfrak{h}_b^{\mathfrak{H}^{\cdot}(X,F)}$ is smooth, and depends smoothly on $b > 0$.

By imitating (1.2.6), set

$$\omega\left(\nabla^{\mathfrak{H}^{\cdot}(X,F)}, \mathfrak{h}_b^{\mathfrak{H}^{\cdot}(X,F)}\right) = \left(\mathfrak{h}_b^{\mathfrak{H}^{\cdot}(X,F)}\right)^{-1} \nabla^{\mathfrak{H}^{\cdot}(X,F)} \mathfrak{h}_b^{\mathfrak{H}^{\cdot}(X,F)}. \tag{5.1.10}$$

Recall that G acts on $\mathfrak{H}^{\cdot}(X,F)$. We define $h_g\left(\nabla^{\mathfrak{H}^{\cdot}(X,F)}, \mathfrak{h}_b^{\mathfrak{H}^{\cdot}(X,F)}\right)$ as in (1.7.3), by making the obvious changes, i.e., by imitating the construction of the form $h_g\left(\nabla^{H^{\cdot}(X,F)}, g^{H^{\cdot}(X,F)}\right)$ in section 1.7.

Definition 5.1.1. For $b > 0$, set

$$\begin{aligned}
u_{b,\infty} &= h_g\left(\nabla^{\mathfrak{H}^{\cdot}(X,F)}, \mathfrak{h}_b^{\mathfrak{H}^{\cdot}(X,F)}\right), \\
v_{b,\infty} &= \pm\varphi \mathrm{Tr_s}^{\mathfrak{H}^{\cdot}(X,F)}\left[g\frac{|p|^2}{b^2}h'\left(\omega\left(\nabla^{\mathfrak{H}^{\cdot}(X,F)}, \mathfrak{h}_b^{\mathfrak{H}^{\cdot}(X,F)}\right)/2\right)\right], \\
w_{b,\infty} &= \varphi \mathrm{Tr_s}^{\mathfrak{H}^{\cdot}(X,F)}\left[g\left(\frac{\lambda_0}{2} \mp \frac{|p|^2}{b^2}\right)h'\left(\omega\left(\nabla^{\mathfrak{H}^{\cdot}(X,F)}, \mathfrak{h}_b^{\mathfrak{H}^{\cdot}(X,F)}\right)/2\right)\right], \\
\underline{w}_{b,\infty} &= \varphi \mathrm{Tr_s}^{\mathfrak{H}^{\cdot}(X,F)}\left[g\left(\frac{N^{T^*X} - n}{2}\right)h'\left(\omega\left(\nabla^{\mathfrak{H}^{\cdot}(X,F)}, \mathfrak{h}_b^{\mathfrak{H}^{\cdot}(X,F)}\right)/2\right)\right].
\end{aligned} \tag{5.1.11}$$

Recall that given F, g^F, b_t was defined in (1.8.5). We write temporarily $b_t = b_t^F$, to emphasize the dependence of b_t on F. Recall that $\chi_g(F), \chi'_g(F)$ were defined in (1.6.1) and (1.7.17), that $\overline{\chi}_g(F), \overline{\chi}'_g(F)$ were defined in (4.2.1), and moreover that (4.2.2), (4.2.3) hold.

Definition 5.1.2. Put

$$\begin{aligned}
c_t &= b_t^F \text{ if } c > 0, \\
&= (-1)^n b_t^{F\otimes o(TX)} \text{ if } c < 0.
\end{aligned} \tag{5.1.12}$$

For $t > 0$, set

$$\begin{aligned}
u_{0,t} &= h_g\left(A', g_t^{\Omega^{\cdot}(X,F)}\right) \text{ if } c > 0, \\
u_{0,t} &= (-1)^n h_g\left(A', g_t^{\Omega^{\cdot}(X,F\otimes o(TX))}\right) \text{ if } c < 0, \\
v_{0,t} &= \pm\frac{n}{2}\chi_g(F), \\
w_{0,t} &= \underline{w}_{0,t} = c_t \mp \frac{n}{4}\chi_g(F).
\end{aligned} \tag{5.1.13}$$

Proposition 5.1.3. *For any $b > 0$,*

$$w_{b,\infty} = \underline{w}_{b,\infty} = \frac{1}{2}\left(\overline{\chi}'_g(F) - n\overline{\chi}_g(F)\right). \tag{5.1.14}$$

Proof. By [B05, Proposition 4.34] applied to a single fiber X as in (4.3.11), or by using directly [B05, Proposition 2.16], we get

$$\frac{1}{2}\left(\left[\overline{d}^{T^*X}_{\phi,2\mathcal{H}^c}, p\right] - \left[d^{T^*X}, i_{\widehat{p}}\right]\right) = \frac{\lambda_0}{2} \mp \frac{|p|^2}{b^2} - \frac{N^{T^*X} - n}{2}. \tag{5.1.15}$$

By (5.1.5), d^{T^*X} and $\overline{d}^{T^*X}_{\phi,2\mathcal{H}^c}$ both vanish on $\mathbb{H}^{\cdot}_c(X, F)$. So by (5.1.15), we get

$$w_{b,\infty} = \underline{w}_{b,\infty}. \tag{5.1.16}$$

Observe that the $\mathfrak{h}^{\mathfrak{H}^{\cdot}(X,F)}_{\mathcal{H}^c}$ are not standard metrics. However, the arguments of [BLo95, Proposition 1.3] or [BG01, Proposition 1.6], can be used to establish the last identity in (5.1.14). □

5.2 THE CONVERGENCE RESULTS

Now we state the essential results of this chapter.

Theorem 5.2.1. *There exists $b_0 > 0$ such that given $t_0 > 0$, there exists $C > 0$ such that for $b \in]0, b_0], t \geq t_0$,*

$$\left|u_{b,t} - u_{b,\infty}\right| \leq \frac{C}{\sqrt{t}}, \qquad \left|v_{b,t} - v_{b,\infty}\right| \leq \frac{C}{\sqrt{t}}, \tag{5.2.1}$$
$$\left|w_{b,t} - w_{b,\infty}\right| \leq \frac{C}{\sqrt{t}}, \qquad \left|\underline{w}_{b,t} - \underline{w}_{b,\infty}\right| \leq \frac{C}{\sqrt{t}}.$$

Moreover, given $t_0 > 0, v \in]0, 1[$, there exists $C_{t_0,v} > 0$ such that for $b \in]0, b_0], t \geq t_0$,

$$\left|u_{b,t} - u_{0,t}\right| \leq C_{t_0,v} b^v, \qquad \left|v_{b,t} - v_{0,t}\right| \leq C_{t_0,v} b^v, \tag{5.2.2}$$
$$\left|w_{b,t} - w_{0,t}\right| \leq C_{t_0,v} b^v, \qquad \left|\underline{w}_{b,t} - \underline{w}_{0,t}\right| \leq C_{t_0,v} b^v.$$

Proof. The remainder of the chapter is devoted to the proof of Theorem 5.2.1. □

Recall that $\mathcal{H} = |p|^2/2$. By (2.8.6), by equations (2.8.9) and (2.8.10) in Theorem 2.8.1 and by (4.3.3), (4.3.8), we get

$$u_{b,t} = (2\pi)^{1/2}\, \varphi \mathrm{Tr_s}\left[gh\left(\mathfrak{F}^{\mathcal{M},t}_{\phi_b,\pm\mathcal{H}-\omega^H,t}\right)\right],$$
$$v_{b,t} = \pm\varphi \mathrm{Tr_s}\left[g\,|p|^2\, h'\left(\mathfrak{F}^{\mathcal{M}}_{\phi_b,\pm\mathcal{H}-\omega^H,t}\right)\right], \tag{5.2.3}$$
$$w_{b,t} = \varphi \mathrm{Tr_s}\left[g\left(e^{-\mu_0}\frac{\lambda_0}{2}e^{\mu_0} \mp |p|^2\right) h'\left(\mathfrak{F}^{\mathcal{M}}_{\phi_b,\pm\mathcal{H}-\omega^H,t}\right)\right], \tag{5.2.4}$$
$$\underline{w}_{b,t} = \varphi \mathrm{Tr_s}\left[g\left(\frac{N^{T^*X} - n}{2} - \frac{b}{t}\omega^H\right) h'\left(\mathfrak{F}^{\mathcal{M}}_{\phi_b,\pm\mathcal{H}-\omega^H,t}\right)\right].$$

Moreover, by (2.1.28), (2.4.8), and (2.4.15), $\mathfrak{C}^{\mathcal{M},2}_{\phi_b,\pm\mathcal{H}-b\omega^H} - \mathfrak{A}'^2_{\phi_b,\pm\mathcal{H}}$ is of positive degree in the variables f^α. Since the f^α supercommute with the other operators, and using also the fact that any nonzero monomial in the $f^\alpha$is of length at most $m = \dim S$, the spectrum of $\mathfrak{C}^{\mathcal{M},2}_{\phi_b,\pm\mathcal{H}-b\omega^H}$ is the same as the spectrum of $\mathfrak{A}'^2_{\phi_b,\pm\mathcal{H}}$, which we considered in section 3.5.

5.3 A CONTOUR INTEGRAL

For $\epsilon > 0$, let $d \subset \mathbf{C}$ be the circle of center 0 and radius ϵ. For $\delta_0 > 0, \delta_1 > 0, \delta_2 = 1/6$, recall that $\mathcal{W}_\delta \in \mathbf{C}$ was defined in (3.4.1). Let γ_δ be its boundary, i.e.

$$\gamma_\delta = \left\{ \lambda \in \mathbf{C}, \operatorname{Re} \lambda = \delta_0 + \delta_1 \left| \operatorname{Im} \lambda \right|^{1/6} \right\}. \tag{5.3.1}$$

We use the results of section 3.5. In particular we choose ϵ small enough as in that section. By (3.5.3), (3.5.11) and by Remark 3.5.2, there exist $b_0 > 0, \delta_0 > 0, \delta_1 > 0$ such that if $b \in]0, b_0]$, the spectrum of $\mathfrak{A}'^2_{\phi,\mathcal{H}^c}$ is included in the union of the disk bounded by d and in the interior of $\mathbf{C} \setminus \mathcal{W}_\delta$, and moreover 0 is possibly the only element of the spectrum included in the disk d.

As we saw in (4.3.12),

$$h'(x) = \left(1 + 2x^2\right) e^{x^2}. \tag{5.3.2}$$

Put

$$r(\lambda) = (1 - 2\lambda) e^{-\lambda}, \tag{5.3.3}$$

so that

$$h'(x) = r\left(-x^2\right). \tag{5.3.4}$$

In what follows, instead of the function $r(\lambda)$, we will use the function $e^{-\lambda}$, simply to make our references to [B97] simpler. However, the analytic details will be exactly the same as for $r(\lambda)$.

Set

$$\begin{aligned} \mathfrak{P}'_b &= \frac{1}{2i\pi} \int_d \left(\lambda - \mathfrak{A}'^2_{\phi_b,\pm\mathcal{H}}\right)^{-1} d\lambda, \\ V_{b,t} &= \frac{1}{2i\pi} \int_{\gamma_\delta} e^{-t\lambda} \left(\lambda - \mathfrak{C}^{\mathcal{M},2}_{\phi_b,\pm\mathcal{H}-b\omega^H}\right)^{-1} d\lambda, \\ W_{b,t} &= \frac{1}{2i\pi} \int_d e^{-\lambda} \left(\lambda - \mathfrak{C}^{\mathcal{M},2}_{\phi_b,\pm\mathcal{H}-b\omega^H,t}\right)^{-1} d\lambda. \end{aligned} \tag{5.3.5}$$

By the above $\mathfrak{P}'_b$ is the natural projector on the finite dimensional vector space $\mathcal{S}^\cdot(T^*X, \pi^*F)_{0,b} = \ker \mathfrak{A}'^2_{\phi_b,\pm\mathcal{H}}$. Using (2.8.6) and the above considerations, for $t \geq 1$, we get

$$\exp\left(-\mathfrak{C}^{\mathcal{M},2}_{\phi_b,\pm\mathcal{H}-b\omega^H,t}\right) = \psi_t^{-1} V_{b,t} \psi_t + W_{b,t}. \tag{5.3.6}$$

Incidentally observe that for $t_0 \geq 0$, by modifying d and δ, we may as well assume that (5.3.6) holds for $t \geq t_0$. This is what we will do in the sequel.

We define $V_{0,t}, W_{0,t}$, by replacing $\mathfrak{A}'^2_{\phi_b,\pm\mathcal{H}}, \mathfrak{C}^{\mathcal{M},2}_{\phi_b,\pm\mathcal{H}-b\omega^H}, \mathfrak{C}^{\mathcal{M},2}_{\phi_b,\pm\mathcal{H}-b\omega^H,t}$ by $\square^X/4, A^{\mathcal{M},2}_{\pm}, A^{\mathcal{M},2}_{\pm,t}$.

Let $\mathcal{N}$ be one of the operators which appear in the right-hand side of the formulas defining $v_{b,t}, w_{b,t}, \underline{w}_{b,t}$, i.e., $\mathcal{N}$ is one of the operators $\pm\frac{t^2}{b^2}|p|^2, \frac{\lambda_0}{2t} \mp \frac{t^2}{b^2}|p|^2, \frac{N^{T^*X}-n}{2} - \omega^H$. Put

$$N = e^{-\mu_0} \tau b^{-1} U_{b,t} \mathcal{N} U_{b,t}^{-1} \tau b e^{\mu_0}. \tag{5.3.7}$$

i.e.,

$$N = \pm |p|^2, \qquad e^{-\mu_0} \frac{\lambda_0}{2} e^{\mu_0} \mp |p|^2, \qquad \frac{N^{T^*X} - n}{2} - \frac{b}{t} \omega^H. \tag{5.3.8}$$

Note that only in the last case does N depend on $t > 0$. We will denote by N_∞ the expression in (5.3.8) for $t = +\infty$.

We now state two results, from which part of Theorem 5.2.1 will follow.

Theorem 5.3.1. *There exists $b_0 > 0$ such that for any $t_0 > 0$, there exist $c > 0, C > 0$ for which if $b \in]0, b_0], t \geq t_0$, then*

$$|\mathrm{Tr_s}[gNV_{b,t}]| \leq Ce^{-ct}. \tag{5.3.9}$$

There exists $b_0 > 0$ such that given $t_0 > 0, v \in]0, 1[$, there exists $C_{t_0,v} > 0, c > 0$ for which if $b \in]0, b_0], t \geq t_0$, then

$$|\mathrm{Tr_s}[gNV_{b,t}] - \mathrm{Tr_s}[gP_\pm NP_\pm V_{0,t}]| \leq C_{t_0,v} e^{-ct} b^v. \tag{5.3.10}$$

Theorem 5.3.2. *There exists $b_0 > 0$ such that for any $t_0 > 0$, there exists $C > 0$ for which if $b \in]0, b_0], t \geq t_0$, then*

$$\Bigg| \mathrm{Tr_s}[gNW_{b,t}] - \mathrm{Tr_s}\Bigg[gP_b e^{\mu_0} K_b^{-1} N_\infty K_b e^{-\mu_0} P_b \\ \exp\left(\omega \left(\nabla^{\mathfrak{H}^{\cdot}(X,F)}, \mathfrak{h}_b^{\mathfrak{H}^{\cdot}(X,F)} \right)^2 /4 \right) \Bigg] \Bigg| \leq \frac{C}{\sqrt{t}}. \tag{5.3.11}$$

Given $t_0 > 0, v \in]0, 1[$, there exists $C_{t_0,v} > 0, c > 0$ such that for $b \in]0, b_0], t \geq t_0$,

$$|\mathrm{Tr_s}[gNW_{b,t}] - \mathrm{Tr_s}[gP_\pm NP_\pm W_{0,t}]| \leq C_{t_0,v} b^v. \tag{5.3.12}$$

Remark 5.3.3. Let us show how to derive Theorem 5.2.1 from equation (5.3.6) and from Theorems 5.3.1 and 5.3.2 for $v_{b,t}, w_{b,t}, \underline{w}_{b,t}$. Indeed the only important change is to replace the function $e^{-\lambda}$ by $r(\lambda)$, but this is very easy. Another possibility is to use instead (4.13.7).

By (2.1.4) and using the notation in (2.8.2), we get

$$e^{-\mu_0} \lambda_0 e^{\mu_0} = \lambda_0 - \mu_0 + N^H - N^V. \tag{5.3.13}$$

From (5.3.13), we obtain

$$P_+ e^{-\mu_0} \lambda_0 e^{\mu_0} P_+ = P_\pm N^H P_\pm, \\ P_- e^{-\mu_0} \lambda_0 e^{\mu_0} P_- = P_- \left(N^H - n\right) P_-. \tag{5.3.14}$$

Moreover, we have the trivial

$$P_\pm |p|^2 P_\pm = \frac{n}{2}. \tag{5.3.15}$$

Using (1.6.5), (1.6.6), by the explicit form of N in (5.3.8), by (5.3.14) and (5.3.15), we get Theorems 5.2.1 for $v_{b,t}, w_{b,t}, \underline{w}_{b,t}$.

Sections 5.4 and 5.5 will be devoted to the proof of Theorems 5.3.1 and 5.3.2.

5.4 A PROOF OF THEOREM 5.3.1

Clearly,

$$V_{b,t} = \frac{(-1)^N N!}{2i\pi t^N} \int_{\gamma_\delta} e^{-t\lambda} \left(\lambda - \mathfrak{C}^{\mathcal{M},2}_{\phi_b,\pm\mathcal{H}-b\omega^H}\right)^{-(N+1)} d\lambda. \tag{5.4.1}$$

By equation (3.4.3) and by the comments we made in section 3.7, when replacing $\mathfrak{A}'^2_{\phi_b,\pm\mathcal{H}}, \square^{X/4}$ by $\mathfrak{C}'^2_{\phi_b,\pm\mathcal{H}-b\omega^H}, A^{\mathcal{M},2}_{\pm}$, as we saw in section 3.7, if $\lambda \in \gamma_\delta$, the obvious analogue of (3.4.3) holds. Using the precise definition of the norms $|||\,|||_\ell$ after equation (17.21.55), there exists $b_0 > 0$ such that for $N \in \mathbf{N}$ large enough, there is $C_N > 0$ for which if $b \in]0, b_0], \lambda \in \gamma_\delta$, the operator $< p >^2 \left(\lambda - \mathfrak{C}^{\mathcal{M},2}_{\phi_b,\pm\mathcal{H}-b\omega^H}\right)^{-(N+1)}$ is trace class, and moreover,

$$\left\| < p >^2 \left(\lambda - \mathfrak{C}^{\mathcal{M},2}_{\phi_b,\pm\mathcal{H}-b\omega^H}\right)^{-(N+1)} \right\|_1 \leq C. \tag{5.4.2}$$

By (5.4.1), (5.4.2), there exists $c > 0, C > 0$ such that under the above conditions,

$$\left\| < p >^2 V_{b,t} \right\|_1 \leq C_N e^{-ct}. \tag{5.4.3}$$

By (5.4.3), we get (5.3.9).

By using the extension of equation (3.4.3) which was described before, from (5.4.1), we get (5.3.10). The proof of Theorem 5.3.1 is completed. □

5.5 A PROOF OF THEOREM 5.3.2

For $i = 0, 1, 2$, let $\mathfrak{F}^{\mathcal{M},(i)}_{\phi_b,\pm\mathcal{H}-b\omega^H}$ be the component of $\mathfrak{F}^{\mathcal{M}}_{\phi_b,\pm\mathcal{H}-b\omega^H}$ of degree i in the variables f^α. Note that by (2.4.8),

$$\mathfrak{F}^{\mathcal{M},(0)}_{\phi_b,\pm\mathcal{H}-b\omega^H} = \mathfrak{B}'_{\phi_b,\pm\mathcal{H}}. \tag{5.5.1}$$

By (2.5.1), $\mathfrak{F}^{\mathcal{M},(1)}_{\phi_b,\pm\mathcal{H}-b\omega^H}, \mathfrak{F}^{\mathcal{M},(2)}_{\phi_b,\pm\mathcal{H}-b\omega^H}$ do not depend on b. By (2.5.1) and (2.8.6), we can expand $\mathfrak{C}^{\mathcal{M},2}_{\phi_b,\pm\mathcal{H},t}$ in the form

$$\begin{aligned} \mathfrak{C}^{\mathcal{M},2}_{\phi_b,\pm\mathcal{H},t} = t\mathfrak{A}'^2_{\phi_b,\pm\mathcal{H}} - \left(\mathfrak{F}^{\mathcal{M},(1)}_{\phi_b,\pm\mathcal{H}-b\omega^H} + \mathfrak{F}^{\mathcal{M},(2)}_{\phi_b,\pm\mathcal{H}-b\omega^H}/\sqrt{t}\right)^2 \\ - \left[\mathfrak{B}'_{\phi_b,\pm\mathcal{H}}, \sqrt{t}\mathfrak{F}^{\mathcal{M},(1)}_{\phi_b,\pm\mathcal{H}-b\omega^H} + \mathfrak{F}^{\mathcal{M},(2)}_{\phi_b,\pm\mathcal{H}-b\omega^H}\right]. \end{aligned} \tag{5.5.2}$$

Now we will use the notation in Theorem 3.5.1, while introducing the extra subscript b. By the third equation in (3.5.12), we find that for $b \in]0, b_0]$,

$$\mathcal{S}^{\cdot}(T^*X, \pi^*F)_{0,b} = \ker \mathfrak{A}'_{\phi_b,\pm\mathcal{H}} = \ker \mathfrak{B}'_{\phi_b,\pm\mathcal{H}}. \tag{5.5.3}$$

Moreover, we have the splitting, which is analogous to (3.2.11),

$$\mathcal{S}^{\cdot}(T^*X, \pi^*F) = \mathcal{S}^{\cdot}(T^*X, \pi^*F)_{0,b} \oplus \mathcal{S}^{\cdot}(T^*X, \pi^*F)_{*,b}. \tag{5.5.4}$$

Using (2.1.20), (3.2.14), (5.5.1), (5.5.4), we find that $\left[\mathfrak{B}'_{\phi_b,\pm\mathcal{H}}, \mathfrak{F}^{\mathcal{M},(1)}_{\phi_b,\pm\mathcal{H}-b\omega^H}\right]$ maps $\mathcal{S}^{\cdot}(T^*X,\pi^*F)_{0,b}$ into $\mathcal{S}^{\cdot}(T^*X,\pi^*F)_{*,b}$.

If u is a matrix written as in (17.1.1) with respect to the splitting (5.5.4) of $\mathcal{S}^{\cdot}(T^*X,\pi^*F)$, we can write formally u^{-1} in the form given in (17.1.3). Namely, let u be given by

$$u = \begin{bmatrix} A & B \\ C & D \end{bmatrix}. \tag{5.5.5}$$

Set

$$H = A - BD^{-1}C. \tag{5.5.6}$$

Then, at least formally,

$$u^{-1} = \begin{bmatrix} H^{-1} & -H^{-1}BD^{-1} \\ -D^{-1}CH^{-1} & D^{-1} + D^{-1}CH^{-1}BD^{-1} \end{bmatrix}. \tag{5.5.7}$$

Now we apply (5.5.5), (5.5.7) to $u = \lambda - \mathfrak{C}^{\mathcal{M},2}_{\phi_b,\mathcal{H}-b\omega^H,t}$. We claim that (5.5.7) is correct. In fact if $u^{(0)}$ is the component of u which has degree 0 in the variables f^α, u is diagonal and (5.5.7) is trivial. Then equation (5.5.7) is obviously a perturbation of $\left(u^{(0)}\right)^{-1}$ by terms of positive degree in the f^α and so it is indeed correct.

By (5.5.1), u is a sum of five terms, which are factors of $t, \sqrt{t}, 1, 1/\sqrt{t}, 1/t$. Using the considerations we made after (5.5.3), we see that $A = \mathcal{O}(1), B = \mathcal{O}\left(\sqrt{t}\right), C = \mathcal{O}\left(\sqrt{t}\right), D = \mathcal{O}(t)$.

Now we will proceed as in [B97, section 9]. For $\alpha \in \mathbf{C}^*, \beta \in \mathbf{C}, \gamma \in \mathbf{C}, b \in]0, b_0]$, set

$$\begin{aligned} \mathfrak{L}^{\mathcal{M}}_{\alpha,\beta,\gamma,b} = {} & \frac{\mathfrak{A}'^2_{\phi_b,\pm\mathcal{H}}}{\alpha^2} - \left(\mathfrak{F}^{\mathcal{M},(1)}_{\phi_b,\pm\mathcal{H}-b\omega^H} + \beta\mathfrak{F}^{\mathcal{M},(2)}_{\phi_b,\pm\mathcal{H}-b\omega^H}\right)^2 \\ & - \left[\mathfrak{B}'_{\phi_b,\pm\mathcal{H}}, \gamma\mathfrak{F}^{\mathcal{M},(1)}_{\phi_b,\pm\mathcal{H}-b\omega^H} + \mathfrak{F}^{\mathcal{M},(2)}_{\phi_b,\pm\mathcal{H}-b\omega^H}\right], \\ \mathfrak{M}^{\mathcal{M}}_{\beta,\gamma,b} = {} & - \left(\mathfrak{F}^{\mathcal{M},(1)}_{\phi_b,\pm\mathcal{H}-b\omega^H} + \beta\mathfrak{F}^{\mathcal{M},(2)}_{\phi_b,\pm\mathcal{H}-b\omega^H}\right)^2 \\ & - \left[\mathfrak{B}'_{\phi_b,\pm\mathcal{H}}, \gamma\mathfrak{F}^{\mathcal{M},(1)}_{\phi_b,\pm\mathcal{H}-b\omega^H} + \mathfrak{F}^{\mathcal{M},(2)}_{\phi_b,\pm\mathcal{H}-b\omega^H}\right]. \end{aligned} \tag{5.5.8}$$

We denote by $\left(\mathfrak{A}'^2_{\phi_b,\pm\mathcal{H}}\right)^{-1}$ the operator acting on $\mathcal{S}^{\cdot}(T^*X,\pi^*F)$ which is 0 on $\mathcal{S}^{\cdot}(T^*X,\pi^*F)_{0,b}$ and acts like the inverse of $\mathfrak{A}'^2_{\phi_b,\pm\mathcal{H}}$ on $\mathcal{S}^{\cdot}(T^*X,\pi^*F)_{*,b}$. Then we have the strict analogue of [B97, Theorem 9.29].

Theorem 5.5.1. *For $\alpha \in \mathbf{C}^*, |\alpha| < 1, \beta \in \mathbf{C}, \gamma \in \mathbf{C}, b \in]0, b_0]$, then*

$$\frac{1}{2i\pi}\int_d \frac{\exp(-\lambda)}{\lambda - \mathfrak{L}^{\mathcal{M}}_{\alpha,\beta,\gamma,b}} d\lambda = \sum_{p=0}^{\dim S} \sum_{\substack{1 \le i_0 \le p+1 \\ 0 \le j_1, \dots, j_{p+1-i_0}, \\ j_1 + \dots + j_{p+1-i_0} \le i_0 - 1}} \frac{(-1)^{p-j_1 \dots - j_{p+1-i_0}}}{(i_0 - 1 - j_1 \dots - j_{p+1-i_0})!} C_1 \mathfrak{M}^{\mathcal{M}}_{\beta,\gamma,b} C_2 \dots \mathfrak{M}^{\mathcal{M}}_{\beta,\gamma,b} C_{p+1}. \tag{5.5.9}$$

In the right-hand side of (5.5.9), i_0 C_j's are equal to $\mathfrak{P}'_b$, and the other C_j's are given by $\left(\alpha^2\left[\mathfrak{A}'^2_{\phi_b,\pm\mathcal{H}}\right]^{-1}\right)^{1+j_1}, \dots, \left(\alpha^2\left(\mathfrak{A}'^2_{\phi_b,\pm\mathcal{H}}\right)^{-1}\right)^{1+j_{p+1-i_0}}$. In particular each term in the right-hand side of (5.5.9) is a monomial in α and a polynomial in β,γ.

Moreover, if $C_1,\dots,C_{p+1}$ are chosen as indicated before, then

$$\deg_\gamma(C_1\mathfrak{M}^{\mathcal{M}}_{\beta,\gamma,b}C_2\dots\mathfrak{M}^{\mathcal{M}}_{\beta,\gamma,b}C_{p+1}) \leq 2(p+1-i_0), \tag{5.5.10}$$
$$\deg_\alpha(C_1\mathfrak{M}^{\mathcal{M}}_{\beta,\gamma,b}C_2\dots\mathfrak{M}^{\mathcal{M}}_{\beta,\gamma,b}C_{p+1}) = 2\left(p+1-i_0+j_1+\dots+j_{p+1-i_0}\right).$$

The above inequality is an equality if and only if $-\left[\mathfrak{B}'_{\phi_b,\pm\mathcal{H}},\gamma\mathfrak{F}^{\mathcal{M},(1)}_{\phi_b,\pm\mathcal{H}-b\omega^H}\right]$ appears exactly $2(p+1-i_0)$ times in sequences of the form

$$\mathfrak{P}'_b\left[-\mathfrak{B}'_{\phi_b,\pm\mathcal{H}},\gamma\mathfrak{F}^{\mathcal{M},(1)}_{\phi_b,\pm\mathcal{H}-b\omega^H}\right]\left(\alpha^2\left(\mathfrak{A}'^2_{\phi_b,\pm\mathcal{H}}\right)^{-1}\right)^{1+j_k}$$
$$\left[-\mathfrak{B}'_{\phi_b,\pm\mathcal{H}},\gamma\mathfrak{F}^{\mathcal{M},(1)}_{\phi_b,\pm\mathcal{H}-b\omega^H}\right]\mathfrak{P}'_b, \tag{5.5.11}$$

the other C_i's being equal to $\mathfrak{P}'_b$.

Proof. The proof is the same as the proof of [B97, Theorem 9.29]. □

First we consider the first two cases in (5.3.8), in which N does not depend on t. By (5.5.9), we get

$$\operatorname{Tr_s}\left[gN\frac{1}{2i\pi}\int_d \frac{\exp(-\lambda)}{\lambda-\mathfrak{L}^{\mathcal{M}}_{\alpha,\beta,\gamma,b}}d\lambda\right] = \sum_{\substack{0\leq 2m\leq \dim S\\ 0\leq m'\leq \ell\leq 2\dim S}} O_{\ell,m,m'}(b)\,\alpha^\ell\beta^m\gamma^{m'}, \tag{5.5.12}$$

where the $O_{\ell,m,m'}(b)$ are smooth even forms on S. Incidentally, note that the condition $2m\leq \dim S$ comes from the fact that β appears as a factor of a term of degree 2 in $\Lambda^\cdot(T^*S)$.

Now we establish the analogue of [B97, Theorem 9.30].

Theorem 5.5.2. *There exist forms $O_{\ell,m,m'}(0)$ such that for any $v\in]0,1]$, there exists $C_v>0$ such that for $b\in]0,b_0]$, if ℓ,m,m' are taken as in (5.5.12),*

$$|O_{\ell,m,m'}(b)-O_{\ell,m,m'}(0)|\leq Cb^v. \tag{5.5.13}$$

Proof. Observe that by Theorem 5.5.1, the the right-hand side of (5.5.12) is a polynomial in the variables α,β,γ. Using Cauchy's residue formula, to prove our theorem, it will be enough to show that there is a smooth form $h(\alpha,\beta,\gamma)$ which is holomorphic in α,β,γ with $\frac{1}{4}\leq|\alpha|\leq 1/2$, $|\beta|\leq 1/2$, $|\gamma|\leq 1/2$ such that for $b\in]0,b_0]$, $\lambda\in d$,

$$\left|\operatorname{Tr_s}\left[gN\frac{1}{2i\pi}\int_d \frac{\exp(-\lambda)}{\lambda-\mathfrak{L}^{\mathcal{M}}_{\alpha,\beta,\gamma,b}}d\lambda\right]-h(\alpha,\beta,\gamma)\right|\leq C_vb^v. \tag{5.5.14}$$

To prove (5.5.14), we need to show only that the analogue of equation (3.4.3) holds in the given range of parameters for $\lambda\in d$. First note that as explained in Remark 17.21.6, the estimates in (3.4.3) are still valid when $\lambda \in d$. Moreover, the operator $\mathfrak{M}^{\mathcal{M}}_{\beta,\gamma,b}$ in (5.5.8) is of order 0 and depends polynomially on p. The whole analysis made in chapters 3, 15, and 17 goes through. The proof of our theorem is completed. $\square$

Observe that by (5.5.1), (5.5.8), we have

$$\mathfrak{E}^{\mathcal{M},2}_{\phi_b,\pm\mathcal{H},t} = \mathfrak{L}^{\mathcal{M}}_{\frac{1}{\sqrt{t}},\frac{1}{\sqrt{t}},\sqrt{t},b}. \tag{5.5.15}$$

By (5.5.12), (5.5.15), we get

$$\mathrm{Tr_s}\left[gNW_{b,t}\right] = \sum_{\substack{0\le 2m\le \dim S\\ 0\le m'\le \ell\le 2\dim S}} O_{\ell,m,m'}\left(b\right)\sqrt{t}^{-\ell-m+m'}. \tag{5.5.16}$$

Now under the conditions on ℓ, m, m' in the right-hand side of (5.5.12), for $t \ge t_0$,

$$\sqrt{t}^{-\ell-m+m'} \le C_{t_0}. \tag{5.5.17}$$

Using Theorem 5.5.2, (5.5.16), and (5.5.17), we get

$$\left|\mathrm{Tr_s}\left[gNW_{b,t}\right] - \sum_{\substack{0\le m\le 2\dim S\\ 0\le m'\le \ell\le 2\dim S}} O_{\ell,m,m'}\left(0\right)\sqrt{t}^{-\ell-m+m'}\right| \le C_{t_0,v}b^{v}. \tag{5.5.18}$$

By using the analogue of equation (3.4.10) for the operator $\mathfrak{E}^{\mathcal{M},2}_{\phi_b,\pm\mathcal{H}-b\omega^H}$ which we described in section 3.7, we find that for a given t, as $b \to 0$,

$$\mathrm{Tr_s}\left[gNW_{b,t}\right] \to \mathrm{Tr_s}\left[gP_\pm NP_\pm W_{0,t}\right]. \tag{5.5.19}$$

From (5.5.18), (5.5.19), we get for $t \ge t_0$,

$$\left|\mathrm{Tr_s}\left[gNW_{b,t}\right] - \mathrm{Tr_s}\left[gP_\pm NP_\pm W_{0,t}\right]d\lambda\right| \le C_{t_0,v}b^{v}, \tag{5.5.20}$$

which is just (5.3.12), in the first two cases in (5.3.8). Establishing this equation in the third case goes along the same line. Details are left to the reader.

Now we establish (5.3.11) in the first two cases for N in (5.3.8). By Theorem 5.5.2, the $O_{l,m,m'}\left(b\right)$ are uniformly bounded. By (5.5.16), we find that for $t \ge t_0$,

$$\left|\mathrm{Tr_s}\left[gNW_{b,t}\right] - \sum_{0\le \ell\le 2\dim S} O_{\ell,0,\ell}\left(b\right)\right| \le \frac{C}{\sqrt{t}}. \tag{5.5.21}$$

Inspection of (5.5.10) shows that only those terms where there is equality in the first line, and where $j_1 = \cdots = j_{p+1-i_0} = 0$, contribute to the sum $\sum_{0\le\ell\le 2\dim S} O_{\ell,0,\ell}\left(b\right)$. Put

$$\begin{aligned}\mathfrak{N}^{\mathcal{M}}_b = \mathfrak{P}'_b\Bigg(\mathfrak{F}^{\mathcal{M}(1),2}_{\phi_b,\pm\mathcal{H}-b\omega^H} - \left[\mathfrak{F}^{\mathcal{M}(1)}_{\phi_b,\pm\mathcal{H}-b\omega^H}, \mathfrak{B}'_{\phi_b,\pm\mathcal{H}}\right]\left(\mathfrak{B}'^2_{\phi_b,\pm\mathcal{H}}\right)^{-1}\\ \left[\mathfrak{F}^{\mathcal{M}(1)}_{\phi_b,\pm\mathcal{H}-b\omega^H}, \mathfrak{B}'_{\phi_b,\pm\mathcal{H}}\right]\Bigg)\mathfrak{P}'_b.\end{aligned} \tag{5.5.22}$$

Then inspection of the right-hand side of (5.5.9) shows that

$$\sum_{0\le\ell\le 2\dim S} O_{\ell,0,\ell}(b) = \mathrm{Tr_s}\left[g\mathfrak{P}'_b N \mathfrak{P}'_b \frac{1}{2i\pi}\int_d \frac{\exp(-\lambda)}{\lambda+\mathfrak{N}_b^{\mathcal{M}}}d\lambda\right]. \tag{5.5.23}$$

Clearly,

$$\frac{1}{2i\pi}\int_d \frac{\exp(-\lambda)}{\lambda+\mathfrak{N}_b^{\mathcal{M}}} = \exp\left(\mathfrak{N}_b^{\mathcal{M}}\right). \tag{5.5.24}$$

Also it is elementary to verify that

$$\mathfrak{N}_b^{\mathcal{M}} = \left(\mathfrak{P}'_b \mathfrak{F}^{\mathcal{M},(1)}_{\phi_b,\pm\mathcal{H}-b\omega^H}\mathfrak{P}'_b\right)^2. \tag{5.5.25}$$

Now by (2.4.5), (2.4.7), and (2.4.15),

$$\begin{aligned}\mathfrak{E}^{\mathcal{M}}_{\phi_b,\pm\mathcal{H}-b\omega^H} &= K_b e^{-\mu_0-(\mathcal{H}^c-\omega^H)} A^{\mathcal{M}}_{\phi,\mathcal{H}^c-\omega^H} e^{\mu_0+\mathcal{H}^c-\omega^H} K_b^{-1}, \\ \mathfrak{F}^{\mathcal{M}}_{\phi_b,\pm\mathcal{H}-b\omega^H} &= K_b e^{-\mu_0-(\mathcal{H}^c-\omega^H)} B^{\mathcal{M}}_{\phi,\mathcal{H}^c-\omega^H} e^{\mu_0+\mathcal{H}^c-\omega^H} K_b^{-1}.\end{aligned} \tag{5.5.26}$$

By (5.5.26), we deduce in particular that

$$\mathfrak{F}^{\mathcal{M},(1)}_{\phi_b,\pm\mathcal{H}-b\omega^H} = K_b e^{-\mu_0-\mathcal{H}^c} B^{\mathcal{M},(1)}_{\phi,\mathcal{H}^c-\omega^H} e^{\mu_0+\mathcal{H}^c} K_b^{-1}. \tag{5.5.27}$$

Incidentally, since (5.5.27) is an identity of operators of order 0, $\mathcal{H}^c$ can be made equal to 0 in the right-hand side of this equation.

As was explained in section 2.4, $A'^{\mathcal{M}}$ is a version of the de Rham operator on the total space of $\mathcal{M}$. By definition, it induces the Gauss-Manin connection $\nabla^{\mathfrak{H}^{\cdot}(X,F)}$ on $\mathfrak{H}^{\cdot}(X,F)$. As we saw after (2.4.10), $\overline{\mathfrak{E}}'^{\mathcal{M}}_{\phi,2(\mathcal{H}^c-\omega^H)}$ is the $\mathfrak{h}^{\Omega^{\cdot}(T^*X,\pi^*F)}_{\mathcal{H}^c-\omega^H}$ adjoint of $A'^{\mathcal{M}}$. Since ω^H is itself of degree 2 in the f^α, $\overline{\mathfrak{E}}'^{\mathcal{M},(1)}_{\phi,2(\mathcal{H}^c-\omega^H)}$ is necessarily the $\mathfrak{h}^{\Omega^{\cdot}(T^*X,\pi^*F)}_{\mathcal{H}^c}$ adjoint of $A'^{\mathcal{M},(1)}$.

By Theorem 3.2.2, the splitting (3.2.11) is $\mathfrak{h}^{S^{\cdot}(T^*X,\pi^*F)}$ orthogonal, and correspondingly, the splitting (5.1.2) is $\mathfrak{h}^{\Omega^{\cdot}(T^*X,\pi^*F)}_{\mathcal{H}^c}$ orthogonal. Using the notation in (5.1.10), it is then elementary to verify that

$$\omega\left(\nabla^{\mathfrak{h}^{\cdot}(X,F)}, \mathfrak{h}^{\mathfrak{h}^{\cdot}_b(X,F)}\right) = P_b 2B^{\mathcal{M},(1)}_{\phi,\mathcal{H}^c-\omega^H}P_b. \tag{5.5.28}$$

Set

$$N' = P_b e^{\mu_0} K_b^{-1} N K_b e^{-\mu_0} P_b. \tag{5.5.29}$$

By (5.5.23)-(5.5.25) and by (5.5.27)-(5.5.29), we get

$$\sum_{0\le\ell\le 2\dim S} O_{\ell,0,\ell}(b) = \mathrm{Tr_s}\left[gN'\exp\left(\omega\left(\nabla^{\mathfrak{h}^{\cdot}(X,F)}, \mathfrak{h}^{\mathfrak{h}^{\cdot}(X,F)}\right)^2/4\right)\right]. \tag{5.5.30}$$

By (5.5.21), (5.5.23), (5.5.29), and (5.5.30), we get (5.3.11) in the first two cases in (5.3.8). The proof of the third case in (5.3.8) follows similar lines, and is left to the reader. This completes the proof of Theorem 5.3.2. □

5.6 A PROOF OF THE FIRST EQUATIONS IN (5.2.1) AND (5.2.2)

Now we will consider the case of $u_{b,t}$. Clearly

$$h\left(\mathfrak{F}^{\mathcal{M}}_{\phi_b,\pm\mathcal{H}-b\omega^H,t}\right) = \mathfrak{F}^{\mathcal{M}}_{\phi_b,\pm\mathcal{H}-b\omega^H,t} \exp\left(-\mathfrak{E}^{\mathcal{M},2}_{\phi_b,\pm\mathcal{H}-b\omega^H,t}\right). \tag{5.6.1}$$

The main difference with respect to what we did before is the appearance of the operator $\mathfrak{F}^{\mathcal{M}}_{\phi_b,\pm\mathcal{H}-b\omega^H,t}$ as a factor in the right-hand side of (5.6.1). By (5.2.3) and (5.6.1), we get

$$u_{b,t} = (2\pi)^{1/2}\,\varphi\mathrm{Tr_s}\left[g\mathfrak{F}^{\mathcal{M}}_{\phi_b,\pm\mathcal{H}-b\omega^H,t} \exp\left(-\mathfrak{E}^{\mathcal{M},2}_{\phi_b,\pm\mathcal{H}-b\omega^H,t}\right)\right]. \tag{5.6.2}$$

We now will prove the analogues of Theorems 5.3.1 and 5.3.2, with N replaced by $\mathfrak{F}^{\mathcal{M}}_{\phi_b,\pm\mathcal{H}-b\omega^H,t}$. The main difficulty is that by equation (2.6.1), the operator $\mathfrak{F}^{\mathcal{M}}_{\phi_b,\pm\mathcal{H}-b\omega^H}$ contains a diverging term when $b \to 0$.

Let z be an extra odd Grassmann variable. If $\alpha \in \Lambda^{\cdot}(T^*S)\widehat{\otimes}\mathbf{R}[z]$, we can write α in the form

$$\alpha = \beta + z\gamma,\ \beta,\gamma \in \Lambda^{\cdot}(T^*S). \tag{5.6.3}$$

Set

$$\alpha^z = \gamma. \tag{5.6.4}$$

Now we use the notation in equation (2.5.1). Put

$$\begin{aligned}\mathfrak{F}^{\mathcal{M}'}_{\phi_b,\pm\mathcal{H}-b\omega^H} &= \mathfrak{F}^{\mathcal{M}}_{\phi_b,\pm\mathcal{H}-b\omega^H} + \frac{1}{2b}\left(\widehat{c}\left(\widehat{e}_i\right)\nabla_{\widehat{e}^i} \pm c\left(\widehat{p}\right)\right), \\ \mathfrak{K}^{\mathcal{M}}_{\phi_b,\pm\mathcal{H}-b\omega^H} &= \mathfrak{E}^{\mathcal{M},2}_{\phi_b,\pm\mathcal{H}-b\omega^H} + \frac{z}{2b}\left(\widehat{c}\left(\widehat{e}_i\right)\nabla_{\widehat{e}^i} \pm c\left(\widehat{p}\right)\right).\end{aligned} \tag{5.6.5}$$

Note that with the notation in (2.6.1),

$$\mathfrak{F}^{\mathcal{M}'}_{\phi_b,\pm\mathcal{H}-b\omega^H} = \mathfrak{H}_{\pm} + b\mathfrak{J}. \tag{5.6.6}$$

Moreover,

$$\mathfrak{F}^{\mathcal{M}'(1)}_{\phi_b,\pm\mathcal{H}-b\omega^H} = \mathfrak{F}^{\mathcal{M}'(1)}_{\phi_b,\pm\mathcal{H}-b\omega^H}, \qquad \mathfrak{F}^{\mathcal{M}'(2)}_{\phi_b,\pm\mathcal{H}-b\omega^H} = \mathfrak{F}^{\mathcal{M}'(2)}_{\phi_b,\pm\mathcal{H}-b\omega^H}. \tag{5.6.7}$$

In the sequel, for $a \geq 0$, the operator ψ_a also maps z into $\sqrt{a}z$. We define the corresponding objects with the extra index t as in (2.8.6). In particular

$$\mathfrak{K}^{\mathcal{M}}_{\phi_b,\pm\mathcal{H}-b\omega^H} = \psi_t^{-1} t\mathfrak{K}^{\mathcal{M}}_{\phi_b,\pm\mathcal{H}-b\omega^H}\psi_t. \tag{5.6.8}$$

Also,

$$\mathfrak{F}^{\mathcal{M}'}_{\phi_b,\pm\mathcal{H}-\omega^H,t} = \sqrt{t}\mathfrak{F}^{\mathcal{M}'(0)}_{\phi_b,\pm\mathcal{H}-\omega^H} + \mathfrak{F}^{\mathcal{M}'(1)}_{\phi_b,\pm\mathcal{H}-\omega^H} + \frac{1}{\sqrt{t}}\mathfrak{F}^{\mathcal{M}'(2)}_{\phi_b,\pm\mathcal{H}-\omega^H}. \tag{5.6.9}$$

Set

$$\begin{aligned}u'_{b,t} &= (2\pi)^{1/2}\,\varphi\mathrm{Tr_s}\left[g\mathfrak{F}^{\mathcal{M}'}_{\phi_b,\pm\mathcal{H}-b\omega^H,t} \exp\left(-\mathfrak{E}^{\mathcal{M},2}_{\phi_b,\pm\mathcal{H}-b\omega^H,t}\right)\right], \\ u''_{b,t} &= (2\pi)^{1/2}\,\varphi\mathrm{Tr_s}\left[g\exp\left(-\mathfrak{K}^{\mathcal{M}}_{\phi_b,\pm\mathcal{H}-b\omega^H,t}\right)\right]^z.\end{aligned} \tag{5.6.10}$$

One verifies easily that

$$u_{b,t} = u'_{b,t} + u''_{b,t}. \tag{5.6.11}$$

We will prove analogues of Theorems 5.3.1 and 5.3.2 for $u'_{b,t}$ and $u''_{b,t}$.

We claim that $u'_{b,t}$ can be handled using the techniques of sections 5.4 and 5.5. Clearly, using (5.3.6), we get

$$\begin{aligned}\mathrm{Tr_s}\left[g\mathfrak{F}^{\mathcal{M}'}_{\phi_b,\pm\mathcal{H}-b\omega^H,t}\exp\left(-\mathfrak{E}^{\mathcal{M},2}_{\phi_b,\pm\mathcal{H}-b\omega^H,t}\right)\right] \\ = \psi_t^{-1}\mathrm{Tr_s}\left[g\sqrt{t}\mathfrak{F}^{\mathcal{M}'}_{\phi_b,\pm\mathcal{H}-b\omega^H}V_{b,t}\right] + \mathrm{Tr_s}\left[g\mathfrak{F}^{\mathcal{M}'}_{\phi_b,\mathcal{H}-b\omega^H,t}W_{b,t}\right].\end{aligned} \tag{5.6.12}$$

Indeed by (3.4.3), besides equation (5.4.2), we also have

$$\left\|\nabla^{\Lambda^{\cdot}(T^*T^*X)\widehat{\otimes}F,u}_{e_i}\left(\lambda - \mathfrak{E}^{\mathcal{M},2}_{\phi_b,\pm\mathcal{H}-b\omega^H}\right)^{-(N+1)}\right\|_1 \leq C. \tag{5.6.13}$$

Then the first term in the right-hand side of (5.6.12) can be handled by the same techniques as the corresponding term in sections 5.4 and 5.5. The factor $\sqrt{t}$ which appears in $\mathfrak{F}^{\mathcal{M}'}_{\phi_b,\pm\mathcal{H}-b\omega^H,t}$ is killed by e^{-ct}. Equivalently, we obtain an analogue of Theorem 5.3.1 for this first term.

The case of the second term in the right-hand side of (5.6.12) is slightly subtler. Indeed by proceeding as in (5.5.16), we get

$$\begin{aligned}\mathrm{Tr_s}\left[g\mathfrak{F}^{\mathcal{M}'}_{\phi_b,\pm\mathcal{H}-b\omega^H,t}W_{b,t}\right] = \sum_{\substack{0\leq 2m\leq \dim S\\ 0\leq m'\leq \ell\leq 2\dim S}}\Big(P_{\ell,m,m'}(b)\sqrt{t} + Q_{\ell,m,m'}(b) \\ + R_{\ell,m,m'}(b)/\sqrt{t}\Big)\sqrt{t}^{-\ell-m+m'}.\end{aligned} \tag{5.6.14}$$

The terms containing $P_{\ell,m,m'}(b)$ are associated with the contribution of $\sqrt{t}\mathfrak{F}^{\mathcal{M}'(0)}_{\phi_b,\pm\mathcal{H}-b\omega^H}$, the terms containing $Q_{\ell,m,m'}(b)$ corresponding to the contribution of $\mathfrak{F}^{\mathcal{M}'(1)}_{\phi_b,\pm\mathcal{H}-\omega^H}$, the terms containing $R_{\ell,m,m'}(b)$ corresponding to $\mathfrak{F}^{\mathcal{M}'(2)}_{\phi_b,\pm\mathcal{H}-\omega^H}$.

Note that $P_{\ell,m,m'}(b), Q_{\ell,m,m'}(b), R_{\ell,m,m'}(b)$ verify estimates similar to (5.5.13).

We claim that

$$\sum_{0\leq\ell\leq 2\dim S} P_{\ell,0,\ell}(b) = 0, \qquad \sum_{1\leq\ell\leq 2\dim S} P_{\ell,0,\ell-1} = 0. \tag{5.6.15}$$

Indeed by proceeding as in (5.5.22)-(5.5.24), we get

$$\sum_{0\leq\ell\leq 2\dim S} P_{\ell,0,\ell}(b) = \mathrm{Tr_s}\left[g\mathfrak{P}'_b\mathfrak{F}^{\mathcal{M}'(0)}_{\phi_b,\pm\mathcal{H}-\omega^H}\mathfrak{P}'_b\exp\left(\mathfrak{N}^{\mathcal{M}}_b\right)\right]. \tag{5.6.16}$$

However, the operator $\mathfrak{F}^{\mathcal{M}'(0)}_{\phi_b,\pm\mathcal{H}-b\omega^H}$ is an odd operator, so that the right-hand side vanishes. Also (5.5.10) shows that $P_{\ell,0,m'}$ is nonzero only if $\ell - m'$ is a multiple of 2. This gives the second equation in (5.6.15).

Equation (5.6.7) and the same arguments as before show that

$$\sum_{0 \le \ell \le 2 \dim S} Q_{\ell,0,\ell}(b) = \mathrm{Tr_s}\left[g \mathfrak{P}_b \mathfrak{F}^{\mathcal{M}(1)}_{\phi_b, \pm\mathcal{H} - b\omega^H} \mathfrak{P}'_b \exp\left(\mathfrak{N}^{\mathcal{M}_b}\right)\right]. \tag{5.6.17}$$

By (2.6.3) in Theorem 2.6.1, (5.5.29), (5.6.6), (5.6.7), and (5.6.14)-(5.6.15), we produce forms $u'_{b,\infty}, u'_{0,t}$ such that under conditions of Theorem 5.2.1, we get

$$\left| u'_{b,t} - u'_{b,\infty} \right| \le \frac{C_{t_0}}{\sqrt{t}}, \qquad \left| u'_{b,t} - u'_{0,t} \right| \le C_{t_0,v} b^v. \tag{5.6.18}$$

Note that $u'_{b,\infty}, u'_{0,t}$ are the sum of two terms. One term is just the contribution of (5.6.17). The other term is the contribution of $\sum_{0 \le \ell \le 2 \dim S} P_{\ell,1,\ell}(b)$, which does not vanish in general.

We claim that

$$u'_{0,t} = u_{0,t}. \tag{5.6.19}$$

Indeed combining the first equation in (2.6.3) with (5.6.6) leads easily to (5.6.19).

Now we will consider $u''_{b,t}$. We claim that the results which are valid for the operator $\mathfrak{E}^{\mathcal{M},2}_{\phi_b,\pm\mathcal{H}-b\omega^H}$ remain valid for the operator $\mathfrak{K}^{\mathcal{M}}_{\phi_b,\pm\mathcal{H}-b\omega^H}$. The point is that the term $\frac{z}{2b}\left(\widehat{c}\left(\widehat{e}_i\right)\nabla_{\widehat{e}^i} \pm c\left(\widehat{p}\right)\right)$ is small compared to $\mathfrak{E}^{\mathcal{M},2}_{\phi_b,\pm\mathcal{H}-b\omega^H}$. Indeed the harmonic oscillator $\alpha_\pm$ which appears in equation (2.6.2) for $\mathfrak{E}^{\mathcal{M},2}_{\phi_b,\pm\mathcal{H}-b\omega^H}$ is minus the square of $\frac{1}{2}\left(\widehat{c}\left(\widehat{e}_i\right)\nabla_{\widehat{e}^i} \pm \widehat{c}\left(\widehat{p}\right)\right)$. Moreover, the spectrum of $\mathfrak{K}^{\mathcal{M}}_{\phi_b,\pm\mathcal{H}-b\omega^H}$ and $\mathfrak{E}^{\mathcal{M},2}_{\phi_b,\pm\mathcal{H}-b\omega^H}$ are identical.

We then write $\exp\left(-\mathfrak{K}^{\mathcal{M}}_{\phi_b,\pm\mathcal{H}-b\omega^H,t}\right)$ in a form similar to (5.3.6). For simplicity we still use the notation $V_{b,t}, W_{b,t}$. The term corresponding to $V_{b,t}$ can be handled exactly as before. So we concentrate on the term corresponding to $W_{b,t}$. We claim that the conclusions in (5.6.14) and the vanishing of (5.6.16) still remain valid in this case. Indeed the argument is essentially the same, and uses the fact that the operator $\widehat{c}\left(\widehat{e}_i\right)\nabla_{\widehat{e}^i} \pm c\left(\widehat{p}\right)$ is an odd operator.

Ultimately, we produce a form $u''_{b,\infty}$ such that under the conditions of Theorem 5.2.1,

$$\left| u''_{b,t} - u''_{b,\infty} \right| \le \frac{C_{t_0}}{\sqrt{t}}, \qquad \left| u''_{b,t} \right| \le C_{t_0,v} b^v. \tag{5.6.20}$$

The reason why in (5.6.20), $u''_{0,t} = 0$ is that the operator $\widehat{c}\left(\widehat{e}_i\right)\nabla_{\widehat{e}^i} \pm c\left(\widehat{p}\right)$ vanishes identically on $\ker \alpha_\pm$. Also $u''_{b,\infty}$ appears as the contribution of a term of the form $\sum_{1 \le \ell \le \dim S} P_{\ell,1,\ell}$.

By (5.6.11), (5.6.18), (5.6.19), and (5.6.20), we get equation (5.2.2) for $u_{b,t}$. Moreover, to establish the first equation in (5.2.1), we need to show only that

$$u_{b,\infty} = u'_{b,\infty} + u''_{b,\infty}. \tag{5.6.21}$$

By (5.5.25) and (5.6.17), we get

$$u_{b,\infty} = (2\pi)^{1/2} \varphi \sum_{0 \le \ell \le \dim S} Q_{\ell,0,\ell}. \tag{5.6.22}$$

So to establish (5.6.21), we need to show only that the sum of the terms of the type $\sum_{0\le \ell\le 2\dim S} P_{\ell,1,\ell}(b)$ which appear in $u'_{b,\infty}$ and $u''_{b,\infty}$ vanishes identically. However, the sum of these two terms involves the operator $\mathfrak{B}'_{\phi_b,\pm\mathcal{H}}$ combined with the terms in the right-hand side of (5.5.9), with $m'=\ell, m=1$. However, inspection of (5.5.11) shows that in the right-hand side of (5.5.9), the corresponding terms are such that $C_1 = C_{p+1} = \mathfrak{P}'_b$. Now by (5.5.3), the operator $\mathfrak{B}'_{\phi,\mathcal{H}}$ vanishes on $\mathcal{S}^{\cdot}(T^*X, \pi^*F)_0$. Therefore we have established the required vanishing result.

This completes the proof of the first equation in (5.2.1).

Remark 5.6.1. One reason that makes the proof of the required estimates on $u_{b,t}$ is difficult is that we have no a priori good understanding of the operator $\mathfrak{A}'_{\phi_b,\pm\mathcal{H}}$. Indeed since it contains differentials of the type $\nabla_{e_i}^{\Lambda^{\cdot}(T^*T^*X)\widehat{\otimes}F,u}$, it cannot be easily dealt with.

Chapter Six

Hypoelliptic torsion and the hypoelliptic Ray-Singer metrics

The purpose of this chapter is to define hypoelliptic torsion forms and corresponding hypoelliptic Ray-Singer metrics on the line $\lambda = \det \mathfrak{H}^{\cdot}(X,F)$. The main result of this chapter is that these objects verify transgression equations very similar to the corresponding equations we gave in chapter 1 for their elliptic counterparts. It is then natural to try to compare the hypoelliptic objects to the elliptic ones. This will in fact be done in chapters 8 and 9.

The present chapter is organized as follows. In section 6.1, we construct the hypoelliptic torsion forms, and we briefly study their dependence on the parameter $b > 0$.

In section 6.2, we study the compatibility of the torsion forms to Poincaré duality.

In section 6.3, we define a generalized Ray-Singer metric on the line $\lambda = \det \mathfrak{H}^{\cdot}(X,F)$ via the analytic torsion of the hypoelliptic Laplacian.

In section 6.4, we introduce a truncation procedure on the spectrum of our Laplacian. This procedure is needed only for large values of the parameter b.

In section 6.5, we show that the hypoelliptic Ray-Singer metric is smooth.

In section 6.6, we extend this procedure to the equivariant determinant already considered in section 1.12 in the elliptic context.

In section 6.7, a key variation formula is given for the hypoelliptic Ray-Singer metric. In particular we show that it does not depend on $b > 0$. The sections which follow are devoted to the proof of this formula.

In section 6.8, we establish an elementary identity.

In section 6.9, we introduce projected connections associated with the truncation procedure.

Finally, in section 6.10, we prove the variation formula.

Throughout the chapter, we make the same assumptions as in chapter 4, and we use the corresponding notation. Also we assume S to be compact.

6.1 THE HYPOELLIPTIC TORSION FORMS

Recall that $Q \in \operatorname{End}(\Lambda^{\cdot}(T^*S))$ was defined in (1.9.5). We take $b_0 > 0$ small enough so that for $b \in]0, b_0]$, the results of sections 3.4, 3.5, and of chapter

5 hold. Also for $b>0$, we set $c=\pm 1/b^2$.

Recall that the map $Q\in \mathrm{End}\left(\Lambda^{\mathrm{even}}\left(T^*S\right)\right)$ was defined in (1.9.5).

Definition 6.1.1. For $g\in G, b\in]0,b_0]$, put

$$\begin{aligned}\mathcal{T}_{h,g,b}\left(T^HM,g^{TX},\nabla^F,g^F\right) &= -\int_0^{+\infty}\left(w_{b,t}-\left(\frac{n}{2}\overline{\chi}_g\left(F\right)-\frac{1}{2}\overline{\chi}'_g\left(F\right)\right)\right.\\ &\left.\left(h'\left(i\sqrt{t}/2\right)h'\left(0\right)\right)\right)\frac{dt}{t}, \qquad (6.1.1)\\ \mathcal{T}_{\mathrm{ch},g,b}\left(T^HM,g^{TX},\nabla^F,g^F\right) &= Q\mathcal{T}_{h,g,b}\left(T^HM,g^{TX},\nabla^F,g^F\right).\end{aligned}$$

Observe that by Theorem 4.4.1, by Proposition 5.1.3, and by Theorem 5.2.1, the integral in the right-hand side of (6.1.1) is well-defined.

The even forms $\mathcal{T}_{h,g,b}\left(T^HM,g^{TX},\nabla^F,g^F\right), \mathcal{T}_{\mathrm{ch},g,b}\left(T^HM,g^{TX},\nabla^F,g^F\right)$ will be called the hypoelliptic torsion forms and the Chern hypoelliptic torsion forms, respectively. These are smooth forms on S, which depend smoothly on $b\in]0,b_0]$.

Theorem 6.1.2. *The following identities hold:*

$$\begin{aligned}d\mathcal{T}_{h,g,b}\left(T^HM,g^{TX},\nabla^F,g^F\right) &= \int_{X_g} e\left(TX_g,\nabla^{TX_g}\right)h_g\left(\nabla^F,g^F\right)\\ &- h_g\left(\nabla^{\mathfrak{H}^{\cdot}(X,F)},\mathfrak{h}_b^{\mathfrak{H}^{\cdot}(X,F)}\right), \qquad (6.1.2)\\ d\mathcal{T}_{\mathrm{ch},g,b}\left(T^HM,g^{TX},\nabla^F,g^F\right) &= \int_{X_g} e\left(TX_g,\nabla^{TX_g}\right)\mathrm{ch}_g^{\circ}\left(\nabla^F,g^F\right)\\ &- \mathrm{ch}_g^{\circ}\left(\nabla^{\mathfrak{H}^{\cdot}(X,F)},\mathfrak{h}_b^{\mathfrak{H}^{\cdot}(X,F)}\right).\end{aligned}$$

Proof. By Theorem 4.3.2,

$$\frac{\partial}{\partial t}u_{b,t}=d\frac{w_{b,t}}{t}. \qquad (6.1.3)$$

Using Theorems 4.4.1 and 5.2.1, by integrating (6.1.3), we get the first equation in (6.1.2). The second equation in (6.1.2) is a consequence of the first one and of the considerations we made in (1.9.1)-(1.9.6). □

Remark 6.1.3. In our definition of $\mathcal{T}_{h,g,b}\left(T^HM,g^{TX},\nabla^F,g^F\right)$, we could as well replace $w_{b,t}$ by $\underline{w}_{b,t}$. A minor difficulty would be that the asymptotic expansion (4.13.19) is not quite enough to produce a converging integral. However by pursuing further along the lines of chapter 4, we find easily that as $t\to 0$, $\underline{w}_{b,t}$ has an asymptotic expansion in powers of $\sqrt{t}$ to arbitrary order. The asymptotics of $\underline{w}_{b,t}$ as $t\to +\infty$ is taken care of by Proposition 5.1.3 and Theorem 5.2.1. To modify the definition of $\mathcal{T}_{h,g,b}\left(T^HM,g^{TX},\nabla^F,g^F\right)$, one just needs to subtract from $\underline{w}_{b,t}$ the constant term in its asymptotic expansion as $t\to 0$ to get an expression which will converge. By Proposition 4.3.4, the obtained expression will differ from $\mathcal{T}_{h,g,b}\left(T^HM,g^{TX},\nabla^F,g^F\right)$ by an exact form, which has no effect on equation (6.1.2). As in [BLo95, BG01], in the sequel, we will be interested in the evaluation of $\mathcal{T}_{h,g,b}\left(T^H M, g^{TX}, \nabla^F, g^F\right)$ modulo exact forms. So replacing $w_{b,t}$ by $\underline{w}_{b,t}$ is indeed permitted. Needless to say, in degree 0, by (4.3.10), $w_{b,t}$ and $\underline{w}_{b,t}$ coincide, and so the two possible definitions of the torsion forms coincide in degree 0.

For $c = 1/b^2$, by (3.1.4), $\mathfrak{H}^{\cdot}(X,F) = H^{\cdot}(X,F)$. Then the first terms of the right-hand sides of (1.8.2), (1.9.8) and of (6.1.2) are the same, while the second terms refer to distinct Hermitian forms on $H^{\cdot}(X,F)$. When $c = -1/b^2$, then $\mathfrak{H}^{\cdot}(X,F) = H^{\cdot-n}(X, F \otimes o(TX))$, and moreover since $e\left(TX_g, \nabla^{TX_g}\right)$ vanishes when $\dim X_g$ is odd, using (1.6.4), we get

$$\int_{X_g} e\left(TX_g, \nabla^{TX_g}\right) h_g\left(\nabla^{F\otimes o(TX)}, g^{F\otimes o(TX)}\right) \\ = (-1)^n \int_{X_g} e\left(TX_g, \nabla^{TX_g}\right) h_g\left(\nabla^F, g^F\right). \quad (6.1.4)$$

It follows that for $c = -1/b^2$, $(-1)^n \mathcal{T}_{h,g}\left(T^H M, g^{TX}, \nabla^{F\otimes o(TX)}, g^{F\otimes o(TX)}\right)$ and $\mathcal{T}_{h,g,b}\left(T^H M, g^{TX}, \nabla^F, g^F\right)$ verify similar equations. In both cases, it is natural to compare the elliptic and hypoelliptic torsion forms.

Also observe that by Theorem 4.3.2, we get

$$\frac{\partial}{\partial b}\frac{w_{b,t}}{t} = \frac{\partial}{\partial t}\frac{v_{b,t}}{b} - d\mathfrak{r}_{b,t}. \quad (6.1.5)$$

By Theorems 4.4.1 and 5.2.1, by (6.1.1) and by (6.1.5), we obtain

$$\frac{\partial}{\partial b}\mathcal{T}_{h,g,b}\left(T^H M, g^{TX}, \nabla^F, g^F\right) = -\frac{v_{b,\infty}}{b} \text{ in } \Omega^{\cdot}(S)/d\Omega^{\cdot}(S). \quad (6.1.6)$$

Equation (6.1.6) can be integrated in b, and so for $0 < b < b'$ and b' small enough, $\mathcal{T}_{h,g,b'}\left(T^H M, g^{TX}, \nabla^F, g^F\right) - \mathcal{T}_{h,g,b}\left(T^H M, g^{TX}, \nabla^F, g^F\right)$ can be evaluated in $\Omega^{\cdot}(S)/d\Omega^{\cdot}(S)$ in terms of the secondary classes of section 1.11. This idea will not be pursued further. In fact in Theorem 8.2.1, we will give a formula comparing $\mathcal{T}_{h,g,b}\left(T^H M, g^{TX}, \nabla^F, g^F\right)$ to the elliptic torsion forms of chapter 1 in $\Omega^{\cdot}(S)/d\Omega^{\cdot}(S)$, from which the above results follow.

In section 6.3, we will establish an analogue of equation (1.8.4), which will relate our definition of the analytic torsion forms in degree 0 to the more classical Ray-Singer torsion, whose definition can be adapted to the hypoelliptic context. Also note that in Theorems 6.7.1 and 6.7.2, in degree 0, we will give another formulation of (6.1.6) which is valid for any $b > 0$, as a result of independence on b of a hypoelliptic Ray-Singer metric.

6.2 HYPOELLIPTIC TORSION FORMS AND POINCARÉ DUALITY

In this section, to distinguish the cases $c > 0$ and $c < 0$, we will temporarily add an index $\pm$ to $\mathcal{T}_{h,g,b}\left(T^H M, g^{TX}, \nabla^F, g^F\right)$.

Proposition 6.2.1. *The following identity holds:*

$$\mathcal{T}_{h,g,b,\pm}\left(T^H M, g^{TX}, \nabla^{\overline{F}^*}, g^{\overline{F}^*}\right) = -\mathcal{T}_{h,g,b,\mp}\left(T^H M, g^{TX}, \nabla^F, g^F\right). \quad (6.2.1)$$

Proof. We use temporarily the notation a^F instead of a in Definition 4.3.1. By (2.7.3), the fact that $\kappa^F_{\mathcal{H}-\omega^H}$ is an even operator and that supertraces vanish on supercommutators, we get

$$a^F_{\pm} = -a^{\overline{F}^*}_{\mp}. \tag{6.2.2}$$

By (4.3.6) and (6.2.2), we obtain

$$w^{\overline{F}^*}_{b,t,\pm} = -w^F_{b,t,\mp}. \tag{6.2.3}$$

Our proposition now follows from (6.1.1) and from the fact that when replacing F by $\overline{F}^*$, by (4.2.2), (4.2.3), $\frac{n}{2}\chi_g(F) - \frac{1}{2}\chi'_g(F)$ is changed into its negative. □

Remark 6.2.2. Of course (6.2.1) still holds for $\mathcal{T}_{\mathrm{ch},g,b}\left(T^H M, g^{TX}, \nabla^F, g^F\right)$. Incidentally it is interesting to establish (6.2.3) directly using (4.3.3). Finally, observe that by (2.7.3) and (4.3.8), the analogue of (6.2.3) holds for $\underline{w}_{b,t}$.

6.3 A GENERALIZED RAY-SINGER METRIC ON THE DETERMINANT OF THE COHOMOLOGY

Recall that the line $\lambda(F)$ was defined in (1.12.3) by

$$\lambda(F) = \det H^{\cdot}(X,F). \tag{6.3.1}$$

Also in (1.12.4), we showed that by Poincaré duality,

$$\lambda(F^* \otimes o(TX)) = (\lambda(F))^{(-1)^{n+1}}. \tag{6.3.2}$$

Set

$$\lambda = \det \mathfrak{H}^{\cdot}(X,F). \tag{6.3.3}$$

Now we will use the notation in section 5.1. By Theorem 3.2.2 and by (5.1.1), we know that

$$\lambda \simeq \det \Omega^{\cdot}(T^*X, \pi^*F)_0. \tag{6.3.4}$$

Also by (3.1.4),

$$\begin{aligned} \lambda &= \lambda(F) \text{ if } c > 0, \\ &= (\lambda(F \otimes o(TX)))^{(-1)^n} \text{ if } c < 0. \end{aligned} \tag{6.3.5}$$

Let $\mathfrak{h}^{\Omega^{\cdot}(T^*X,\pi^*F)_0}_{\mathcal{H}^c}$ be the restriction of $\mathfrak{h}^{\Omega^{\cdot}(T^*X,\pi^*F)}_{\mathcal{H}^c}$ to $\Omega^{\cdot}(T^*X,\pi^*F)_0$. By Theorem 3.2.2, the Hermitian form $\mathfrak{h}^{\Omega^{\cdot}(T^*X,\pi^*F)_0}_{\mathcal{H}^c}$ is nondegenerate.

Definition 6.3.1. We denote by $|\,|^2_{\det \Omega^{\cdot}(T^*X,\pi^*F)_0}$ the generalized metric on the line $\det \Omega^{\cdot}(T^*X,\pi^*F)_0$ which is induced by $\mathfrak{h}^{\Omega^{\cdot}(T^*X,\pi^*F)_0}_{\mathcal{H}^c}$. Let $|\,|^2_{\lambda}$ be the corresponding generalized metric on λ via the isomorphism (6.3.4).

Note that the above objects depend explicitly on c.

In Definition 6.3.1, we follow the terminology in [B05, subsection 1.4]. Indeed if $\mathfrak{h}_{\mathcal{H}^c}^{\Omega^{\cdot}(T^*X,\pi^*F)_0}$ turns out to be a standard Hermitian metric, then $|\,|_\lambda^2$ is the corresponding Hermitian metric on λ. The above notation is just the obvious extension to the general case. Note here that the square in $|\,|_\lambda^2$ is not meant to indicate any positivity. To the contrary, as explained in detail in [B05, subsection 1.4], the generalized metric $|\,|_\lambda^2$ has a definite sign $\epsilon\left(|\,|_\lambda^2\right)$. If $\mathfrak{h}_{\mathcal{H}^c}^{\Omega^{\cdot}(T^*X,\pi^*F)_0}$ has signature (p,q), this sign is $(-1)^q$.

Put

$$D_{\phi,\mathcal{H}^c} = 2A_{\phi,\mathcal{H}^c}, \tag{6.3.6}$$

which, by (2.1.20), is equivalent to

$$D_{\phi,\mathcal{H}^c} = d^{T^*X} + \overline{d}^{T^*X}_{\phi,2\mathcal{H}^c}. \tag{6.3.7}$$

Let $\left(D^2_{\phi,\mathcal{H}^c}\right)^{-1}$ be the inverse of $D_{\phi,\mathcal{H}^c}$ on $\Omega^{\cdot}\left(T^*X,\pi^*F\right)_*$. For $g\in G, s\in$ $\mathbf{C}, \mathrm{Re} s >> 0$, set

$$\vartheta_g(s) = -\mathrm{Tr_s}^{\Omega^{\cdot}(T^*X,\pi^*F)_*}\left[gN^{T^*X}\left(D^2_{\phi,\mathcal{H}^c}\right)^{-s}\right]. \tag{6.3.8}$$

When $g=1$, we will write $\vartheta(s)$ instead of $\vartheta_1(s)$.

It is not even clear that (6.3.8) makes sense. However, we claim that $e^{\vartheta'(0)}$ is indeed well-defined.

As we saw after equation (3.3.18), there are only a finite number of $\lambda\in \mathrm{Sp}A^2_{\phi,\mathcal{H}^c}$ such that $\mathrm{Re}\,\lambda\le 0$. Let $P_{<0}$ be the obvious projector on the finite dimensional complex

$$\Omega^{\cdot}\left(T^*X,\pi^*F\right)_{<0} = \oplus_{\mathrm{Re}\lambda<0}\Omega^{\cdot}\left(T^*X,\pi^*F\right)_\lambda. \tag{6.3.9}$$

We define the projector $P_{\le 0}$ in the same way. Set

$$P_{>0} = 1 - P_{\le 0}. \tag{6.3.10}$$

We define $\vartheta_{<0}(s)$ and $\vartheta_{>0}(s)$ by inserting in (6.3.8) the operators $P_{<0}$ and $P_{>0}$, so that in principle,

$$\exp\left(\vartheta'(0)\right) = \exp\left(\vartheta'_{<0}(0)\right)\exp\left(\vartheta'_{>0}(0)\right). \tag{6.3.11}$$

Let $\mathfrak{h}_{\mathcal{H}^c}^{\Omega^{\cdot}(T^*X,\pi^*F)_{<0}}$ be the restriction of $\mathfrak{h}_{\mathcal{H}^c}^{\Omega^{\cdot}(T^*X,\pi^*F)}$ to $\Omega^{\cdot}\left(T^*X,\pi^*F\right)_{<0}$. By Theorem 3.2.3, $\mathfrak{h}_{\mathcal{H}^c}^{\Omega^{\cdot}(T^*X,\pi^*F)_{<0}}$ is nondegenerate.

Any definition of $\exp\left(\vartheta'_{<0}(0)\right)$ leads to the fact that this should be an alternate product of determinants of the restriction $D^2_{\phi,\mathcal{H}^c,<0}$ of $D^2_{\phi,\mathcal{H}^c}$ to $\Omega^{\cdot}\left(T^*X,\pi^*F\right)_{<0}$, which is always well-defined. Note that since the spectrum of $D^2_{\phi,\mathcal{H}^c}$ is conjugation-invariant, this quantity is always real. Contrary to what the notation seems to indicate, $\exp\left(\vartheta'_{<0}(0)\right)$ has a sign. By [B05, Theorem 1.9 and Remark 1.10], this sign is just $(-1)^{\mathrm{sign}\,\mathfrak{h}_{\mathcal{H}^c}^{\Omega^{\cdot}(T^*X,\pi^*F)_{<0}}}$.

Now we will make sense of $\exp\left(\vartheta'_{>0}(0)\right)$. Indeed, by equation (3.3.9) or by equation (15.7.5) in Theorem 15.7.1, we know that for $s\in\mathbf{R}, s>>0$, the operator $\left(D^2_{\phi,\mathcal{H}^c}\right)^{-s}$ is trace class. Therefore the operator $P_{>0}\left(D^2\right)^{-s}_{\phi,\mathcal{H}^c}$ is also trace class. Using Proposition 3.2.1, it is clear that if $s \in \mathbf{R}, s >> 0$, then $\vartheta_{>0}(s) \in \mathbf{R}$.

Now we show that $\vartheta_{>0}(s)$ extends to a holomorphic function near $s = 0$. First we claim that for $s \in \mathbf{R}, \mathrm{Re}\, s >> 0$,

$$\mathrm{Tr_s}\left[P_{>0}\left(D^2_{\phi,\mathcal{H}^c}\right)^{-s}\right] = 0. \tag{6.3.12}$$

Indeed,

$$\mathrm{Tr_s}\left[P_{>0}\left(D^2_{\phi,\mathcal{H}^c}\right)^{-s}\right] = \sum_{\substack{\lambda \in \mathrm{Sp} D^2_{\phi\mathcal{H}^c} \\ \mathrm{Re}\lambda>0}} \frac{\chi_\lambda}{\lambda^s}, \tag{6.3.13}$$

where χ_λ is the Euler characteristic of the complex $\left(\Omega^{\cdot}(T^*X, \pi^*F)_\lambda, d^{T^*X}\right)$. Note that (6.3.13) is a consequence of Weyl's inequality [ReSi78, Theorem XIII.10.3, p. 318], which guarantees that the series in the right-hand side of (6.3.13) is absolutely convergent, and also of Lidskii's theorem [ReSi78, Corollary, p. 328]. Since the complexes $\left(\Omega^{\cdot}(T^*X, \pi^*F)_\lambda, d^{T^*X}\right)$ are exact, the χ_λ vanish identically, so that (6.3.12) holds.

We deduce from (6.3.12) that for $s \in \mathbf{R}, s >> 0$,

$$\vartheta_{>0}(s) = -\mathrm{Tr_s}\left[\left(N^{T^*X} - n\right) P_{>0}\left(D^2_{\phi,\mathcal{H}^c}\right)^{-s}\right]. \tag{6.3.14}$$

Now we use the Mellin transform to rewrite $\vartheta_{>0}(s)$ in the form

$$\vartheta_{>0}(s) = -\frac{1}{\Gamma(s)} \int_0^{+\infty} t^{s-1} \mathrm{Tr_s}\left[\left(N^{T^*X} - n\right) P_{>0} \exp\left(-tD^2_{\phi,\mathcal{H}^c}\right)\right] dt. \tag{6.3.15}$$

The integral in (6.3.15) splits as $\int_0^1 + \int_1^{+\infty}$. By using a contour integral similar to the contour integral in (5.3.5) for the definition of $V_{b,t}$, and using an estimate similar to (5.4.3), we find that the integrand after t^{s-1} decays exponentially as $t \to +\infty$. Therefore the integral $\int_1^{+\infty}$ extends to a holomorphic function of $s \in \mathbf{C}$.

We already know that $P_{\leq 0}$ is a projector on a finite dimensional vector space. Moreover,

$$\begin{aligned}
&\frac{1}{\Gamma(s)} \int_0^1 t^{s-1} \mathrm{Tr_s}\left[\left(N^{T^*X} - n\right) P_{>0} \exp\left(-tD^2_{\phi,\mathcal{H}^c}\right)\right] dt \\
&\qquad = \frac{1}{\Gamma(s)} \int_0^1 t^{s-1} \mathrm{Tr_s}\left[\left(N^{T^*X} - n\right) \exp\left(-tD^2_{\phi,\mathcal{H}^c}\right)\right] dt \\
&\qquad - \frac{1}{\Gamma(s)} \int_0^1 t^{s-1} \mathrm{Tr_s}\left[\left(N^{T^*X} - n\right) P_{\leq 0} \exp\left(-tD^2_{\phi,\mathcal{H}^c}\right)\right] dt.
\end{aligned} \tag{6.3.16}$$

Clearly, as $t \to 0$,

$$\mathrm{Tr_s}\left[\left(N^{T^*X} - n\right) P_{\leq 0} \exp\left(-tD^2_{\phi,\mathcal{H}^c}\right)\right] = \mathrm{Tr_s}\left[\left(N^{T^*X} - n\right) P_{\leq 0}\right] + \mathcal{O}(t). \tag{6.3.17}$$

By (6.3.17), we see that the second integral in the right-hand side of (6.3.16) extends to a holomorphic function near $s = 0$.

Using the notation in section 2.8 and in chapter 4, from (6.3.6), we get

$$\mathrm{Tr_s}\left[\left(N^{T^*X} - n\right)\exp\left(-tD^2_{\phi,\mathcal{H}^c}\right)\right] = \mathrm{Tr_s}\left[\left(N^{T^*X} - n\right)\exp\left(-4A^2_{b,t}\right)\right]. \tag{6.3.18}$$

By equation (4.13.18) in Proposition 4.13.1, by Remark 4.13.2, and by the above, there are $a \in \mathbf{C}, a' \in \mathbf{C}$ such that as $t \to 0$,

$$\mathrm{Tr_s}\left[\left(N^{T^*X} - n\right)\exp\left(-tD^2_{\phi,\mathcal{H}^c}\right)\right] = \frac{a}{\sqrt{t}} + a' + \mathcal{O}\left(\sqrt{t}\right). \tag{6.3.19}$$

Note that by the first equation in (4.13.18) and also by Remark 4.13.2, which guarantees that in degree 0, the left-hand sides in (4.13.18) coincide, we have indeed $a' = 0$ instead of in the right-hand side of (6.3.19), but (6.3.19) is enough for our purpose.

By (6.3.19), it is clear that the first integral in the right-hand side of (6.3.16) also extends to a holomorphic function near $s = 0$. This function is still real for $s \in \mathbf{R}$.

The conclusion is that $\theta_{>0}(s)$ extends to a holomorphic function near $s = 0$.

Definition 6.3.2. Put

$$S_b\left(g^{TX}, \nabla^F, g^F\right) = \exp\left(\vartheta'(0)\right). \tag{6.3.20}$$

The right-hand side of (6.3.20) is unambiguously defined as a nonzero real number. In particular, its sign is well-defined. Only the real negative eigenvalues of $D^2_{\phi,\mathcal{H}^c}$ contribute to this sign.

Definition 6.3.3. Let $\| \|^2_\lambda$ be the generalized metric on λ,

$$\| \|^2_\lambda = S_b\left(g^{TX}, \nabla^F, g^F\right) | |^2_\lambda. \tag{6.3.21}$$

The metric $\| \|^2_\lambda$ will be called a generalized Ray-Singer metric.

Let $\epsilon\left(\| \|^2_\lambda\right)$ be the sign of the generalized metric $\| \|^2_\lambda$. By (6.3.21), we get

$$\epsilon\left(\| \|^2_\lambda\right) = \mathrm{sign}\left(S_b\left(g^{TX}, \nabla^F, g^F\right)\right)\epsilon\left(| |^2_\lambda\right). \tag{6.3.22}$$

Now we adapt Definition 6.1.1 to the case where S is a point, so that $M = X$. Here $T^HM = \{0\}$. For $b \in \mathbf{R}^*_+$ small enough, and $c = \pm 1/b^2$, we have defined the real number $\mathcal{T}_{h,b}\left(g^{TX}, \nabla^F, g^F\right)$, which is just $\mathcal{T}_{h,g,b}\left(g^{TX}, \nabla^F, g^F\right)$ in the case where $g = 1$. Here the notation T^HM has been dropped. Also we may as well replace the subscript h by ch, since Q has no effect in degree 0.

For $b > 0$ small enough, by Theorem 3.5.1, there is no ambiguity in the definition of $\vartheta(s)$, since the nonzero eigenvalues have positive real part.

Proposition 6.3.4. *For $b > 0$ small enough, the following identity holds:*

$$\mathcal{T}_{h,b}\left(g^{TX}, \nabla^F, g^F\right) = \frac{1}{2}\vartheta'(0). \tag{6.3.23}$$

Proof. By (4.3.10), when defining $\mathcal{T}_{h,b}\left(g^{TX},\nabla^F,g^F\right)$, we can replace $w_{b,t}$ by $\underline{w}_{b,t}$. Using the asymptotic expansion in (6.3.19), the proof of our proposition is exactly the same as the proof of [BLo95, Theorem 3.29]. □

We claim that the above result can be extended to the case of a general b. Indeed the projectors $P_{<0}, P_{>0}$ were defined before. By inserting $P_{>0}$ in the definition of $\underline{w}_{b,t}$, we obtain now the real number $\underline{w}_{b,t}^{>0}$. Moreover, using Proposition 4.3.4 and Theorem 4.4.1, we find that as $t \to 0$, we have an asymptotic expansion of the type

$$\underline{w}_{b,t}^{>0} = \underline{w}_{b,0}^{>0} + \mathcal{O}\left(\sqrt{t}\right). \tag{6.3.24}$$

Put

$$\mathcal{T}_{h,b}\left(g^{TX},\nabla^F,g^F\right)_{>0} = -\int_0^{+\infty}\left(\underline{w}_{b,t}^{>0} - \underline{w}_{b,0}^{>0}h'\left(i\sqrt{t}/2\right)\right)\frac{dt}{t}. \tag{6.3.25}$$

The same arguments as in the proof of Proposition 6.3.4 show that

$$\mathcal{T}_{h,b}\left(g^{TX},\nabla^F,g^F\right)_{>0} = \frac{1}{2}\vartheta'_{>0}(0). \tag{6.3.26}$$

By (6.3.20), we find that

$$S_b\left(g^{TX},\nabla^F,g^F\right) = \exp\left(2\mathcal{T}_{h,b}\left(g^{TX},\nabla^F,g^F\right)_{>0}\right)\exp\left(\vartheta'_{<0}(0)\right). \tag{6.3.27}$$

The critical fact which ensures that (6.3.23) and (6.3.27) are compatible is that if $\mu \in \mathbf{C}$ is such that $\operatorname{Re}\mu^2 > 0$, then

$$-\int_0^{+\infty}\left(h'\left(i\sqrt{t}\mu/2\right) - h'\left(i\sqrt{t}/2\right)\right)\frac{dt}{t} = \log\left(\mu^2\right). \tag{6.3.28}$$

6.4 TRUNCATION OF THE SPECTRUM AND RAY-SINGER METRICS

Here we will adapt arguments of Quillen [Q85a] to the present situation. We use the arguments in [B05, subsection 1.6].

Let $r \in \mathbf{R}$ be such that if $\lambda \in \operatorname{Sp} D^2_{\phi,\mathcal{H}^c}$, then $|\lambda| \neq r$.

Definition 6.4.1. Let $r \in \mathbf{R}_+^*$ be such that if $\lambda \in \operatorname{Sp}\mathfrak{A}^2_{\phi,\mathcal{H}^c}$, then $|\lambda| \neq r$. Put

$$\mathcal{S}^\cdot\left(T^*X,\pi^*F\right)_{<r} = \bigoplus_{\substack{\lambda\in\operatorname{Sp}\mathfrak{A}^2_{\phi,\mathcal{H}^c}\\ |\lambda|<r}} \mathcal{S}^\cdot\left(T^*X,\pi^*F\right)_\lambda. \tag{6.4.1}$$

We define the projector $\mathfrak{P}_{<r}$ on $\mathcal{S}^\cdot\left(T^*X,\pi^*F\right)_{<r}$ as in (3.2.9), the contour δ being now the circle of center 0 and radius r. Set $\mathfrak{P}_{>r} = 1-\mathfrak{P}_{<r}$. Then $\mathfrak{P}_{<r}$ is a projector on the finite dimensional vector space $\mathcal{S}^\cdot\left(T^*X,\pi^*F\right)_{<r}$, and $\mathfrak{P}_{>r}$ projects on a supplementary subspace $\mathcal{S}^\cdot\left(T^*X,\pi^*F\right)_{*>r}$. As in (5.1.1), by multiplying by $\exp\left(c|p|^2/2\right)$, we obtain vector spaces $\Omega^{\cdot}(T^*X,\pi^*F)_{<r}$ and $\Omega^{\cdot}(T^*X,\pi^*F)_{*>r}$.

Clearly $\Omega^{\cdot}(T^*X,\pi^*F)_{<r}$ and $\Omega^{\cdot}(T^*X,\pi^*F)_{*>r}$ are themselves complexes, and moreover, instead of (5.1.2), we have now

$$\Omega^{\cdot}(T^*X,\pi^*F) = \Omega^{\cdot}(T^*X,\pi^*F)_{<r} \oplus \Omega^{\cdot}(T^*X,\pi^*F)_{*>r}. \tag{6.4.2}$$

Proposition 6.4.2. *If $r>0$ is taken as before, then*

$$\begin{aligned} H^{\cdot}\left(\Omega^{\cdot}(T^*X,\pi^*F)_{>r}, d^X\right) &= 0, \\ H^{\cdot}\left(\Omega^{\cdot}(T^*X,\pi^*F)_{<r}, d^X\right) &= \mathfrak{H}^{\cdot}(X,F). \end{aligned} \tag{6.4.3}$$

The subcomplexes $\Omega^{\cdot}(T^*X,\pi^*F)_{<r}$ and $\Omega^{\cdot}(T^*X,\pi^*F)_{*>r}$ are $\mathfrak{h}_{\mathcal{H}^c}^{\Omega^{\cdot}(T^*X,\pi^*F)}$ orthogonal, and the restriction of $\mathfrak{h}_{\mathcal{H}^c}^{\Omega^{\cdot}(T^*X,\pi^*F)}$ to each of these subcomplexes is nondegenerate.

Proof. This follows from Theorems 3.2.2 and 3.2.3. □

For $r>0$, by [KMu76] and by Proposition 6.4.2, we have the canonical isomorphism

$$\lambda \simeq \det \Omega^{\cdot}(T^*X,\pi^*F)_{<r}. \tag{6.4.4}$$

Note that the canonical isomorphisms in (6.3.4), (6.4.4) are compatible with the isomorphism $\det\Omega^{\cdot}(T^*X,\pi^*F)_{<r} \simeq \det\Omega^{\cdot}(T^*X,\pi^*F)_0$ one also obtains via [KMu76]. Let $\mathfrak{h}_{\mathcal{H}^c}^{\Omega^{\cdot}(T^*X,\pi^*F)_{<r}}$ be the restriction of $\mathfrak{h}_{\mathcal{H}^c}^{\Omega^{\cdot}(T^*X,\pi^*F)}$ to $\Omega^{\cdot}(T^*X,\pi^*F)_{<r}$ and let $|\,|^2_{\det\Omega^{\cdot}(T^*X,\pi^*F)_{<r}}$ be the induced generalized metric on $\det\Omega^{\cdot}(T^*X,\pi^*F)_{<r}$. Let $|\,|^2_{\lambda,<r}$ be the corresponding generalized metric on λ via the isomorphism (6.4.4).

For $r>0$ such that $r/4$ verifies the previous assumptions, $s\in\mathbf{C}$, $\operatorname{Re}s>>0$, set

$$\vartheta_{>r}(s) = -\operatorname{Tr_s}^{\Omega^{\cdot}(T^*X,\pi^*F)_{>r/4}}\left[N^{T^*X}\left(D^2_{\phi,\mathcal{H}^c}\right)^{-s}\right]. \tag{6.4.5}$$

In (6.4.5), $r/4$ appears in the right-hand side because $A^2_{\phi,\mathcal{H}^c} = D^2_{\phi,\mathcal{H}^c}/4$. Again, it is not clear that (6.4.5) is well-defined.

As we saw after (3.3.18), the set $\left\{\lambda\in\operatorname{Sp}D^2_{\phi\mathcal{H}^c}, \operatorname{Re}\lambda\le 0\right\}$ is bounded, and so it is finite. Also for $r>0$ large enough, if $\lambda\in\operatorname{Sp}D^2_{\phi,\mathcal{H}^c}$, $|\lambda|>r$, then $\operatorname{Re}\lambda>0$. So the definition of $\vartheta_{>r}(s)$ is unambiguous.

Set

$$S_b\left(g^{TX},\nabla^F,g^F\right)_{>r} = \exp\left(\vartheta'_{>r}(0)\right). \tag{6.4.6}$$

By proceeding as in section 6.3, we find that $S_b\left(g^{TX},\nabla^F,g^F\right)_{>r}$ is a well-defined nonzero real number.

As in (6.3.25), we can define $T_{h,b}\left(g^{TX},\nabla^F,g^F\right)_{>r}$. The obvious analogue of (6.3.27) is

$$S_b\left(g^{TX},\nabla^F,g^F\right)_{>r} = \exp\left(2T_{h,b}\left(g^{TX},\nabla^F,g^F\right)_{>r}\right). \tag{6.4.7}$$

Proposition 6.4.3. *The following identities hold:*

$$\| \|_\lambda^2 = S_b \left(g^{TX}, \nabla^F, g^F\right)_{>r} | |_{\lambda,<r}^2 \,. \tag{6.4.8}$$

Proof. The restriction of $\overline{d}_{\phi,2\mathcal{H}^c}^{T^*X}$ to $\Omega^{\cdot}(T^*X, \pi^*F)_{<r}$ is the $\mathfrak{h}_{\mathcal{H}^c}^{\Omega^{\cdot}(T^*X,\pi^*F)_{<r}}$ adjoint of d^{T^*X}. Our proposition now follows from [B05, Theorems 1.9 and 1.14]. This is an analogue for our main result in a finite dimensional context, which is here clearly enough. □

Remark 6.4.4. Recall that $c = \pm 1/b^2$. We have not noted explicitly the dependence of $\| \|_\lambda$ on $b \in \mathbf{R}_+^*$. Still keeping track of the dependence will be important in the sequel. Temporarily, we will denote these metrics $\| \|_{\lambda,b}$. Of course the generalized metric also depends on the sign of c, which is not explicitly written.

6.5 A SMOOTH GENERALIZED METRIC ON THE DETERMINANT BUNDLE

We make the same assumptions as in chapter 4. In particular g^{TX} is a metric on TX, and g^F is a Hermitian metric on F. Otherwise, we use the notation of this chapter.

Recall that $\mathfrak{H}^{\cdot}(X, F)$ is equipped with the flat connection $\nabla^{\mathfrak{H}^{\cdot}(X,F)}$.

We then define the complex lines $\lambda_s, s \in S$ as in (6.3.3). Clearly these lines patch into a smooth line bundle λ on S, which is canonically equipped with a flat connection ∇^λ.

We replace temporarily S by $S' = S \times \mathbf{R}_+^*$. The line bundle λ lifts to S', together with the flat connection ∇^λ. For $s' = (s, b) \in S'$, we can equip the fibers λ_s with the generalized metric $\| \|_{\lambda_s,b}^2$. For simplicity, this metric will be denoted as $\| \|_\lambda^2$.

Theorem 6.5.1. *The metric $\| \|_\lambda^2$ is a smooth generalized metric on λ over S'.*

Proof. Take $r > 0$, and define the open set $V_r \subset S$ as in [B05, eq. (1.68)], i.e.,

$$V_r = \left\{s' \in S', \text{if } \lambda \in \operatorname{Sp} D_{\phi,\mathcal{H}^c}^2, \text{ then } |\lambda| \neq r\right\}. \tag{6.5.1}$$

Then S' is covered by the V_r. Moreover, given $s' \in S', M > 0$, there is $r > M$ such that $s' \in V_r$. Also $\Omega^{\cdot}(T^*X, \pi^*F)_{<r}$ is a smooth $\mathbf{Z}$-graded finite dimensional vector bundle on V_r, equipped with the smooth generalized metric $\mathfrak{h}_{\mathcal{H}^c}^{\Omega^{\cdot}(T^*X,\pi^*F)_{<r}}$. If $s \in V_r$, and $r > 0$ is large enough, $T_{h,b}\left(g^{TX}, g^F, \mathcal{H}^c\right)_{>r}$ is a smooth function on V_r. By Proposition 6.4.3, it is now clear that $\| \|_\lambda^2$ is a smooth generalized metric on λ. □

6.6 THE EQUIVARIANT DETERMINANT

In this section, we use the formalism of section 1.12. We still make the same assumptions as in chapter 4. Then G acts naturally on $H^{\cdot}(X,F)$ or $H^{\cdot}(X,F\otimes o(TX))$.

Clearly the action of G lifts to T^*X. Also G acts on $\mathfrak{H}^{\cdot}(TX)$. Then (3.1.4) is an identification of G-spaces.

Now we use the notation in (1.12.6)-(1.12.8). If $\in \widehat{G}$, set

$$\lambda_W = \det\left(\mathrm{Hom}_G\left(W, \mathfrak{H}^{\cdot}(X,F)\right)\otimes W\right). \tag{6.6.1}$$

Put

$$\lambda = \oplus_{W\in\widehat{G}}\lambda_W. \tag{6.6.2}$$

Recall that $\lambda(F)$ was defined in (1.12.8). By (3.1.4),

$$\begin{aligned}\lambda =&\lambda(F) \text{ if } c>0, \\ =&\left(\lambda\left(F\otimes o(TX)\right)\right)^{(-1)^n} \text{ if } c<0.\end{aligned} \tag{6.6.3}$$

Clearly G commutes with $D^2_{\phi,\mathcal{H}^c}$. The splitting (5.1.2) of $\Omega^{\cdot}(T^*X,\pi^*F)$ is preserved by G. The generalized metric $\mathfrak{h}^{\Omega^{\cdot}(T^*X,\pi^*F)}_{\mathcal{H}^c}$ is also G-invariant. Moreover, the identifications in Theorem 3.2.2 are identifications of G-spaces.

Now we proceed as in [B05, subsection 1.12]. If $W\in\widehat{G}$, set

$$\Omega^{\cdot}(T^*X,\pi^*F)_{0,W} = \mathrm{Hom}_G\left(W,\Omega^{\cdot}(T^*X,\pi^*F)_0\right)\otimes W. \tag{6.6.4}$$

Then we have the isotypical decomposition of $\Omega^{\cdot}(T^*X,\pi^*F)_0$,

$$\Omega^{\cdot}(T^*X,\pi^*F)_0 = \bigoplus_{W\in\widehat{G}}\Omega^{\cdot}(T^*X,\pi^*F)_{0,W}. \tag{6.6.5}$$

By [B05, Proposition 1.24], the decomposition (6.6.5) is $\mathfrak{h}^{\Omega^{\cdot}(T^*X,\pi^*F)_0}_{\mathcal{H}^c}$ orthogonal, and the restriction of $\mathfrak{h}^{\Omega^{\cdot}(T^*X,\pi^*F)_0}_{\mathcal{H}^c}$ to each term in the right-hand side of (6.6.5) is nondegenerate.

The $\Omega^{\cdot}(T^*X,\pi^*F)_{0,W}$ are subcomplexes of $\Omega^{\cdot}(T^*X,\pi^*F)_0$. Set

$$\mu_W = \det\Omega^{\cdot}(T^*X,\pi^*F)_{0,W}, \qquad \mu = \bigoplus_{W\in\widehat{G}}\mu_W. \tag{6.6.6}$$

By Theorem 3.2.2, there are canonical isomorphisms

$$\lambda_W \simeq \mu_W. \tag{6.6.7}$$

By (6.6.7), we have the canonical isomorphism

$$\lambda\simeq\mu. \tag{6.6.8}$$

Now we use the notation in [B05, Definition 1.25].

Definition 6.6.1. Let $\log\left(|\,|^2_\mu\right)$ be the logarithm of the generalized equivariant metric on μ which is associated to $\mathfrak{h}^{\Omega^{\cdot}(T^*X,\pi^*F)_0}_{\mathcal{H}^c}$, and let $\log\left(|\,|^2_\lambda\right)$ be the corresponding object on λ via the canonical isomorphism (6.6.8).

Let $\log|\ |^2_{\det \mu_W}$ be the logarithm of the generalized metric on μ_W associated to the restriction of $\mathfrak{h}^{\Omega^\cdot(T^*X,\pi^*F)_0}_{\mathcal{H}^c}$ to $\Omega^\cdot(T^*X,\pi^*F)_{0,W}$. Recall that χ_W is the character of the representation associated to W. Then we have the identity

$$\log\left(|\ |^2_\mu\right) = \sum_{W\in\widehat{G}} \log\left(|\ |^2_{\mu_W}\right) \otimes \frac{\chi_W}{\operatorname{rk} W}. \tag{6.6.9}$$

As before, some care has to be given to the fact that the generalized metrics in (6.6.9) are not necessarily positive. Therefore each term $\log\left(|\ |^2_{\mu_W}\right)$ contains implicitly the logarithm of the sign of the corresponding metric. The logarithm of the sign is just $ki\pi$, with $k\in\mathbf{Z}$ determined modulo 2.

Recall that $\vartheta_g(s)$ was formally defined in (6.3.8). Note that there are still ambiguities in the definition of $\vartheta_g(s)$, due to the negative part of the spectrum. These ambiguities are lifted as before, by still splitting the spectrum of $D^2_{\phi,\mathcal{H}^c}$. There remain ambiguities of the type

$$\sum k_W i\pi\chi_W, \tag{6.6.10}$$

with $k_W\in\mathbf{Z}$ being unambiguously determined mod 2. In the case where G is trivial, we got rid of the ambiguity by taking the exponential.

Now we imitate Definition 1.12.1.

Definition 6.6.2. Put

$$\log\left(\|\ \|^2_\lambda\right) = \log\left(|\ |^2_\lambda\right) + \vartheta'_\cdot(0). \tag{6.6.11}$$

The object in (6.6.8) will be called the logarithm of the generalized equivariant Ray-Singer metric on λ.

Note that the techniques and results of section 6.4 apply without any change to these new metrics. This adaptation is left to the reader.

For $b>0$ small enough, as in (6.3.20), we can define $S_{g,b}\left(g^{TX},\nabla^F,g^F\right)$ by the formula

$$S_{g,b}\left(g^{TX},\nabla^F,g^F\right) = \exp\left(\vartheta'_g(0)\right). \tag{6.6.12}$$

Assume temporarily that S is just a point. For $b>0$ small enough, we will write $\mathcal{T}_{h,g,b}\left(g^{TX},\nabla^F,g^F\right)$ for the torsion form which is concentrated in degree 0. The same arguments as in Proposition 6.3.4 show that for $b>0$ small enough,

$$\mathcal{T}_{h,g,b}\left(g^{TX},\nabla^F,g^F\right) = \frac{1}{2}\vartheta'_g(0). \tag{6.6.13}$$

We can construct the generalized equivariant line bundle λ on S, which is now a direct sum of line bundles over S. The definition of $\vartheta_{>r}(s)$ in (6.4.5) is now changed into

$$\vartheta_{>r,g}(s) = -\operatorname{Tr}_s^{\Omega^\cdot(T^*X,\pi^*F)_{>r/4}}\left[gN^{T^*X}\left(D^2_{\phi,\mathcal{H}^c}\right)^{-s}\right]. \tag{6.6.14}$$

We define $S_{g,b}\left(g^{TX},g^{F},\mathcal{H}^{c}\right)_{>r}$ as in (6.4.6) and $T_{h,g,b}\left(g^{TX},\nabla^{F},g^{F}\right)_{>r}$ by a formula similar to (6.3.25). The analogue of (6.3.27) is the identity

$$S_{g,b}\left(g^{TX},g^{F},\mathcal{H}^{c}\right)_{>r}=\exp\left(2T_{h,g,b}\left(g^{TX},\nabla^{F},g^{F}\right)_{>r}\right). \tag{6.6.15}$$

Proposition 6.4.3 takes the following form.

Proposition 6.6.3. *The following identity holds:*

$$\log\left(\|\ \|^{2}\right)=\log\left(|\ |^{2}_{\lambda,<r}\right)+\log S_{b,\cdot}\left(g^{TX},g^{F},g^{F}\right)_{>r}. \tag{6.6.16}$$

Needless to say, the equality in (6.6.16) is valid for all choices of $g\in G$. As before, to emphasize dependence on $b\in\mathbf{R}^{*}_{+}$, we will sometimes write instead $\log\left(\|\ \|^{2}_{\lambda,b}\right)$.

The obvious analogue of Theorem 6.5.1 is that the generalized equivariant Quillen metric on λ is "smooth" over $S'=S\times\mathbf{R}^{*}$. In the case where G is trivial, this was already proved in Theorem 6.5.1. In the general case, this means in particular that the sign of the various metrics remains constant or, equivalently, that the integers which express the logarithms of these signs remain constant modulo 2. The proof is exactly the same as the proof of Theorem 6.5.1.

6.7 A VARIATION FORMULA

The main result of this chapter is as follows.

Theorem 6.7.1. *The generalized metric* $\|\ \|^{2}_{\lambda,b}$ *does not depend on* b. *Moreover, the following identity of closed* 1*-forms holds on* S*:*

$$\begin{aligned}\frac{1}{2}d\log\|\ \|^{2}_{\lambda}(g)&=\int_{X_g}e\left(TX_g,\nabla^{TX_g}\right)\mathrm{Tr}^{F}\left[g\frac{1}{2}\omega\left(\nabla^{F},g^{F}\right)\right]\\ &\quad\text{for } c>0, \qquad\qquad (6.7.1)\\ &=(-1)^{n}\int_{X_g}e\left(TX_g,\nabla^{TX_g}\right)\mathrm{Tr}^{F\otimes o(TX)}\left[g\frac{1}{2}\omega\left(\nabla^{F\otimes o(TX)},g^{F\otimes o(TX)}\right)\right] \qquad (6.7.2)\\ &\quad\text{for } c<0.\end{aligned}$$

Theorem 6.7.2. *The following identity of closed* 1*-forms holds on* $S'=S\times\mathbf{R}^{*}$*:*

$$\frac{1}{2}d\log\|\ \|^{2}_{\lambda}(g)=\int_{X_g}e\left(TX_g,\nabla^{TX_g}\right)\mathrm{Tr}^{F}\left[g\frac{1}{2}\omega\left(\nabla^{F},g^{F}\right)\right]. \tag{6.7.3}$$

Proof. The remainder of the chapter is devoted to the proof of Theorem 6.7.2. □

Remark 6.7.3. By (6.7.3), we see that the metric $\| \|^2_{\lambda,b}$ is locally constant in the b variable. Then Theorem 6.7.1 follows from (1.6.4), from Theorem 6.7.2 and from the fact that only even dimensional X_g contribute to the right-hand side of (6.7.3).

Again it is remarkable that the right-hand sides of (1.12.11) and (6.7.3) coincide.

6.8 A SIMPLE IDENTITY

Proposition 6.8.1. *For any $b \in \mathbf{R}^*, t > 0$, the following identity holds:*

$$w^{(0)}_{b,t} = \underline{w}^{(0)}_{b,t} = \mathrm{Tr_s}\left[g\frac{1}{2}\left(N^{T^*X} - n\right) h'\left(\sqrt{t}B_{\phi,\mathcal{H}^c}\right)\right]$$
$$= \left(1 + 2t\frac{\partial}{\partial t}\right)\mathrm{Tr_s}\left[g\frac{1}{2}\left(N^{T^*X} - n\right)\exp\left(-tA^2_{\phi,\mathcal{H}^c}\right)\right]. \quad (6.8.1)$$

Moreover,

$$\frac{\partial}{\partial b}\frac{w^{(0)}_{b,t}}{t} = \frac{\partial}{\partial t}\frac{v^{(0)}_{b,t}}{b}. \quad (6.8.2)$$

Proof. The first part of our proposition follows from equation (2.8.8) in Theorem 2.8.1, from (4.3.8), from Proposition 4.3.4, and from (4.13.7). Moreover, (6.8.2) is a consequence of Theorem 4.3.2. □

6.9 THE PROJECTED CONNECTIONS

We denote by $\mathcal{H}^*$ the smooth function which coincides with $\mathcal{H}^c$ on $\mathcal{M} \times \{b\}$.

Now we follow [B05, section 4]. If $U \in TS$, let $U^H \in T^H\mathcal{M}$ be its horizontal lift. Then the Lie derivative operator L_{U^H} acts naturally on smooth sections of $\Omega^{\cdot}(T^*X, \pi^*F)$. If s is such a smooth section, set

$$\nabla^{\Omega^{\cdot}(T^*X,\pi^*F)}_U s = L_{U^H} s. \quad (6.9.1)$$

Then $\nabla^{\Omega^{\cdot}(T^*X,\pi^*F)}$ is a connection on $\Omega^{\cdot}(T^*X, \pi^*F)$. Put

$$\omega\left(\Omega^{\cdot}(T^*X, \pi^*F), \mathfrak{h}^{\Omega^{\cdot}(T^*X,\pi^*F)}_{\mathcal{H}^*}\right) = -E \pm \frac{2}{b^2} f^\alpha \left\langle T\left(f^H_\alpha, p\right), p\right\rangle \pm 2\frac{db}{b^3}|p|^2$$
$$+ f^\alpha \omega\left(\nabla^F, g^F\right)\left(f^H_\alpha\right),$$
$$\nabla^{\Omega^{\cdot}(T^*X,\pi^*F)*} = \nabla^{\Omega^{\cdot}(T^*X,\pi^*F)} + \omega\left(\Omega^{\cdot}(T^*X, \pi^*F), \mathfrak{h}^{\Omega^{\cdot}(T^*X,\pi^*F)}_{\mathcal{H}^*}\right), \quad (6.9.2)$$
$$\nabla^{\Omega^{\cdot}(T^*X,\pi^*F),u} = \frac{1}{2}\left(\nabla^{\Omega^{\cdot}(T^*X,\pi^*F)*} + \nabla^{\Omega^{\cdot}(T^*X,\pi^*F)}\right).$$

By [B05, eq. (4.27), Propositions 4.21 and 4.24, and eq. (4.109)], the connection $\nabla^{\Omega^{\cdot}(T^*X,\pi^*F)*}$ is the $\mathfrak{h}^{\Omega^{\cdot}(T^*X,\pi^*F)}_{\mathcal{H}^*}$-adjoint of $\nabla^{\Omega^{\cdot}(T^*X,\pi^*F)}$ over $S \times \mathbf{R}^*$. Therefore $\nabla^{\Omega^{\cdot}(T^*X,\pi^*F),u}$ preserves the generalized metric $\mathfrak{h}^{\Omega^{\cdot}(T^*X,\pi^*F)}_{\mathcal{H}^*}$.

Let $V_r \subset S \times \mathbf{R}^*$ be the open set defined as in (6.5.1). Recall that on V_r, we have the canonical isomorphism in (6.4.4).

Definition 6.9.1. Let $\nabla^{\Omega^{\cdot}(T^*X,\pi^*F)_{<r}}$ and $\nabla^{\Omega^{\cdot}(T^*X,\pi^*F)_{<r},u}$ denote the connections on $\Omega^{\cdot}(T^*X,\pi^*F)_{<r}$ over V_r which are obtained by projection of $\nabla^{\Omega^{\cdot}(T^*X,\pi^*F)}$ and $\nabla^{\Omega^{\cdot}(T^*X,\pi^*F),u}$ on $\Omega^{\cdot}(T^*X,\pi^*F)_{<r}$ with respect to the splitting (6.4.2).

By (6.4.4), the connections $\nabla^{\Omega^{\cdot}(T^*X,\pi^*F)_{<r}}, \nabla^{\Omega^{\cdot}(T^*X,\pi^*F)_{<r},u}$ induce connections $\nabla^{\lambda}_{<r}, \nabla^{\lambda,u}_{<r}$ on λ over V_r.

Recall that after (6.4.4), we defined the generalized metric $|\;|^2_{\lambda,<r}$ on the restriction of λ to V_r. Now we establish an analogue of [B05, Proposition 1.21 and eq. (1.96)]. Recall that if $g \in G$, $T_{h,g,b}\left(g^{TX},\nabla^F,g^F\right)_{>r}$ is a smooth function on V_r.

Proposition 6.9.2. *The following identity of connections holds on V_r:*

$$\nabla^{\lambda} = \nabla^{\lambda}_{<r}. \tag{6.9.3}$$

Moreover,

$$\nabla^{\lambda,u}_{<r} = \nabla^{\lambda} + \frac{1}{2}\nabla^{\lambda}\log|\;|^2_{\lambda,<r}. \tag{6.9.4}$$

Finally,

$$d\log\|\;\|^2_{\lambda} = d\log|\;|^2_{\lambda,<r} + 2dT_{h,\cdot,b}\left(g^{TX},\nabla^F,g^F\right)_{>r}. \tag{6.9.5}$$

Proof. The proof of the first part of our proposition is the same as the proof of [B05, Propositions 1.17 and 1.21]. Since the splitting (6.4.2) is $\mathfrak{h}^{\Omega^{\cdot}(T^*X,\pi^*F)}_{\mathcal{H}^c}$ orthogonal, the considerations which follow (6.9.2) show that (6.9.4) holds. Identity (6.9.5) follows from (6.6.15) and from Proposition 6.6.3. □

6.10 A PROOF OF THEOREM 6.7.2

We may and we will assume that S has dimension 1, so that $\omega^H = 0$. Also we fix $g \in G$. Given $b > 0, t > 0$, we have the identity of 1-forms on S,

$$u_{b,t} = \mathrm{Tr_s}\left[gh\left(B^{\mathcal{M}}_{b,t}\right)\right]. \tag{6.10.1}$$

By (2.8.8) and (6.9.2), we get the equality of 1-forms on S,

$$u_{b,t} = \mathrm{Tr_s}\left[g\frac{1}{2}\omega\left(\Omega^{\cdot}(T^*X,\pi^*F),\mathfrak{h}^{\Omega^{\cdot}(T^*X,\pi^*F)}_{\mathcal{H}^c}\right)h'\left(\sqrt{t}B_{\phi,\mathcal{H}^c}\right)\right]. \tag{6.10.2}$$

Since $B_{\phi,\mathcal{H}^c}$ commutes with $A^2_{\phi,\mathcal{H}^c}$, it preserves the splitting (6.4.2) of the vector space $\Omega^{\cdot}(T^*X,\pi^*F)$. Let $P_{<r}, P_{>r}$ be the spectral projectors on $\Omega^{\cdot}(T^*X,\pi^*F)_{<r}, \Omega^{\cdot}(T^*X,\pi^*F)_{>r}$. We denote with a superscript $< r$ or

$> r$ the restriction of $B_{\phi,\mathcal{H}^c}$ to one of these two vector spaces. Put

$$\begin{aligned}(6.10.3)\quad &\omega\left(\Omega^{\cdot}(T^*X,\pi^*F),\mathfrak{h}_{\mathcal{H}^c}^{\Omega^{\cdot}(T^*X,\pi^*F)}\right)_{<r}\\ =&P_{<r}\omega\left(\Omega^{\cdot}(T^*X,\pi^*F),\mathfrak{h}_{\mathcal{H}^c}^{\Omega^{\cdot}(T^*X,\pi^*F)}\right)P_{<r},\\ &\omega\left(\Omega^{\cdot}(T^*X,\pi^*F),\mathfrak{h}_{\mathcal{H}^c}^{\Omega^{\cdot}(T^*X,\pi^*F)}\right)_{>r}\\ =&P_{>r}\omega\left(\Omega^{\cdot}(T^*X,\pi^*F),\mathfrak{h}_{\mathcal{H}^c}^{\Omega^{\cdot}(T^*X,\pi^*F)}\right)P_{>r}.\end{aligned}$$

Set

$$\begin{aligned}(6.10.4)\quad u_{b,t}^{<r}&=\mathrm{Tr_s}\left[g\frac{1}{2}\omega\left(\Omega^{\cdot}(T^*X,\pi^*F),\mathfrak{h}_{\mathcal{H}^c}^{\Omega^{\cdot}(T^*X,\pi^*F)}\right)_{<r}h'\left(\sqrt{t}B_{\phi,\mathcal{H}^c}^{<r}\right)\right],\\ u_{b,t}^{>r}&=\mathrm{Tr_s}\left[g\frac{1}{2}\omega\left(\Omega^{\cdot}(T^*X,\pi^*F),\mathfrak{h}_{\mathcal{H}^c}^{\Omega^{\cdot}(T^*X,\pi^*F)}\right)_{>r}h'\left(\sqrt{t}B_{\phi,\mathcal{H}^c}^{>r}\right)\right].\end{aligned}$$

It follows from the above that

$$u_{b,t}=u_{b,t}^{<r}+u_{b,t}^{>r}. \tag{6.10.5}$$

Recall that $h'(0)=1$. Therefore, as $t\to 0$,

$$u_{b,t}^{<r}\to u_{b,0}^{<r}=\mathrm{Tr_s}\left[g\frac{1}{2}\omega\left(\Omega^{\cdot}(T^*X,\pi^*F),\mathfrak{h}_{\mathcal{H}^c}^{\Omega^{\cdot}(T^*X,\pi^*F)}\right)_{<r}\right]. \tag{6.10.6}$$

By Proposition 6.9.2, it is clear that

$$u_{b,0}^{<r}=\frac{1}{2}\nabla^{\lambda}\log\left(|\ |_{\lambda,<r}^{2}\right). \tag{6.10.7}$$

By Theorem 4.4.1, we know that as $t\to 0$,

$$u_{b,t}\to u_{b,0}=\int_{X_g}e\left(TX_g,\nabla^{TX_g}\right)\mathrm{Tr}^F\left[g\frac{1}{2}\omega\left(\nabla^F,g^F\right)\right]. \tag{6.10.8}$$

By (6.10.5), (6.10.6), (6.10.8), we conclude that as $t\to 0$, there is a 1-form $u_{b,0}^{>r}$ on V_r such that

$$u_{b,t}^{>r}\to u_{b,0}^{>r}, \tag{6.10.9}$$

so that

$$u_{b,0}=u_{b,0}^{<r}+u_{b,0}^{>r}. \tag{6.10.10}$$

Proposition 6.10.1. *The following identity holds:*

$$u_{b,0}^{>r}=dT_{h,b,g}\left(g^{TX},g^F,\mathcal{H}^{1/b^2}\right)_{>r}. \tag{6.10.11}$$

Proof. We define $\underline{w}_{b,t}^{>r}$ as in Definition 4.3.3:

$$\underline{w}_{b,t}^{>r}=\mathrm{Tr_s}\left[g\frac{1}{2}\left(N^{T^*X}-n\right)h'\left(\sqrt{t}B_{\phi,\mathcal{H}^c}^{>r}\right)\right]. \tag{6.10.12}$$

Note that here, $\underline{w}_{b,t}^{>r}$ is a smooth function on V_r. We claim that

$$\frac{\partial}{\partial t} u_{b,t}^{>r} = d \frac{\underline{w}_{b,t}^{>r}}{t}. \tag{6.10.13}$$

Indeed if the superscript $> r$ was omitted, this identity would just be a consequence of Theorem 4.3.2. Recall that since d^{T^*X} commutes with $B^2_{\phi,\mathcal{H}^{1/b^2}}$, d^{T^*X} acts on $\Omega^{\cdot}(T^*X, \pi^*F)_{>r}$. Let $d^{T^*X>r}$ be the restriction of d^{T^*X} to $\Omega^{\cdot}(T^*X, \pi^*F)_{>r}$. Since S is one dimensional,

$$A'^{\mathcal{M}>r} = d^{T^*X>r} + \nabla^{\Omega^{\cdot}(T^*X,\pi^*F)_{>r}}$$

is still a flat superconnection on $\Omega^{\cdot}(T^*X, \pi^*F)_{>r}$. The proof of (6.10.13) continues as the proof of Theorem 4.3.2. Needless to say, we could as well replace $> r$ by $< r$ and obtain (6.10.13) by difference.

Recall that we can take $r > 0$ large enough so that if $\lambda \in \mathrm{Sp} A^2_{\phi,\mathcal{H}^c}$, then $\mathrm{Re}\,\lambda \geq 1$. By proceeding as in (5.4.3), there are $c > 0, C > 0$ such that if $s \in V_r$, near s, for $t \geq 1$,

$$\left\| \exp\left(-tA^{>r,2}_{\phi,\mathcal{H}^c}\right)\right\|_1 \leq Ce^{-ct}. \tag{6.10.14}$$

From (6.10.4), (6.10.12), (6.10.14), we deduce that as $t \to +\infty$, $u_{b,t}^{>r}$ and $\underline{w}_{b,t}^{>r}$ tend to 0 exponentially fast near $s \in V_r$. By (6.10.13), we get, for $t > 0$,

$$u_{b,t}^{>r} = -d \int_t^{\infty} \underline{w}_{b,u}^{>r} \frac{du}{u}. \tag{6.10.15}$$

Now we will make $t \to 0$ in (6.10.15). As we saw in (6.10.9), the left-hand side has a limit as $t \to 0$. We can define $\underline{w}_{b,t}^{<r}$ as in (6.10.12), and we have an analogue of (6.10.5) for $\underline{w}_{b,t}$. Put

$$\underline{w}_{b,0}^{<r} = \mathrm{Tr_s}\left[g\frac{1}{2}\left(N^{T^*X} - n\right)P_{<r}\right]. \tag{6.10.16}$$

Then $\underline{w}_{b,0}^{<r}$ is locally constant on V_r. Moreover, as $t \to 0$,

$$\underline{w}_{b,t}^{<r} = \underline{w}_{b,0}^{<r} + \mathcal{O}(t). \tag{6.10.17}$$

By using Theorem 4.4.1, (4.3.10) and (6.10.17), we find that as $t \to 0$,

$$\underline{w}_{b,t}^{>r} = -\underline{w}_{b,0}^{<r} + \mathcal{O}\left(\sqrt{t}\right). \tag{6.10.18}$$

From (6.10.17), we find that in the integral in the right-hand side of (6.10.15), there is a logarithmic divergence as $t \to 0$, which, being a locally constant term, is killed by the operator d.

By (4.13.7),

$$\mathrm{Tr_s}\left[gh'\left(\sqrt{t}B^{>r}_{\phi,\mathcal{H}^c}\right)\right] = \left(1 + 2t\frac{\partial}{\partial t}\right)\mathrm{Tr_s}\left[g\exp\left(tB^{>r,2}_{\phi,\mathcal{H}^c}\right)\right]. \tag{6.10.19}$$

By proceeding as in the proof of Theorem 4.2.1, $\mathrm{Tr_s}\left[g\exp\left(tB^{>r,2}_{\phi,\mathcal{H}^c}\right)\right]$ does not depend on t. Using (6.10.14) and making $t \to +\infty$, we find that this quantity is just 0. So by (6.10.19), we get

$$\mathrm{Tr_s}\left[gh'\left(\sqrt{t}B^{>r}_{\phi,\mathcal{H}^c}\right)\right] = 0. \tag{6.10.20}$$

By (6.3.19) and using the observation which follows, which guarantees that $a' = 0$ in the right-hand side of (6.3.19), we find that as $t \to 0$,

$$\frac{1}{2}\mathrm{Tr_s}\left[gN^{T^*X}\exp\left(tB^{2,>r}_{\phi,\mathcal{H}^c}\right)\right] = \frac{\alpha}{\sqrt{t}} + \beta + \mathcal{O}\left(\sqrt{t}\right), \tag{6.10.21}$$

and $\beta = -\underline{w}^{<r}_{b,0}$ is locally constant on V_r. Also observe that by (6.10.14), there is $c > 0$ such that as $t \to +\infty$,

$$\frac{1}{2}\mathrm{Tr_s}\left[gN^{T^*X}\exp\left(tB^{2,>r}_{\phi,\mathcal{H}^c}\right)\right] = \mathcal{O}\left(e^{-ct}\right). \tag{6.10.22}$$

By (4.13.7), we get

$$\underline{w}^{>r}_{b,t} = \left(1 + 2t\frac{\partial}{\partial t}\right)\frac{1}{2}\mathrm{Tr_s}\left[gN^{T^*X}\exp\left(tB^{2,>r}_{\phi,\mathcal{H}^c}\right)\right]. \tag{6.10.23}$$

By (6.10.15), (6.10.21), (6.10.22) and (6.10.23), we obtain

$$u^{>r}_{b,t} = -d\Big(\int_t^{+\infty}\frac{1}{2}\mathrm{Tr_s}\left[gN^{T^*X}\exp\left(uB^{2,>r}_{\phi,\mathcal{H}^{1/b^2}}\right)\right]\frac{du}{u} - \mathrm{Tr_s}\left[gN^{T^*X}\exp\left(tB^{2,>r}_{\phi,\mathcal{H}^c}\right)\right]\Big). \tag{6.10.24}$$

By (6.10.9), (6.10.21), (6.10.24), and using the fact that β is locally constant, we get

$$u^{>r}_{b,0} = -d\Big(\int_0^1\left(\frac{1}{2}\mathrm{Tr_s}\left[gN^{T^*X}\exp\left(uB^{2,>r}_{\phi,\mathcal{H}^{1/b^2}}\right)\right] - \frac{\alpha}{\sqrt{u}} - \beta\right)\frac{du}{u} - 2\alpha - \int_1^{+\infty}\frac{1}{2}\mathrm{Tr_s}\left[gN^{T^*X}\exp\left(uB^{2,>r}_{\phi,\mathcal{H}^c}\right)\right]\frac{du}{u}\Big). \tag{6.10.25}$$

Using (1.8.10), (6.10.14), (6.10.21), and (6.10.23), we have the easy formula

$$\mathcal{T}_{h,g,b}\left(g^{TX}, g^F, g^F\right)_{>r} = -\int_0^1\left(\frac{1}{2}\mathrm{Tr_s}\left[gN^{T^*X}\exp\left(uB^{2,>r}_{\phi,\mathcal{H}^{1/b^2}}\right)\right] - \frac{\alpha}{\sqrt{u}} - \beta\right)\frac{du}{u} + 2\alpha - \int_1^{+\infty}\frac{1}{2}\mathrm{Tr_s}\left[gN^{T^*X}\exp\left(uB^{2,>r}_{\phi,\mathcal{H}^{1/b^2}}\right)\right]\frac{du}{u}\Big) + \left(\Gamma'(1) + 2\log(2)\right)\beta. \tag{6.10.26}$$

Since β is locally constant, (6.10.11) follows from (6.10.25) and (6.10.26). The proof of our proposition is completed. □

By (6.9.5), (6.10.5)-(6.10.11), we get (6.7.1).

Now we briefly show how to establish that $\| \|^2_{\lambda,b}$ does not depend on b. Here we can take S to be a point. The discussion is then exactly the same as before. We still use Theorem 4.3.2 in the form given in (6.8.2) and we exploit the second identity in (4.4.1). The combination of these two facts permits us to complete the proof of Theorem 6.7.2. □

Chapter Seven

The hypoelliptic torsion forms of a vector bundle

The purpose of this chapter is to calculate the hypoelliptic torsion forms of a real Euclidean vector bundle equipped with a Euclidean connection. These explicit computations will play a key role in establishing the formula which compares the elliptic to the hypoelliptic torsion forms. The fact that our computations are closely related to computations in [BG01, section 4] in the elliptic case can be considered as a microlocal version of our comparison formula.

The present chapter is also related in spirit with similar computations which were done in a holomorphic context in [B90, B94].

This chapter is organized as follows. In section 7.1, we introduce a key function $\tau(c, \eta, x)$.

In section 7.2, we give the Weitzenböck formula for the hypoelliptic curvature in the case of a vector bundle E.

In section 7.3, we establish a translation invariance property of the hypoelliptic Laplacian.

In section 7.4, we equip E with a flat isometry g, and we consider the corresponding eigenbundle decomposition of E.

In section 7.5, we define a von Neumann supertrace of the corresponding heat kernel on $E \oplus E^*$.

In section 7.6, we give a probabilistic expression for the heat kernel.

In section 7.7, certain finite dimensional supertraces are expressed in terms of infinite determinants.

In section 7.8, we complete the evaluation of the supertraces of the heat kernel in terms of the traces of certain operators acting on the circle.

In section 7.9, some extra computations are done on related supertraces.

In section 7.10, the Mellin transforms of certain Fourier series are introduced.

Finally, in section 7.11, the hypoelliptic torsion forms of the vector bundle E are evaluated.

7.1 THE FUNCTION $\tau(\mathbf{c}, \eta, \mathbf{x})$

Take $x \in \mathbf{C}, c \in \mathbf{R}^*$. Let $P_{c,x}(\lambda)$ be the polynomial

$$P_{c,x}(\lambda) = -\lambda^3 + \frac{c^2}{4}(\lambda + x). \qquad (7.1.1)$$

Note that

$$P_{c,-x}(-\lambda) = -P_{c,x}(\lambda). \tag{7.1.2}$$

Let $\lambda_1, \lambda_2, \lambda_3$ be the roots of $P_{c,x}(\lambda)$. Take $\eta \in \mathbf{C}$.

Definition 7.1.1. Set

$$\tau(c,\eta,x) = \prod_{i=1}^{3} 2\sinh\left(\frac{\lambda_i+\eta}{2}\right). \tag{7.1.3}$$

By (7.1.2),

$$\tau(c,-\eta,-x) = -\tau(c,\eta,x). \tag{7.1.4}$$

Also observe that when changing η into $\eta + 2i\pi$, $\tau(c,\eta,x)$ is changed into $-\tau(c,\eta,x)$.

Let L^2 be the vector space of square integrable complex functions on $[0,1]$. The operator $J = \frac{d}{dt}$ with periodic boundary conditions acts as an unbounded antisymmetric operator on L^2, with simple eigenvalues $2ik\pi, k \in \mathbf{Z}$.

Take $z \in \mathbf{C}, |z| = 1$, so that $z = e^{i\theta}, \theta \in \mathbf{R}/2\pi\mathbf{Z}$. Let $\mathcal{C}_z([0,1],\mathbf{C})$ be the vector space of smooth functions f defined on $[0,1]$ with values in $\mathbf{C}$ such that $f_1 = zf_0$. Then the operator $\frac{d}{dt}$ acting on $\mathcal{C}_z([0,1],\mathbf{C})$ has a unique skew-adjoint extension, which will be denoted J_z. Observe that if $T_\theta \in \operatorname{End}(L^2)$ is given by $f_t \to e^{it\theta} f$, then

$$T_\theta^{-1} J_z T_\theta = J + i\theta, \tag{7.1.5}$$

where the domain of the operator in the right-hand side of (7.1.5) consists of the standard periodic functions on $[0,1]$. Moreover, the spectrum of J_z is just $i(2k\pi+\theta), k \in \mathbf{Z}$. Finally, note that when $z = 1$, we will use the notation J instead of J_1.

In the sequel, we will consider determinants of operators acting on L^2. These determinants will always be infinite products over $k \in \mathbf{Z}$. The considered products either will be obviously convergent or will converge when considering products of the type $\prod_{-M\le k\le M}$ and making M tend to $+\infty$. Also note that expressions like J^{-1} or $J_{e^{i\theta}}^{-1}$ will appear. It will be implicitly assumed that such operators are restricted to the direct sums of eigenspaces where J or $J_{e^{i\theta}}$ is invertible. In the case where such restricted determinants appear, they will be signaled by $\det^*$, instead of the usual det. Similarly, the trace of operators of this kind will be written as Tr^*.

Theorem 7.1.2. *If $\theta \in \mathbf{R}$, the following identity holds:*

$$\tau(c,i\theta,x) = P_{c,x}(-i\theta)\prod_{k\in\mathbf{Z}^*}\frac{P_{c,x}(-i(2k\pi+\theta))}{(2ik\pi)^3}. \tag{7.1.6}$$

Moreover, if $k \in \mathbf{Z}$,

$$\tau(c,2ik\pi,x) = (-1)^k \frac{c^2}{4}x \det{}^*\left(1 - \frac{c^2}{4}J^{-2} + \frac{c^2}{4}xJ^{-3}\right). \tag{7.1.7}$$

If $\theta \notin 2\pi\mathbf{Z}$, then

$$\tau(c, i\theta, x) = \left(2\sinh\left(\frac{i\theta}{2}\right)\right)^3 \det\left(1 - \frac{c^2}{4} J_{e^{i\theta}}^{-2} + \frac{c^2}{4} x J_{e^{i\theta}}^{-3}\right). \tag{7.1.8}$$

Proof. We have the well-known formula

$$2\sinh(y/2) = y \prod_{k \in \mathbf{Z}^*} \frac{2ik\pi + y}{2ik\pi}. \tag{7.1.9}$$

By (7.1.1), (7.1.3), (7.1.9), we get (7.1.6). Equation (7.1.7) follows trivially when $k = 0$, and by antiperiodicity, we get (7.1.7) in full generality.

If $\theta \notin 2\pi\mathbf{Z}$, then

$$\prod_{k \in \mathbf{Z}^*} \frac{P_{c,x}(-i(2k\pi + \theta))}{(2ik\pi)^3} = \prod_{k \in \mathbf{Z}^*} \left(1 + \frac{\theta}{2k\pi}\right)^3 \prod_{k \in \mathbf{Z}^*} \left(1 - \frac{c^2}{4(2ik\pi + i\theta)^2} + \frac{c^2 x}{4(2ik\pi + i\theta)^3}\right). \tag{7.1.10}$$

Moreover,

$$P_{c,x}(-i\theta) = (i\theta)^3 \left(1 - \frac{c^2}{4(i\theta)^2} + \frac{c^2 x}{4(i\theta)^3}\right). \tag{7.1.11}$$

By combining (7.1.6) with (7.1.9)-(7.1.11), we get (7.1.8). The proof of our theorem is completed. □

7.2 HYPOELLIPTIC CURVATURE FOR A VECTOR BUNDLE

Let S be a smooth manifold. Let $\pi : E \to S$ be a real vector bundle of dimension n on S. Let g^E be a Euclidean metric on E, let ∇^E be a Euclidean connection on E, and let R^E be the curvature of ∇^E.

Let M^E be the total space of E. Now, we will use the formalism of section 2.4, with $M = M^E$, and $X = E$. Indeed observe that the connection ∇^E induces a horizontal vector bundle $T^H M^E \subset TM^E$, so that

$$TM^E = T^H M^E \oplus E. \tag{7.2.1}$$

Moreover, $TX = E$ is equipped with the metric g^E. So the assumptions of section 2.4 are verified.

Let y be the generic element of E. One verifies easily that the tensor T of (1.3.1) is purely horizontal, and that if $U, V \in TS$,

$$T\left(U^H, V^H\right) = R^E(U, V)\, y. \tag{7.2.2}$$

Clearly,

$$T^*X = E \oplus E^*. \tag{7.2.3}$$

Then $\mathcal{M}^E$, the total space of T^*X, is just the total space of $E \oplus E^*$.

The connection ∇^E induces the dual connection ∇^{E^*} on E^*, so that $E \oplus E^*$ is equipped with the connection $\nabla^E \oplus \nabla^{E^*}$, which is metric compatible. It induces a horizontal subbundle $T^H \mathcal{M}^E$ on $\mathcal{M}^E$, which is just the one which was considered in [B05, subsection 4.5] and in section 2.4.

Let $e_1, \ldots, e_n$ be a basis of E, and let $e^1, \ldots, e^n$ be the corresponding dual basis. Let $\widehat{e}^1, \ldots, \widehat{e}^n$ denote the associated basis of E^* (in $E \oplus E^*$), and let $\widehat{e}_1, \ldots, \widehat{e}_n$ be the corresponding dual basis. The fiberwise symplectic form on $T^*X = E \oplus E^*$ is just

$$\omega^V = \widehat{e}_i \wedge e^i. \tag{7.2.4}$$

We will identify the $e^i, \widehat{e}_i$ with the corresponding vertical 1-forms on $\mathcal{M}^E$.

Let p be the generic element of E^*, so that (y, p) is the generic element of $T^*X = E \oplus E^*$. The form θ of section 2.4 is just the horizontal form on M^E associated to the canonical section p. Using (2.4.2), (2.4.3) or a simple direct computation,

$$\omega = \omega^V + \left\langle R^E y, p \right\rangle . \tag{7.2.5}$$

From now on, $e_1, \ldots, e_n$ will be supposed to be an orthonormal basis of E. Let $c \in \mathbf{R}$. We will give formulas for some of the operators considered in section 2.5. Here we will take F to the trivial flat Euclidean vector bundle $\mathbf{R}$ on S, so that $\omega\left(\nabla^F, g^F\right) = 0$.

Let $\Omega^{\cdot}(E \oplus E^*)$ be the vector space of smooth sections of $\Lambda^{\cdot}(E^* \oplus E^*) \otimes \Lambda^n(E)$ along the fibers $E \oplus E^*$. The operators which we will consider act on smooth sections of $\Omega^{\cdot}(E \oplus E^*)$.

If $p \in E^*$, recall that p can be identified with a corresponding element in E. Then ∇^E_p denotes the associated fiberwise differentiation operator along the fibers of E. Also Δ^V still denotes the Laplacian along the fibers of E^*. Here that $c \in \mathbf{R}^*$ is allowed to vary, so that $dc \in \Lambda^1(\mathbf{R})$.

Theorem 7.2.1. *The following identities hold:*

$$\begin{aligned} \underline{\widehat{\mathfrak{C}}}^{\mathcal{M}^E,2}_{\phi,\mathcal{H}^c-\omega^H} &= \frac{1}{4}\left(-\Delta^V + c^2 |p|^2 + c\left(2 i_{\widehat{e}_i} \widehat{e}^i - n\right)\right) + \frac{1}{4}\left\langle e_i, R^E e_j \right\rangle \widehat{e}^i \widehat{e}^j \\ &- \frac{c}{2}\left(\nabla^E_p + i_{\widehat{e}_i}\left(e^i + i_{e_i}\right) + \left\langle R^E y, p\right\rangle\right) + \frac{1}{2} dc\left(\widehat{p} - p - i_p\right), \qquad (7.2.6) \\ \underline{\widehat{\mathfrak{G}}}^{\mathcal{M}^E}_{\phi,\mathcal{H}^c-\omega^H} &= -\frac{c}{2}\left(\widehat{p} + 6 i_{\widehat{p}}\right) - \frac{3}{2} dc\, |p|^2, \\ \underline{\widehat{\mathfrak{G}}}^{\mathcal{M}^E}_{\phi,\mathcal{H}^c-\omega^H} &+ \left[\underline{\widehat{\mathfrak{C}}}'^{\mathcal{M}^E}_{\mathcal{H}^c-\omega^H}, \frac{3c}{2}|p|^2\right] = -\frac{c}{2}\widehat{p}. \end{aligned}$$

Proof. Our theorem follows from (2.5.14), (2.5.17), from Theorem 2.5.4, and from (7.2.2). □

7.3 TRANSLATION INVARIANCE OF THE CURVATURE

Take $y_0 \in E$. Let T_{y_0} be the operator

$$s(y, p) \to T_{y_0} s(y, p) = s(y + y_0, p) . \tag{7.3.1}$$

Proposition 7.3.1. *The following identity holds:*

$$T_{y_0}\underline{\widehat{\mathfrak{L}}}^{\mathcal{M}^E,2}_{\phi,\mathcal{H}^c-\omega^H}T_{y_0}^{-1} = \exp\left(-\left\langle R^E y_0, y\right\rangle\right)\underline{\widehat{\mathfrak{L}}}^{\mathcal{M}^E,2}_{\phi,\mathcal{H}^c-\omega^H}\exp\left(\left\langle R^E y_0, y\right\rangle\right). \quad (7.3.2)$$

Proof. This is a trivial consequence of Theorem 7.2.1. □

7.4 AN AUTOMORPHISM OF E

Let g be an automorphism of the vector bundle E, which preserves the metric g^E, and is parallel with respect to ∇^E. Let $e^{\pm i\theta}, 0 < \theta < \pi$ be the distinct nonreal eigenvalues of g, the other possible eigenvalues being 1 and -1. These eigenvalues are locally constant on S. Let $E^{e^{\pm i\theta}}, E^1, E^{-1}$ be the corresponding eigenbundles. We get the orthogonal splitting

$$E\otimes_{\mathbf{R}}\mathbf{C} = E^1\otimes_{\mathbf{R}}\mathbf{C}\oplus E^{-1}\otimes_{\mathbf{R}}\mathbf{C}\oplus\bigoplus_{0<\theta<\pi}E^{e^{i\theta}}\oplus E^{e^{-i\theta}}. \quad (7.4.1)$$

Note that E^1, E^{-1} are real vector bundles, and that $E^{e^{i\theta}}\oplus E^{e^{-i\theta}}$ is the complexification of a real vector bundle $E_{\mathbf{R}}^{e^{\pm i\theta}}$. By (7.4.1), we have the real splitting

$$E = E^1\oplus E^{-1}\oplus\bigoplus_{0<\theta<\pi}E_{\mathbf{R}}^{e^{\pm i\theta}}. \quad (7.4.2)$$

Finally, observe that the connection ∇^E preserves the splittings (7.4.1), (7.4.2).

Let $\Omega^{\cdot}(E\oplus E^*)$ be the vector space of smooth sections of $\Lambda^{\cdot}(E^*\oplus E)$ along $E\oplus E^*$.

We will now assume that in the operators $\underline{\widehat{\mathfrak{L}}}^{\mathcal{M}^E,2}_{\phi,\mathcal{H}^c-\omega^H}, \underline{\widehat{\mathfrak{L}}}^{\prime\mathcal{M}^E}_{\mathcal{H}^c-\omega^H}, \underline{\widehat{\mathfrak{G}}}^{\mathcal{M}^E}_{\phi,\mathcal{H}^c-\omega^H}$, the Grassmann variables $\widehat{e}^i, i_{\widehat{e}_i}$ have been replaced by $i_{\widehat{e}^i}, \widehat{e}_i$. Still these operators will be denoted as before. These operators now act on $\Omega^{\cdot}(E\oplus E^*)$.

Clearly, g acts on $\Omega^{\cdot}(E\oplus E^*)$ and commutes with $\underline{\widehat{\mathfrak{L}}}^{\mathcal{M}^E,2}_{\phi,\mathcal{H}^c-\omega^H}$. Let z be an odd Grassmann variable, which anticommutes with all the other odd variables. Set

$$\mathfrak{L}_c^E = \underline{\widehat{\mathfrak{L}}}^{\mathcal{M}^E,2}_{\phi,\mathcal{H}^c-\omega^H} - z\left(\underline{\widehat{\mathfrak{G}}}^{\mathcal{M}^E}_{\phi,\mathcal{H}^c-\omega^H} + \left[\underline{\widehat{\mathfrak{L}}}^{\prime\mathcal{M}^E}_{\mathcal{H}^c-\omega^H}, \frac{3c}{2}|p|^2\right]\right). \quad (7.4.3)$$

Using Theorem 7.2.1, we get

$$\mathfrak{L}_c^E = \underline{\widehat{\mathfrak{L}}}^{\mathcal{M}^E,2}_{\phi,\mathcal{H}^c-\omega^H} + z\frac{c}{2}i_{\widehat{p}}. \quad (7.4.4)$$

Let dy, dp be the volumes along the fibers of E, E^*. Let $Q_c((y,p),(y',p'))$ be the smooth kernel associated to $\exp\left(-\underline{\widehat{\mathfrak{L}}}^{\mathcal{M}^E,2}_{\phi,\mathcal{H}^c-\omega^H}\right)$ and the volume $dydp$.

The kernel associated to $g\exp\left(-\underline{\widehat{\mathfrak{L}}}^{\mathcal{M}^E,2}_{\phi,\mathcal{H}^c-\omega^H}\right)$ is $g_*Q_c\left(g^{-1}(y,p),(y',p')\right)$.

Now we will rewrite the operator $\mathfrak{L}_c^E$ using the notation of section 1.1. By Theorem 6.1.2, we get

$$\mathfrak{L}_c^E = \frac{1}{4}\left(-\Delta^V + c^2 |p|^2 + c\left(2\widehat{e}_i i_{\widehat{e}^i} - n\right)\right) + \frac{1}{4}\left\langle e_i, R^E e_j\right\rangle i_{\widehat{e}^i} i_{\widehat{e}^j}$$
$$- \frac{c}{2}\left(\nabla_p^E + \widehat{e}_i \widehat{c}(e_i) + \left\langle R^E y, p\right\rangle\right) + \frac{1}{2} dc\left(i_{\widehat{p}} - \widehat{c}(p)\right) + z\frac{c}{2} i_{\widehat{p}}. \quad (7.4.5)$$

Note that in (7.4.5), the Clifford variables $\widehat{c}(e_i)$ anticommute with the $\widehat{e}_i, i_{\widehat{e}^i}$. The fact they both wear hats should not make them related in any way.

7.5 THE VON NEUMANN SUPERTRACE OF $\exp\left(-\mathfrak{L}_c^E\right)$

In the sequel, we suppose $c \neq 0$. A first crucial observation is that in general, the operator $\exp\left(-\mathfrak{L}_c^E\right)$ is not trace class. One evidence for this is that, by Proposition 7.3.1, $\mathfrak{L}_c^E$ is essentially translation-invariant by translations in E. Also observe that for $1 \leq i \leq n$,

$$c(e_i)\widehat{c}(e_i) = 2e^i i_{e_i} - 1. \quad (7.5.1)$$

By (7.5.1), we find that among the monomials in the $c(e_i), \widehat{c}(e_i)$ acting on $\Lambda^{\cdot}(E^*)$, up to permutation, only $c(e_1)\widehat{c}(e_1)\dots c(e_n)\widehat{c}(e_n)$ has a nonzero supertrace, and moreover,

$$\mathrm{Tr_s}\left[c(e_1)\widehat{c}(e_1)\dots c(e_n)\widehat{c}(e_n)\right] = (-2)^n. \quad (7.5.2)$$

By (7.5.2), since no Clifford variable $c(e_i)$ appears in the right-hand side of (7.4.5),

$$\mathrm{Tr_s}^{\Lambda^{\cdot}(E^* \oplus E)}\left[Q_c\left((y,p),(y,p)\right)\right] = 0. \quad (7.5.3)$$

To overcome the above difficulties, we describe a recipe to produce a "natural" von Neumann supertrace of the operator $g\exp\left(-\mathfrak{L}_c^E\right)$. This recipe is inspired from [B90, section 4], [B94, section 2].

Let $o(E)$ be the orientation line bundle of E. Let $\widehat{c}(E)$ be the algebra spanned by the $\widehat{c}(e_i), 1 \leq i \leq n$. If $A \in \widehat{c}(E)$, A can be expanded as a sum of monomials in the $\widehat{c}(e_i)$. Let $\lambda(A) \in o(E)$ be the coefficient of $\widehat{c}(e_1)\dots\widehat{c}(e_n)$ in this expansion. The map $A \in \widehat{c}(E) \to \lambda(A)$ is a supertrace in the sense that if $A, B \in \widehat{c}(E)$, then $[A, B]$ maps to 0.

First suppose that $g = 1$, so that $E = E^1$. We expand $Q\left((y,p),(y,p)\right)$ as a sum of monomials in the $\widehat{c}(e_i)$. We denote by $Q_c^\circ\left((y,p),(y,p)\right)$ the coefficient of $(-1)^{n(n+1)/2}\widehat{c}(e_1)\dots\widehat{c}(e_n)$ so that

$$Q_c\left((y,p),(y,p)\right) = \dots + Q_c^\circ\left((y,p),(y,p)\right)(-1)^{n(n+1)/2}\widehat{c}(e_1)\dots\widehat{c}(e_n). \quad (7.5.4)$$

Note that

$$Q_c^\circ\left((y,p),(y,p)\right) \in \mathrm{End}\left(\Lambda^{\cdot}(E)\right)\widehat{\otimes}\Lambda^{\cdot}(T^*S)\widehat{\otimes}\mathbf{R}[z,dc]\widehat{\otimes}o(E),$$

and so

$$\mathrm{Tr_s}^{\Lambda^{\cdot}(E)}\left[Q_c^\circ\left((y,p),(y,p)\right)\right] \in \Lambda^{\cdot}(T^*S)\widehat{\otimes}\mathbf{R}[z,dc]\widehat{\otimes}o(E).$$

By Proposition 7.3.1, $\mathrm{Tr_s}^{\Lambda^{\cdot}(E)}\left[Q_c^{\circ}\left((y,p),(y,p)\right)\right]$ is invariant by translations in $y \in E$, so that we can as well take $y = 0$.

Finally, it is easy to verify that as $|p| \to +\infty$, $Q_c\left((0,p),(0,p)\right)$ decays like a Gaussian as $|p| \to +\infty$.

Definition 7.5.1. We define the von Neumann supertrace $\mathrm{Tr_s}\left[\exp\left(-\mathfrak{L}_c^E\right)\right]$ by the formula

$$\mathrm{Tr_s}\left[\exp\left(-\mathfrak{L}_c^E\right)\right] = \int_{E^*} \mathrm{Tr_s}^{\Lambda^{\cdot}(E)}\left[Q_c^{\circ}\left((0,p),(0,p)\right)\right] dp. \tag{7.5.5}$$

Note that $\mathrm{Tr_s}\left[\exp\left(-\mathfrak{L}_c^E\right)\right]$ is a smooth section of $\Lambda^{\cdot}(T^*S)\widehat{\otimes}\mathbf{R}[z,dc]\widehat{\otimes}o(E)$.

Assume now that no eigenvalue of g is equal to 1. In this case,

$$\mathrm{Tr_s}^{\Lambda^{\cdot}(E^*\oplus E)}\left[gQ_c\left(g^{-1}(y,p),(y,p)\right)\right]$$

no longer vanishes identically for trivial reasons, since, in general, g contains the missing $c(e_i), 1 \le i \le n$.

Moreover, one verifies that as $|(y,p)| \to +\infty$, $Q_c\left(g^{-1}(y,p),(y,p)\right)$ still exhibits a Gaussian-like decay. We will give a short probabilistic proof for that. Many arguments have already been used in chapter 4. Indeed the dampening factor $c^2|p|^2$ ensures the proper Gaussian decay as $|p| \to +\infty$.

Gaussian decay as $|y| \to +\infty$ is subtler and still uses this dampening factor. Indeed for large $|y|$, $|gy - y|$ is large. Because the dynamics of the underlying path integral is determined by an equation like the one we will write in (7.5.1), this means that $\sup_{0\le t\le 1}|p_t| \simeq |y|$. If $|p| \simeq |y|$, Gaussian decay in p guarantees Gaussian decay in y. If $|p| \le c|y|/2$, an estimate like (4.7.22) still guarantees Gaussian decay when $|y| \to +\infty$.

Definition 7.5.2. If no eigenvalue of g is equal to 1, we define the von Neumann trace $\mathrm{Tr_s}\left[g\exp\left(-\mathfrak{L}_c^E\right)\right]$ by the formula

$$\mathrm{Tr_s}\left[g\exp\left(-\mathfrak{L}_c^E\right)\right] = \int_{E\oplus E^*} \mathrm{Tr_s}^{\Lambda^{\cdot}(E^*\oplus E)}\left[gQ_c\left(g^{-1}(y,p),(y,p)\right)\right] dydp. \tag{7.5.6}$$

In this case, $\mathrm{Tr_s}\left[g\exp\left(-\mathfrak{L}_c^E\right)\right]$ is a smooth section of $\Lambda^{\cdot}(T^*S)\widehat{\otimes}\mathbf{R}[z,dc]$.

In the general case, we use the splitting (7.4.1) of $E\otimes_{\mathbf{R}}\mathbf{C}$ to define the von Neumann supertrace $\mathrm{Tr_s}\left[g\exp\left(-\mathfrak{L}_c^E\right)\right]$ by combining the above techniques. Indeed if $e_1,\dots,e_n$ is such that $e_1,\dots,e_m$ is an orthonormal basis of E^1, only the Clifford variables $\widehat{c}(e_i), 1 \le i \le m$ receive a special treatment. Similarly, if $E^{1,\perp}$ denotes the orthogonal bundle to E^1 in E, in the obvious extension of (7.5.5), (7.5.6), the variable (y,p) will be integrated on $E^{1,\perp}\oplus E^*$. Details are easy to fill in, and are left to the reader. Note that in the general case, $\mathrm{Tr_s}\left[g\exp\left(-\mathfrak{L}_c^E\right)\right]$ is a smooth section of $\Lambda^{\cdot}(T^*S)\widehat{\otimes}\mathbf{R}[z,dc]\widehat{\otimes}o\left(E^1\right)$.

Let $\mathfrak{L}_c^1, \mathfrak{L}_c^{-1}, \mathfrak{L}_c^{e^{\pm i\theta}}$ be the operators $\mathfrak{L}_c^E$ attached to $E^1, E^{-1}, E_{\mathbf{R}}^{e^{\pm i\theta}}$.

Proposition 7.5.3. *The following identity holds:*

$$\mathrm{Tr_s}\left[g\exp\left(-\mathfrak{L}_c^E\right)\right] = \mathrm{Tr_s}\left[\exp\left(-\mathfrak{L}_c^1\right)\right]\mathrm{Tr_s}\left[g\exp\left(-\mathfrak{L}_c^{-1}\right)\right] \prod_{0<\theta<\pi} \mathrm{Tr_s}\left[g\exp\left(-\mathfrak{L}_c^{e^{\pm i\theta}}\right)\right]. \tag{7.5.7}$$

Proof. This is a trivial consequence of the fact that the operator $\mathfrak{L}_c^E$ splits according to (7.4.2). □

If $a \in \mathbf{R}$, let $H_a \in \operatorname{End}(\Omega^{\cdot}(E \oplus E^*))$ be given by

$$s(y,p) \to H_a s(y,p) = s(y,ap). \tag{7.5.8}$$

Put

$$\mathfrak{L}_c^{E\prime} = H_{1/\sqrt{2}} \mathfrak{L}_c^E H_{\sqrt{2}}. \tag{7.5.9}$$

By (7.4.5),

$$\mathfrak{L}_c^{E\prime} = \frac{1}{2}\left(-\Delta^V + \frac{c^2}{4}|p|^2 + \frac{c}{2}\left(2\widehat{e}_i i_{\widehat{e}^i} - n\right)\right) + \frac{1}{4}\left\langle e_i, R^E e_j\right\rangle i_{\widehat{e}^i} i_{\widehat{e}^j}$$
$$- \frac{c}{2}\left(\frac{1}{\sqrt{2}}\nabla_p^E + \widehat{e}_i \widehat{c}(e_i) + \frac{1}{\sqrt{2}}\left\langle R^E y, p\right\rangle\right) + \frac{1}{2\sqrt{2}} dc\left(i_{\widehat{p}} - \widehat{c}(p)\right) + z\frac{c}{2\sqrt{2}} i_{\widehat{p}}. \tag{7.5.10}$$

Let $Q_c'((y,p),(y',p'))$ be the smooth kernel associated to $\exp\left(-\mathfrak{L}_c^{E\prime}\right)$. We define $\operatorname{Tr_s}\left[g\exp\left(-\mathfrak{L}_c^{E\prime}\right)\right]$ just as before, using instead the kernel Q_c'. One verifies easily that

$$\operatorname{Tr_s}\left[g\exp\left(-\mathfrak{L}_c^{E}\right)\right] = \operatorname{Tr_s}\left[g\exp\left(-\mathfrak{L}_c^{E\prime}\right)\right]. \tag{7.5.11}$$

Replacing $\mathfrak{L}_c^E$ by $\mathfrak{L}_c^{E\prime}$ is done simply for the convenience of later references.

7.6 A PROBABILISTIC EXPRESSION FOR Q_c'

Let $q((y,p),(y',p'))$ be the smooth kernel for $\exp\left(\frac{\Delta^V}{2} + \frac{c}{2\sqrt{2}}\nabla_p^E\right)$. Given $(y_0,p_0) \in E \oplus E^*$, let $R_{(y_0,p_0)}$ be the probability law on $\mathcal{C}([0,1], E \oplus E^*)$ of the hypoelliptic diffusion process (y_t,p_t), whose infinitesimal generator is $\frac{\Delta^V}{2} + \frac{c}{2\sqrt{2}}\nabla_p^E$. Let $S_{(y_0,p_0)}$ be the probability law on $\mathcal{C}([0,1], E \oplus E^*)$ of the corresponding bridge (y_t,p_t), which starts at (y_0,p_0) at time 0, and is such that $(y_1,p_1) = g(y_0,p_0)$. Let $E^{S_{(y_0,p_0)}}$ be the expectation operator with respect to $S_{(y_0,p_0)}$.

Note that under $R_{(y_0,p_0)}$ or $S_{(y_0,p_0)}$,

$$\frac{dy}{dt} = \frac{c}{2\sqrt{2}} p, \tag{7.6.1}$$

so that

$$\frac{c}{2\sqrt{2}}\int_0^1 p_t dt = y_1 - y_0. \tag{7.6.2}$$

Equivalently,

$$\frac{c}{2\sqrt{2}}\int_0^1 p_t dt = (g-1)\, y_0. \tag{7.6.3}$$

In particular, if $g = 1$, so that $E = E^1$, then

$$\int_0^1 p_t dt = 0, \tag{7.6.4}$$

i.e., $p \in L_0^2$, where L_0^2 is the subvector space of L^2 orthogonal to the constants.

Now we fix the path $p_t \in \mathcal{C}([0,1], E^*)$. Let $U_t, 0 \leq t \leq 1$ be the solution of the differential equation

$$\begin{aligned} \frac{dU}{dt} &= U\Big[\frac{c}{2}\widehat{e}_i\left(\widehat{c}\left(e_i\right) - i_{\widehat{e}^i}\right) + \frac{cn}{4} - \frac{1}{4}\left\langle e_i, R^E e_j\right\rangle i_{\widehat{e}^i} i_{\widehat{e}^j} \\ &\qquad - \frac{dc}{2\sqrt{2}}\left(i_{\widehat{p}} - \widehat{c}\left(p\right)\right) - \frac{cz}{2\sqrt{2}} i_{\widehat{p}}\Big], \\ U_0 &= 1. \end{aligned} \tag{7.6.5}$$

Proposition 7.6.1. *For any $(y_0, p_0) \in E \oplus E^*$, the following identity holds:*

$$\begin{aligned} Q'_c\left(\left(y_0, p_0\right), g\left(y_0, p_0\right)\right) = E^{S(y_0,p_0)}\Bigg[\exp\Bigg(-\frac{c^2}{8}\int_0^1 |p|^2\, dt \\ + \frac{c}{2\sqrt{2}}\int_0^1 \left\langle R^E y, p\right\rangle dt\Bigg) U_1\Bigg] q\left(\left(y_0, p_0\right), g\left(y_0, p_0\right)\right). \end{aligned} \tag{7.6.6}$$

Proof. Our proposition follows easily from (7.5.10) and from the Feynman-Kac formula. □

7.7 FINITE DIMENSIONAL SUPERTRACES AND INFINITE DETERMINANTS

In this section, we will give an expression for the supertrace of gU_1. Let $so_g(E)$ be the bundle of antisymmetric sections of $\mathrm{End}(E)$ which commute with g. Temporarily, we replace R^E by $A \in so_g(E)$.

So let V_t be the solution of the differential equation

$$\begin{aligned} \frac{dV}{dt} &= V\Big[\frac{c}{2}\widehat{e}_i\left(\widehat{c}\left(e_i\right) - i_{\widehat{e}^i}\right) + \frac{cn}{4} - \frac{1}{4}\left\langle e_i, A e_j\right\rangle i_{\widehat{e}^i} i_{\widehat{e}^j} \\ &\qquad - \frac{dc}{2\sqrt{2}}\left(i_{\widehat{p}} - \widehat{c}\left(p\right)\right) - \frac{cz}{2\sqrt{2}} i_{\widehat{p}}\Big], \\ V_0 &= 1. \end{aligned} \tag{7.7.1}$$

Note that we will adopt the conventions of section 7.5 concerning supertraces, that is, the Clifford variables $\widehat{c}(e_i), 1 \leq i \leq m$, will play a special role. Note that in principle, $\mathrm{Tr_s}[gV_1]$ is a section of $\Lambda^{\cdot}(T^*S)\widehat{\otimes}\mathbf{R}[z, dc]\widehat{\otimes}o(E^1)$. Still, we will often be interested here in the part of the supertrace which is even in z, dc, that is, either it does not contain z, dc or it contains the factor zdc. We will denote this restricted supertrace by the notation $\mathrm{Tr_s}^{\mathrm{even}}$. The part of the supertrace which only contains either z or dc will be denoted $\mathrm{Tr_s}^{\mathrm{odd}}$.

Let $\mathcal{C}_g^\infty([0,1],E)$ be the vector space of smooth functions f defined on $[0,1]$ with values in E, such that

$$f_1 = gf_0. \tag{7.7.2}$$

The operator $\frac{d}{dt}$ acts as a real antisymmetric operator on $\mathcal{C}_g^\infty([0,1],E)$. We denote by J_g the corresponding skew-adjoint extension of $\frac{d}{dt}$. If $g=1$, we will use the notation J instead of J_g.

Note that $\ker J_g$ is in one to one correspondence with $\ker(g-1)$. We will denote by J_g^{-1} the inverse of J_g restricted to the orthogonal space $\ker J_g^\perp$ to $\ker J_g$ in $L^2([0,1],E)$. Equivalently, J_g^{-1} is the inverse of J_g restricted to $\ker J_g^\perp$.

We will denote Fredholm determinants of operators where J_g^{-1} appears with the notation $\det^*$, to remind the reader that the zero eigenvalue of J_g is excluded.

Let $E^{1,\perp} \subset E$ be the orthogonal vector bundle to E^1 in E. Since A commutes with g, E^1 and $E^{1,\perp}$ are stable by A. In particular $A|_{E^1}$ is an antisymmetric endomorphism of E^1. By definition, $\mathrm{Pf}[A|_{E^1}]$ vanishes if E^1 is odd dimensional. In general, it is a section of $o(E^1)$.

We claim that $\det^*\left(1-\frac{c^2}{4}J_g^{-2}+\frac{c^2}{4}AJ_g^{-3}\right)$ has a natural square root. Indeed the operator $1-\frac{c^2}{4}J_g^{-2}$ is positive definite, so that $\det\left(1-\frac{c^2}{4}J_g^{-2}\right)$ is positive. For A close enough to 0, we can define the "natural" square root $\det^{*1/2}\left(1-\frac{c^2}{4}J_g^{-2}+\frac{c^2}{4}AJ_g^{-3}\right)$ of this determinant, which is an analytic function of A.

Although this will not be used in the next chapters, we show briefly how to define this square root as an analytic function of $A \in so_g(E)$. Indeed the operators A and J_g are antisymmetric and real. Complex conjugation maps the corresponding eigenspaces into eigenspaces associated with the negative of the eigenvalues. Since we exclude the zero eigenvalue of J_g, this makes that in the infinite product which defines the Fredholm determinant, the same factor appears necessarily twice. By picking just one of these factors, we can then define the square root $\det^{*1/2}\left(1-\frac{c^2}{4}J_g^{-2}+\frac{c^2}{4}AJ_g^{-3}\right)$. It is obviously an analytic function of $A \in so(E)$, which extends the previously defined function, since it coincides with that function at $A=0$.

Observe that if Pf^* denotes the Pfaffian of an operator acting on the orthogonal of $\ker J_g$, at least formally,

$$\det{}^{*1/2}\left(1-\frac{c^2}{4}J_g^{-2}+\frac{c^2}{4}AJ_g^{-3}\right) = \frac{\mathrm{Pf}^*\left[J_g-\frac{c^2}{4}J_g^{-1}+\frac{c^2}{4}AJ_g^{-2}\right]}{\mathrm{Pf}^*[J_g]}. \tag{7.7.3}$$

Note that the ratio of two Pfaffians is a real number. By splitting $\ker J_g^\perp$ into a direct sum of eigenspaces associated to the nonzero eigenvalues of J_g, and writing the Pfaffians in (7.7.3) as a product of Pfaffians, one can make easily sense of (7.7.3).

Definition 7.7.1. Set

$$\chi(c,g,A) = \mathrm{Pf}\left[\frac{c^2}{8}A|_{E^1}\right] \det{}^{E^{1,\perp}}(1-g)^2 \det{}^{*1/2}\left(1 - \frac{c^2}{4}J_g^{-2} + \frac{c^2}{4}AJ_g^{-3}\right). \quad (7.7.4)$$

We will give a more concrete expression for $\chi(c,g,A)$. Assume first that $E = E^1$ is even dimensional and oriented by the choice of the basis $e_1, \ldots, e_n$. Then there is an oriented orthonormal basis of E such that the matrix of A is a union of semidiagonal blocks

$$\begin{pmatrix} 0 & -y_j \\ y_j & 0 \end{pmatrix}, 1 \le j \le n/2,$$

so that if $x_j = iy_j, 1 \le j \le n/2$, the $\pm x_j$ are the eigenvalues of A. In particular the Pfaffian of A is just

$$\mathrm{Pf}[A] = \prod_1^{n/2}(-y_j). \quad (7.7.5)$$

Then one verifies easily that

$$\chi(c,1,A) = \prod_{j=1}^{n/2}\left(\frac{ic^2}{8}x_j\right)\prod_{j=1}^{n/2}\det{}^*\left(1 - \frac{c^2}{4}J^{-2} + \frac{c^2}{4}x_jJ^{-3}\right). \quad (7.7.6)$$

Assume now that $E = E_{\mathbf{R}}^{e^{\pm i\theta}}$, with $0 < \theta < \pi$, so that E is even dimensional. Let $B \in \mathrm{End}(E)$ be the antisymmetric endomorphism with semidiagonal blocks $\begin{pmatrix} 0 & -\theta \\ \theta & 0 \end{pmatrix}$, so that

$$g = e^B. \quad (7.7.7)$$

We may and we will assume that A is also reduced in semidiagonal blocks

$$\begin{pmatrix} 0 & -y_j \\ y_j & 0 \end{pmatrix}, 1 \le j \le n/2$$

on the same basis as B. Put again $x_j = iy_j$. Then we have the identity

$$\chi(c,g,A) = \left(2\sin\left(\frac{\theta}{2}\right)\right)^n \prod_{j=1}^{n/2}\det\left(1 - \frac{c^2}{4}J_{e^{i\theta}}^{-2} + \frac{c^2}{4}x_jJ_{e^{i\theta}}^{-3}\right). \quad (7.7.8)$$

We will not be more specific in the case $E = E^{-1}$, i.e., when $g = -1$.
We will also consider expressions of the type

$$\chi(c,g,A)\exp\left(\frac{c^2zdc}{8}\left\langle\frac{J_g}{J_g^3 - \frac{c^2}{4}J_g + \frac{c^2}{4}A}p,p\right\rangle\right). \quad (7.7.9)$$

Equation (7.7.9) should be properly interpreted. Indeed recall that in (7.6.5) and in (7.7.1), p_t is such that $p_1 = gp_0$.

If A is close enough to 0, the operator $J_g^3 - \frac{c^2}{4}J_g + \frac{c^2}{4}A$ is invertible. However, if A is arbitrary, this operator may well be noninvertible. Still, the expression (7.7.9) remains an analytic function of A because of the determinant appearing in the definition of $\chi(c,g,A)$ given in (7.7.4). For similar considerations in a simpler situation, we refer to Mathai-Quillen [MatQ86] and to [B90, section 5].

Theorem 7.7.2. *The following identity holds:*

$$\mathrm{Tr_s}^{\mathrm{even}}[gV_1] = \chi(c,g,A)\exp\left(\frac{c^2zdc}{8}\left\langle\frac{J_g}{J_g^3 - \frac{c^2}{4}J_g + \frac{c^2}{4}A}p, p\right\rangle\right). \tag{7.7.10}$$

Proof. We will consider in succession the cases $E = E^1, E = E_{\mathbf{R}}^{e^{\pm i\theta}}, E = E^{-1}$.

• **The case where $\mathbf{E} = \mathbf{E^1}$.** Here, we assume that $E = E^1$, i.e., $g = 1$. First we will make $z = 0, dc = 0$. We claim that if E is odd dimensional, $\mathrm{Tr_s}[V^1]$ vanishes. Indeed V_1 is even in the monomials containing the $\widehat{e}_i, i_{\widehat{e}^i}, \widehat{c}(e_i)$. Moreover, only even monomials in the $\widehat{e}_i, i_{\widehat{e}^i}$ can have a nonzero supertrace when acting on $\Lambda^\cdot(E)$. In the expansion of V_1, such even monomials will always be factors of even monomials in the $\widehat{c}(e_i)$. If E is odd dimensional, the monomial $\widehat{c}(e_1)\dots\widehat{c}(e_n)$ never appears in this expansion, so that indeed $\mathrm{Tr_s}[V_1]$ vanishes.

We can now assume that E is even dimensional. Note that

$$\widehat{e}_i = \frac{1}{2}(\widehat{c}(\widehat{e}_i) + c(\widehat{e}_i)), \qquad i_{\widehat{e}^i} = \frac{1}{2}(\widehat{c}(\widehat{e}_i) - c(\widehat{e}_i)). \tag{7.7.11}$$

In the sequel, we will use the notation for $1 \le i \le n$,

$$\widehat{c}(e_i) = i\overline{c}(e_i), \qquad \widehat{c}(\widehat{e}_i) = i\overline{c}(\widehat{e}_i). \tag{7.7.12}$$

The $\overline{c}(e_i), \overline{c}(\widehat{e}_i)$ are now standard Clifford variables, which anticommute, and also anticommute with the other Clifford variables. Observe that since n is even,

$$\begin{gathered}\prod_{1\le i\le n} c(\widehat{e}_i)\widehat{c}(\widehat{e}_i) = \prod_{1\le i\le n} c(\widehat{e}_i)\prod_{1\le i\le n}\overline{c}(\widehat{e}_i),\\ (-1)^{n(n+1)/2}\widehat{c}(e_1)\dots\widehat{c}(e_n) = \overline{c}(e_1)\dots\overline{c}(e_n).\end{gathered} \tag{7.7.13}$$

We assume temporarily that the Euclidean vector bundle E is oriented and spin, and we denote by $S^E = S_+^E \oplus S_-^E$ the corresponding $\mathbf{Z}_2$-graded vector bundle of spinors. Then S^E is a $c(E)$ Clifford module. Among the monomials in the $c(e_i)$, only $c(e_1)\dots c(e_n)$ has a nonzero supertrace, and moreover,

$$\mathrm{Tr_s}^{S^E}[c(e_1)\dots c(e_n)] = (-2i)^{n/2}. \tag{7.7.14}$$

In the sequel we make $c(E)^{\otimes 3}$ act on $S^E\widehat{\otimes}S^E\widehat{\otimes}S^E$. The corresponding three copies of $c(E)$ are generated by the $\overline{c}(e_i), c(\widehat{e}_i), \overline{c}(\widehat{e}_i)$. Using (7.7.13), (7.7.14), we find easily that we have the identity of functionals on $c(E)^{\otimes 3}$,

$$\mathrm{Tr_s} = \frac{1}{(2i)^{n/2}}\mathrm{Tr_s}^{S^E\widehat{\otimes}S^E\widehat{\otimes}S^E}. \tag{7.7.15}$$

Using (7.7.1), the fact that $z = 0, dc = 0$ and also (7.7.11), (7.7.12), we get

$$V_1 = \exp\Big(\frac{ic}{4}\overline{c}(\widehat{e}_i)\, c(\widehat{e}_i) + \frac{ic}{4}(c(\widehat{e}_i) + i\overline{c}(\widehat{e}_i))\,\overline{c}(e_i) \\ + \frac{1}{16}\langle Ae_i, e_j\rangle (c(\widehat{e}_i) - i\overline{c}(\widehat{e}_i))(c(\widehat{e}_j) - i\overline{c}(\widehat{e}_j))\Big). \quad (7.7.16)$$

Let V be a finite dimensional real Euclidean vector space, and let $so(V)$ be the Lie algebra of antisymmetric elements in $\mathrm{End}(V)$. Let $H \in so(V)\otimes_{\mathbf{R}}\mathbf{C}$. Let $c(V)$ be the Clifford algebra of V. Let $v_1, \ldots, v_p$ be an orthonormal basis of V. Then the image $c(H)$ of H in $c(V)\otimes_{\mathbf{R}}\mathbf{C}$ is given by

$$c(H) = \frac{1}{4}\langle Hv_i, v_j\rangle\, c(v_i)\, c(v_j). \quad (7.7.17)$$

Assume that V has even dimension p and also that it is oriented. Let $S^V = S^V_+ \oplus S^V_-$ be the corresponding $\mathbf{Z}_2$-graded vector space of spinors. Then S^V is a $c(V)$-Clifford module. Moreover, $c^{\mathrm{even}}(V)$ preserves the $\mathbf{Z}_2$-grading of S^V.

If $a \in c(V)\otimes_{\mathbf{R}}\mathbf{C}$, let $\mathrm{Tr_s}^{S^V}[a]$ be the supertrace of a when acting on $c(V)$. If $H \in so(V)$, then there is an oriented orthogonal basis of V such that the matrix of H has semidiagonal blocks $\begin{pmatrix} 0 & -\gamma_j \\ \gamma_j & 0 \end{pmatrix}, 1 \le j \le p/2$. By (7.7.14) and (7.7.17), one finds easily that

$$\mathrm{Tr_s}^{S^V}[\exp(c(H))] = \prod_{j=1}^{p/2}(-2i\sin(\gamma_j/2)). \quad (7.7.18)$$

Clearly,

$$\mathrm{Pf}[H] = \prod_{j=1}^{p/2}(-\gamma_j). \quad (7.7.19)$$

The function Pf extends to an analytic function on $so(V)\otimes_{\mathbf{R}}\mathbf{C}$, which is such that

$$\mathrm{Pf}^2 = \det. \quad (7.7.20)$$

Moreover, the function

$$\widehat{A}(H) = \prod_{j=1}^{p/2}\left(\frac{\gamma_j/2}{\sin(\gamma_j/2)}\right) \quad (7.7.21)$$

is an analytic symmetric function in the γ_j^2. Therefore it is an ad-invariant analytic function of the coefficients of the characteristic polynomial of H. It can be expressed as an analytic function of the $\mathrm{Tr}\left[H^{2k}\right], k \in \mathbf{N}$, and so it extends uniquely as an invariant analytic function on $\mathrm{End}(V)\otimes_{\mathbf{R}}\mathbf{C}$. If

$H \in so(V) \otimes_{\mathbf{R}} \mathbf{C}$, it has eigenvalues $\pm\delta_1, \ldots, \pm\delta_{p/2}$. Then

$$\mathrm{Pf}^2[H] = \det[H] = \prod_{j=1}^{p/2} \left(-\delta_j^2\right), \tag{7.7.22}$$

$$\widehat{A}(H) = \prod_{j=1}^{p/2} \frac{\delta_j/2}{\sinh(\delta_j/2)}.$$

By analyticity, we deduce from the above that if $H \in so(V) \otimes_{\mathbf{R}} \mathbf{C}$,

$$\mathrm{Tr_s}^{S^V}\left[\exp\left(c(H)\right)\right] = \mathrm{Pf}\left[iH\right] \widehat{A}^{-1}(H). \tag{7.7.23}$$

Let $M \in \mathrm{End}\left(E^3\right) \otimes_{\mathbf{R}} \mathbf{C}$ be given by

$$M = \begin{pmatrix} 0 & i\frac{c}{2} & -\frac{c}{2} \\ -i\frac{c}{2} & \frac{A}{4} & -i\frac{A}{4} + i\frac{c}{2} \\ \frac{c}{2} & -i\frac{A}{4} - i\frac{c}{2} & -\frac{A}{4} \end{pmatrix}. \tag{7.7.24}$$

Then M is antisymmetric. We make the first columns and rows be associated with the $\overline{c}(e_i)$, the second with the $c(\widehat{e}_i)$, the third with the $\overline{c}(\widehat{e}_i)$. By (7.7.16), and using (7.7.17) with $V = E^3$, we get

$$V_1 = \exp\left(c(M)\right). \tag{7.7.25}$$

By (7.7.23),

$$\mathrm{Tr_s}^{S^E \widehat{\otimes} S^E \widehat{\otimes} S^E}\left[V_1\right] = \mathrm{Pf}\left[iM\right] \widehat{A}^{-1}(M). \tag{7.7.26}$$

Let $x \in \mathbf{C}$. Let C be the $(3,3)$ matrix

$$C = \begin{pmatrix} 0 & i\frac{c}{2} & -\frac{c}{2} \\ -i\frac{c}{2} & \frac{x}{4} & -i\frac{x}{4} + i\frac{c}{2} \\ \frac{c}{2} & -i\frac{x}{4} - i\frac{c}{2} & -\frac{x}{4} \end{pmatrix}. \tag{7.7.27}$$

Let $Q(\lambda) = \det(C - \lambda)$ be the characteristic polynomial of C. By a straightforward computation, we get

$$Q(\lambda) = P_{c,x}(\lambda). \tag{7.7.28}$$

Let $\lambda_1, \lambda_2, \lambda_3$ be the roots of $Q(\lambda)$. By (7.1.3), (7.7.28),

$$\tau(c, 0, x) = \prod_1^3 2\sinh\left(\frac{\lambda_i}{2}\right). \tag{7.7.29}$$

We fix an orientation of E. Then there is an oriented orthonormal basis of E such that the matrix of A on this basis has semidiagonal blocks $\begin{pmatrix} 0 & -y_i \\ y_i & 0 \end{pmatrix}$. Put $x_i = iy_i$. Then $\pm x_1, \ldots, \pm x_{n/2}$ are the eigenvalues of A. By (7.7.23), by (7.7.26)-(7.7.29), and by a careful computation of signs, we get

$$\mathrm{Tr_s}^{S^E \widehat{\otimes} S^E \widehat{\otimes} S^E}\left[V_1\right] = (-1)^{n/2} \prod_1^{n/2} \tau(c, 0, x_i). \tag{7.7.30}$$

By (7.1.7) in Theorem 7.1.2, by (7.7.6), (7.7.15), and (7.7.30), we get (7.7.10) when $E = E^1$ and z and dc are made equal to 0.

Now we consider the case where $E = E^1$, but z, dc do not vanish. Essentially the same arguments as before show that if n is odd, then $\mathrm{Tr_s}^{\mathrm{even}}[V_1]$ still vanishes. Assume now that n is even. Note that we can rewrite the differential equation (7.7.1) in the form

$$\begin{aligned}\frac{dV}{dt} &= V\Big[c(M) - \frac{dc}{2\sqrt{2}}\left(\frac{1}{2}\left(i\overline{c}(\widehat{p}) - c(\widehat{p})\right) - i\overline{c}(p)\right) \\ &\qquad - \frac{z}{2\sqrt{2}}\frac{c}{2}\left(i\overline{c}(\widehat{p}) - c(\widehat{p})\right)\Big], \\ V_0 &= 1.\end{aligned} \tag{7.7.31}$$

Let $\vartheta_t \in E^3 \otimes \mathbf{C}[z, dc]$ be given by

$$\vartheta_t = -\begin{bmatrix} \frac{i}{2\sqrt{2}} dcp_t \\ \frac{1}{4\sqrt{2}}(cz + dc)\, p_t \\ \frac{-i}{4\sqrt{2}}(cz + dc)\, p_t \end{bmatrix}. \tag{7.7.32}$$

We denote by $c(\vartheta_t)$ the element of $c(E^3)\widehat{\otimes}\mathbf{C}(z, dc)$ which is obtained by replacing each of the three components in (7.7.32) by the corresponding Clifford multiplications, and introducing the obvious minus sign when taking z or dc out of the Clifford multiplication. Then (7.7.31) can be rewritten in the form

$$\begin{aligned}\frac{dV}{dt} &= V\left[c(M) + c(\vartheta_t)\right], \\ V_0 &= 1.\end{aligned} \tag{7.7.33}$$

By [B90, Theorem 5.1], by (7.7.31), (7.7.32), we get

$$\mathrm{Tr_s}[V_1] = \mathrm{Tr_s}\left[\exp\left(c(M)\right)\right] \exp\left(\left\langle\left(\frac{d}{dt} + M\right)^{-1}\vartheta, \vartheta\right\rangle\right). \tag{7.7.34}$$

In (7.7.34), the boundary conditions are the standard periodic boundary conditions on [0, 1]. Note that to make sense of the right-hand side of (7.7.34), considerations similar to the ones we made after (7.7.9) should be made. For details we refer to [B90, Chapter 5].

The first factor in the right-hand side of (7.7.34) was evaluated before. So we now concentrate on the evaluation of the second factor. We still take C as in (7.7.27). Observe that if λ is not a root of the polynomial $P_{c,x}$, then

$$(C - \lambda)^{-1} = \frac{1}{P_{c,x}(\lambda)} \begin{pmatrix} \lambda^2 - \frac{c^2}{4} & \frac{ic}{2}\lambda + \frac{ic}{4}x + \frac{ic^2}{4} & -\frac{c}{2}\lambda + \frac{c}{4}x - \frac{c^2}{4} \\ -\frac{ic}{2}\lambda - \frac{ic}{4}x + \frac{ic^2}{4} & \lambda^2 + \frac{x}{4}\lambda + \frac{c^2}{4} & i\left(-\frac{x}{4} + \frac{c}{2}\right)\lambda + i\frac{c^2}{4} \\ \frac{c}{2}\lambda - \frac{c}{4}x - \frac{c^2}{4} & -i\left(\frac{x}{4} + \frac{c}{2}\right)\lambda + i\frac{c^2}{4} & \lambda^2 - \frac{x}{4}\lambda - \frac{c^2}{4} \end{pmatrix}. \tag{7.7.35}$$

Let us temporarily consider p_t in (7.7.32) as a real number, and ϑ_t as an element of $\mathbf{C}(z, dc)$. By (7.7.35), we get

$$\left\langle (C-\lambda)^{-1}\vartheta, \vartheta \right\rangle = -zdc \frac{c^2\lambda}{8P_{c,x}(\lambda)} |p_t|^2 . \tag{7.7.36}$$

By (7.7.24), (7.7.27), (7.7.36), we get easily

$$\left\langle \left(\frac{d}{dt} + M\right)^{-1} \vartheta, \vartheta \right\rangle = \frac{c^2}{8} zdc \left\langle \frac{J}{J^3 - \frac{c^2}{4}J + \frac{c^2}{4}A} p, p \right\rangle . \tag{7.7.37}$$

By (7.7.34), (7.7.37), we get (7.7.10) when $E = E^1$.

• **The case where $\mathbf{E} = \mathbf{E}_{\mathbf{R}}^{\mathbf{e}^{\pm i\theta}}$.** Now, we assume that $E = E_{\mathbf{R}}^{e^{i\pm\theta}}$, so that E is even dimensional. Moreover, there is $B \in \mathrm{End}(E)$ which is antisymmetric, commutes with A, and is such that (7.7.7) holds. Then the action of B on $\Lambda^{\cdot}(E^* \oplus E)$ is given by

$$B|_{\Lambda^{\cdot}(E^*\oplus E)} = \frac{1}{4} \langle Be_i, e_j \rangle \Big(c(e_i) c(e_j) - \widehat{c}(e_i) \widehat{c}(e_j) \\ + c(\widehat{e}_i) c(\widehat{e}_j) - \widehat{c}(\widehat{e}_i) \widehat{c}(\widehat{e}_j) \Big) . \tag{7.7.38}$$

As before, we replace in (7.7.38) the $\widehat{c}(e_j), \widehat{c}(\widehat{e}_j)$ by $i\overline{c}(e_i), i\overline{c}(\widehat{e}_j)$, the effect being to change the $-$ signs into $+$ signs in (7.7.38).

Recall that E is oriented. We will temporarily assume that E is spin, and we denote by $S^E = S^E_+ \oplus S^E_-$ the corresponding spinors. Then $\left(S^E\right)^{\otimes 4}$ is a $c(E)^{\otimes 4}$ Clifford module. It is then elementary to verify that in our computation of the supertrace, we can as well replace $\Lambda^{\cdot}(E^* \oplus E)$ by $\left(S^E\right)^{\otimes 4}$.

First we assume that $z = 0, dc = 0$. Let $N \in \mathrm{End}\left(E^4\right) \otimes_{\mathbf{R}} \mathbf{C}$ be given by

$$N = \begin{pmatrix} B & 0 & 0 & 0 \\ 0 & B & i\frac{c}{2} & -\frac{c}{2} \\ 0 & -i\frac{c}{2} & \frac{A}{4} + B & -i\frac{A}{4} + i\frac{c}{2} \\ 0 & \frac{c}{2} & -i\frac{A}{4} - i\frac{c}{2} & -\frac{A}{4} + B \end{pmatrix} . \tag{7.7.39}$$

Note that N is antisymmetric. It follows from (7.7.31) and (7.7.38) that we have the identity of operators acting on $\Lambda^{\cdot}(E^* \oplus E)$,

$$gV_1 = \exp(c(N)) . \tag{7.7.40}$$

Let $\pm\sigma_j, 1 \leq j \leq 2n$ be the eigenvalues of N. Then by (7.7.22), (7.7.23),

$$\mathrm{Tr_s}\left[\exp(c(N))\right] = \prod_{j=1}^{2n} 2\sinh\left(\frac{\sigma_j}{2}\right) . \tag{7.7.41}$$

The explicit sign in (7.7.41) is evaluated using explicitly the orientation of E^4.

If $\eta \in \mathbf{C}$, set

$$D = \begin{pmatrix} \eta & 0 & 0 & 0 \\ 0 & \eta & i\frac{c}{2} & -\frac{c}{2} \\ 0 & -i\frac{c}{2} & \frac{x}{4}+\eta & -i\frac{x}{4}+i\frac{c}{2} \\ 0 & \frac{c}{2} & -i\frac{x}{4}-i\frac{c}{2} & -\frac{x}{4}+\eta \end{pmatrix}. \tag{7.7.42}$$

Let $R(\lambda)$ be the characteristic polynomial of D. By comparing with (7.7.27), (7.7.28), we get

$$R(\lambda) = (\eta - \lambda) P_{c,x}(\lambda - \eta). \tag{7.7.43}$$

Assume that the eigenvalues of A acting on $E^{e^{i\theta}}$ are given by $x_1, \ldots, x_{n/2}$. By (7.1.3), (7.7.41), (7.7.43), we get

$$\mathrm{Tr_s}[gV_1] = (2i\sin(\theta/2))^{n/2} \prod_1^{n/2} \tau(c, i\theta, x_j). \tag{7.7.44}$$

Using (7.1.8) in Theorem 7.1.2, (7.7.8), and (7.7.44), we get (7.7.10) when z, dc are made equal to 0.

Consider now the general case, where z, dc do not vanish. Put

$$\varpi_t = -\begin{bmatrix} 0 \\ \frac{i}{2\sqrt{2}} dc p_t \\ \frac{1}{4\sqrt{2}}(cz+dc)p_t \\ \frac{-i}{4\sqrt{2}}(cz+dc)p_t \end{bmatrix}. \tag{7.7.45}$$

Consider the differential equation

$$\frac{dW}{dt} = W\left[c(N) + c(\exp(-tB)\varpi_t)\right], \qquad W_0 = 1. \tag{7.7.46}$$

If we identify B to the corresponding matrix acting on E^4, and M to the matrix acting like 0 on the first copy of E, and like the given M on the last E^3, from (7.7.39), we get

$$N = B + M. \tag{7.7.47}$$

By (7.7.33), (7.7.46), (7.7.47), we find that

$$V_1 g = W_1. \tag{7.7.48}$$

Now by using again [B90, Theorem 5.1] as in (7.7.34) and by (7.7.48), we get

$$\mathrm{Tr_s}[gV_1] = \mathrm{Tr_s}[W_1] = \mathrm{Tr_s}[\exp(c(N))] \exp\left(\left\langle \left(\frac{d}{dt}+N\right)^{-1} \exp(-tB)\varpi_t, \exp(-tB)\varpi_t \right\rangle\right). \tag{7.7.49}$$

Again, standard periodic boundary conditions are used in the right-hand side of (7.7.49). However, since e^{-tB} is an orthogonal matrix,

$$\left\langle \left(\frac{d}{dt}+N\right)^{-1} \exp(-tB)\varpi_t, \exp(-tB)\varpi_t \right\rangle = \left\langle \left(\frac{d}{dt}+M\right)^{-1} \varpi_t, \varpi_t \right\rangle, \tag{7.7.50}$$

where the operator $\left(\frac{d}{dt}+M\right)^{-1}$ should now act on smooth functions f on $[0,1]$ with values in E^4 such that $f_1 = gf_0$, which, incidentally, is the case for ϖ_t. By using (7.7.36) and proceeding as in (7.7.37), we get

$$\left\langle \left(\frac{d}{dt}+M\right)^{-1}\vartheta,\vartheta\right\rangle = \frac{c^2}{8}zdc\left\langle \frac{J_g}{J_g^3-\frac{c^2}{4}J_g+\frac{c^2}{4}A}p,p\right\rangle. \tag{7.7.51}$$

From (7.7.49),(7.7.51), we get (7.7.10) in full generality when $E = E^{i\theta}$.

• **The case where $\mathbf{E} = \mathbf{E}^{-1}$.** Now we assume that $E = E^{-1}$, i.e., $g = -1$. We replace temporarily E by $E \oplus E$, so that g acts like -1 on both copies of E. Then $\mathrm{Tr_s}^{\mathrm{even}}[gV_1]$ is replaced by its square. Let $B \in \mathrm{End}(E\oplus E)$ be given by

$$B = \begin{pmatrix} 0 & -\pi \\ \pi & 0 \end{pmatrix}, \tag{7.7.52}$$

so that $g = e^B$. Since $E\oplus E$ is even dimensional, we can then proceed as before, and get the identity corresponding to (7.7.10) for $E \oplus E$, which is the square of the one we are looking for. The expressions we consider are analytic in A, c, z, dc. When $A = 0, c = 0, z = 0, dc = 0$, then $V_1 = 1$, so that

$$\mathrm{Tr_s}[gV_1] = 2^{2n}, \tag{7.7.53}$$

which coincides in this case with (7.7.10). Therefore we have established (7.7.10) also in this case. The proof of our theorem is completed. □

7.8 THE EVALUATION OF THE FORM $\mathrm{Tr_s}\left[\mathbf{g}\exp\left(-\mathfrak{L}_{\mathbf{c}}^{\mathbf{E}}\right)\right]$

Here we establish the main result of this chapter.

Theorem 7.8.1. *The following identity holds:*

$$\mathrm{Tr_s}\left[g\exp\left(-\mathfrak{L}_c^E\right)\right] = e\left(E^1,\nabla^{E^1}\right) \left(1-\frac{c^2}{8}zdc\mathrm{Tr}^*\left[\frac{1}{\left(-J_g^2+\frac{c^2}{4}\left(1-R^EJ_g^{-1}\right)\right)^2}\right]\right). \tag{7.8.1}$$

Proof. We split the proof of our theorem into two parts.

• **The case where $\mathbf{E} = \mathbf{E}^1$.** First, we assume that $g = 1$, so that $J_g = J = \frac{d}{dt}$ with periodic boundary conditions. Note that by (7.1.9),

$$\left(2\sinh\left(\sqrt{u}/2\right)\right)^n = u^{n/2}\det{}^{*1/2}\left(1-uJ^{-2}\right). \tag{7.8.2}$$

If $p \in E^*$, let P_p be the probability law on $\mathcal{C}([0,1],E^*)$ of the Brownian bridge $t \in [0,1] \to p_t \in E^*$, with $p_0 = p_1 = p$. For $u > 0$, let Q_u be the probability law on $\mathcal{C}([0,1],E^*)$ of the Gaussian process with values in E^* with covariance $\left(-J^2+u\right)^{-1}$. By (7.8.2) and by [B90, eq. (7.36)], we have the equality of positive finite measures on $\mathcal{C}([0,1],E^*)$,

$$\exp\left(-\frac{u}{2}\int_0^1 |p_t|^2\,dt\right)dP_p\frac{dp}{(2\pi)^{n/2}} = \frac{dQ_u}{\left(2\sinh\left(\sqrt{u}/2\right)\right)^n}. \tag{7.8.3}$$

Set

$$h = \int_0^1 p_t dt, \qquad q_t = p_t - h. \tag{7.8.4}$$

By construction, $q \in L_0^2$. Then under Q_u, h and $q_\cdot$ are independent random variables, the probability law of h is a centered Gaussian with values in E with covariance $1/u$, and q is a Gaussian process concentrated on L_0^2, with covariance $\left(-J^2+u\right)^{-1}$, whose corresponding probability law will be denoted Q_u^0.

Using (7.8.2) and the above, we find that under the finite positive measure in (7.8.3), the law of $h, q_\cdot$ is just

$$\frac{1}{(2\pi)^{n/2} \det{}^{*1/2}\left(1-uJ^{-2}\right)} \exp\left(-u\left|h\right|^2/2\right) dh \otimes dQ_u^0\left(q\right). \tag{7.8.5}$$

Let $\sigma \in \operatorname{End}\left(\Omega^\cdot\left(E \oplus E^*\right)\right)$ be given by $f\left(y,p\right) \to f\left(-y,-p\right)$. By (7.4.5), the part of $\mathfrak{L}_c^E$ which does not contain z, dc is invariant under conjugation by σ, and the part which contains z, dc is antiinvariant. We then deduce easily that

$$\operatorname{Tr_s}^{\mathrm{odd}}\left[\exp\left(-\mathfrak{L}_c^E\right)\right] = 0. \tag{7.8.6}$$

So we concentrate now on $\operatorname{Tr_s}^{\mathrm{even}}\left[\exp\left(-\mathfrak{L}_c^E\right)\right]$. By Proposition 7.6.1 and by Theorem 7.7.2, this expression vanishes if E is odd dimensional. So we assume now that E is even dimensional.

Note that y_t is such that (7.6.1) holds. So by (7.6.2), (7.8.4),

$$y_1 - y_0 = \frac{c}{2\sqrt{2}} h. \tag{7.8.7}$$

By (7.6.4), under $S_{(0,p)}$, $p_\cdot \in L_2^0$, so that $J^{-1}p \in L_2^0$ is well defined. By (7.5.5), (7.6.1) and by (7.6.6) in Proposition 7.5.3, we get

$$\operatorname{Tr_s}\left[\exp\left(-\mathfrak{L}_c^E\right)\right] = \int_{E^*} E^{S_{(0,p)}}\Bigg[\exp\left(-\frac{c^2}{8}\int_0^1 \left|p\right|^2 dt + \frac{c^2}{8}\left\langle R^E J^{-1}p, p\right\rangle\right) \\ \operatorname{Tr_s}\left[U_1\right]\Bigg] q\left(\left(0,p\right),\left(0,p\right)\right) dp. \tag{7.8.8}$$

Also by (7.8.5), (7.8.7), (7.8.8) and using Fubini's theorem, which allows us to condition on $y_1 = 0$ after having done the integration in the variable $p \in E^*$, we get

$$\operatorname{Tr_s}\left[\exp\left(-\mathfrak{L}_c^E\right)\right] = \frac{2^n}{\det{}^{*1/2}\left(1-\frac{c^2}{4}J^{-2}\right)\left|c\right|^n \pi^{n/2}} \\ E^{Q^0_{c^2/4}}\left[\exp\left(\frac{c^2}{8}\left\langle R^E J^{-1}p, p\right\rangle\right)\operatorname{Tr_s}\left[U_1\right]\right]. \tag{7.8.9}$$

Using now Theorem 7.7.2 and (7.8.9), we obtain

$$\operatorname{Tr_s}\left[\exp\left(-\mathfrak{L}_c^E\right)\right] = \frac{2^n \chi\left(c,1,R^E\right)}{\det{}^{*1/2}\left(1-\frac{c^2}{4}J^{-2}\right)\left|c\right|^n \pi^{n/2}} \\ E^{Q^0_{c^2/4}}\left[\exp\left(\frac{c^2}{8}\left\langle\left(R^E J^{-1} + z\,dc\frac{J}{J^3-\frac{c^2}{4}J+\frac{c^2}{4}R^E}\right)p, p\right\rangle\right)\right]. \tag{7.8.10}$$

Trivial properties of Gaussians show that

$$E^{Q^0_{c^2/4}}\left[\exp\left(\frac{c^2}{8}\left\langle\left(R^E J^{-1}+zdc\frac{J}{J^3-\frac{c^2}{4}J+\frac{c^2}{4}R^E}\right)p,p\right\rangle\right)\right]$$
$$=\det{}^{*1/2}\left(\frac{-J^2+\frac{c^2}{4}}{-J^2+\frac{c^2}{4}-\frac{c^2}{4}\left(R^E J^{-1}+zdc\frac{J}{J^3-\frac{c^2}{4}J+\frac{c^2}{4}R^E}\right)}\right). \quad (7.8.11)$$

Moreover, we have the obvious,

$$\det{}^{*1/2}\left(\frac{-J^2+\frac{c^2}{4}}{-J^2+\frac{c^2}{4}-\frac{c^2}{4}\left(R^E J^{-1}+zdc\frac{J}{J^3-\frac{c^2}{4}J+\frac{c^2}{4}R^E}\right)}\right)$$
$$=\frac{\det{}^{*1/2}\left(1-\frac{c^2}{4}J^{-2}\right)}{\det{}^{*1/2}\left(1-\frac{c^2}{4}J^{-2}+\frac{c^2}{4}R^E J^{-3}\right)}$$
$$\left(1-\frac{zdc}{8}c^2\mathrm{Tr}^*\left[\frac{1}{\left(-J^2+\frac{c^2}{4}\left(1-R^E J^{-1}\right)\right)^2}\right]\right). \quad (7.8.12)$$

By (7.7.4), (7.8.10)-(7.8.12), we get (7.8.1) when $E=E^1$.

• **The case where 1 is not an eigenvalue of g.** Now we assume that 1 is not an eigenvalue of g. For $p\in E^*$, let $P_{g,p}$ be the probability law on $\mathcal{C}([0,1],E^*)$ of the Brownian bridge $t\in[0,1]\to p_t\in E^*$, with $p_0=p, p_1=gp$. For $u>0$, let $Q_{u,g}$ be the probability law on $\mathcal{C}([0,1],E^*)$ of the Gaussian process with covariance $\left(-J_g^2+u\right)^{-1}$. Note that under $Q_{u,g}$, $p_\cdot$ is a continuous process such that $p_1=gp_0$. We claim that we have the equality of positive finite measures on $\mathcal{C}([0,1],E^*)$ which generalizes (7.8.3),

$$\exp\left(-\frac{u}{2}\int_0^1 |p_t|^2\,dt-\frac{1}{2}\left|(1-g)\,p\right|^2\right)dP_{g,p}\frac{dp}{(2\pi)^{n/2}}$$
$$=\frac{dQ_{u,g}}{\left|\det\left(e^{\sqrt{u}/2}-e^{-\sqrt{u}/2}g\right)\right|}. \quad (7.8.13)$$

The proof of (7.8.13) is exactly the same as the proof of (7.8.3) given in [B90, proof of Theorem 7.3], which we repeat. Take $f\in L^2([0,1],E^*)$. Set

$$I=\int_{E^*}E^{P_{g,p}}\left[\exp\Big(\int_0^1\langle f,dp-\sqrt{u}pdt\rangle-\frac{u}{2}\int_0^1|p_t|^2\,dt\right.$$
$$\left.-\frac{1}{2}\left|(1-g)\,p\right|^2\Big)\right]\frac{dp}{(2\pi)^{n/2}}. \quad (7.8.14)$$

Using the fact that g is an isometry, the obvious analogues of [B90, eq. (7.27) and (7.28)] remain valid.

Let $w_{\cdot}$ be a Brownian motion with values in E^*. Given $p \in E^*$, let $p'_{\cdot}$ be the solution of the stochastic differential equation

$$dp' = \left(\sqrt{u}p' + f\right) dt + dw, \qquad p'_0 = p. \tag{7.8.15}$$

Let δp be the Itô differential of p. The same use of the Girsanov transformation as in [B90] shows that

$$\frac{1}{(2\pi)^{n/2}} E^{P_{g,p}} \left[\exp\left(\int_0^1 \left\langle f + \sqrt{u}p, \delta p \right\rangle - \frac{1}{2} \left| f + \sqrt{u}p \right|^2 \right) \right] \exp\left(-\frac{1}{2} \left| (1-g) p \right|^2 \right) \tag{7.8.16}$$

is exactly the value of the density of the probability law of p'_1 with respect to dp at $p' = gp$. This probability law was obtained in [B90, eq. 7.32)] by an easy explicit computation. Equation (7.8.13) then follows.

Using (7.1.9), instead of (7.8.2), we have the more general

$$\det\left(e^{\sqrt{u}/2} - e^{-\sqrt{u}/2} g\right) = \frac{\det^{1/2}\left(-J_g^2 + u\right)}{\det^{*1/2}\left(-J^2\right)}. \tag{7.8.17}$$

By (7.8.17), we get in particular

$$\det(1-g) = \frac{\det^{1/2}\left(-J_g^2\right)}{\det^{*1/2}\left(-J^2\right)}. \tag{7.8.18}$$

By (7.8.17), (7.8.18), we obtain

$$\det\left(e^{\sqrt{u}/2} - e^{-\sqrt{u}/2} g\right) = \det(1-g) \det^{1/2}\left(1 - uJ_g^{-2}\right). \tag{7.8.19}$$

We now use the notation in (7.8.4). Under $Q_{u,g}$, the probability law of h is a centered Gaussian on E^* whose variance $\sigma^2(u) > 0$ can be calculated explicitly, its exact value being irrelevant.

Under $S_{(y,p)}$, (7.6.1), (7.6.2), (7.8.4), and (7.8.7) still hold. Since $y_1 = gy_0$, we get

$$y = \frac{c}{2\sqrt{2}} J_g^{-1} p. \tag{7.8.20}$$

By (7.5.6), (7.6.6), and (7.8.20), we obtain

$$\operatorname{Tr_s}\left[g \exp\left(-\mathfrak{L}_c^E\right)\right] = \int_{E \oplus E^*} E^{S_{(y,p)}} \left[\exp\left(-\frac{c^2}{8} \int_0^1 |p|^2 dt + \frac{c^2}{8} \left\langle J_g^{-1} R^E p, p \right\rangle \right) \operatorname{Tr_s}\left[gU_1\right] \right] q\left((y,p), g(y,p)\right) dy dp. \tag{7.8.21}$$

Using (7.8.13), (7.8.19)-(7.8.21), and also Fubini's theorem, we get

$$\operatorname{Tr_s}\left[g \exp\left(-\mathfrak{L}_c^E\right)\right] = \frac{1}{\det(1-g)^2 \det^{1/2}\left(1 - \frac{c^2}{4} J_g^{-2}\right)} E^{Q_{g,c^2/4}} \left[\exp\left(\frac{c^2}{8} \left\langle R^E J_g^{-1} p, p \right\rangle \right) \operatorname{Tr_s}\left[gU_1\right] \right]. \tag{7.8.22}$$

By Theorem 7.7.2 and (7.8.22), we obtain

$$\operatorname{Tr}_{\mathrm{s}}\left[g\exp\left(-\mathfrak{L}_c^E\right)\right]=\frac{\chi\left(c,g,R^E\right)}{\det\left(1-g\right)^2\det{}^{1/2}\left(1-\frac{c^2}{4}J_g^{-2}\right)}$$
$$E^{Q^0_{g,c^2/4}}\left[\exp\left(\frac{c^2}{8}\left\langle\left(R^EJ_g^{-1}+zdc\frac{J_g}{J_g^3-\frac{c^2}{4}J_g+\frac{c^2}{4}R^E}\right)p,p\right\rangle\right)\right]. \tag{7.8.23}$$

Now the obvious analogue of (7.8.11), (7.8.12) just says that

$$E^{Q^0_{g,c^2/4}}\left[\exp\left(\frac{c^2}{8}\left\langle\left(R^EJ_g^{-1}+zdc\frac{J_g}{J_g^3-\frac{c^2}{4}J_g+\frac{c^2}{4}R^E}\right)p,p\right\rangle\right)\right]$$
$$=\frac{\det{}^{1/2}\left(1-\frac{c^2}{4}J_g^{-2}\right)}{\det{}^{1/2}\left(1-\frac{c^2}{4}J_g^{-2}+\frac{c^2}{4}R^EJ_g^{-3}\right)}$$
$$\left(1-\frac{zdc}{8}c^2\operatorname{Tr}\left[\frac{1}{\left(-J_g^2+\frac{c^2}{4}\left(1-R^EJ_g^{-1}\right)\right)^2}\right]\right). \tag{7.8.24}$$

By (7.7.4), (7.8.23), (7.8.24), we obtain

$$\operatorname{Tr}_{\mathrm{s}}\left[g\exp\left(-\mathfrak{L}_c^E\right)\right]=1-\frac{zdc}{8}c^2\operatorname{Tr}\left[\frac{1}{\left(-J_g^2+\frac{c^2}{4}\left(1-R^EJ_g^{-1}\right)\right)^2}\right], \tag{7.8.25}$$

which is just (7.8.1). The proof of our theorem is completed. □

7.9 SOME EXTRA COMPUTATIONS

Put

$$\mathfrak{M}_c^E=\widehat{\underline{\mathfrak{C}}}^{\mathcal{M}^E,2}_{\phi,\mathcal{H}^c-\omega^H}-z\widehat{\underline{\mathfrak{G}}}^{\mathcal{M}^E}_{\phi,\mathcal{H}^c-\omega^H}. \tag{7.9.1}$$

The von Neumann supertraces

$$\operatorname{Tr}_{\mathrm{s}}\left[g\left|p\right|^2\exp\left(-\widehat{\underline{\mathfrak{C}}}^{\mathcal{M}^E,2}_{\phi,\mathcal{H}^c-\omega^H}\right)\right],\ \operatorname{Tr}_{\mathrm{s}}\left[g\exp\left(-\mathfrak{M}_c^E\right)\right]$$

are defined as in Definitions 7.5.1 and 7.5.2.

In the sequel, we write $\widehat{\underline{\mathfrak{D}}}_c^{\mathcal{M}^E}$ instead of $\widehat{\underline{\mathfrak{D}}}^E_{\phi,\mathcal{H}^c-\omega^H}$ when the value c has been fixed, so that $dc=0$.

Recall that $h(x)$ is given by (4.3.1), h' by (4.3.12). In particular $h'(x)$ is an even function of x, so that $h'\left(\widehat{\underline{\mathfrak{D}}}_c^{\mathcal{M}^E}\right)$ is in fact a function of $\widehat{\underline{\mathfrak{D}}}_c^{\mathcal{M}^E,2}$. In particular, we can define the generalized supertrace $\operatorname{Tr}_{\mathrm{s}}\left[g\left|p\right|^2h'\left(\widehat{\underline{\mathfrak{D}}}_c^{\mathcal{M}^E,2}\right)\right]$ as before.

If α is a section of $\Lambda^\cdot\left(T^*S\right)\widehat{\otimes}\mathbf{C}\left[dc\right]$, we denote $\alpha^{dc=0}$ the section of $\Lambda^\cdot\left(T^*S\right)$ which is obtained by making $dc=0$. If $\beta\in\Lambda^\cdot\left(T^*S\right)\widehat{\otimes}\mathbf{C}\left[z,dc\right]$, β^{zdc} denotes the section of $\Lambda^\cdot\left(T^*S\right)$ which is a factor of zdc in the obvious expansion of β.

Theorem 7.9.1. *The following identities hold:*

$$\mathrm{Tr_s}\left[g\left|p\right|^2\exp\left(-\widehat{\underline{\mathfrak{C}}}^{\mathcal{M}^E,2}_{\phi,\mathcal{H}^c-\omega^H}\right)\right]=e\left(E^1,\nabla^{E^1}\right)$$
$$\frac{1}{2}\mathrm{Tr}^*\left[\frac{1}{-J_g^2+\frac{c^2}{4}\left(1-R^EJ_g^{-1}\right)}\right], \tag{7.9.2}$$
$$\mathrm{Tr_s}\left[g\exp\left(-\mathfrak{M}_c^E\right)\right]=e\left(E^1,\nabla^{E^1}\right)$$
$$\left(1-\frac{1}{8}zdc\mathrm{Tr}^*\left[\frac{-6J_g^2+\frac{c^2}{2}\left(3R^EJ_g^{-1}-1\right)}{\left(-J_g^2+\frac{c^2}{4}\left(1-R^EJ_g^{-1}\right)\right)^2}\right]\right), \tag{7.9.3}$$
$$-\frac{1}{2}\mathrm{Tr_s}\left[g\left|p\right|^2h'\left(\widehat{\underline{\mathfrak{D}}}_c^{\mathcal{M}^E}\right)\right]=\mathrm{Tr_s}\left[g\exp\left(-\mathcal{M}_c^E\right)\right]^{zdc}.$$

Moreover,

$$\begin{aligned}\mathrm{Tr_s}&\left[g\exp\left(-\mathfrak{L}_c^E\right)\right]^{zdc}\\&=\mathrm{Tr_s}\left[g\exp\left(-\mathfrak{M}_c^E\right)\right]^{zdc}+\frac{\partial}{\partial c}\mathrm{Tr_s}\left[g\frac{3c}{2}\left|p\right|^2\exp\left(-\widehat{\underline{\mathfrak{C}}}^{\mathcal{M}^E,2}_{\phi,\mathcal{H}^c-\omega^H}\right)\right]\\&=-e\left(E^1,\nabla^{E^1}\right)\frac{c^2}{8}\mathrm{Tr}^*\left[\frac{1}{\left(-J_g^2+\frac{c^2}{4}\left(1-R^EJ_g^{-1}\right)\right)^2}\right].\end{aligned} \tag{7.9.4}$$

Proof. Let η be an even Grassmann variable which is such that $\eta^2=0$. We claim that we have the identity

$$\begin{aligned}\mathrm{Tr_s}\left[g\exp\left(-\widehat{\underline{\mathfrak{C}}}^{\mathcal{M}^E,2}_{\phi,\mathcal{H}^c-\omega^H}+\eta\left|p\right|^2\right)\right]&=\mathrm{Tr_s}\left[g\exp\left(-\widehat{\underline{\mathfrak{C}}}^{\mathcal{M}^E,2}_{\phi,\mathcal{H}^c-\omega^H}\right)\right]\\&+\eta\mathrm{Tr_s}\left[g\left|p\right|^2\exp\left(-\widehat{\underline{\mathfrak{C}}}^{\mathcal{M}^E,2}_{\phi,\mathcal{H}^c-\omega^H}\right)\right].\end{aligned} \tag{7.9.5}$$

Note that since the objects appearing in (7.9.5) are only generalized supertraces, the identity in (7.9.5) is not entirely trivial. However, it can be easily proved by the methods used in the proof of [B90, Theorem 4.6]. Namely, we can express the left-hand side of (7.9.5) by using a kernel version of Duhamel's formula, which only contains terms of degree 0 and 1 in the variable η. Then we use the fact that the finite dimensional version of our generalized supertrace vanishes on supercommutators, together with equation (7.3.2) in Proposition 7.3.1. Details are left to the reader.

By proceeding as in the proof of Theorem 7.8.1 on the left-hand side of (7.9.5), we easily get the first identity in (7.9.2). To establish the second identity, we use equation (7.2.6), and also we proceed as in the proof of Theorem 7.8.1. By the first identity in (7.9.2), the contribution of $-\frac{3}{2}dc\left|p\right|^2$ to $\mathrm{Tr_s}\left[g\exp\left(-\mathfrak{M}_c^E\right)\right]$ is given by

$$-e\left(E^1,\nabla^{E^1}\right)\frac{3}{4}zdc\mathrm{Tr}^*\left[\frac{1}{-J_g^2+\frac{c^2}{4}\left(1-R^EJ_g^{-1}\right)}\right]. \tag{7.9.6}$$

To obtain the contribution of the first terms in formula (7.2.6) for $\underline{\widehat{\mathfrak{G}}}^{\mathcal{M}}_{\phi,\mathcal{H}^c-\omega^H}$, we proceed again as in the proof Theorem 7.8.1. Instead of (7.7.32), we have now

$$\vartheta_t = -\begin{pmatrix} \frac{i}{2\sqrt{2}} dcp_t \\ \frac{1}{4\sqrt{2}} \left(-5cz + dc\right) p_t \\ \frac{-i}{4\sqrt{2}} \left(7cz + dc\right) p_t \end{pmatrix}. \tag{7.9.7}$$

Instead of (7.7.36), we get

$$\left\langle \left(C - \lambda\right)^{-1} \theta, \theta \right\rangle = zdc \frac{c^2 \left(2\lambda + 3x\right)}{8P_{c,x}\left(\lambda\right)}. \tag{7.9.8}$$

Also (7.7.45) is correspondingly modified. By proceeding as in the proof of Theorems 7.7.2 and 7.8.1, and using (7.9.8), we find that the contribution of the first two terms in equation (7.2.6) for $\underline{\widehat{\mathfrak{G}}}^{\mathcal{M}^E}_{\phi,\mathcal{H}^c-\omega^H}$ is given by

$$e\left(E^1, \nabla^{E^1}\right) \frac{c^2}{8} zdc\mathrm{Tr}^* \left[\frac{2 - 3R^E J_g^{-1}}{\left(-J_g^2 + \frac{c^2}{4}\left(1 - R^E J_g^{-1}\right)\right)^2} \right]. \tag{7.9.9}$$

By summing (7.9.6) and (7.9.9), we get the second identity in (7.9.2). Using Theorem 7.8.1 and the second identity in (7.9.2), we get (7.9.4).

We will establish the last identity in (7.9.2). In the sequel, we make $dc = 0$. We use (4.3.12) and we get

$$\mathrm{Tr_s}\left[g\left|p\right|^2 h'\left(\underline{\widehat{\mathfrak{D}}}_c^{\mathcal{M}^E}\right)\right] = \left(1 + 2\frac{\partial}{\partial a}\right) \mathrm{Tr_s}\left[g\left|p\right|^2 \exp\left(-a\underline{\widehat{\mathfrak{C}}}_c^{\mathcal{M}^E,2}\right)\right]\Big|_{a=1}. \tag{7.9.10}$$

Now the right-hand side of (7.9.10) can be evaluated by the same method as the first identity in (7.9.2). Instead we will use another method. Let R_a be the map $s\left(y,p\right) \to s\left(\sqrt{a}y, \sqrt{a}p\right)$. Using the first identity in (7.2.6), we observe that when conjugating $a\underline{\widehat{\mathfrak{C}}}_c^{\mathcal{M}^E,2}$ by R_a, we get the operator $\psi_a \underline{\widehat{\mathfrak{C}}}_{ac}^{\mathcal{M}^E,2} \psi_a^{-1}$, so that c has been replaced by ac, and R^E by aR^E. Also, especially when 1 is an eigenvalue of g, one has to be careful, because the effect of the conjugation is not trivial on the generalized supertrace. Ultimately we get

$$\mathrm{Tr_s}\left[g\left|p\right|^2 \exp\left(-a\underline{\widehat{\mathfrak{C}}}_c^{\mathcal{M}^E,2}\right)\right] = \frac{a}{2} e\left(E^1, \nabla^{E^1}\right) \mathrm{Tr}^*\left[\frac{1}{-J_g^2 + \frac{a^2c^2}{4}\left(1 - aR^E J_g^{-1}\right)}\right]. \tag{7.9.11}$$

By the second equality in (7.9.2), by (7.9.10) and (7.9.11), we obtain the last equality in (7.9.2). The proof of our theorem is completed. □

Remark 7.9.2. Although we have given a direct proof of the identities in Theorem 7.9.1, some of them should be viewed as almost tautological. Indeed one can give a direct proof of the first identity in (7.9.4) similar to the proof of (4.3.17) in Theorem 4.3.6. Similarly the third identity in (7.9.2) simply comes from two related evaluations of the components of an odd form, very similar to the identity (4.5.2) for $u_{b,t}$.

7.10 THE MELLIN TRANSFORM OF CERTAIN FOURIER SERIES

If $z \in \mathbf{C}$, $\sqrt{z}$ denotes an arbitrary (but fixed) square root of z. Now we follow [B94, Definition 4.1]. For $u, \eta, x \in \mathbf{C}$, put

$$\sigma(u,\eta,x) = 2\sinh\left(\frac{x-2\eta+\sqrt{x^2+4u}}{4}\right) 2\sinh\left(\frac{-x+2\eta+\sqrt{x^2+4u}}{4}\right). \tag{7.10.1}$$

Observe that $-\sigma(u,\eta,x)$ is the analogue of $\tau(c,\eta,x)$ in (7.1.3), where the polynomial of degree $P_{c,x}(\lambda)$ in (7.1.1) is replaced by the polynomial of degree 2,

$$Q_{u,x}(\lambda) = \lambda^2 + x\lambda - u. \tag{7.10.2}$$

Clearly,

$$\sigma\left(c^2, i\theta, 0\right) = 2\left(\cosh(c) - \cos(\theta)\right). \tag{7.10.3}$$

Using (7.10.3) and an easy computation given in [BZ94, eq. (5.45)], we get for $c > 0$,

$$\frac{\frac{\partial}{\partial c}\sigma\left(c^2,i\theta,0\right)}{\sigma\left(c^2,i\theta,0\right)} = \frac{\sinh(c)}{\cosh(c)-\cos(\theta)} = 1 + 2\sum_{n\geq 1} e^{-nc}\cos(n\theta). \tag{7.10.4}$$

By (7.10.4), we find that as $c \to 0$,

$$\begin{aligned}\frac{\frac{\partial}{\partial c}\sigma\left(c^2,i\theta,0\right)}{\sigma\left(c^2,i\theta,0\right)} &= \frac{2}{c} + \mathcal{O}(c) \text{ if } \theta \in 2\pi\mathbf{Z},\\ &= \mathcal{O}(c) \text{ if } \theta \notin 2\pi\mathbf{Z},\end{aligned} \tag{7.10.5}$$

and that as $c \to +\infty$,

$$\frac{\frac{\partial}{\partial c}\sigma\left(c^2,i\theta,0\right)}{\sigma\left(c^2,i\theta,0\right)} = 1 + \mathcal{O}\left(e^{-c}\right). \tag{7.10.6}$$

By [B94, Proposition 4.2],

$$\sigma(u,i\theta,0) = \left(\theta^2 + i\theta x + u\right)\prod_{k\in\mathbf{Z}^*}\left(\frac{(\theta+2k\pi)^2 + i(\theta+2k\pi)x + u}{4k^2\pi^2}\right). \tag{7.10.7}$$

From (7.10.7), if $c \in \mathbf{R}^*$, we get as in [BG01, Proposition 4.14],

$$\frac{\frac{\partial}{\partial c}\sigma\left(c^2,i\theta,0\right)}{\sigma\left(c^2,i\theta,0\right)} = \sum_{k\in\mathbf{Z}}\frac{2c}{(\theta+2k\pi)^2+c^2}. \tag{7.10.8}$$

Definition 7.10.1. For $\theta \in \mathbf{R}, s \in \mathbf{C}, \mathrm{Re}(s) > 1$, put

$$\zeta(\theta,s) = \sum_{n=1}^{+\infty}\frac{\cos(n\theta)}{n^s}, \qquad \eta(\theta,s) = \sum_{n=1}^{+\infty}\frac{\sin(n\theta)}{n^s}. \tag{7.10.9}$$

Then $\zeta(\theta,s), \eta(\theta,s)$ are the real and imaginary parts of the Lerch series [Le88] $L(\theta,s) = \sum_{n=1}^{+\infty} \frac{e^{in\theta}}{n^s}$. If $\theta \notin 2\pi\mathbf{Z}$, $s \mapsto \zeta(\theta,s)$ extends to a holomorphic function on $\mathbf{C}$, if $\theta \in 2\pi\mathbf{Z}$, $s \mapsto \zeta(y,s)$ extends to a meromorphic function on $\mathbf{C}$ with a simple pole at $s=1$. Also $s \to \eta(\theta,s)$ extends to a holomorphic function on $\mathbf{C}$.

By (7.10.4), we find that if $s \in \mathbf{C}, \operatorname{Re}(s) > 1$,

$$\frac{1}{\Gamma(s)} \int_0^{+\infty} c^{s-1} \frac{1}{2} \left(\frac{\frac{\partial}{\partial c} \sigma(c^2, i\theta, 0)}{\sigma(c^2, i\theta, 0)} - 1 \right) dc = \zeta(\theta, s). \tag{7.10.10}$$

By the above, it is clear that the left-hand side of (7.10.10) extends to a meromorphic function of $s \in \mathbf{C}$, with a simple pole at $s=1$, which is holomorphic if $\theta \notin 2\pi\mathbf{Z}$. In particular (7.10.10) is an equality of meromorphic or holomorphic functions on $\mathbf{C}$.

By (7.10.5), (7.10.6), we find that as $c \to 0$,

$$\begin{aligned} -\frac{1}{4} c \frac{\partial}{\partial c} \frac{1}{c} \frac{\frac{\partial}{\partial c} \sigma(c^2, i\theta, 0)}{\sigma(c^2, i\theta, 0)} &= \frac{1}{c^2} + \mathcal{O}(c^2) \text{ if } \theta \in 2\pi\mathbf{Z}, \\ &= \mathcal{O}(c^2) \text{ if } \theta \notin 2\pi\mathbf{Z}, \end{aligned} \tag{7.10.11}$$

and that as $c \to +\infty$,

$$-\frac{1}{4} c \frac{\partial}{\partial c} \frac{1}{c} \frac{\frac{\partial}{\partial c} \sigma(c^2, i\theta, 0)}{\sigma(c^2, i\theta, 0)} = \frac{1}{4c} + \mathcal{O}(e^{-c}). \tag{7.10.12}$$

By (7.10.8),

$$-\frac{1}{4} c \frac{\partial}{\partial c} \frac{1}{c} \frac{\frac{\partial}{\partial c} \sigma(c^2, i\theta, 0)}{\sigma(c^2, i\theta, 0)} = \sum_{k \in \mathbf{Z}} \frac{c^2}{\left((\theta + 2k\pi)^2 + c^2 \right)^2}. \tag{7.10.13}$$

Using (7.10.10) and integration by parts, we find that for $s \in \mathbf{C}, \operatorname{Re}(s) > 1$,

$$-\frac{1}{\Gamma(s)} \int_0^{+\infty} c^s \frac{1}{4} c \frac{\partial}{\partial c} \frac{1}{c} \left(\frac{\frac{\partial}{\partial c} \sigma(c^2, i\theta, 0)}{\sigma(c^2, i\theta, 0)} - 1 \right) dc = \frac{1}{2} (s+1) \zeta(\theta, s). \tag{7.10.14}$$

If $f(x)$ is an analytic function of $x \in \mathbf{C}$, we denote by $f^{(>0)}(x)$ the function $f(x) - f(0)$.

Definition 7.10.2. For $c \in \mathbf{R}^*, x \in \mathbf{C}, \theta \in \mathbf{R}^*$, put

$$K^\theta(c, x) = \frac{2c}{\theta^2 + c^2 \left(1 - \frac{ix}{\theta}\right)}, \quad L^\theta(c, x) = \frac{c^2}{\left(\theta^2 + c^2 \left(1 - \frac{ix}{\theta}\right)\right)^2}. \tag{7.10.15}$$

An elementary computation using finite increments shows that given $M > 0$, there is a constant $C > 0$ such that if c, θ are taken as before and $x \in \mathbf{C}, |x| \leq \inf(M, |\theta|/2)$, then

$$\left| K^{\theta(>0)}(c, x) \right| \leq C \frac{c^3}{|\theta| (\theta^2 + c^2)^2}, \quad \left| L^{\theta(>0)}(c, x) \right| \leq C \frac{c^4}{|\theta| (\theta^2 + c^2)^3}. \tag{7.10.16}$$

In the sequel, we denote by $\sum'_{k\in\mathbf{Z}} K^{2k\pi+\theta}(c,x)$, $\sum'_{k\in\mathbf{Z}} L^{2k\pi+\theta}(c,x)$ the sum of the corresponding series, where we take as a convention that if $2k\pi+\theta$ vanishes, the corresponding term is omitted. Similar conventions will be used with other functions as well.

Using (7.10.16) and an obvious integral bound, we find that given $\theta \in \mathbf{R}$, there exists $C'>0$ such that if $x\in\mathbf{C}, |x|<1/2$ if $\theta\in 2\pi\mathbf{Z}$, $|x|<\frac{1}{2}\inf_{k\in\mathbf{Z}}|\theta+2k\pi|$ if $\theta\notin 2\pi\mathbf{Z}$, for $c\geq 1$,

$$\left|\sum_{k\in\mathbf{Z}}{}' K^{2k\pi+\theta(>0)}(c,x)\right| \leq \frac{C'}{c}, \quad \left|\sum_{k\in\mathbf{Z}}{}' L^{2k\pi+\theta(>0)}(c,x)\right| \leq \frac{C'}{c^2}. \tag{7.10.17}$$

By (7.10.6), (7.10.8), (7.10.12), (7.10.13), as $c\to+\infty$,

$$\sum_{k\in\mathbf{Z}}{}' K^{2k\pi+\theta}(c,0) = 1+\mathcal{O}\left(\frac{1}{c}\right), \quad \sum_{k\in\mathbf{Z}}{}' L^{2k\pi+\theta}(c,0) = \frac{1}{4c}+\mathcal{O}\left(\frac{1}{c^2}\right). \tag{7.10.18}$$

By (7.10.17), (7.10.18), we conclude that as $c\to+\infty$,

$$\sum_{k\in\mathbf{Z}}{}' K^{2k\pi+\theta}(c,x) = 1+\mathcal{O}\left(\frac{1}{c}\right), \quad \sum_{k\in\mathbf{Z}}{}' L^{2k\pi+\theta}(c,x) = \frac{1}{4c}+\mathcal{O}\left(\frac{1}{c^2}\right). \tag{7.10.19}$$

By splitting the integral

$$\frac{1}{\Gamma(s)}\int_0^{+\infty} c^s\left(\sum_{k\in\mathbf{Z}}{}' L^{2k\pi+\theta}(c,x)-\frac{1}{4c}\right)dc$$

into two pieces $\int_0^1$ and $\int_1^{+\infty}$, the first piece is holomorphic in $s\in\mathbf{C}, \mathrm{Re}(s)>1$, and extends to a holomorphic function near $s=0$, and the second piece is holomorphic on $s\in\mathbf{C}, \mathrm{Re}(s)<1$, so that the integral itself is holomorphic near $s=0$.

Recall that if $f(x)$ is a holomorphic function, we defined the holomorphic function $Qf(x)$ in (1.9.1).

Definition 7.10.3. If $x\in\mathbf{C}, \theta\in\mathbf{R}, |x|<2\pi$ when $\theta\in 2\pi\mathbf{Z}$, $|x|<\inf_{k\in\mathbf{Z}}|\theta+2k\pi|$ when $\theta\notin 2\pi\mathbf{Z}$, put

$$\mathbf{I}(\theta,x) = \frac{\partial}{\partial s}\left[\frac{1}{\Gamma(s)}\int_0^{+\infty} c^s\left(\sum_{k\in\mathbf{Z}}{}' L^{2k\pi+\theta}(c,x)-\frac{1}{4c}\right)dc\right]\Big|_{s=0}, \tag{7.10.20}$$

$$\mathbf{J}(\theta,x) = Q\mathbf{I}(\theta,x).$$

Now we recall a few definitions in [BG01, Definitions 4.28 and 4.33].

Definition 7.10.4. For $\theta\in\mathbf{R}^*, x\in\mathbf{C}, |x|<|\theta|$, put

$$I^\theta(x) = \frac{\pi}{4}\frac{1}{|\theta|}\left(1-\frac{ix}{\theta}\right)^{-3/2}, \quad J^\theta(x) = \frac{\pi}{4}\frac{1}{|\theta|}\left(1-\frac{ix}{\theta}\right)^{-1}. \tag{7.10.21}$$

By [BG01, Proposition 4.34],

$$QI^{\theta}(x) = J^{\theta}(x). \tag{7.10.22}$$

Similarly, in [BG01, Definitions 4.21 and 4.25], for $x \in \mathbf{C}, \theta \in \mathbf{R}, |x| < 2\pi$ if $\theta \in 2\pi\mathbf{Z}$, $|x| < \inf_{k\in\mathbf{Z}} |2k\pi + \theta|$ if $\theta \notin 2\pi\mathbf{Z}$, functions $I(\theta, x), J(\theta, x)$ were defined such that

$$J(\theta, x) = QI(\theta, x). \tag{7.10.23}$$

In the sequel, we use the notation

$${}^{0}I(\theta, x) = I(\theta, x) - I(0,0), \quad {}^{0}J(\theta, x) = J(\theta, x) - J(0,0). \tag{7.10.24}$$

By (7.10.23), (7.10.24), we get

$${}^{0}J(\theta, x) = Q^{0}I(\theta, x). \tag{7.10.25}$$

In the sequel, sums like $\sum'_{k\in\mathbf{Z}} \left(I^{2k\pi+\theta}(x) - I^{2k\pi}(0)\right)$ do appear. It is understood that if $2k\pi + \theta$ or $2k\pi$ vanishes, the corresponding term $I^{2k\pi+\theta}(x)$ or $I^{2k\pi}(0)$ is not counted in the sum. Similar conventions will be used with other functions as well.

By [BG01, Theorems 4.30 and 4.35],

$$\begin{aligned} I(\theta, x) = \frac{1}{2}\Bigg[& \sum_{\substack{p\in\mathbf{N} \\ p \text{ even}}} \frac{(2p+1)!}{(p!)^3} \frac{\partial\zeta}{\partial s}(\theta, -p) \left(\frac{x}{4}\right)^p \\ & + i \sum_{\substack{p\in\mathbf{N} \\ p \text{ odd}}} \frac{(2p+1)!}{(p!)^3} \frac{\partial\eta}{\partial s}(\theta, -p) \left(\frac{x}{4}\right)^p \Bigg], \end{aligned}$$

$${}^{0}I(\theta, x) = \sum_{k\in\mathbf{Z}}{}' \left(I^{2k\pi+\theta}(x) - I^{2k\pi}(0)\right), \tag{7.10.26}$$

$$J(\theta, x) = \frac{1}{2}\left[\sum_{\substack{p\in\mathbf{N} \\ p \text{ even}}} \frac{\partial\zeta}{\partial s}(\theta, -p) \frac{x^p}{p!} + i \sum_{\substack{p\in\mathbf{N} \\ p \text{ odd}}} \frac{\partial\eta}{\partial s}(\theta, -p) \frac{x^p}{p!} \right],$$

$${}^{0}J(\theta, x) = \sum_{k\in\mathbf{Z}}{}' \left(J^{2k\pi+\theta}(x) - J^{2k\pi}(0)\right).$$

By [BG01, Theorem 4.37], if $\theta \notin 2\pi\mathbf{Z}$,

$$J(\theta, x) = \frac{1}{2}\frac{\partial\zeta}{\partial s}(\theta - ix, 0), \tag{7.10.27}$$

and by [BG01, Theorem 4.38], if $\theta' \in]-2\pi, 2\pi[\setminus \{0\}, |x| < \inf_{k\in\mathbf{Z}} |\theta' + 2k\pi|$,

$$J(\theta', x) = J(0, x + i\theta') + J^{\theta'}(x). \tag{7.10.28}$$

Theorem 7.10.5. *The following identity holds:*

$$\mathbf{I}(\theta, x) = I(\theta, x) - \frac{1}{4}, \qquad \mathbf{J}(\theta, x) = J(\theta, x) - \frac{1}{4}. \tag{7.10.29}$$

Proof. By (7.10.20), (7.10.23), the second identity in (7.10.29) follows from the first one.

We use (7.10.13) to obtain a more explicit expression for the left-hand side of (7.10.14). We claim that the sum $\sum_{k\in\mathbf{Z}}$ in (7.10.13) can be replaced by the truncated sum $\sum'_{k\in\mathbf{Z}}$. Indeed these two sums differ at most by the function $1/c^2$, whose contribution to a Mellin transform vanishes identically. By comparing (7.10.13) and (7.10.15) with (7.10.20), we get

$$\mathbf{I}(\theta,0)=\frac{1}{2}\frac{\partial}{\partial s}\left((s+1)\,\zeta(\theta,s)\right)|_{s=0}. \tag{7.10.30}$$

By Lerch's formula [W76, chapter 7, eqs. (15)-(23)] as used in [BZ94, eqs. (5.51)-(5.54)], we get

$$\zeta(\theta,0)=-\frac{1}{2}. \tag{7.10.31}$$

By (7.10.26), (7.10.30), (7.10.31), we get equation (7.10.29) for $x=0$.

By (7.10.19), (7.10.20), we get

$$\mathbf{I}^{(>0)}(\theta,x)=\int_0^{+\infty}\sum_{k\in\mathbf{Z}}{}' L^{2k\pi+\theta(>0)}(c,x)\,dc. \tag{7.10.32}$$

Clearly, for $\theta\neq 0$,

$$\int_0^{+\infty}L^{\theta}(c,x)\,dc=\frac{1}{|\theta|}\left(1-\frac{ix}{\theta}\right)^{-3/2}\int_0^{+\infty}\frac{c^2}{\left(1+c^2\right)^2}dc. \tag{7.10.33}$$

By differentiating the equality valid for $y>0$,

$$\int_0^{+\infty}\frac{1}{1+yc^2}dc=\frac{\pi}{2\sqrt{y}}, \tag{7.10.34}$$

we get

$$\int_0^{+\infty}\frac{c^2}{\left(1+c^2\right)^2}dc=\frac{\pi}{4}. \tag{7.10.35}$$

By (7.10.21), and by (7.10.32), (7.10.33), (7.10.35), we get

$$\mathbf{I}^{(>0)}(\theta,x)=\sum_{k\in\mathbf{Z}}{}' I^{2k\pi+\theta(>0)}(x). \tag{7.10.36}$$

Comparing with the second identity in (7.10.26), we see that

$$\mathbf{I}^{(>0)}(\theta,x)=I^{(>0)}(\theta,x). \tag{7.10.37}$$

Since (7.10.29) has already been established for $x=0$, by (7.10.37), we get (7.10.29) in full generality. □

Remark 7.10.6. The identity in (7.10.29) is stunning. Indeed the function $I(\theta,x)$ was obtained in [BG01] by a construction which is very different from the present one, based on the family of Witten Laplacians along the fibers E. The fact that the present construction gives essentially the same answer demonstrates the extraordinary rigidity of the quantities we are considering. It should be pointed out that the functions $I^{\theta}(x), J^{\theta}(x)$ also appeared in another context in [BG04, Definition 4.5], when evaluating the defect in the behavior of certain currents with respect to Morse-Bott functions.

7.11 THE HYPOELLIPTIC TORSION FORMS FOR VECTOR BUNDLES

By (7.10.19), we find easily that as $c \to +\infty$,

$$\begin{aligned}\mathrm{Tr}^*\left[\frac{2c}{-J_g^2 + c^2\left(1 - R^E J_g^{-1}\right)}\right] &= \dim E + \mathcal{O}\left(\frac{1}{c}\right), \\ \mathrm{Tr}^*\left[\frac{c^2}{\left(-J_g^2 + c^2\left(1 - R^E J_g^{-1}\right)\right)^2}\right] &= \frac{\dim E}{4c} + \mathcal{O}\left(\frac{1}{c^2}\right).\end{aligned} \tag{7.11.1}$$

Also observe that by equation (7.8.1) in Theorem 7.8.1, the second form in (7.11.1) is intimately related to $\mathrm{Tr_s}\left[g \exp\left(-\mathfrak{L}_c^E\right)\right]$. In chapter 8, it will precisely appear in this way.

Recall that the map $\varphi : \Lambda^{\cdot}\left(T^*S\right) \to \Lambda^{\cdot}\left(T^*S\right)$ was defined in section 1.7. Also the map $Q \in \mathrm{End}\left(\Lambda^{\cdot}\left(T^*S\right)\right)$ was defined in (1.9.5).

Definition 7.11.1. Set

$$\begin{aligned}\mathbf{I}_g\left(E, \nabla^E\right) = \frac{\partial}{\partial s}\Bigg[\frac{1}{\Gamma(s)}\int_0^{+\infty} c^s \Bigg(\varphi \mathrm{Tr}^*\left[\frac{c^2}{\left(-J_g^2 + c^2\left(1 - R^E J_g^{-1}\right)\right)^2}\right] \\ - \frac{\dim E}{4c}\Bigg) dc\Bigg]\Big|_{s=0},\end{aligned} \tag{7.11.2}$$

$$\mathbf{J}_g\left(E, \nabla^E\right) = Q\mathbf{I}_g\left(E, \nabla^E\right).$$

As we saw before, g has eigenvalues $1, e^{\pm i\theta_j}$, and -1. Let $B \in \mathrm{End}\left(E\right) \otimes_{\mathbf{R}} \mathbf{C}$ commuting with g, which is 0 on E^1, which has semidiagonal blocks $\begin{pmatrix} 0 & -\theta_j \\ \theta_j & 0 \end{pmatrix}$ on $E^{e^{\pm i\theta_j}}$, and which is $i\pi$ on E^{-1}. Note that B is real if E^{-1} is reduced to 0. Moreover,

$$g = e^B. \tag{7.11.3}$$

Finally, since R^E commutes with g, it also commutes with B.

Proposition 7.11.2. *The following identity of real closed differential forms holds:*

$$\mathbf{I}_g\left(E, \nabla^E\right) = \mathrm{Tr}\left[\mathbf{I}\left(-iB, -\frac{R^E}{2\pi}\right)\right], \quad \mathbf{J}_g\left(E, \nabla^E\right) = \mathrm{Tr}\left[\mathbf{J}\left(-iB, -\frac{R^E}{2\pi}\right)\right]. \tag{7.11.4}$$

Proof. This is a trivial consequence of (7.10.15), (7.10.20), (7.10.23), and (7.11.2). □

Remark 7.11.3. By using equation (7.8.1) and comparing with (7.11.2), it is legitimate to call $\mathbf{I}_g\left(E, \nabla^E\right), \mathbf{J}_g\left(E, \nabla^E\right)$ the hypoelliptic torsion forms for the vector bundle E.

In [BG01, eqs. (4.65) and (4.69)], the following closed forms were introduced:

$$I_g\left(E,\nabla^E\right)=\operatorname{Tr}\left[I\left(-iB,-\frac{R^E}{2\pi}\right)\right],\quad J_g\left(E,\nabla^E\right)=\operatorname{Tr}\left[J\left(-iB,-\frac{R^E}{2\pi}\right)\right].\tag{7.11.5}$$

Note that with respect to [BG01], we replaced $-\frac{R^E}{2i\pi}$ by $-\frac{R^E}{2\pi}$. The forms $I_g\left(E,\nabla^E\right),J_g\left(E,\nabla^E\right)$ were obtained in [BG01] as elliptic torsion forms of the vector bundle E.

By Theorem 7.10.5 and Proposition 7.11.2, we get

$$\mathbf{I}_g\left(E,\nabla^E\right)=I_g\left(E,\nabla^E\right)-\frac{1}{4}\dim E,\quad \mathbf{J}_g\left(E,\nabla^E\right)=J_g\left(E,\nabla^E\right)-\frac{1}{4}\dim E.\tag{7.11.6}$$

This is still a version of the extraordinary coincidences which were alluded to in Remark 7.10.6. The content of (7.11.6) is that the elliptic and hypoelliptic torsion forms of E are essentially equivalent.

When replacing I,J by ${}^0I,{}^0J$, the forms ${}^0I_g\left(E,\nabla^E\right),{}^0J_g\left(E,\nabla^E\right)$ were also considered in [BG01, section 7.1], where they play a critical role. By (7.10.25),

$${}^0J_g\left(E,\nabla^E\right)=Q^0I_g\left(E,\nabla^E\right).\tag{7.11.7}$$

By Lerch's formula [W76, chapter 7, eqs. (15)-(23)], we know that

$$\frac{\partial\zeta}{\partial s}(0,0)=-\frac{1}{2}\log(2\pi).\tag{7.11.8}$$

So by Theorem 7.10.5 and by (7.11.8), we get

$$\begin{aligned}\mathbf{I}_g\left(E,\nabla^E\right)&={}^0I_g\left(E,\nabla^E\right)-\frac{1}{4}\left(\log(2\pi)+1\right)\dim E,\\ \mathbf{J}_g\left(E,\nabla^E\right)&={}^0J_g\left(E,\nabla^E\right)-\frac{1}{4}\left(\log(2\pi)+1\right)\dim E.\end{aligned}\tag{7.11.9}$$

Incidentally observe that by (7.8.1), (7.9.4), (7.11.1), as $c\to+\infty$,

$$\operatorname{Tr_s}\left[g\exp\left(-\mathfrak{M}_c^E\right)\right]^{zdc}=-\frac{\dim E}{4c}e\left(E^1,\nabla^{E^1}\right)+\mathcal{O}\left(\frac{1}{c^2}\right).\tag{7.11.10}$$

Chapter Eight

Hypoelliptic and elliptic torsions: a comparison formula

In this chapter, we establish the main result of the book. Namely, we give an explicit formula relating the hypoelliptic torsion forms to the corresponding elliptic torsion forms. The proofs of several intermediate results are deferred to the following chapters.

This chapter is organized as follows. In section 8.1, we construct natural secondary Chern classes attached to two couples of generalized metrics on $\mathfrak{H}^{\cdot}(X,F)$.

In section 8.2, we state our main result. The next sections are devoted to the proof of this result.

In section 8.3, we introduce a rectangular contour Γ in $\mathbf{R}_{+}^{*2}$ on which the form a of section 4.2 is integrated. Our main formula will be obtained by pushing Γ to the boundary of $\mathbf{R}_{+}^{*2}$.

In section 8.4, we state four intermediate results, which will be used in the proof of our main formula. The proofs of these results are deferred to chapters 10, 11, and 12.

In section 8.5, the asymptotics of the integral of a on the four sides of the rectangle Γ is studied under the deformation of Γ.

In section 8.6, the divergences of the integrals of a on the four sides of the rectangle are matched.

In section 8.7, our final identity is shown to be our main result.

In this chapter, S is assumed to be compact.

8.1 ON SOME SECONDARY CHERN CLASSES

We use the notation of chapters 5 and 6.

By (3.1.4), $\mathfrak{H}^{\cdot}(X,F)$ is $H^{\cdot}(X,F)$ for $c>0$, and $H^{\cdot-n}(X,F\otimes o(TX))$ for $c<0$. Moreover, as we saw after (1.2.5), each of these last $\mathbf{Z}$-graded vector spaces inherits a standard Hermitian metric via the Hodge theory of X. We denote by $\mathfrak{h}_0^{\mathfrak{H}^{\cdot}(X,F)}$ the corresponding Hermitian metric on $\mathfrak{H}^{\cdot}(X,F)$ for $c>0$, and the product by $(-1)^n$ of the associated Hermitian metric for $c<0$.

Recall that $\mathfrak{H}^{\cdot}(X,F)$ is a $\mathbf{Z}$-graded flat vector bundle on S equipped with the flat Gauss-Manin connection $\nabla^{\mathfrak{H}^{\cdot}(X,F)}$. Also G acts on the fibers of $\mathfrak{H}^{\cdot}(X,F)$ and preserves the connection $\nabla^{\mathfrak{H}^{\cdot}(X,F)}$.

Now we use the notation of section 1.11. If $g_0^{\mathfrak{H}^{\cdot}(X,F)}, g_1^{\mathfrak{H}^{\cdot}(X,F)}$ are two split Hermitian metrics on $\mathfrak{H}^{\cdot}(X,F)$, set

$$\begin{aligned}
&\widetilde{h}_g\left(\nabla^{\mathfrak{H}^{\cdot}(X,F)}, g_0^{\mathfrak{H}^{\cdot}(X,F)}, g_1^{\mathfrak{H}^{\cdot}(X,F)}\right) \\
&=\sum_{i=0}^{2n}(-1)^i \widetilde{h}_g\left(\nabla^{\mathfrak{H}^i(X,F)}, g_0^{\mathfrak{H}^i(X,F)}, g_1^{\mathfrak{H}^i(X,F)}\right), \qquad (8.1.1)\\
&\widetilde{\mathrm{ch}}_g^{\circ}\left(\nabla^{\mathfrak{H}^{\cdot}(X,F)}, g_0^{\mathfrak{H}^{\cdot}(X,F)}, g_1^{\mathfrak{H}^{\cdot}(X,F)}\right) \\
&=\sum_{i=0}^{2n}(-1)^i \widetilde{\mathrm{ch}}_g^{\circ}\left(\nabla^{\mathfrak{H}^i(X,F)}, g_0^{\mathfrak{H}^i(X,F)}, g_1^{\mathfrak{H}^i(X,F)}\right).
\end{aligned}$$

As was observed in section 1.11, in the above, we may as well replace some of the metrics $g_0^{\mathfrak{H}^i(X,F)}, g_1^{\mathfrak{H}^i(X,F)}$ by their negatives, the point being that the sign of the corresponding objects should be the same for any i.

8.2 THE MAIN RESULT

We take $\underline{b_0} > 0$ small enough so that the results in section 3.5, of chapter 5 and of section 6.1 hold for $b \in]0, \underline{b}_0]$. Also we still take $c = \pm 1/b^2$. We will distinguish these two cases as the case $+$ and the case $-$.

Put

$$\begin{aligned}
\mathcal{T}_{h,g,0}\left(T^H M, g^{TX}, \nabla^F, g^F\right) &= \mathcal{T}_{h,g}\left(T^H M, g^{TX}, \nabla^F, g^F\right) \text{ if } c>0, \qquad (8.2.1)\\
&(-1)^n \mathcal{T}_{h,g}\left(T^H M, g^{TX}, \nabla^{F\otimes o(TX)}, g^{F\otimes o(TX)}\right) \text{ if } c<0.
\end{aligned}$$

Note that by (1.10.1), for $c<0$, we can rewrite (1.10.1) in the form

$$\mathcal{T}_{h,g,0}\left(T^H M, g^{TX}, \nabla^F, g^F\right) = -\mathcal{T}_{h,g}\left(T^H M, g^{TX}, \nabla^{\overline{F}^*}, g^{\overline{F}^*}\right). \qquad (8.2.2)$$

Theorem 8.2.1. *Assume that* $b_0 \in]0, \underline{b}_0]$. *For any* $i, 0 \leq i \leq 2n$, *the Hermitian form* $\mathfrak{h}_{b_0}^{\mathfrak{H}^i(X,F)}$ *or its negative is a Hermitian metric, and moreover* $\mathfrak{h}_{b_0}^{\mathfrak{H}^i(X,F)}$ *has the same sign as* $\mathfrak{h}_0^{\mathfrak{H}^i(X,F)}$.

The following identities hold:

$$\begin{aligned}
&-\mathcal{T}_{h,g,b_0}\left(T^H M, g^{TX}, \nabla^F, g^F\right) + \mathcal{T}_{h,g,0}\left(T^H M, g^{TX}, \nabla^F, g^F\right)\\
&-\widetilde{h}_g\left(\nabla^{\mathfrak{H}^{\cdot}(X,F)}, \mathfrak{h}_0^{\mathfrak{H}^{\cdot}(X,F)}, \mathfrak{h}_{b_0}^{\mathfrak{H}^{\cdot}(X,F)}\right) \pm \int_{X_g} e\left(TX_g\right)^0 I_g\left(TX|_{M_g}\right) \mathrm{Tr}^F[g] = 0\\
&\quad \text{in } \Omega^{\cdot}(S)/d\Omega^{\cdot}(S), \qquad (8.2.3)\\
&-\mathcal{T}_{\mathrm{ch},g,b_0}\left(T^H M, g^{TX}, \nabla^F, g^F\right) + \mathcal{T}_{\mathrm{ch},g,0}\left(T^H M, g^{TX}, \nabla^F, g^F\right)\\
&\quad -\widetilde{\mathrm{ch}}_g^{\circ}\left(\nabla^{\mathfrak{H}^{\cdot}(X,F)}, \mathfrak{h}_0^{\mathfrak{H}^{\cdot}(X,F)}, \mathfrak{h}_{b_0}^{\mathfrak{H}^{\cdot}(X,F)}\right)\\
&\quad \pm \int_{X_g} e\left(TX_g\right)^0 J_g\left(TX|_{M_g}\right) \mathrm{Tr}^F[g] = 0 \text{ in } \Omega^{\cdot}(S)/d\Omega^{\cdot}(S).
\end{aligned}$$

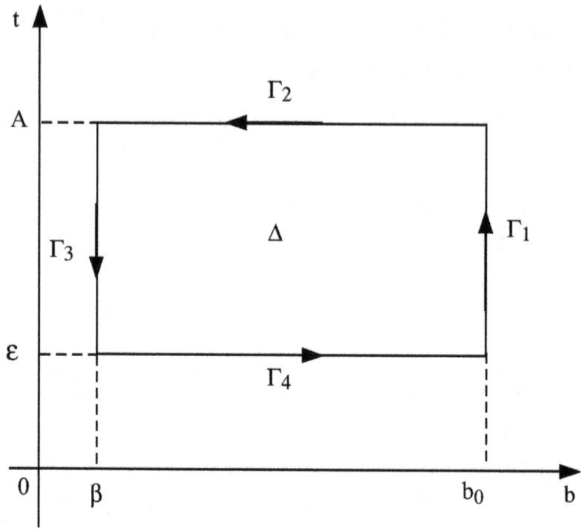

Figure 8.1

Proof. First we observe that by (1.9.7), (6.1.1), and (7.11.7), the second equation in (8.2.3) follows from the first one. The remaining sections are devoted to the proof of the first equation in (8.2.3). □

Remark 8.2.2. Observe that when applying the d operator to (8.2.1), we get an identity which itself follows from (1.8.2), (1.11.1), and (6.1.2). Also we observe that by (1.10.1), by Proposition 6.2.1, and by (8.2.2), (8.2.3) is compatible to Poincaré duality. Equivalently, it would be enough to prove our theorem in the case $c > 0$, the case $c < 0$ being simply a consequence.

8.3 A CONTOUR INTEGRAL

Here we use the notation of chapter 4. In particular the even differential form a on $S \times \mathbf{R}_+^{*2}$ was defined in Definition 4.3.1, and a formula for a was given in (4.3.6). Let β, ϵ, A be such that $0 < \beta < b_0, 0 < \epsilon < 1 < A$. Let Γ be the oriented rectangular contour in $\mathbf{R}_+^{2*}$ indicated in Figure 8.1. The contour Γ is made of four oriented pieces $\Gamma_1, \ldots, \Gamma_4$. It bounds a domain Δ.

Proposition 8.3.1. *The following identity of even forms holds on S:*

$$\int_\Gamma a = -d \int_\Delta a. \tag{8.3.1}$$

Proof. Since a is an odd closed form on $S \times \mathbf{R}_+^{2*}$, equation (8.3.1) follows from Stokes's formula. □

For $1 \leq k \leq 4$, set

$$I_k^0 = \int_{\Gamma_k} a. \tag{8.3.2}$$

Then by (8.3.1), we get

$$\sum_{k=1}^{4} I_k^0 = -d \int_{\Delta} a. \tag{8.3.3}$$

To obtain Theorem 8.2.1, we will make $A \to +\infty, \beta \to 0, \epsilon \to 0$ in this order in (8.3.3). We will study in succession each of the terms in the left-hand side of (8.3.3).

8.4 FOUR INTERMEDIATE RESULTS

We use the notation of Definitions 5.1.1 and 5.1.2. Put

$$w_{0,0} = \mp \frac{n}{4} \chi_g(F), \qquad w_{0,\infty} = \pm \frac{1}{2} \left(\chi'_g(F) - n\chi_g(F)\right). \tag{8.4.1}$$

By (1.8.7), (1.8.8),

$$\begin{aligned} w_{0,t} &= w_{0,0} + \mathcal{O}\left(\sqrt{t}\right) \text{ as } t \to 0, \\ &= w_{0,\infty} + \mathcal{O}\left(1/\sqrt{t}\right) \text{ as } t \to +\infty. \end{aligned} \tag{8.4.2}$$

We use the notation of chapter 7, with $S = M_g, E = TX|_{M_g}$.

Definition 8.4.1. For $c \in \mathbf{R}^*$, put

$$m_c = -\varphi \int_{X_g} \mathrm{Tr_s} \left[g \exp\left(-\mathfrak{M}_c^{TX|_{M_g}}\right)\right]^{zdc} \mathrm{Tr}^F[g]. \tag{8.4.3}$$

By (1.6.3) and (7.11.10), we know that as $c \to +\infty$,

$$m_c = \frac{n}{4c} \chi_g(F) + \mathcal{O}\left(\frac{1}{c^2}\right). \tag{8.4.4}$$

Theorem 8.4.2. *If $c = \pm 1/b^2$, for any $v \in]0,1[$, when $b \to 0$,*

$$\mathfrak{h}_b^{\mathfrak{H}^{\cdot}(X,F)} = \left(b\sqrt{\pi}\right)^{\pm n} \left(\mathfrak{h}_0^{\mathfrak{H}^{\cdot}(X,F)} + \mathcal{O}\left(b^v\right)\right). \tag{8.4.5}$$

Theorem 8.4.3. *For $b > 0$, as $\epsilon \to 0$,*

$$v_{\sqrt{\epsilon}b,\epsilon} \to \pm 2 \frac{m_{1/b^2}}{b^2}. \tag{8.4.6}$$

There exist $C > 0, \alpha \in]0,1]$ such that for $\epsilon \in]0,1], b \in]0,1]$,

$$\left|v_{\sqrt{\epsilon}b,\epsilon} - v_{0,\epsilon}\right| \leq C b^{\alpha}. \tag{8.4.7}$$

For any $b_0 \geq 1$, there exist $C > 0$ such that for $\epsilon \in]0,1], b \in [\sqrt{\epsilon}, b_0]$,

$$\left|v_{b,\epsilon}\right| \leq C \frac{\epsilon}{b^2}. \tag{8.4.8}$$

Remark 8.4.4. Observe that (8.4.4), (8.4.6) are compatible with (8.4.7), (8.4.8), and that (4.4.1) and (8.4.8) are also compatible. Theorem 8.4.2 will be proved in chapter 10, Theorem 8.4.3 in chapters 11, 12, and 13. More precisely equation (8.4.6) will be proved in chapter 11, equation (8.4.8) in chapter 12, and equation (8.4.7) in chapter 13.

Remark 8.4.5. By Theorem 8.4.2, we know that for $0 < b \leq \underline{b}_0$ and b small enough, for $c > 0$, for any $i, 0 \leq i \leq 2n$, $\mathfrak{h}_b^{\mathfrak{H}^i(X,F)}$ is a Hermitian metric. Moreover, as we saw in section 5.1, for $b \in]0, \underline{b}_0]$, the Hermitian forms $\mathfrak{h}_b^{\mathfrak{H}^\cdot(X,F)}$ are nondegenerate. It follows that for $0 < b \leq \underline{b}_0$, the signature of these Hermitian forms remains constant. Therefore for $b \in]0, \underline{b}_0]$, the $\mathfrak{h}_b^{\mathfrak{H}^i(X,F)}$ are Hermitian metrics. The same argument can be used for $c < 0$, except that if n is odd, the considered Hermitian forms are the negative of standard Hermitian metrics. Therefore the first part of Theorem 8.2.1 has been proved.

8.5 THE ASYMPTOTICS OF THE $\mathbf{I_k^0}$

We start from identity (8.3.3), which asserts that

$$\sum_{k=1}^{4} I_k^0 = 0 \text{ in } \Omega^\cdot(S)/d\Omega^\cdot(S). \tag{8.5.1}$$

Note that if α_n is a sequence of smooth exact forms on S which converges uniformly to a smooth form α, then α is still exact.

1) The term $\mathbf{I_1^0}$

Clearly,

$$I_1^0 = \int_\epsilon^A w_{b_0,t}\frac{dt}{t}. \tag{8.5.2}$$

- $A \to +\infty$

By Theorem 5.2.1, as $A \to +\infty$,

$$I_1^0 - w_{b_0,\infty}\log(A) \to I_1^1 = \int_\epsilon^1 w_{b_0,t}\frac{dt}{t} + \int_1^{+\infty}\left(w_{b_0,t} - w_{b_0,\infty}\right)\frac{dt}{t}. \tag{8.5.3}$$

- $\beta \to 0$

The term I_1^1 remains constant and equal to I_1^2.

- $\epsilon \to 0$

By Theorem 4.4.1, as $\epsilon \to 0$,

$$I_1^2 \to I_1^3 = \int_0^1 w_{b_0,t}\frac{dt}{t} + \int_1^{+\infty}\left(w_{b_0,t} - w_{b_0,\infty}\right)\frac{dt}{t}. \tag{8.5.4}$$

- Evaluation of I_1^3

Proposition 8.5.1. *The following identity holds:*

$$I_1^3 = -\mathcal{T}_{h,g,b_0}\left(T^H M, g^{TX}, \nabla^F, g^F\right) - \left(\Gamma'(1) + 2\left(\log(2) - 1\right)\right)w_{b_0,\infty}. \tag{8.5.5}$$

Proof. This follows from (1.8.10), (6.1.1), and (8.5.4). □

2) The term $\mathbf{I_2^0}$

We have the identity

$$I_2^0 = -\int_\beta^{b_0} v_{b,A} \frac{db}{b}. \tag{8.5.6}$$

• $A \to +\infty$

By Theorem 5.2.1, as $A \to +\infty$,

$$I_2^0 \to I_2^1 = -\int_\beta^{b_0} v_{b,\infty} \frac{db}{b}. \tag{8.5.7}$$

Also using [BLo95, Definition 1.12] or [BG01, Definition 1.10] or by (1.11.2), we get

$$I_2^1 = -\widetilde{h}_g \left(\nabla^{\mathfrak{H}^\cdot(X,F)}, \mathfrak{h}_\beta^{\mathfrak{H}^\cdot(X,F)}, \mathfrak{h}_{b_0}^{\mathfrak{H}^\cdot(X,F)} \right) \text{ in } \Omega^\cdot(S)/d\Omega^\cdot(S). \tag{8.5.8}$$

• $\beta \to 0$

Using obvious properties of the classes in (8.5.8) following from (1.11.2), we get

$$\begin{aligned} I_2^1 = \widetilde{h}_g &\left(\nabla^{\mathfrak{H}^\cdot(X,F)}, \mathfrak{h}_0^{\mathfrak{H}^\cdot(X,F)}, \mathfrak{h}_\beta^{\mathfrak{H}^\cdot(X,F)} \right) \\ &- \widetilde{h}_g \left(\nabla^{\mathfrak{H}^\cdot(X,F)}, \mathfrak{h}_0^{\mathfrak{H}^\cdot(X,F)}, \mathfrak{h}_{b_0}^{\mathfrak{H}^\cdot(X,F)} \right) \text{ in } \Omega^\cdot(S)/d\Omega^\cdot(S). \end{aligned} \tag{8.5.9}$$

By (1.11.2) and by Theorem 8.4.2, as $\beta \to 0$, for any $v \in]0,1[$,

$$\begin{aligned} \widetilde{h}_g &\left(\nabla^{\mathfrak{H}^\cdot(X,F)}, \mathfrak{h}_0^{\mathfrak{H}^\cdot(X,F)}, \mathfrak{h}_\beta^{\mathfrak{H}^\cdot(X,F)} \right) \\ &= \pm \frac{1}{2} \left(n \log(\beta) + \frac{n}{2} \log(\pi) \right) \chi_g(F) + \mathcal{O}(\beta^v). \end{aligned} \tag{8.5.10}$$

So by (8.5.9), (8.5.10), we find that as $\beta \to 0$,

$$\begin{aligned} I_2^1 &\mp \frac{1}{2} n \chi_g(F) \log(\beta) \to I_2^2 \\ &= -\widetilde{h}_g \left(\nabla^{\mathfrak{H}^\cdot(X,F)}, \mathfrak{h}_0^{\mathfrak{H}^\cdot(X,F)}, \mathfrak{h}_{b_0}^{\mathfrak{H}^\cdot(X,F)} \right) \pm \frac{n}{4} \log(\pi) \chi_g(F). \end{aligned} \tag{8.5.11}$$

• $\epsilon \to 0$

As $\epsilon \to 0$, I_2^2 remains constant and equal to I_2^3.

3) The term $\mathbf{I_3^0}$

We have the identity

$$I_3^0 = -\int_\epsilon^A w_{\beta,t} \frac{dt}{t}. \tag{8.5.12}$$

• $A \to +\infty$

By Theorem 5.2.1, as $A \to +\infty$,

$$I_3^0 + w_{\beta,\infty} \log(A) \to I_3^1 = -\int_\epsilon^1 w_{\beta,t} \frac{dt}{t} - \int_1^{+\infty} (w_{\beta,t} - w_{\beta,\infty}) \frac{dt}{t}. \tag{8.5.13}$$

• $\beta \to 0$

By Theorem 5.2.1, as $\beta \to 0$,

$$I_3^1 \to I_3^2 = -\int_\epsilon^1 w_{0,t}\frac{dt}{t} - \int_1^{+\infty}\left(w_{0,t} - w_{0,\infty}\right)\frac{dt}{t}. \tag{8.5.14}$$

• $\epsilon \to 0$

By (8.4.1), (8.4.2), we find that as $\epsilon \to 0$,

$$I_3^2 \pm \frac{n}{4}\chi_g(F)\log(\epsilon) \to I_3^3 = -\int_0^1 \left(w_{0,t} - w_{0,0}\right)\frac{dt}{t} - \int_1^{+\infty}\left(w_{0,t} - w_{0,\infty}\right)\frac{dt}{t}. \tag{8.5.15}$$

• Evaluation of I_3^3

Proposition 8.5.2. *The following identity holds:*

$$I_3^3 = \mathcal{T}_{h,g,0}\left(T^H M, g^{TX}, \nabla^F, g^F\right) + \left(\Gamma'(1) + 2\left(\log(2) - 1\right)\right)\left(w_{0,\infty} - w_{0,0}\right). \tag{8.5.16}$$

Proof. Our proposition follows from (1.8.11), (5.1.12), (5.1.13) and from (8.5.15). □

4) The term $\mathbf{I_4^0}$

Clearly,

$$I_4^0 = \int_\beta^{b_0} v_{b,\epsilon}\frac{db}{b}. \tag{8.5.17}$$

• $A \to +\infty$

The term I_4^0 remains constant and equal to I_4^1.

• $\beta \to 0$

By Theorem 5.2.1, as $\beta \to 0$,

$$I_4^1 + v_{0,\epsilon}\log(\beta) \to I_4^2 = \int_0^{b_0}\left(v_{b,\epsilon} - v_{0,\epsilon}\right)\frac{db}{b} + v_{0,\epsilon}\log(b_0). \tag{8.5.18}$$

• $\epsilon \to 0$

Take $\epsilon > 0$ small enough so that $b_0/\sqrt{\epsilon} > 1$. Set

$$J_1^0 = \int_0^1 \left(v_{\sqrt{\epsilon}b,\epsilon} - v_{0,\epsilon}\right)\frac{db}{b}, \qquad J_2^0 = \int_1^{b_0/\sqrt{\epsilon}} v_{\sqrt{\epsilon}b,\epsilon}\frac{db}{b}. \tag{8.5.19}$$

Clearly,

$$I_4^2 = J_1^0 + J_2^0 + v_{0,\epsilon}\log\left(\sqrt{\epsilon}\right). \tag{8.5.20}$$

By (8.4.4), as $b \to 0$,

$$2\frac{m_{1/b^2}}{b^2} = \frac{n}{2}\chi_g(F) + \mathcal{O}\left(b^2\right). \tag{8.5.21}$$

By (8.4.6), by equation (8.4.7) in Theorem 8.4.3, and by (8.5.21), as $\epsilon \to 0$,

$$J_1^0 \to J_1^1 = \pm\int_0^1 \left(2\frac{m_{1/b^2}}{b^2} - \frac{n}{2}\chi_g(F)\right)\frac{db}{b} = \pm\int_1^{+\infty}\left(m_c - \frac{n}{4c}\chi_g(F)\right)dc. \tag{8.5.22}$$

Moreover, by (8.4.6) and (8.4.8) in Theorem 8.4.3, as $\epsilon \to 0$,

$$J_2^0 \to J_2^1 = \pm \int_1^{+\infty} 2 \frac{m_{1/b^2}}{b^2} \frac{db}{b} = \pm \int_0^1 m_c dc. \tag{8.5.23}$$

So by (5.1.13), (8.5.20)-(8.5.23), we find that as $\epsilon \to 0$,

$$I_4^2 \mp \frac{n}{4} \chi_g (F) \log (\epsilon) \to I_4^3 = J_1^1 + J_2^1. \tag{8.5.24}$$

• Evaluation of I_4^3

Proposition 8.5.3. *The following identity holds:*

$$I_4^3 = \pm \left(\int_{X_g} e \left(TX_g, \nabla^{TX_g} \right) \mathbf{I}_g \left(TX|_{M_g}, \nabla^{TX|_{M_g}} \right) \mathrm{Tr}^F [g] + \left(3 - \Gamma' (1) - \log (2)\right) \frac{n}{4} \chi_g (F) \right). \tag{8.5.25}$$

Proof. Set

$$l_c = -\varphi \int_{X_g} \mathrm{Tr_s} \left[g \exp \left(-\mathfrak{L}_c^{TX|_{M_g}} \right) \right]^{zdc} \mathrm{Tr}^F [g] . \tag{8.5.26}$$

By the first identity in (7.9.2), (7.9.4), (7.11.1) and by (8.5.22)-(8.5.24), we get

$$\pm I_4^3 = \int_0^1 l_c dc + \int_1^{+\infty} \left(l_c - \frac{1}{4c} \chi_g (F) \right) dc + \frac{3}{4} n \chi_g (F) . \tag{8.5.27}$$

Moreover, by (7.9.4), (7.11.1), (7.11.2), we get

$$\int_{X_g} e (TX_g) \mathbf{I}_g (TX|_{M_g}) \mathrm{Tr}^F [g] = \int_0^1 l_c dc + \int_1^{+\infty} \left(l_c - \frac{n}{4c} \chi_g (F) \right) dc + \left(\Gamma' (1) + \log (2)\right) \frac{n}{4} \chi_g (F) . \tag{8.5.28}$$

By (8.5.27), (8.5.28), we get (8.5.25). □

8.6 MATCHING THE DIVERGENCES

Proposition 8.6.1. *The following identity holds:*

$$\sum_{k=1}^4 I_k^3 = 0 \text{ in } \Omega^{\cdot} (S) / d\Omega^{\cdot} (S) . \tag{8.6.1}$$

Proof. We start from equation (8.5.1). As $A \to +\infty$, by Proposition 5.1.3, and by (8.5.3), (8.5.13), we have the diverging terms

$$\left(w_{\beta,\infty} - w_{b_0,\infty} \right) \log (A) = 0. \tag{8.6.2}$$

From (8.5.1), (8.6.2), we get

$$\sum_{k=1}^{4} I_k^1 = 0 \text{ in } \Omega^{\cdot}(S)/d\Omega^{\cdot}(S). \tag{8.6.3}$$

By (5.1.13), (8.5.11), (8.5.18), as $\beta \to 0$, we have the diverging terms

$$\pm\Big(\frac{n}{2}\chi_g(F) - \frac{n}{2}\chi_g(F)\Big)\log(\beta) = 0. \tag{8.6.4}$$

So we get

$$\sum_{k=1}^{4} I_k^2 = 0 \text{ in } \Omega^{\cdot}(S)/d\Omega^{\cdot}(S). \tag{8.6.5}$$

By (8.5.15), (8.5.24), as $\epsilon \to 0$, we have the diverging terms

$$\Big(\pm\frac{n}{4}\chi_g(F) \mp \frac{n}{4}\chi_g(F)\Big)\log(\epsilon) = 0. \tag{8.6.6}$$

By (8.6.5), (8.6.6), we get (8.6.1). □

8.7 A PROOF OF THEOREM 8.2.1

Now we establish Theorem 8.2.1. Using Propositions 8.5.1, (8.5.11), Propositions 8.5.2 and 8.5.3, we get

$$\begin{aligned} &-\mathcal{T}_{h,g,b_0}\left(T^H M, g^{TX}, \nabla^F, g^F\right) + \mathcal{T}_{h,g,0}\left(T^H M, g^{TX}, \nabla^F, g^F\right) \\ &\quad - \widetilde{h}_g\Big(\nabla^{\mathfrak{H}^{\cdot}(X,F)}, h_0^{\mathfrak{H}^{\cdot}(X,F)}, h_{b_0}^{\mathfrak{h}^{\cdot}(X,F)}\Big) \pm \int_{X_g} e\left(TX_g, \nabla^{TX_g}\right) \\ &\quad \mathbf{I}_g\Big(TX|_{M_g}, \nabla^{TX|_{M_g}}\Big)\operatorname{Tr}^F[g] \\ &\quad - \left(\Gamma'(1) + 2(\log(2) - 1)\right)\left(w_{b_0,\infty} - w_{0,\infty} + w_{0,0}\right) \\ &\pm \left(3 + \log(\pi) - \Gamma'(1) - \log(2)\right)\frac{n}{4}\chi_g(F) = 0 \text{ in } \Omega^{\cdot}(S)/d\Omega^{\cdot}(S). \end{aligned} \tag{8.7.1}$$

As we saw before Theorem 5.2.1, $w_{0,\infty} = w_{b_0,\infty}$. Using (8.4.1), we get

$$\begin{aligned} -\left(\Gamma'(1) + 2(\log(2) - 1)\right)w_{0,0} \pm \left(3 + \log(\pi) - \Gamma'(1) - \log(2)\right)\frac{n}{4}\chi_g(F) \\ = \pm\left(\log(2\pi) + 1\right)\frac{n}{4}\chi_g(F). \end{aligned} \tag{8.7.2}$$

By (1.6.3), (8.7.1), (8.7.2), we get

$$\begin{aligned} &-\mathcal{T}_{h,g,b_0}\left(T^H M, g^{TX}, \nabla^F, g^F\right) + \mathcal{T}_{h,g,0}\left(T^H M, g^{TX}, \nabla^F, g^F\right) \\ &\quad - \widetilde{h}_g\Big(\nabla^{\mathfrak{H}^{\cdot}(X,F)}, h_0^{\mathfrak{H}^{\cdot}(X,F)}, h_{b_0}^{\mathfrak{h}^{\cdot}(X,F)}\Big) \pm \int_{X_g} e\left(TX_g, \nabla^{TX_g}\right) \\ &\Big(\mathbf{I}_g\Big(TX|_{M_g}, \nabla^{TX|_{M_g}}\Big) + \frac{n}{4}\left(\log(2\pi) + 1\right)\Big)\operatorname{Tr}^F[g] = 0 \text{ in } \Omega^{\cdot}(S)/d\Omega^{\cdot}(S). \end{aligned} \tag{8.7.3}$$

By (7.11.9) and (8.7.3), we get the first equation in (8.2.3). The proof of Theorem 8.2.1 is completed. □

Chapter Nine

A comparison formula for the Ray-Singer metrics

We make the same assumptions as in chapter 6, and we use the corresponding notation. Also we assume here that S is reduced to a point.

Recall that $b \in \mathbf{R}_+^*$ and that $c = \pm 1/b^2$. By Theorem 6.7.1, the generalized metric $\| \|_{\lambda,b}^2$ does not depend on b. Recall that a priori, $\| \|_{\lambda,b}^2$ is only a generalized equivariant metric, in the sense that the sign of the $\| \|_{\lambda_W}^2, W \in \widehat{G}$ is not necessarily positive.

When n is even, or when n is odd and $c > 0$, we denote by $\| \|_{\lambda,0}^2$ the corresponding more classical Ray-Singer metric on λ which was constructed in section 1.12. When n is odd and $c < 0$, we use the same notation for the generalized equivariant metric on λ, in which $g^{H^\cdot(X,F)}$ is replaced by $(-1)^n g^{H^\cdot(X,F)}$. Observe that in this case, the Euler characteristic $\chi(F)$ vanishes identically, so that if $G = 1$, this is again the Ray-Singer metric on λ.

Theorem 9.0.1. *For any $b > 0$, $g \in G$, the following identity of positive equivariant Hermitian metrics holds:*

$$\log\left(\frac{\| \|_{\lambda,b}^2}{\| \|_{\lambda,0}^2}\right)(g) = \pm 2 \int_{X_g} e\left(TX_g\right)^0 J_g\left(TX|_{X_g}\right) \operatorname{Tr}^F[g]. \tag{9.0.1}$$

In particular, if G is reduced to a point, we have the identity of Hermitian metrics on the complex line λ,

$$\| \|_{\lambda,b}^2 = \| \|_{\lambda,0}^2. \tag{9.0.2}$$

Proof. By Theorem 6.7.1, we know that the generalized metric $\| \|_{\lambda,b}^2$ does not depend on $b > 0$. Therefore we only need to establish (9.0.1) for $b > 0$ small enough.

We claim that our theorem is a consequence of Theorem 8.2.1. Indeed if n is even, for $1 \le i \le 2n$, the $\mathfrak{h}_0^{\mathfrak{H}^i(X,F)}$ or their negative are Hermitian metrics. By Theorem 8.2.1, we know that for $b > 0$ small enough, the $\mathfrak{h}_b^{\mathfrak{H}^i(X,F)}$ have the same type as $\mathfrak{h}^{\mathfrak{H}^\cdot(X,F)}$. Moreover, by (1.11.2), (1.11.4), we get

$$\widehat{\mathrm{ch}}_g^\circ\left(\nabla^{\mathfrak{H}^\cdot(X,F)}, \mathfrak{h}_0^{\mathfrak{H}^\cdot(X,F)}, \mathfrak{h}_b^{\mathfrak{H}^\cdot(X,F)}\right) = \frac{1}{2}\log\left(\frac{| |_{\lambda,b}^2}{| |_{\lambda,0}^{0,2}}\right)(g). \tag{9.0.3}$$

Using (1.8.4), (1.12.10), (6.6.11), (6.6.13), (8.2.3), and (9.0.3), we get (9.0.2).

When G is reduced to a point, if n is even, $\| \|_\lambda^{0,2}$ is a Hermitian metric. When n is negative, since the Euler characteristic $\chi(F)$ vanishes, $\| \|_{\lambda,0}^2$ is still a Hermitian metric. Finally, (9.0.2) follows from (9.0.1). The proof of our theorem is completed. □

Chapter Ten

The harmonic forms for $b \to 0$ and the formal Hodge theorem

The purpose of this chapter is twofold.

On the one hand, in section 10.1, we prove Theorem 8.4.2, i.e., we compute the asymptotics of the generalized metrics $\mathfrak{h}_b^{\mathfrak{H}^{\cdot}(X,T)}$ as $b \to 0$.

On the other hand, in section 10.2, we give a direct proof of a formal Hodge theorem as $b \to 0$. Namely, we prove that up to some trivial scaling, the space of formal power series in the variable $b > 0$ which lies in $r_b^* \mathbb{H}_b^{\cdot}(X, F)$ is in one to one correspondence with $\mathfrak{H}^{\cdot}(X, F)$. More precisely, given a fixed class in $\mathfrak{H}^{\cdot}(X, F)$, we compute the formal power series in the variable b of the closed form in $r_b^* \mathbb{H}_b^{\cdot}(X, F)$ which represents this cohomology class.

In section 10.3, we show that the above formal power series also vanishes under a scaled version of $\overline{d}_{\phi, 2\mathcal{H}^c}^{T^*X}$.

Finally, in section 10.4, we show that this formal power series is the Taylor expansion near $b = 0$ of the harmonic form in $\mathbb{H}_b^{\cdot}(X, F)$ which represents the given cohomology class.

10.1 A PROOF OF THEOREM 8.4.2

Take $b > 0$. Recall that $c = \pm 1/b^2$ and that $\mathcal{H} = \frac{1}{2}|p|^2$.

In the $+$ case, $\square^X$ is the standard elliptic Laplacian acting on $\Omega^{\cdot}(X, F)$; in the $-$ case, it is the elliptic Laplacian acting on $\Omega^{\cdot}(X, F \otimes o(TX))$. Set

$$\mathbf{H}^{\cdot}(X, F) = \ker \square^X. \tag{10.1.1}$$

Then by (1.2.5), $\mathbf{H}^{\cdot}(X, F)$ is canonically identified to $H^{\cdot}(X, F)$ in the $+$ case and to $H^{\cdot}(X, F \otimes o(TX))$ in the $-$ case. Equivalently,

$$\begin{aligned} \mathbf{H}^{\cdot}(X, F) &\simeq \mathfrak{H}^{\cdot}(X, F) \text{ for } c > 0, \\ &\simeq \mathfrak{H}^{\cdot - n}(X, F) \text{ for } c < 0. \end{aligned} \tag{10.1.2}$$

For $b > 0$, recall that in (5.1.7), we defined $\mathbb{H}_b^{\cdot}(X, F)$ as

$$\mathbb{H}_b^{\cdot}(X, F) = \ker A_{\phi, \mathcal{H}^c}^2. \tag{10.1.3}$$

By (5.1.9), for $b \in]0, b_0]$,

$$\mathbb{H}_b^{\cdot}(X, F) \simeq \mathfrak{H}^{\cdot}(X, F). \tag{10.1.4}$$

Recall that by (5.1.8), for $b \in]0, b_0]$,

$$\Omega^{\cdot}(T^*X, \pi^* F)_{b,0} = \mathbb{H}_b^{\cdot}(X, F). \tag{10.1.5}$$

In section 10.1, for $b > 0$, we defined P_b as the projector on the vector space $\Omega^{\cdot}(T^*X, \pi^*F)_{b,0}$ with respect to the splitting (5.1.2) of $\Omega^{\cdot}(T^*X, \pi^*F)$. By (10.1.5), for $b \in]0, b_0]$, P_b is a projector on $\mathbb{H}^{\cdot}_b(X, F)$. The projector P_b can be defined by a contour integral similar to (5.3.5). Since $A^2_{\phi,\mathcal{H}^c}$ commutes with d^{T^*X}, the projector P_b also commutes with d^{T^*X}. By (5.1.5), we get

$$d^{T^*X}P_b = P_b d^{T^*X} = 0. \tag{10.1.6}$$

In the $+$ case, we identify $s \in \Omega^{\cdot}(X, F)$ to $\overline{s} = \pi^* s \in \Omega^{\cdot}(T^*X, \pi^*F)$. In the $-$ case, we identify $s \in \Omega^{\cdot}(X, F \otimes o(TX))$ to $\overline{s} = \pi^* s \wedge r^*_{1/b}\Phi^{T^*X} \in \Omega^{\cdot}(T^*X, \pi^*F)$. In both cases, we obtain this way closed forms on T^*X which represent the corresponding cohomology classes in $\mathfrak{H}^{\cdot}(X, F)$.

It follows from the above that the map $s \in \mathbf{H}^{\cdot}(X, F) \to P_b\overline{s} \in \mathbb{H}^{\cdot}(X, F)$ is an isomorphism which is compatible with the canonical identifications in (10.1.1) and (10.1.2).

As we saw in section 5.1, the restriction of $\mathfrak{h}^{\Omega^{\cdot}(T^*X,\pi^*F)}_{\mathcal{H}^c}$ to $\mathbb{H}^{\cdot}_b(X, F)$ is nondegenerate and induces the Hermitian form $\mathfrak{h}^{\mathfrak{H}^{\cdot}(X,F)}_b$ on $\mathfrak{H}^{\cdot}(X, F)$.

By (2.1.21), (2.1.22), (2.1.28),

$$\mathfrak{A}'_{\phi_b,\pm\mathcal{H}} = e^{\mp\mathcal{H}-\mu_0}K_b A_{\phi,\mathcal{H}^c}K_b^{-1}e^{\pm\mathcal{H}+\mu_0}. \tag{10.1.7}$$

Recall that the projector $\mathfrak{P}'_b$ from $\mathcal{S}^{\cdot}(T^*X, \pi^*F)$ on $\mathcal{S}^{\cdot}(T^*X, \pi^*F)_{0,b} = \ker \mathfrak{A}'^2_{\phi_b,\pm\mathcal{H}}$ was defined in (5.3.5). By (10.1.7), we get

$$\mathfrak{P}'_b = e^{\mp\mathcal{H}-\mu_0}K_b P_b K_b^{-1}e^{\pm\mathcal{H}+\mu_0}. \tag{10.1.8}$$

The map $s \in \mathbb{H}_b(X, F) \to e^{\mp\mathcal{H}}K_b e^{-\mu_0}s \in \ker \mathfrak{A}'^2_{\phi_b,\pm H}$ provides the canonical identity of these two spaces. By (2.1.25), if $s, s' \in \Omega^{\cdot}(T^*X, \pi^*F)$ have compact support, then

$$\left\langle e^{\mp\mathcal{H}-\mu_0}K_b s, e^{\mp\mathcal{H}-\mu_0}K_b s'\right\rangle_{h^{\Omega^{\cdot}(T^*X,\pi^*F)}} = \frac{1}{b^n}\left\langle s, s'\right\rangle_{\mathfrak{h}^{\Omega^{\cdot}(T^*X,\pi^*F)}_{\mathcal{H}^c}}. \tag{10.1.9}$$

Take $s, s' \in \mathbf{H}^{\cdot}(X, F)$. By (10.1.8), (10.1.9), we get

$$\left\langle P_b\overline{s}, P_b\overline{s}'\right\rangle_{\mathfrak{h}^{\Omega^{\cdot}(T^*X,\pi^*F)}_{\mathcal{H}^c}} = b^n\left\langle \mathfrak{P}'_b e^{\mp\mathcal{H}-\mu_0}K_b\overline{s}, \mathfrak{P}'_b e^{\mp\mathcal{H}-\mu_0}K_b\overline{s}'\right\rangle_{h^{\Omega^{\cdot}(T^*X,\pi^*F)}}. \tag{10.1.10}$$

Let P be the orthogonal projector from $\Omega^{\cdot}(X, F)$ or $\Omega^{\cdot}(X, F \otimes o(TX))$ on $\mathbf{H}^{\cdot}(\mathbf{X}, \mathbf{F})$. By using the estimate in (3.5.9) with $\lambda = 0$, we find that for any $v \in]0, 1[$ there is $C_v > 0$ such that for $b \in]0, b_0]$,

$$\|\mathfrak{P}'_b - i_\pm P P_\pm\|_1 \leq C_v b^v. \tag{10.1.11}$$

We assume first that we are in the $+$ case, i.e., $c = 1/b^2$. Then P_+ projects on forms which have fiberwise degree 0. Also $e^{-\mathcal{H}}$ spans fiberwise the image of P_+. Using (10.1.10), (10.1.11), we get

$$\begin{aligned}\left\langle P_b\overline{s}, P_b\overline{s}'\right\rangle_{\mathfrak{h}^{\Omega^{\cdot}(T^*X,\pi^*F)}_{\mathcal{H}^c}} \\ = b^n\left(\left\langle e^{-\mathcal{H}}\pi^*s, e^{-\mathcal{H}}\pi^*s'\right\rangle_{h^{\Omega^{\cdot}(T^*X,\pi^*F)}} + \mathcal{O}(b^v)|s||s'|\right).\end{aligned} \tag{10.1.12}$$

The restriction of the Hermitian form $h^{\Omega^{\cdot}(T^*X,\pi^*F)}$ to forms of the type $e^{-\mathcal{H}}\pi^*s$ is just the standard Hermitian product. From (10.1.12), we get

$$\left\langle P_b\overline{s}, P_b\overline{s}'\right\rangle_{\mathfrak{h}_{\mathcal{H}^c}^{\Omega^{\cdot}(T^*X,\pi^*F)}} = b^n\left(\pi^{n/2}\left\langle s,s'\right\rangle_{g^{\Omega^{\cdot}(X,F)}} + \mathcal{O}\left(b^v\right)|s|\left|s'\right|\right). \quad (10.1.13)$$

By (10.1.10)-(10.1.13), we get (8.4.5) in the + case.

Let us now consider the − case. Then

$$K_b\overline{s} = \pi^*s\wedge K_b r_{1/b}^*\Phi^{T^*X}. \quad (10.1.14)$$

When acting on functions, the operator $K_b r_{1/b}^*$ is just the identity. Let η be a fiberwise volume form of norm 1 in T^*X. Using (2.3.4), (10.1.11), (10.1.14), and the fact that $\exp\left(-|p|^2/2\right)\eta$ spans fiberwise the image of P_-, we get

$$\mathfrak{P}_b' e^{\mathcal{H}-\mu_0}K_b\overline{s} = b^{-n}\left(\pi^*s\frac{1}{\pi^{n/2}}\exp\left(-|p|^2/2\right)\eta + \mathcal{O}\left(b^v\right)|s|\right). \quad (10.1.15)$$

Also the restriction of $h^{\Omega^{\cdot}(T^*X,\pi^*F)}$ to forms of the type which appear in the right-hand side of (10.1.15) is $(-1)^n$ times the usual Hermitian product on these forms. Therefore,

$$\left\langle\pi^*s\exp\left(-|p|^2/2\right)\eta, \pi^*s'\exp\left(-|p|^2/2\right)\eta\right\rangle_{h^{\Omega^{\cdot}(T^*X,\pi^*F)}} = \pi^{n/2}(-1)^n\left\langle s,s'\right\rangle_{g^{\Omega^{\cdot}(X,F\otimes o(TX))}}. \quad (10.1.16)$$

By (10.1.12), (10.1.15), (10.1.16), we get (8.4.5) in the − case. The proof of Theorem 8.4.2 is completed. □

Remark 10.1.1. Let δ_X be the current of integration on X viewed as the zero section of T^*X. Then δ_X can be viewed as a compactly supported current of degree n with values in $o(TX)$. In the − case, we could as well have taken $\overline{s} = s\delta_X$. This is of course a current, but it is permitted in our theory. Note in this case that

$$K_b\overline{s} = b^{-n}\overline{s}. \quad (10.1.17)$$

Also

$$P_-\overline{s} = \pi^{-n/2}\pi^*s\exp\left(-|p|^2/2\right)\eta. \quad (10.1.18)$$

Then (10.1.10), (10.1.17), (10.1.18) also lead to (8.4.5) in the − case.

10.2 THE KERNEL OF $\mathbf{A}^2_{\phi,\mathcal{H}^c}$ AS A FORMAL POWER SERIES

Here we use the notation of section 2.3, in particular for the definition of $\Omega^{\cdot}(T^*X,\pi^*F)$.

We will assume $c>0$. Still all the arguments we will give in this case, the + case, are also valid for the case where $c<0$. This is why we will not write the subscript + explicitly.

By [B05, Proposition 3.5], the operators $\mathfrak{a}, \mathfrak{b}$ commute with the de Rham operator d^{T^*X}. Let $\Omega^{\cdot\dagger}(T^*X, \pi^*F), \Omega^{\cdot\perp}(T^*X)$ be the ± 1 eigenspaces for the action of r^*, so that

$$\Omega^{\cdot}(T^*X, \pi^*F) = \Omega^{\cdot\dagger}(T^*X, \pi^*F) \oplus \Omega^{\cdot\perp}(T^*X, \pi^*F). \tag{10.2.1}$$

Then $\mathfrak{a}$ preserves this splitting, and $\mathfrak{b}$ exchanges the vector spaces in this splitting. We can then write these two operators in matrix form as

$$\mathfrak{a} = \begin{pmatrix} \mathfrak{a}^\dagger & 0 \\ 0 & \mathfrak{a}^\perp \end{pmatrix}, \qquad \mathfrak{b} = \begin{pmatrix} 0 & \mathfrak{b}^\perp \\ \mathfrak{b}^\dagger & 0 \end{pmatrix}. \tag{10.2.2}$$

As we saw in section 2.3, $\ker \mathfrak{a}$ is generated by the even form 1. In particular the operator $\mathfrak{a}^\perp$ is invertible. Moreover, $\Omega^{\cdot\dagger}(T^*X, \pi^*F)$ splits as

$$\Omega^{\cdot\dagger}(T^*X, \pi^*F) = \ker \mathfrak{a}^\dagger \oplus \operatorname{Im} \mathfrak{a}^\dagger. \tag{10.2.3}$$

We denote by $\left(\mathfrak{a}^\dagger\right)^{-1}$ the inverse of $\mathfrak{a}^\dagger$ acting on $\operatorname{Im} \mathfrak{a}^\dagger$.

Recall that $\Omega^{\cdot}(X, F)$ embeds as a vector subspace of $\ker \mathfrak{a}^\dagger \subset \Omega^{\cdot\dagger}(X, F)$ by the map $s \to \pi^* s$. Put

$$L = \frac{1}{2} \square^X. \tag{10.2.4}$$

Then $\Omega^{\cdot}(X, F)$ splits orthogonally as

$$\Omega^{\cdot}(X, F) = \ker L \oplus \operatorname{Im} L. \tag{10.2.5}$$

Let L^{-1} be the inverse of L acting on $\operatorname{Im} L$.

By Theorem 2.3.1, we have the identity of operators acting on $\Omega^{\cdot}(X, F)$,

$$-Q^{T^*X} \mathfrak{b} \mathfrak{a}^{-1} \mathfrak{b} Q^{T^*X} = L. \tag{10.2.6}$$

Recall that by equation (2.3.2), for $c > 0$,

$$2A^2_{\phi_b, \mathcal{H}} = \frac{\mathfrak{a}}{b^2} + \frac{\mathfrak{b}}{b}. \tag{10.2.7}$$

Consider the equation for $s \in \mathbb{H}_b(X, F) = \ker A^2_{\phi, \mathcal{H}^c}$,

$$A^2_{\phi, \mathcal{H}^c} s = 0. \tag{10.2.8}$$

Put

$$\sigma = r_b^* s. \tag{10.2.9}$$

By (2.1.28), (10.2.8), (10.2.9), we get

$$A^2_{\phi_b, \mathcal{H}} \sigma = 0. \tag{10.2.10}$$

By (10.2.7), we can rewrite (10.2.10) in the form

$$\left(\frac{\mathfrak{a}}{b} + \mathfrak{b}\right) \sigma = 0. \tag{10.2.11}$$

In (10.2.11), we split σ using the splitting (10.2.1) of $\Omega^{\cdot}(T^*X, \pi^*F)$,

$$\sigma = \sigma^\dagger + \sigma^\perp. \tag{10.2.12}$$

Similarly, we will also split $\sigma^{\dagger}$ using the splitting (10.2.3),

$$\sigma^{\dagger} = \tau + \upsilon. \tag{10.2.13}$$

By (10.2.2), (10.2.11), we get

$$\sigma^{\perp} = -b\left(\mathfrak{a}^{\perp}\right)^{-1}\mathfrak{b}^{\dagger}\sigma^{\dagger}. \tag{10.2.14}$$

Equation (10.2.11) is then equivalent to

$$\left(\frac{\mathfrak{a}^{\dagger}}{b^2} - \mathfrak{b}^{\perp}\left(\mathfrak{a}^{\perp}\right)^{-1}\mathfrak{b}^{\dagger}\right)\sigma^{\dagger} = 0. \tag{10.2.15}$$

We will expand $\sigma^{\dagger} = \sigma_c^{\dagger}$ as a formal power series in the variable b^2, of the form

$$\sigma_c^{\dagger} = \sigma_0^{\dagger} + \sigma_1^{\dagger} b^2 + \cdots \tag{10.2.16}$$

By (10.2.15), (10.2.16), we get for $i \geq 0$,

$$\mathfrak{a}^{\dagger}\sigma_i^{\dagger} - \mathfrak{b}^{\perp}\left(\mathfrak{a}^{\perp}\right)^{-1}\mathfrak{b}^{\dagger}\sigma_{i-1}^{\dagger} = 0 \tag{10.2.17}$$

Equation (10.2.17) for $i=0$ says that $\sigma_0^{\dagger} \in \ker \mathfrak{a}^{\dagger}$, so that $\sigma_0^{\dagger} \in \Omega^{\cdot}(X,F)$. By (10.2.6), equation (10.2.17) for $i=1$ says that

$$L\sigma_0^{\dagger} = 0, \tag{10.2.18}$$

i.e., $\sigma_0^{\dagger}$ is harmonic on X, and then closed on X. Using the splitting (10.2.13) for $\sigma_1^{\dagger}$, we get

$$\upsilon_1 = \left(\mathfrak{a}^{\dagger}\right)^{-1}\mathfrak{b}^{\perp}\left(\mathfrak{a}^{\perp}\right)^{-1}\mathfrak{b}^{\dagger}\sigma_0. \tag{10.2.19}$$

Note that τ_1 is not determined yet.

A consequence of [B05, Proposition 3.5] is that, as we shall see in (10.3.5), the operators $\mathfrak{a}, \mathfrak{b}$ can be written as anticommutators of d^{T^*X} with other odd operators, so that in particular they commute with d^{T^*X}. Since the form σ_0 is closed, it follows that υ_1 is an exact form.

For $i=2$, equation (10.2.17) says that

$$\mathfrak{a}^{\dagger}\sigma_2^{\dagger} - \mathfrak{b}^{\perp}\left(\mathfrak{a}^{\perp}\right)^{-1}\mathfrak{b}^{\dagger}\sigma_1^{\dagger} = 0. \tag{10.2.20}$$

Using (10.2.6), we find that if (10.2.20) holds,

$$L\tau_1 = Q_+^{T^*X}\mathfrak{b}^{\perp}\left(\mathfrak{a}^{\perp}\right)^{-1}\mathfrak{b}^{\dagger}\upsilon_1. \tag{10.2.21}$$

Since $Q_+^{T^*X}$ commutes with the de Rham operator, and since υ_1 is exact, the form in the right-hand side of (10.2.21) is exact in $\Omega^{\cdot}(X,F)$. Therefore the right-hand side of (10.2.21) lies indeed in $\operatorname{Im} L$. If we insist on the fact that the $\sigma_i, i \geq 1$ are exact, the only possibility in (10.2.21) is to take

$$\tau_1 = L^{-1}Q_+\mathfrak{b}^{\perp}\left(\mathfrak{a}^{\perp}\right)^{-1}\mathfrak{b}^{\dagger}\upsilon_1. \tag{10.2.22}$$

Then υ_1, τ_1 are exact, and $\sigma_1^{\dagger}$ is exact.

It should now be clear that we can proceed by recursion to solve all the equations (10.2.20), so that for $i \geq 1$, the $\sigma_i^\dagger$ are exact, and the τ_i lie in $\operatorname{Im} L$. By (10.2.14), (10.2.16), we have the formal expansion

$$\sigma^\perp = -b\left(\mathfrak{a}^\perp\right)^{-1}\mathfrak{b}^\dagger\left(\sigma_0^\dagger + \sigma_1^\dagger b^2 + \cdots\right). \tag{10.2.23}$$

The same arguments as before show that all the terms which appear in the right-hand side of (10.2.23) are themselves exact.

From the above, we find that in the space of formal power series, the form σ_c is closed, and its cohomology class is equal to the class of $\sigma_0^\dagger$. By the above we get a formal expansion of σ of the form

$$\sigma = \sum_{i=0}^{+\infty} \sigma_i b^i. \tag{10.2.24}$$

In (10.2.24),

$$\begin{aligned} \sigma_i &= \sigma_{i/2}^\dagger \text{ if } i \text{ is even,} \\ &= -\left(\mathfrak{a}^\perp\right)^{-1}\mathfrak{b}^\dagger\sigma_{(i-1)/2}^\dagger \text{ if } i \text{ is odd.} \end{aligned} \tag{10.2.25}$$

Finally, inspection of equations (2.2.3) and (2.3.1) and a trivial recursion argument shows that for any $i \in \mathbf{N}$, σ_i is a polynomial of degree at most $2i$ in the variable p.

This last fact is true only in the $+$ case. In the $-$ case, the σ_i are the product of $e^{-|p|^2}$ by a polynomial of degree at most $2i$.

10.3 A PROOF OF THE FORMAL HODGE THEOREM

Again we take $c = \pm 1/b^2$. Recall that by (2.1.28),

$$\overline{d}^{T^*X}_{\phi_b,\pm 2\mathcal{H}} = r_b^* \overline{d}^{T^*X}_{\phi,2\mathcal{H}^c} r_b^{*-1}. \tag{10.3.1}$$

By [B05, Proposition 2.33], we get

$$\overline{d}^{T^*X}_{\phi_b,\pm 2\mathcal{H}} = \frac{\overline{d}^{T^*X} \mp i_{2Y^{\mathcal{H}}}}{b} + \frac{\delta^{T^*X,V} \pm i_{2\widehat{p}}}{b^2}. \tag{10.3.2}$$

In (10.3.2), the operators $\overline{d}^{T^*X}$, $\delta^{T^*X,V}$ are explicitly determined odd operators whose exact form is irrelevant. Recall that $\overline{d}^{T^*X,2}_{\phi_b,\pm 2\mathcal{H}} = 0$. From (10.3.2), we deduce that

$$\begin{aligned} &\left(\overline{d}^{T^*X} \mp i_{2Y^{\mathcal{H}}}\right)^2 = 0, \qquad \left(\delta^{T^*X,V} \pm i_{2\widehat{p}}\right)^2 = 0, \\ &\left[\overline{d}^{T^*X} \mp 2i_{Y^{\mathcal{H}}}, \delta^{T^*X} \pm 2i_{\widehat{p}}\right] = 0. \end{aligned} \tag{10.3.3}$$

Of course, the identities in (10.3.3) can be proved directly. Let us just mention a proof of the third equation. Recall that λ_0 was introduced in (2.1.4). By [B05, Proposition 2.16],

$$\delta^{T^*X} \pm 2i_{\widehat{p}} = -\left[\overline{d}^{T^*X} \mp 2i_{Y^{\mathcal{H}}}, \lambda_0\right]. \tag{10.3.4}$$

The third equation in (10.3.3) then follows from the first two equations and from (10.3.4).

Also by [B05, Proposition 3.5],

$$\mathfrak{a}_\pm = \frac{1}{2}\left[d^{T^*X}, \delta^{T^*X,V} \pm 2i_{\widehat{p}}\right], \quad \mathfrak{b}_\pm = \frac{1}{2}\left[d^{T^*X}, \overline{d}^{T^*X} \mp 2i_{Y^{\mathcal{H}}}\right]. \tag{10.3.5}$$

In section 10.2, we showed that equation (10.2.11) has a unique solution as a power series σ in (10.2.24), such that the cohomology class of σ_0 is fixed and the σ_i are exact. In particular

$$d^{T^*X}\sigma = 0. \tag{10.3.6}$$

To establish that the Hodge Theorem holds in the sense of formal power series, we must check that $\overline{d}^{T^*X}_{\phi_b,\pm 2\mathcal{H}}\sigma = 0$.

Using (10.3.2), this is equivalent to the following result.

Theorem 10.3.1. *The following identity of formal power series holds:*

$$\left(\frac{\delta^{T^*X,V} \pm 2i_{\widehat{p}}}{b} + \overline{d}^{T^*X} \mp 2i_{Y^{\mathcal{H}}}\right)\sigma = 0. \tag{10.3.7}$$

Equivalently, for $i \geq 0$,

$$\left(\delta^{T^*X,V} \pm 2i_{\widehat{p}}\right)\sigma_i + \left(\overline{d}^{T^*X} \mp 2i_p\right)\sigma_{i-1} = 0. \tag{10.3.8}$$

Proof. As before, we will consider only the + case, the − case being similar. We will prove (10.3.8) by recursion. By construction, (10.3.8) holds for $i=0$. Assume that $i'=2i$ is even, and that (10.3.8) holds for $j \leq i'$. We will prove that (10.3.8) holds for $i'+1$. Equivalently, we will show that

$$\left(\overline{d}^{T^*X} - 2i_{Y^{\mathcal{H}}}\right)\sigma_i^\dagger - \left(\delta^{T^*X,V} + 2i_{\widehat{p}}\right)\left(\mathfrak{a}^\perp\right)^{-1}\mathfrak{b}^\dagger\sigma_i^\dagger = 0. \tag{10.3.9}$$

Using the second identity in (10.3.3) and (10.3.5) , we find that $\delta^{T^*X,V} + 2i_{\widehat{p}}$ commutes with $\mathfrak{a}^{-1}$. Since $\sigma_i^\dagger$ is d^{T^*X} closed, using again (10.3.5), we get

$$\begin{aligned}\left(\delta^{T^*X,V} + 2i_{\widehat{p}}\right)&\left(\mathfrak{a}^\perp\right)^{-1}\mathfrak{b}^\dagger\sigma_i^\dagger \\ &= \frac{1}{2}\left(\mathfrak{a}^\perp\right)^{-1}\left(\delta^{T^*X,V} + 2i_{\widehat{p}}\right)d^{T^*X}\left(\overline{d}^{T^*X} - 2i_{Y^{\mathcal{H}}}\right)\sigma_i^\dagger.\end{aligned} \tag{10.3.10}$$

Using (10.3.3) and equation (10.3.8) for $i'=2i$, we obtain

$$\left(\delta^{T^*X,V} + 2i_{\widehat{p}}\right)\left(\overline{d}^{T^*X} - 2i_{Y^{\mathcal{H}}}\right)\sigma_i^\dagger = 0. \tag{10.3.11}$$

By (10.3.5) and (10.3.11), we find that in the right-hand side of (10.3.10), we can as well replace $\left(\delta^{T^*X,V} + 2i_{\widehat{p}}\right)d^{T^*X}$ by $\mathfrak{a}^\perp$, i.e., we get (10.3.9).

Now we assume that $i'=2i-1$ is odd, and that (10.3.8) holds for $j \leq i'$. We will show that (10.3.7) holds for $i'+1$, i.e.,

$$\left(\delta^{T^*X,V} + 2i_{\widehat{p}}\right)\sigma_i^\dagger + \left(\overline{d}^{T^*X} - 2i_p\right)\sigma_{2i-1} = 0. \tag{10.3.12}$$

Now, by (10.2.17) and (10.2.25), we get

$$\left(\delta^{T^*X,V}+2i_{\widehat{p}}\right)\sigma_i^{\dagger}=-\left(\delta^{T^*X,V}+2i_{\widehat{p}}\right)\left(\mathfrak{a}^{\dagger}\right)^{-1}\mathfrak{b}^{\perp}\sigma_{2i-1}. \tag{10.3.13}$$

The same arguments as in (10.3.10) show that

$$\left(\delta^{T^*X,V}+2i_{\widehat{p}}\right)\left(\mathfrak{a}^{\dagger}\right)^{-1}\mathfrak{b}^{\perp}\sigma_{2i-1}=\frac{1}{2}\left(\mathfrak{a}^{\dagger}\right)^{-1}\left(\delta^{T^*X,V}+i_{2i\widehat{p}}\right) d^{T^*X}\left(\overline{d}^{T^*X}-2i_p\right)\sigma_{2i-1}. \tag{10.3.14}$$

Now we use the third identity in (10.3.3) and also (10.3.8) for $i'=2i-1$, and we get

$$\left(\delta^{T^*X,V}+2i_{2\widehat{p}}\right)\left(\overline{d}^{T^*X}-2i_{Y^{\mathcal{H}}}\right)\sigma_{2i-1}=0. \tag{10.3.15}$$

By (10.3.5), (10.3.13)-(10.3.15), we get (10.3.12). The proof of our theorem is completed. □

10.4 TAYLOR EXPANSION OF HARMONIC FORMS NEAR b = 0

We use the notation of section 10.2. For $m\in\mathbf{N}$, set

$$\sigma^m=\sum_{i=0}^{m}\sigma_i b^i. \tag{10.4.1}$$

Put

$$\tau^m=A^2_{\phi_b,\pm\mathcal{H}}\sigma^m. \tag{10.4.2}$$

A trivial computation shows that

$$\tau^m=\frac{1}{2}\mathfrak{b}_{\pm}\sigma_m b^{m-1}. \tag{10.4.3}$$

By (2.1.21), (2.1.28), and(2.8.10),

$$\mathfrak{A}'^2_{\phi_b,\pm\mathcal{H}}=e^{-\mu_0\mp\mathcal{H}}\tau_b^{-1}A^2_{\phi_b,\pm\mathcal{H}}\tau_b e^{\mu_0\pm\mathcal{H}}. \tag{10.4.4}$$

Let $d\in\mathbf{C}$ be the small circle of center 0 which was considered in section 5.3. Then if $b\in]0,b_0]$, $\lambda\in d$, the resolvent $\left(\lambda-A^2_{\phi_b,\pm\mathcal{H}}\right)$ exists. Moreover, the natural projector $\mathcal{P}_b$ from $\mathcal{S}^{\cdot}\left(T^*X,\pi^*F\right)$ on $\mathcal{S}^{\cdot}\left(T^*X,\pi^*F\right)_{0,b}=\ker A^2_{\phi_b,\pm\mathcal{H}}$ is given

$$\mathcal{P}_b=\frac{1}{2i\pi}\int_d\left(\lambda-A^2_{\phi_b,\pm\mathcal{H}}\right)^{-1}d\lambda. \tag{10.4.5}$$

Also since d^{T^*X} commutes with $A^2_{\phi_b,\pm\mathcal{H}}$, $\mathcal{P}_b$ also commutes with d^{T^*X}. Moreover, using (5.3.5), (10.4.4), and (10.4.5), we get

$$\mathfrak{P}_b=e^{-\mu_0\mp\mathcal{H}}\tau_b^{-1}\mathcal{P}_b\tau_b e^{\mu_0\pm\mathcal{H}}. \tag{10.4.6}$$

If $\lambda \in d$, by (10.4.2),

$$\left(\lambda - A^2_{\phi_b,\pm\mathcal{H}}\right)\sigma^m = \lambda\sigma^m - \tau^m. \tag{10.4.7}$$

From (10.4.7), we get

$$\left(\lambda - A^2_{\phi_b,\pm\mathcal{H}}\right)^{-1}\sigma^m = \lambda^{-1}\sigma^m + \left(\lambda - A^2_{\phi_b,\pm\mathcal{H}}\right)^{-1}\lambda^{-1}\tau^m. \tag{10.4.8}$$

By (10.4.5), (10.4.8), we obtain

$$\mathcal{P}_b\sigma_m = \sigma_m + \frac{1}{2i\pi}\int_d \left(\lambda - A^2_{\phi_b,\pm\mathcal{H}}\right)^{-1}\lambda^{-1}d\lambda\tau^m. \tag{10.4.9}$$

Now as we saw in section 10.2, $\sigma_m - \sigma_0$ is d^{T^*X} exact. Also by Theorem 3.5.1, for $b \in]0, b_0]$, d^{T^*X} vanishes on $\ker A^2_{\phi_b,\pm\mathcal{H}}$. Since $\mathcal{P}_b$ commutes with d^{T^*X}, we get

$$\mathcal{P}_b\sigma_m = \mathcal{P}_b\sigma_0. \tag{10.4.10}$$

By (10.4.9), (10.4.10), we obtain

$$\mathcal{P}_b\sigma_0 = \sigma_m + \frac{1}{2i\pi}\int_d \left(\lambda - A^2_{\phi_b,\pm\mathcal{H}}\right)^{-1}\lambda^{-1}d\lambda\tau^m. \tag{10.4.11}$$

By (10.4.4), if $\lambda \in d$,

$$\left(\lambda - A^2_{\phi_b,\pm\mathcal{H}}\right)^{-1} = e^{\pm\mathcal{H}}\tau_b e^{\mu_0}\left(\lambda - \mathfrak{A}'^2_{\phi_b,\pm\mathcal{H}}\right)^{-1}e^{-\mu_0}\tau_b^{-1}e^{\mp\mathcal{H}}. \tag{10.4.12}$$

Moreover, it follows from equation (17.21.23) in Theorem 17.21.3 and from Remark 17.21.6 that for $b \in]0, b_0]$, $\lambda \in d$, the operators $\left(\lambda - \mathfrak{A}'^2_{\phi_b,\pm\mathcal{H}}\right)^{-1}$ have uniformly bounded norm when acting on the standard L^2 space over T^*X.

If $s \in \mathcal{S}^{\cdot}(T^*X, \pi^*F)$, set

$$\|s\|_{\pm\mathcal{H}} = \left\|e^{\mp\mathcal{H}}s\right\|_{L^2}. \tag{10.4.13}$$

By noting that $\tau^{b^{-1}}$ introduces a singularity as $b \to 0$ which is at most $\mathcal{O}(b^{-n})$, from (10.4.11), (10.4.12), we claim that

$$\|\mathcal{P}_b\sigma_0 - \sigma_m\|_{\pm\mathcal{H}} \le C_m b^{m-1-n}\|\sigma_0\|_{\pm\mathcal{H}}. \tag{10.4.14}$$

Indeed in the + case, as we saw at the end of section 10.2, the σ_i are polynomials in p of degree at most $2i$, which depend continuously on σ_0 by construction. The estimate (10.4.14) is then obvious. In the − case, σ_i is the product of $e^{-2\mathcal{H}}$ by a polynomial of degree at most $2i$, which also depends continuously on σ_0. Equation (10.4.14) still follows.

It follows from (10.4.14) that $\mathcal{P}_b\sigma_0$ is approximated by the polynomial σ_m to arbitrary order.

Chapter Eleven

A proof of equation (8.4.6)

The purpose of this chapter is to establish equation (8.4.6) in Theorem 8.4.3. We will thus compute explicitly the limit as $t \to 0$ of $v_{\sqrt{t}b,t}$. The techniques are closely related to the ones we used in chapter 4. However, a direct application of the results of that chapter would lead to spurious divergences. This forces us to modify our trivializations, very much in the spirit of the local version of the families index theorem of [B86].

This chapter is organized as follows. In section 11.1, we introduce our new trivialization and rescaling of the creation and annihilation variables.

In section 11.2, we prove the convergence of certain supertraces.

In section 11.3, we establish equation (8.4.6).

11.1 THE LIMIT OF THE RESCALED OPERATOR AS t → 0

In this chapter, we use the notation and conventions of chapter 4. However, we do not replace c by t/b^2, but given $b > 0$, we fix $c = \pm 1/b^2 \in \mathbf{R}^*$. With respect to chapter 4, we make $z = 0, dc = 0, dt = 0, db = 0$.

Recall that $h(x)$ was defined in (1.7.1) and $r(\lambda)$ in (5.3.3), and also that (5.3.4) holds.

As in (4.5.6), we get

$$\mathrm{Tr_s}\left[gt\,|p|^2\,h'\left(\mathfrak{D}^{\mathcal{M}}_{\phi,\mathcal{H}^c-\omega^H}\right)\right] = (-1)^n\,\mathrm{Tr_s}\left[gt\,|p|^2\,r\left(-\underline{\widehat{\mathfrak{L}}}^{\mathcal{M},2}_{\phi,\mathcal{H}^c-\omega^H}\right)\right]. \quad (11.1.1)$$

As in (4.6.2), set

$$\underline{\widehat{\mathfrak{M}}}_{c,t} = r^*_{1/\sqrt{t}}\underline{\widehat{\mathfrak{L}}}^{\mathcal{M},2}_{\phi,\mathcal{H}^c-\omega^H}r^*_{\sqrt{t}}. \quad (11.1.2)$$

Note that with respect to (4.6.2), the subscripts are now (c,t) instead of (b,t). Moreover, the operators which appear in (11.1.2) are not the same as in (4.6.2), even though the notation suggests otherwise.

Let ${}^1\nabla_t^{\Lambda^{\cdot}(T^*X)\widehat{\otimes}\Lambda^{\cdot}(T^*X)\widehat{\otimes}\Lambda^n(TX)\widehat{\otimes}F,u}$ be the connection on

$$\Lambda^{\cdot}(T^*X)\,\widehat{\otimes}\Lambda^{\cdot}(T^*X)\,\widehat{\otimes}\Lambda^n(TX)\,\widehat{\otimes}F$$

along the fibers X,

$${}^1\nabla_t^{\Lambda^{\cdot}(T^*X)\widehat{\otimes}\Lambda^{\cdot}(T^*X)\widehat{\otimes}\Lambda^n(TX)\widehat{\otimes}F,u} = \nabla_{\cdot}^{\Lambda^{\cdot}(T^*X)\widehat{\otimes}\Lambda^{\cdot}(T^*X)\widehat{\otimes}\Lambda^n(TX)\widehat{\otimes}F,u}$$
$$+\frac{1}{t}\left\langle T(f_\alpha,e_i)\,f^\alpha\left(e^i - t i_{e_i}\right) + T^H,\cdot\right\rangle. \quad (11.1.3)$$

By equation (2.5.19) in Theorem 2.5.4 and by equation (4.6.4) in Proposition 4.6.1, we get

$$\begin{aligned}\underline{\widehat{\mathfrak{M}}}_{c,t} = \frac{1}{4}\Big(-\Delta^V + c^2 |p|^2 + c\Big(2 i_{\widehat{e}_i}\big(\widehat{e}^i - e^i/\sqrt{t} - \sqrt{t} i_{e_i}\big) - n\Big)\Big) \\ + \frac{1}{4}\langle e_i, R^{TX} e_j\rangle \widehat{e}^i\widehat{e}^j - \frac{\sqrt{t}}{4}\omega\left(\nabla^F, g^F\right)(e_i)\nabla_{\widehat{e}^i} \\ - \frac{\sqrt{t}}{4}\nabla^{\Lambda^{\cdot}(T^*T^*X)\widehat{\otimes}F}\widehat{\omega}\left(\nabla^F, g^F\right) - \frac{1}{4}\omega\left(\nabla^F, g^F\right)^2 \\ - \frac{c\sqrt{t}}{2}\left({}^1\nabla^{\Lambda^{\cdot}(T^*X)\widehat{\otimes}\Lambda^{\cdot}(T^*X)\widehat{\otimes}\Lambda^n(TX)\widehat{\otimes}F,u}_{t,Y^{\mathcal{H}}} + \left\langle R^{TX}(\cdot,p)e_i, p\right\rangle \widehat{e}^i\right). \quad (11.1.4)\end{aligned}$$

Take $x \in X^g$. We trivialize X near x using the fiberwise geodesic coordinate system centered at x. Namely, we identify a neighborhood of 0 in T_xX with a neighborhood of x in X by the map $y \in T_xX \to \exp_x^X(y) \in X$. Also we trivialize $\Lambda^{\cdot}(T^*X)\widehat{\otimes}\Lambda^{\cdot}(T^*X)\widehat{\otimes}\Lambda^n(TX)\widehat{\otimes}F$ along geodesics centered at x by parallel transport with respect to the fiberwise connection ${}^1\nabla_t^{\Lambda^{\cdot}(T^*X)\widehat{\otimes}\Lambda^{\cdot}(T^*X)\widehat{\otimes}\Lambda^n(TX)\widehat{\otimes}F,u}$.

Using the above trivialization, we define the operator $\underline{\widehat{\mathfrak{N}}}_{c,t}$ as in section 4.8, with b is replaced by $b\sqrt{t}$. For $a > 0$, we define I_a as in (4.8.8).

Set

$$\underline{\widehat{\mathfrak{O}}}_{c,t} = I_{\sqrt{t}}\underline{\widehat{\mathfrak{N}}}_{c,t}I_{1/\sqrt{t}}. \quad (11.1.5)$$

Observe that when $b = t^2$, the operator $\underline{\widehat{\mathfrak{O}}}_{b,t}$ in (4.8.9) is just $\underline{\widehat{\mathfrak{O}}}_{c,t}$ written in a different trivialization.

Let $\widehat{c}(e_i), 1 \le i \le \ell$ be a family of Clifford variables such that if $U, V \in T_xX_g$,

$$[\widehat{c}(U), \widehat{c}(V)] = 2\langle U, V\rangle. \quad (11.1.6)$$

These $\widehat{c}(e_i)$ anticommute with all the other Clifford variables. Still, one should keep in mind that contrary to what may be suggested by the notation in section 1.1, for $1 \le i \le \ell$, $\widehat{c}(e_i)$ and $e^i + i_{e_i}$ are for the moment unrelated.

Definition 11.1.1. Let $\underline{\widehat{\mathfrak{P}}}_{c,t}$ be the operator obtained from $\underline{\widehat{\mathfrak{O}}}_{c,t}$ by making the following replacements for $1 \le i \le \ell$:

- e^i is unchanged.
- i_{e_i} is replaced by $-e^i/t + i_{e_i} + \widehat{c}(e_i)/\sqrt{t}$.
- $\widehat{e}^i, i_{\widehat{e}_i}$ are unchanged.

And for $\ell + 1 \le i \le n$:

- e^i is replaced by $\sqrt{t}e^i$.
- i_{e_i} is replace by $i_{e_i}/\sqrt{t}$.
- $\widehat{e}^i, i_{\widehat{e}_i}$ are unchanged.

Incidentally, observe that as we just saw, for $1 \le i \le \ell$, $\widehat{c}(e_i)$ is a Clifford variable having nothing to do with e^i, i_{e_i}, while for $\ell + 1 \le i \le n$, $\widehat{c}(e_i)$ is taken as in (1.1.2). This adds an element of extra confusion, which the reader has to accept for a while.

Still, it is important to observe that when $b = \sqrt{t}$, the rescaling of Definition 4.8.4 coincides with the ones in Definition 11.1.1, as long as one makes the following conventions for $1 \le i \le \ell$:

- $\widehat{c}(e_i) = \mathfrak{e}_i + i_{\mathfrak{e}^i}$.
- $\widehat{\mathfrak{e}}_i = 0$.

The above formulas make clear that indeed the $\widehat{c}(e_i), 1 \le i \le \ell$ should be considered as independent Clifford variables.

Let $r\left(-\widehat{\underline{\mathfrak{P}}}_{c,t}\right)((y,p),(y',p'))$ be the smooth kernel for $r\left(-\widehat{\underline{\mathfrak{P}}}_{c,t}\right)$ with respect to $dy'dp'$. We can define $\widehat{\mathrm{Tr_s}}\left[g\,|p|^2 r\left(-\widehat{\underline{\mathfrak{P}}}_{c,t}\right)\left(\left(g^{-1}(y,p)\right),(y,p)\right)\right]$ by following rules similar to the ones we used in section 7.5. We explain these rules in more detail in the present context.

First we concentrate on the case where $g = 1$ and F is trivial, so that $\ell = n$. Then the kernel $r\left(-\widehat{\underline{\mathfrak{P}}}_{c,t}\right)((y,p),(y,p))$ can be expanded in monomials in the $e^i, -e^i/t + i_{e_i} + \widehat{c}(e_i)/\sqrt{t}, \widehat{e}^i, i_{\widehat{e}_i}$. We denote by

$$\widehat{\mathrm{Tr_s}}\left[|p|^2 r\left(-\widehat{\underline{\mathfrak{P}}}_{c,t}\right)((y,p),(y,p))\right] \in \Lambda^{\cdot}(T^*M)$$

the object which one obtains first writing $r\left(-\widehat{\underline{\mathfrak{P}}}_{c,t}\right)((y,p),(y,p))$ using a normal ordering as in section 4.8, i.e., by putting all the annihilation operators i_{e_i} to the right, and then, after ignoring any term containing any of the i_{e_i}, by taking the standard supertrace in the variables $\widehat{e}^i, i_{\widehat{e}_i}$, and by selecting only those terms containing the terms to the left of the monomial $\widehat{c}(e_1)\ldots\widehat{c}(e_n)$ with a correcting sign $(-1)^{n(n+1)/2}$. In the general case, we combine the above conventions for the indices $1 \le i \le \ell$ with taking a classical supertrace in the remaining variables.

Let $\widehat{\mathrm{Tr_s}}\left[g\,|p|^2 r\left(-\widehat{\underline{\mathfrak{P}}}_{c,t}\right)\left(\left(g^{-1}(y,p)\right),(y,p)\right)\right]^{\max}$ be the form in $\Lambda^{\cdot}(T^*S)$ which comes to the left of $e^1 \wedge \ldots \wedge e^\ell$ in the expansion of

$$\widehat{\mathrm{Tr_s}}\left[g\,|p|^2 r\left(-\widehat{\underline{\mathfrak{P}}}_{c,t}\right)\left(\left(g^{-1}(y,p)\right),(y,p)\right)\right] \in \Lambda^{\cdot}(T^*M_g).$$

Proposition 11.1.2. *The following identity holds:*

$$t^{-\ell/2}\mathrm{Tr_s}\left[g\,|p|^2 r\left(-\widehat{\underline{\mathfrak{D}}}_{c,t}\right)\left(\left(g^{-1}(y,p)\right),(y,p)\right)\right] = \widehat{\mathrm{Tr_s}}\left[g\,|p|^2 r\left(-\widehat{\underline{\mathfrak{P}}}_{c,t}\right)\left(\left(g^{-1}(y,p)\right),(y,p)\right)\right]^{\max}. \quad (11.1.7)$$

Proof. We make the replacements which were described in Definition 11.1.1. For $1 \le i \le \ell$, $e^i i_{e_i}$ is replaced by

$$e^i\left(i_{e_i} + \widehat{c}(e_i)/\sqrt{t}\right). \quad (11.1.8)$$

Also recall that for $1 \le i \le \ell$, each $e^i i_{e_i}$ contributes to the initial supertrace by a factor -1. When making the above replacements, we find that when expanding the kernel $g \exp\left(-\widehat{\mathfrak{D}}_{c,t}\right)\left(g^{-1}(y,p),(y,p)\right)$ in the above Grassmann or Clifford variables, the original local supertrace will be obtained from the coefficient of

$$e^1 \dots e^\ell \widehat{c}(e_1)\widehat{c}(e_\ell)$$

by taking the supertrace in the remaining variables $e^i, i_{e_i}, \ell+1 \le i \le n$ and $\widehat{e}^i, i_{\widehat{e}_i}, 1 \le i \le n$ with the correcting factor $(-1)^{\ell(\ell+1)/2} t^{\ell/2}$. This last factor overcomes the singularity $t^{-\ell/2}$. This concludes the proof of Proposition 11.1.2. □

Set

$$\begin{aligned}\underline{\widehat{\mathfrak{P}}}_c = &\frac{1}{4}\left(-\Delta^V + c^2|p|^2 + c\left(2i_{\widehat{e}_i}\left(\widehat{e}^i - \widehat{c}(e_i)\right) - n\right)\right)\\ &+\frac{1}{4}\left\langle e_i, i^* R^{TX} e_j\right\rangle \widehat{e}^i\widehat{e}^j - \frac{1}{4} i^*\omega\left(\nabla^F, g^F\right)^2 - \frac{c}{2}\left(\nabla_p + \left\langle i^* R^{TX} y, p\right\rangle\right).\end{aligned} \tag{11.1.9}$$

Note that $\underline{\widehat{\mathfrak{P}}}_c$ depends on $x \in X_g$.

In the asymptotic expansion of operators, we follow the same rules as in Theorem 4.9.1. However, the various $\mathcal{O}$ which appear are taken only with respect to a given value of b.

Theorem 11.1.3. *As $t \to 0$,*

$$\underline{\widehat{\mathfrak{P}}}_{c,t} \to \underline{\widehat{\mathfrak{P}}}_c. \tag{11.1.10}$$

More precisely, for $t \in]0,1]$, $|y| \le 2\eta_0/\sqrt{t}$,

$$\underline{\widehat{\mathfrak{P}}}_{c,t} = \underline{\widehat{\mathfrak{P}}}_c + \sqrt{t}\mathcal{O}\left(1 + |y| + \widehat{\nabla} + |y|\nabla_p + |p|^2\widehat{\nabla} + |p|^2 + |y|^2 p\right). \tag{11.1.11}$$

Proof. Even though the connection ${}^1\nabla_t^{\Lambda^\cdot(T^*X)\widehat{\otimes}\Lambda^\cdot(T^*X)\widehat{\otimes}\Lambda^n(TX)\widehat{\otimes}F,u}$ contains a term diverging like $1/t$, the only part of this connection which contains annihilation operators i_{e_i} does not depend on t. The consequence is that when acting on terms which only contain creation operators, the trivialization using ${}^1\nabla_t^{\Lambda^\cdot(T^*X)\widehat{\otimes}\Lambda^\cdot(T^*X)\widehat{\otimes}\Lambda^n(TX)\widehat{\otimes}F,u}$ does not introduce spurious divergences.

Also observe that

$$\left[e^i/\sqrt{t} - \sqrt{t}i_{e_i}, e^i/\sqrt{t} + \sqrt{t}i_{e_i}\right] = 0. \tag{11.1.12}$$

Using (11.1.3) and (11.1.12), we get

$$\left[{}^1\nabla_t^{\Lambda^\cdot(T^*X)\widehat{\otimes}\Lambda^\cdot(T^*X)\widehat{\otimes}\Lambda^n(TX)\widehat{\otimes}F,u}, 2i_{\widehat{e}_i}\left(\widehat{e}^i - e^i/\sqrt{t} - \sqrt{t}i_{e_i}\right)\right] = 0, \tag{11.1.13}$$

so that the operator which appears in the first line of (11.1.4) is parallel with respect to the connection ${}^1\nabla_t^{\Lambda^\cdot(T^*X)\widehat{\otimes}\Lambda^\cdot(T^*X)\widehat{\otimes}\Lambda^n(TX)\widehat{\otimes}F,u}$. Also observe for

$1 \leq i \leq \ell$, $e^i/\sqrt{t}+\sqrt{t}i_{e_i}$ is changed into $\sqrt{t}i_{e_i}+\widehat{c}(e_i)$. We thus handle easily all the terms which appear in the right-hand side of (11.1.4), except for the difficult term in the fourth line which starts with $-\frac{c\sqrt{t}}{2}(\ldots)$.

Let $\left(E,\nabla^E\right)$ be a vector bundle on a fiber X, let $R^E = \nabla^{E,2}$ be the curvature of ∇^E. By [ABP73, Proposition 3.7], if E is trivialized on a small neighborhood of $x \in X$ by parallel transport with respect to the connection ∇^E along geodesics centered at x, if Γ is the corresponding connection form, if $y \in T_xX$ is close enough to 0,

$$\Gamma_y = \frac{1}{2}R^E_x(y,\cdot) + \mathcal{O}\left(|y|^2\right). \tag{11.1.14}$$

We assume temporarily that F is the trivial line bundle, so that we can drop the superscript u in the definition of $\nabla^{\Lambda^\cdot(T^*X)\widehat{\otimes}\Lambda^\cdot(T^*X)\widehat{\otimes}\Lambda^n(TX),u}$. Also we will temporarily underline sections of TX or T^*X when they will be considered as vectors or 1-forms along the fibers. Along these lines, we denote temporarily by $\underline{\nabla}^{TX}$ the restriction of ∇^{TX} to the fiber X, and by $\underline{R}^{TX}$ its curvature. Also we use temporarily the notation ${}^1\underline{\nabla}^{\Lambda^\cdot(T^*X)\widehat{\otimes}\Lambda^\cdot(T^*X)\widehat{\otimes}\Lambda^n(TX)}$ instead of $\nabla^{\Lambda^\cdot(T^*X)\widehat{\otimes}\Lambda^\cdot(T^*X)\widehat{\otimes}\Lambda^n(TX)}$.

Put

$$\begin{aligned}
\underline{T}^H &= \frac{1}{2}\left\langle T\left(f^H_\alpha, f^H_\beta\right), e_i\right\rangle \underline{e}^i \wedge f^\alpha \wedge f^\beta,\\
\underline{T}^0 &= f^\alpha \wedge \underline{e}^i \wedge e^j \left\langle T\left(f^H_\alpha, e_i\right), e_j\right\rangle, \\
\left|\underline{T}^0\right|^2 &= \sum_{j=1}^n \left(\sum_{\substack{1\leq i\leq n\\ 1\leq\alpha\leq m}} \left\langle T\left(f^H_\alpha, e_i\right), e_j\right\rangle f^\alpha \wedge \underline{e}^i \right)^2 .
\end{aligned} \tag{11.1.15}$$

In the sequel our tensors will be evaluated at $x \in X$. A straightforward computation, which uses in particular (1.3.2), shows that

$$\begin{aligned}
{}^1\underline{\nabla}_t^{\Lambda^\cdot(T^*X)\widehat{\otimes}\Lambda^\cdot(T^*X)\widehat{\otimes}\Lambda^n(TX),2} &= -\left\langle \underline{R}^{TX} e_i, e_j\right\rangle \left(e^i i_{e_j} + \widehat{e}^i i_{\widehat{e}_j}\right) + \underline{\nabla}^{TX}\underline{T}^H/t \\
&- \underline{\nabla}^{TX}\underline{T}^0\left(f^H_\alpha, e_i\right) f^\alpha\left(e^i - t i_{e_i}\right)/t - \left|\underline{T}^0\right|^2/t.
\end{aligned} \tag{11.1.16}$$

Now we use the same arguments as in the proof of Theorem 4.9.1 on the splitting of the scalar part of $\sqrt{t}\,{}^1\nabla^{\Lambda^\cdot(T^*X)\widehat{\otimes}\Lambda^\cdot(T^*X)\widehat{\otimes}\Lambda^n(TX)}_{t,Y^\mathcal{H}}$. By (11.1.14), (11.1.16), we find that after having done the replacements in the Clifford variables indicated in Definition 11.1.1, by considering y,p as sections of TX which can be contracted with underlined exterior variables along TX, we get

$$\begin{aligned}
I_{\sqrt{t}}\sqrt{t}\,{}^1\nabla^{\Lambda^\cdot(T^*X)\widehat{\otimes}\Lambda^\cdot(T^*X)\widehat{\otimes}\Lambda^n(TX)}_{t,Y^\mathcal{H}} I_{1/\sqrt{t}} &= \nabla_p \\
&+ \frac{1}{2}\sum_{1\leq i,j\leq \ell} \left\langle \underline{R}^{TX}\left(\underline{y},\underline{p}\right) e_i, e_j\right\rangle e^i e^j \\
&+ \frac{1}{2}\underline{\nabla}^{TX}\underline{T}^H\left(\underline{y},\underline{p}\right) - i^*\underline{\nabla}^{TX}\underline{T}^0\left(\underline{y},\underline{p}\right) - \frac{1}{2}\left|\underline{T}^0\right|^2\left(\underline{y},\underline{p}\right) \\
&+ \sqrt{t}\mathcal{O}\left(1 + |y|\nabla_p + |p|^2\widehat{\nabla} + |y|^2 p\right).
\end{aligned} \tag{11.1.17}$$

In (11.1.17), we temporarily underlined y, p to emphasize the fact that they were contracted with Grassmann variables like $\underline{e}^i$. Also i^* in $i^*\underline{\nabla}^{TX}\underline{T}^0(\underline{y}, \underline{p})$ refers to the Grassmann variables $e^i, 1 \leq i \leq n$. Note that the coefficient just before this expression is -1 and not $-1/2$. Indeed, as we saw at the beginning of the proof, for $1 \leq i \leq \ell$, $e^i - ti_{e_i}$ is replaced by $2e^i - ti_{e_i} - \sqrt{t}\widehat{c}(e_i)$, whose limit is $2e^i$. Using now [BG04, Theorem 3.26] or [B05, Theorem 4.15], we can rewrite (11.1.17) as

$$I_{\sqrt{t}}\sqrt{t}^1\nabla^{\Lambda^{\cdot}(T^*X)\widehat{\otimes}\Lambda^{\cdot}(T^*X)\widehat{\otimes}\Lambda^n(TX)}_{t,Y^{\mathcal{H}}} I_{1/\sqrt{t}} = \nabla_p + \langle i^*R^{TX}y, p\rangle + \sqrt{t}\mathcal{O}\left(1 + |y|\nabla_p + |p|^2\widehat{\nabla} + |y|^2|p|\right). \tag{11.1.18}$$

When F is nontrivial, the asymptotics of

$$I_{\sqrt{t}}{}^1\nabla^{\Lambda^{\cdot}(T^*X)\widehat{\otimes}\Lambda^{\cdot}(T^*X)\widehat{\otimes}\Lambda^n(TX),u}_{\sqrt{t}p} I_{1/\sqrt{t}}$$

is the same as in (11.1.18), by still using (11.1.14) and the above computations. The proof of our theorem is completed. □

Remark 11.1.4. Consider the vector bundle $\left(TX|_{M_g}, g^{TX|_{M_g}}, \nabla^{TX|_{M_g}}\right)$ on M_g. It satisfies the assumptions which are used in the constructions of chapter 7. We define the operator $\widehat{\underline{\mathfrak{C}}}^{\mathcal{M}^{TX|_{M_g}},2}_{\phi,\mathcal{H}^c-\omega^H}$ as in (7.2.6). When $\omega\left(\nabla^F, g^F\right) = 0$, and comparing (7.2.6) and (11.1.10), we get

$$\widehat{\underline{\mathfrak{P}}}_c = \widehat{\underline{\mathfrak{C}}}^{\mathcal{M}^{TX|_{M_g}},2}_{\phi,\mathcal{H}^c-\omega^H}. \tag{11.1.19}$$

Note that the identification (11.1.19) relies on the equality (1.1.2),

$$\widehat{c}(e_i) = e^i + i_{e_i}. \tag{11.1.20}$$

Identity (11.1.19) can be very confusing if one considers the e^i as forms on X_g. However, in the formalism of (7.2.6), there is no relation between E^1 and the base manifold S. Hence the $\widehat{c}(e_i)$ are treated in (7.2.6) as exogenous Clifford variables coming from E and not from S. This is precisely what we did before when introducing the $\widehat{c}(e_i)$. The identification (11.1.19) is then legitimate.

Identity (11.1.19) is a hypoelliptic version of the local families index theorem of [B86]. It plays a crucial role in the sequel.

11.2 THE LIMIT OF THE SUPERTRACE AS $t \to 0$

By taking (11.1.19) into account, we define the form $\mathrm{Tr_s}\left[g|p|^2 r\left(-\widehat{\underline{\mathfrak{P}}}_c\right)\right]$ on M_g by using the same conventions as in section 7.5, i.e., it is defined as a von Neumann supertrace. We take the operator $\mathfrak{M}_c^{TX|_{M_g}}$ as in (7.9.1).

Theorem 11.2.1. *Given $c \in \mathbf{R}^*$, as $t \to 0$,*

$$\mathrm{Tr_s}\left[gt|p|^2 h'\left(\mathfrak{D}^{\mathcal{M}}_{\phi,\mathcal{H}^c-\omega^H}\right)\right] \to (-1)^n \int_{X_g} \mathrm{Tr_s}\left[g|p|^2 r\left(-\widehat{\underline{\mathfrak{P}}}_c\right)\right]. \tag{11.2.1}$$

Moreover, the following identities hold:

$$(-1)^n \int_{X_g} \mathrm{Tr_s}\left[g\,|p|^2 r\left(-\widehat{\mathfrak{P}}_c\right)\right]$$
$$= -2\int_{X_g} \mathrm{Tr_s}\left[g\,|p|^2 \exp\left(-\mathfrak{M}_c^{TX|_{M_g}}\right)\right]^{zdc} \mathrm{Tr}^F[g]. \quad (11.2.2)$$

Proof. Using (11.1.1) and (11.1.2), we get

$$\mathrm{Tr_s}\left[gt\,|p|^2 h'\left(\mathfrak{D}^{\mathcal{M}}_{\phi,\mathcal{H}^c-\omega^H}\right)\right] = (-1)^n\,\mathrm{Tr_s}\left[g\,|p|^2 r\left(-\widehat{\mathfrak{M}}_{c,t}\right)\right]. \quad (11.2.3)$$

Moreover,

$$\mathrm{Tr_s}\left[g\,|p|^2 r\left(-\widehat{\mathfrak{M}}_{c,t}\right)\right]$$
$$= \int_{T^*X} \mathrm{Tr_s}\left[g\,|p|^2 r\left(-\widehat{\mathfrak{M}}_{c,t}\right)\left(g^{-1}(x,p),(x,p)\right)\right] dv_{T^*X}. \quad (11.2.4)$$

Using Proposition 4.7.1 and proceeding as in Remark 4.7.2, it is clear that for $\beta > 0$, the integral in the right-hand side of (11.2.4) localizes on $\pi^{-1}\mathcal{U}_\beta$. By Proposition 4.8.2, it is also clear that when evaluating the limit, we can as well replace $\widehat{\mathfrak{M}}_{c,t}$ by $\widehat{\mathfrak{N}}_{c,t}$.

Now we use the obvious analogue of (4.13.1). Moreover, by proceeding as in (4.13.2) and using (11.1.7), for $x \in X_g$, we get

$$\int_{\left\{y\in N_{X_g/X},|y|\le\eta_0\right\}\times T^*_xX} \mathrm{Tr_s}\left[g\,|p|^2 r\left(-\widehat{\mathfrak{N}}_{c,t}\right)\left(g^{-1}(y,p),(y,p)\right)\right] dv_{N_{X_g/X}}(y)\,dp$$
$$= \int_{\left\{y\in N_{X_g/X},|y|\le\eta_0/\sqrt{t}\right\}\times T^*_xX} \widehat{\mathrm{Tr}}_{\mathrm{s}}\left[g\,|p|^2 r\left(-\widehat{\mathfrak{P}}_{c,t}\right)\left(g^{-1}(y,p),(y,p)\right)\right] dv_{N_{X_g/X}}(y)\,dp. \quad (11.2.5)$$

Now by equation (4.10.1) in Theorem 4.10.1, for $t \in]0,1], x \in X_g, y \in N_{X_g/X}, |y| \le \eta_0/\sqrt{t}, p \in T^*_xX$, at least when $b \ge 1$,

$$\left|\widehat{\mathrm{Tr}}_{\mathrm{s}}\left[g\,|p|^2 r\left(-\widehat{\mathfrak{P}}_{c,t}\right)\left(\left(g^{-1}(y,p)\right),(y,p)\right)\right]\right|$$
$$\le C\left(1+\frac{1}{t^{m/2}}\right)\exp\left(-c\left(|y|^{1/3}+|p|^{2/3}\right)\right). \quad (11.2.6)$$

However, inspection of the proof of Theorem 4.10.1 shows trivially that given $b > 0$, the estimate (11.2.6) is always valid. Also note that when writing the uniform bound in (11.2.6), we took into account the fact that the rescaling on the variable y differs by a factor b^2. Still the uniform bound in (11.2.6) is inadequate because of the diverging factor $\frac{1}{t^{m/2}}$.

However, for a given $b > 0$, the above divergence is a ghost divergence. Indeed the estimate in (11.2.6) was obtained in Theorem 4.10.1 using a trivialization via the connection $\nabla^{\Lambda^\cdot(T^*X)\widehat{\otimes}\Lambda^\cdot(T^*X)\widehat{\otimes}\Lambda^n(TX)\widehat{\otimes}F,u}$. As explained before (11.1.3), we use here a different trivialization. As is clear from Theorem

11.1.3, this makes the above divergence disappear. Therefore the estimate (11.2.6) is valid when deleting the term $1/t^{m/2}$.

By using (11.1.11) and the appropriate uniform estimates on the heat kernels, we find that as $t \to 0$,

$$\widehat{\mathrm{Tr}}_{\mathrm{s}}\left[g\,|p|^{2}\,r\left(-\underline{\widehat{\mathfrak{P}}}_{c,t}\right)\left(g^{-1}(y,p),(y,p)\right)\right] \to \widehat{\mathrm{Tr}}_{\mathrm{s}}\left[g\,|p|^{2}\,r\left(-\underline{\widehat{\mathfrak{P}}}_{c}\right)\left(g^{-1}(y,p),(y,p)\right)\right]. \qquad (11.2.7)$$

By combining (11.2.3)-(11.2.7), it is clear that (11.2.1) holds.

When $\omega\left(\nabla^{F}, g^{F}\right)$ vanishes, i.e., when g^{F} is parallel, equation (11.2.2) follows from (7.9.2) in Theorem 7.9.1 and from (11.1.19). Note that the replacement of $\widehat{e}^{i}, i_{\widehat{e}_{i}}$ by $i_{\widehat{e}^{i}}, \widehat{e}_{i}$ which is done in section 7.4 makes the sign $(-1)^{n}$ ultimately disappear in the right-hand side of (11.2.2). In the general case, we also use (4.13.4). The proof of our theorem is completed. □

11.3 A PROOF OF EQUATION (8.4.6)

By (4.3.3) and by Theorem 11.2.1, it is clear that for $b>0$, as $t \to 0$,

$$v_{\sqrt{t}b,t} \to \mp\frac{2}{b^{2}}\varphi\int_{X_{g}}\mathrm{Tr}_{\mathrm{s}}\left[g\exp\left(-\mathfrak{M}_{c}^{TX|_{M_{g}}}\right)\right]^{zdc}. \qquad (11.3.1)$$

Comparing with (8.4.3), we find that (11.3.1) is just (8.4.6).

Remark 11.3.1. In our proof of (11.3.1), we could have instead used the expression for $v_{b,t}$ which was suggested before (4.13.6). This would have led us more directly to the operator $\mathfrak{M}_{c}^{TX|_{M_{g}}}$.

Chapter Twelve

A proof of equation (8.4.8)

The purpose of this chapter is to establish equation (8.4.8) in Theorem 8.4.3. We thus establish a uniform bound on $|v_{b,\epsilon}|$ for $\epsilon \in]0,1], b \in [\sqrt{\epsilon}, b_0]$. The proof relies on techniques already used in chapters 4 and 11.

This chapter is organized as follows. In section 12.1, we combine the techniques of chapters 4 and 11 to obtain a uniform expansion of the rescaled operator in the considered range of parameters.

In section 12.2, we prove the required estimate.

12.1 UNIFORM RESCALINGS AND TRIVIALIZATIONS

In this chapter, we fix $b_0 \geq 1$. We take $t \in]0,1], b \in [\sqrt{t}, b_0]$. We use the notation of section 4.7, while making in the whole chapter $z = 0, dc = 0, db = 0, dt = 0$.

We will use the expression for $v_{b,t}$ given in (4.13.8),

$$v_{b,t} = \pm (-1)^n \varphi \left(1 + 2\frac{\partial}{\partial a}\right) \mathrm{Tr_s}\left[g\frac{t}{b^2}|p|^2 \exp\left(-a\widehat{\underline{\mathfrak{M}}}_{b,t}\right)\right]|_{a=1}. \tag{12.1.1}$$

The point of equation (12.1.1) is that we have expressed $v_{b,t}$ in terms of supertraces of heat kernels. We can rewrite (12.1.1) in the form

$$v_{b,t} = \left(1 + 2\frac{\partial}{\partial a}\right) \left(\pm (-1)^n \varphi \int_{T^*X} \mathrm{Tr_s}\left[g\frac{t}{b^2}|p|^2 \exp\left(-a\widehat{\underline{\mathfrak{M}}}_{b,t}\right)(z, gz)\right] dv_{T^*X}(z)\right)|_{a=1}. \tag{12.1.2}$$

Using Proposition 4.7.1, which covers precisely the range of parameters which is considered here, and proceeding as in Remark 4.7.2, we find that for $a \in [1/2, 1], \beta > 0$,

$$\left|\int_{\pi^{-1}(X\setminus\mathcal{U}_\beta)} \mathrm{Tr_s}\left[g\frac{t}{b^2}|p|^2 \exp\left(-a\widehat{\underline{\mathfrak{M}}}_{b,t}\right)(z, gz)\right] dv_{T^*X}\right| \leq C\exp\left(-c\beta^2/t\right). \tag{12.1.3}$$

Recall that the connection ${}^1\nabla_t^{\Lambda^\cdot(T^*X)\widehat{\otimes}\Lambda^\cdot(T^*X)\widehat{\otimes}\Lambda^n(TX)\widehat{\otimes}F,u}$ was defined in

(11.1.3). By (4.6.4) in Proposition 4.6.1 and by (11.1.3), we get

$$
\begin{aligned}
\widehat{\underline{\mathfrak{M}}}_{b,t} = \frac{1}{4}\Bigg(-\Delta^V + \frac{t^2}{b^4}|p|^2 \pm \frac{t}{b^2}\left(2i_{\widehat{e}_i}\left(\widehat{e}^i - e^i/\sqrt{t} - \sqrt{t}i_{e_i}\right) - n\right)\Bigg) \\
+ \frac{1}{4}\left\langle e_i, R^{TX} e_j\right\rangle \widehat{e}^i\widehat{e}^j - \frac{\sqrt{t}}{4}\omega\left(\nabla^F, g^F\right)(e_i)\nabla_{\widehat{e}^i} \\
- \frac{\sqrt{t}}{4}\nabla^{\Lambda^\cdot(T^*T^*X)\widehat{\otimes}F}\widehat{\omega}\left(\nabla^F, g^F\right) - \frac{1}{4}\omega\left(\nabla^F, g^F\right)^2 \\
\mp \frac{t^{3/2}}{2b^2}\left({}^1\nabla_{t,Y^{\mathcal{H}}}^{\Lambda^\cdot(T^*X)\widehat{\otimes}\Lambda^\cdot(T^*X)\widehat{\otimes}\Lambda^n(TX)\widehat{\otimes}F,u} + \left\langle R^{TX}(\cdot, p)e_i, p\right\rangle \widehat{e}^i\right). \quad (12.1.4)
\end{aligned}
$$

Take now $x \in X_g$. As in section 11.1, we will use the trivialization of $\Lambda^\cdot(T^*X)\widehat{\otimes}\Lambda^\cdot(T^*X)\widehat{\otimes}\Lambda^n(TX)\widehat{\otimes}F$ along radial geodesics centered at x by parallel transport with respect to ${}^1\nabla_t^{\Lambda^\cdot(T^*X)\widehat{\otimes}\Lambda^\cdot(T^*X)\widehat{\otimes}\Lambda^n(TX)\widehat{\otimes}F,u}$. We define $\widehat{\underline{\mathfrak{N}}}_{b,t}$ as in section 4.8. The basic difference with respect to chapter 4 is that we use the connection ${}^1\nabla_t^{\Lambda^\cdot(T^*X)\widehat{\otimes}\Lambda^\cdot(T^*X)\widehat{\otimes}\Lambda^n(TX)\widehat{\otimes}F,u}$ for the trivialization of the considered vector bundles.

By Proposition 4.8.2, over $\pi^{-1}\mathcal{U}_\beta$, we can as well replace the operator $\widehat{\underline{\mathfrak{M}}}_{b,t}$ by $\widehat{\underline{\mathfrak{N}}}_{b,t}$. Incidentally, note that we can indeed use Proposition 4.8.2 even though our choice of trivialization is now different.

We define the operator $\widehat{\underline{\mathfrak{O}}}_{b,t}$ as in (4.8.9). From $\widehat{\underline{\mathfrak{O}}}_{b,t}$, we obtain the operator $\widehat{\underline{\mathfrak{P}}}_{b,t}$ as in Definition 4.8.4, that is, by making the replacements indicated there. As in section 4.8, the kernels for $\exp\left(-a\widehat{\underline{\mathfrak{O}}}_{b,t}\right), \exp\left(-\widehat{\underline{\mathfrak{P}}}_{b,t}\right)$ will be calculated with respect to the volume $dydp$.

Proposition 12.1.1. *The following identity holds:*

$$
\begin{aligned}
\left(\frac{t^{3/2}}{b^2}\right)^{n-\ell} \operatorname{Tr_s}\left[g\frac{t}{b^2}|p|^2\exp\left(-\widehat{\underline{\mathfrak{N}}}_{b,t}\right)\left(g^{-1}\left(\frac{t^{3/2}}{b^2}y, p\right), \left(\frac{t^{3/2}}{b^2}y, p\right)\right)\right] \\
= (-1)^\ell \widehat{\operatorname{Tr_s}}\left[g\frac{t}{b^2}|p|^2\exp\left(-\widehat{\underline{\mathfrak{P}}}_{b,t}\right)\left(g^{-1}(y,p), (y,p)\right)\right]. \quad (12.1.5)
\end{aligned}
$$

Proof. The proof of our proposition is the same as the proof of Proposition 4.8.5. □

Recall that the operator $\widehat{\underline{\mathfrak{P}}}$ was defined in (4.9.1). We give a better version of Theorem 4.9.1, in which the potentially diverging terms will have disappeared. The whole point of this new theorem is that we use a different trivialization.

Theorem 12.1.2. *As $t \to 0$,*

$$
\widehat{\underline{\mathfrak{P}}}_{b,t} \to \widehat{\underline{\mathfrak{P}}}. \quad (12.1.6)
$$

More precisely, for $t \in]0,1], b \in \left[\sqrt{t}, b_0\right], |y| \leq 2b^2\eta_0/t^{3/2}$,

$$\begin{aligned}\underline{\widehat{\mathfrak{P}}}_{b,t} = \underline{\widehat{\mathfrak{P}}} + \frac{t^2}{4b^4}|p|^2 + \mathcal{O}\left(\frac{\sqrt{t}}{b}\right)\left(\mathcal{O}\left(1+\frac{t}{b}|y|\right)\right) \\ + \mathcal{O}\left(\sqrt{t}\widehat{\nabla} + \frac{t^{3/2}}{b^2}|y|\nabla_p + \frac{t^{3/2}}{b^2}|p|^2\widehat{\nabla} + \frac{t^{3/2}}{b^2}|p|^2 + \frac{t^2}{b^4}p|y|\right). \quad (12.1.7)\end{aligned}$$

Proof. We will combine the methods used in the proof of Theorems 4.9.1 and 11.1.3. Theorem 11.1.3 will be especially useful, since its proof uses the same trivialization as ours.

We start from equation (12.1.4). We make the preliminary observation that using the connection ${}^1\nabla_t^{\Lambda^\cdot(T^*X)\widehat{\otimes}\Lambda^\cdot(T^*X)\widehat{\otimes}\Lambda^n(TX)\widehat{\otimes}F,u}$ does not add extra divergences when applied to expressions containing only creation operator $e^i, \widehat{e}^i$. Also to handle the matrix terms in the first line in the right-hand side of (12.1.4), we use (11.1.13). One finds easily that they contribute to $\underline{\widehat{\mathfrak{P}}}_{b,t} - \underline{\widehat{\mathfrak{P}}}$ by a term $\mathcal{O}\left(\frac{\sqrt{t}}{b}\right)$. The terms in the next two following lines can be estimated as in the proof of Theorem 4.9.1. This way, we get in particular the term $\mathcal{O}\left(\frac{t^{3/2}}{b^2}|y|\right)$ in the right-hand side of (12.1.7).

The terms $\mathcal{O}\left(\sqrt{t}\widehat{\nabla} + \frac{t^{3/2}}{b^2}|p|^2\right)$ appear for the same reason as in the proof of Theorem 4.9.1. What remains is the contribution of

$$\frac{t^{3/2}}{b^2}{}^1\nabla_{t,Y^{\mathcal{H}}}^{\Lambda^\cdot(T^*X)\widehat{\otimes}\Lambda^\cdot(T^*X)\widehat{\otimes}\Lambda^n(TX)\widehat{\otimes}F,u}$$

to the expansion of $\underline{\widehat{\mathfrak{P}}}_{b,t}$. In fact by (11.1.17), we get, once the replacements of Definition 4.8.4 have been done,

$$\begin{aligned}I_{t^{3/2}/b^2}\frac{t^{3/2}}{b^2}{}^1\nabla_{t,Y^{\mathcal{H}}}^{\Lambda^\cdot(T^*X)\widehat{\otimes}\Lambda^\cdot(T^*X)\widehat{\otimes}\Lambda^n(TX)\widehat{\otimes}F,u}I_{b^2/t^{3/2}} \\ = \nabla_p + \mathcal{O}\left(\frac{t^{3/2}}{b^2}|y|\nabla_p + \frac{t^{3/2}}{b^2}|p|^2\widehat{\nabla} + \frac{t^2}{b^4}p|y|\right). \quad (12.1.8)\end{aligned}$$

The proof of our theorem is completed. □

12.2 A PROOF OF (8.4.8)

Now we establish a better version of Theorem 4.10.1.

Theorem 12.2.1. *For* $\eta_0 > 0$ *small enough, there exist* $c > 0, C > 0, m \in \mathbf{N}$ *such that for* $a \in \left[\frac{1}{2}, 1\right], t \in]0,1], b \in \left[\sqrt{t}, b_0\right], x \in X_g, y \in N_{X_g/X,x}, |y| \leq b^2\eta_0/t^{3/2}, p \in T_x^*X$, *then*

$$\left|g\exp\left(-a\underline{\widehat{\mathfrak{P}}}_{b,t}\right)\left(g^{-1}(y,p),(y,p)\right)\right| \leq C\exp\left(-c\left(|y|^{1/3} + |p|^{2/3}\right)\right). \quad (12.2.1)$$

Proof. To establish our theorem, we will use the expansion (12.1.7) in Theorem 12.1.2. Indeed the expansion makes clear that in the given range of parameters, the coefficients of the operator $\underline{\widehat{\mathfrak{P}}}_{b,t}$ remain uniformly controlled. It is then easy to adapt the techniques of the proof of Theorem 4.10.1 so as to get (12.2.1). The proof of our theorem is completed. □

By using (12.1.5) and proceeding as in (4.13.2), we get

$$\int_{\{y\in N_{X_g/X},|y|\leq\eta_0\}} \mathrm{Tr_s}\left[g\frac{t}{b^2}|p|^2\exp\left(-a\underline{\mathfrak{N}}_{b,t}\right)\left(g^{-1}(y,p),(y,p)\right)\right]dydp$$
$$= (-1)^{\ell}\int_{\{y\in N_{X_g/X},|y|\leq b^2\eta_0\}} \widehat{\mathrm{Tr_s}}\left[g\frac{t}{b^2}\exp\left(-\underline{\widehat{\mathfrak{P}}}_{b,t}\right)\left(g^{-1}(y,p),(y,p)\right)\right] dydp. \quad (12.2.2)$$

From (12.2.1), (12.2.2), we get

$$\left|\int_{y\in N_{X_g/X},|y|\leq b^2\eta_0/t^{3/2}} \widehat{\mathrm{Tr_s}}\left[g\frac{t}{b^2}\exp\left(-a\underline{\widehat{\mathfrak{P}}}_{b,t}\right)\left(\left(g^{-1}(y,p)\right),(y,p)\right)\right]dydp\right| \leq C\frac{t}{b^2}. \quad (12.2.3)$$

By combining the previous estimates with (12.2.3), we get (8.4.8).

Chapter Thirteen

A proof of equation (8.4.7)

The purpose of this chapter is to establish the estimate (8.4.7), which gives an estimate for $\left|v_{\sqrt{t}b,t} - v_{0,t}\right|$ which is uniform in $b \in]0,1], t \in]0,1]$. The idea is to combine the local index techniques of chapter 11 with the functional analytic machine which is extensively developed in chapter 17 to prove that in the proper sense, as $b \to 0$, the hypoelliptic Laplacian converges to the standard Laplacian. Indeed our estimate is compatible with the convergence result for $v_{\sqrt{t}b,t}$ as $t \to 0$, which is established in chapter 11. This explains why the methods of that chapter play an important role in the whole proof. The main point of the present chapter is actually to show that the convergence result of chapter 11 can be made uniform in $b \in]0,1]$.

This chapter is organized as follows. In section 13.1, we establish our estimate in the range $t \leq b^{\beta}$.

In section 13.2, we show that the estimate can be localized near $\pi^{-1}X_g$.

In section 13.3, we give a new approach to the local index theoretic techniques of chapter 11, which will permit us to obtain the required uniformity.

In section 13.4, we evaluate the limit as $t \to 0$ of the properly rescaled Laplacian.

In section 13.5, we replace the fiber X by one of the tangent spaces $T_xX, x \in X_g$, and we consider a corresponding localized operator.

Finally, in section 13.6, we complete the proof of (8.4.7).

13.1 THE ESTIMATE IN THE RANGE $\mathbf{t} \geq \mathbf{b}^{\beta}$

By (2.2.5), we find easily that $A^{2}_{b,t}$ is conjugate to $tA'^{2}_{\phi_b,\mathcal{H}}$. Note that this is also a consequence of (2.8.8). Finally, $\mathfrak{C}^{\mathcal{M},2}_{b,t} - A^{2}_{b,t}$ has positive degree in $\Lambda^{\cdot}(T^*S)$. As explained in section 3.7, this implies that $\mathfrak{C}^{\mathcal{M},2}_{b,t}$ and $A^{2}_{b,t}$ have the same spectrum.

The set $\mathcal{W}_{\delta',b,r}$ is defined in (3.4.2) and in (17.20.1). By taking δ', r as in Theorem 17.21.3, for $b \in]0,b_0], \lambda \in \mathcal{W}_{\delta',b,r}$, the resolvent $\left(A^{2}_{\phi_b,\pm\mathcal{H}} - \lambda\right)^{-1}$ exists and verifies appropriate estimates. Since $t \in]0,1]$, for $b \in]0,b_0], \lambda \in \mathcal{W}_{\delta',b,r}$, the resolvent $\left(A^{2}_{\phi_{b\sqrt{t}},\mathcal{H}} - \lambda\right)^{-1}$ exists.

By the above considerations, it follows that for $b \in]0,b_0], t \in]0,1], \lambda \in \mathcal{W}_{\delta',b,r}$, the resolvent $\left(\mathfrak{C}^{\mathcal{M}\times\mathbf{R}^{*2}_{+}}_{b,t} - \lambda\right)^{-1}$ exists.

Take $c_0 > 0, c_1 > 0$. Let γ be the contour in $\mathbf{C}$ given by

$$\gamma = \left\{ \lambda \in \mathbf{C}, \operatorname{Re}\lambda = -c_0 + c_1 \left|\operatorname{Im}\lambda\right|^{1/6} \right\}. \tag{13.1.1}$$

We orient γ downward. By taking c_0 large enough and c_1 small enough, we find that for $b \in]0,1]$, $\gamma \subset \mathbf{C} \setminus W_{\delta',b,r}$.

By equation (2.8.9) and by the above,

$$U_{b,t} r\left(-\mathfrak{C}_{b,t}^{\mathcal{M},2}\right) U_{b,t}^{-1} = \psi_t^{-1} r\left(-t\mathfrak{C}_{\phi_b,\pm\mathcal{H}-b\omega^H}^{\mathcal{M},2}\right)\psi_t. \tag{13.1.2}$$

Proposition 13.1.1. *Take $v \in]0,1[$. There exist $C_v > 0, u > 0$ such that for $t \in]0,1], b \in]0,1]$,*

$$\left| v_{\sqrt{t}b,t} - v_{0,t} \right| \le C_v t^{-u} b^v. \tag{13.1.3}$$

Proof. By (4.3.3) and (13.1.2), we get

$$v_{\sqrt{t}b,t} = \pm\psi_t^{-1}\varphi \operatorname{Tr_s}\left[g\left|p\right|^2 r\left(-t\mathfrak{C}_{\phi_{\sqrt{t}b},\pm\mathcal{H}-\sqrt{t}b\omega^H}^{\mathcal{M},2}\right)\right]. \tag{13.1.4}$$

Now observe that for $t \in]0,1], b \in]0,1]$, if λ lies to the left of γ, then $\lambda/t \in \mathcal{W}_{\delta',\sqrt{t}b,r}$. It follows that

$$r\left(-t\mathfrak{C}_{\phi_{\sqrt{t}b},\pm\mathcal{H}-\sqrt{t}b\omega^H}^{2}\right) = \frac{1}{2i\pi}\int_\gamma r\left(-\lambda\right)\left(\lambda - t\mathfrak{C}_{\phi_{\sqrt{t}b},\pm\mathcal{H}-\sqrt{t}b\omega^H}^{\mathcal{M},2}\right)^{-1} d\lambda. \tag{13.1.5}$$

For $N \in \mathbf{N}$, let $r_N\left(-\lambda\right)$ be the Nth integral of $r\left(-\lambda\right)$ which vanishes at $\operatorname{Re}\lambda = +\infty$. By (13.1.5), we get

$$r\left(-t\mathfrak{C}_{\phi_{\sqrt{t}b},\pm\mathcal{H}-\sqrt{t}b\omega^H}^{2}\right) = \frac{(N-1)!}{2i\pi}\int_\gamma r_{N-1}\left(-\lambda\right)\left(\lambda - t\mathfrak{C}_{\phi_{\sqrt{t}b},\pm\mathcal{H}-\sqrt{t}b\omega^H}^{\mathcal{M},2}\right)^{-N} d\lambda. \tag{13.1.6}$$

On the other hand, one has the trivial

$$r\left(-tA_\pm^2\right) = \frac{(N-1)!}{2i\pi}\int_\gamma r_{N-1}\left(-\lambda\right)\left(\lambda - tA_\pm^2\right)^{-N} d\lambda. \tag{13.1.7}$$

Now we take $L \in \mathbf{N}$. By equation (3.4.3) and Theorem 17.21.5, we know that given $v \in]0,1[$, for $N \in \mathbf{N}$ large enough, for $b \in]0,1], t \in]0,1], \lambda \in \gamma$,

$$\left|\left|\left|\left(\lambda - t\mathfrak{C}_{\phi_{\sqrt{t}b},\pm\mathcal{H}-\sqrt{t}b\omega^H}^{\mathcal{M},2}\right)^{-N} - i_\pm\left(\lambda - tA_\pm^2\right)^{-N} P_\pm\right|\right|\right|_L \le C_{v,N} t^{-N} b^v. \tag{13.1.8}$$

Note that in degree 0, equation (13.1.8) follows directly from Theorem 17.21.5. However, as explained in section 3.7, the arguments of chapter 17 can be easily extended to the hypoelliptic curvature $E_{\phi_{\sqrt{t}b},\pm\mathcal{H}-\sqrt{t}b\omega^H}^{\mathcal{M},2}$.

As explained in (3.4.10), (13.1.8) guarantees at the same time a uniform bound of the difference of the corresponding kernels and also the adequate uniform decay of the difference of kernels as $|p| \to +\infty$.

By (5.3.15),

$$P_{\pm}|p|^2 P_{\pm} = \frac{n}{2}. \tag{13.1.9}$$

Finally, by (1.7.5),

$$\operatorname{Tr_s}\left[gr\left(-tA_{+}^{2}\right)\right] = \chi_g(F), \qquad \operatorname{Tr_s}\left[gr\left(-tA_{-}^{2}\right)\right] = \chi_g(F \otimes o(TX)). \tag{13.1.10}$$

Using (1.6.5), (1.6.6), (13.1.9) and taking into account the fact that P_+ projects on forms of fiberwise degree 0 and P_- on forms of fiberwise degree n, we finally get

$$\operatorname{Tr_s}\left[gP_{\pm}|p|^2 P_{\pm} r\left(-tA_{\pm}^{2}\right)\right] = \frac{n}{2}\chi_g(F). \tag{13.1.11}$$

By (5.1.13) and (13.1.4)-(13.1.11), we get (13.1.3). The proof of our proposition is completed. □

By (13.1.3), we find that if $v' < v$, there exists $C > 0$ such that for $t \in]0,1], b \in]0,1], t \geq b^{v'/u}$,

$$\left|v_{\sqrt{t}b,t} - v_{0,t}\right| \leq Cb^{v-v'}, \tag{13.1.12}$$

i.e., we have established the estimate (8.4.7) in the considered range of parameters. Therefore to establish (8.4.7) in full generality, we may as well assume that $\beta > 0$ is given and that $t \leq b^{\beta}$.

Ultimately to establish (8.4.7), it will be enough to show that given $\beta > 0$, there exists $C > 0, \alpha > 0$ such that for $b \in]0,1], t \in]0,1], t \leq b^{\beta}$,

$$\left|v_{\sqrt{t}b,t} - v_{0,t}\right| \leq C\left(t^{\alpha} + b^{\alpha}\right). \tag{13.1.13}$$

Now we concentrate on the proof of (13.1.13).

13.2 LOCALIZATION OF THE ESTIMATE NEAR $\pi^{-1}X_g$

By (13.1.4),

$$v_{\sqrt{t}b,t} = \pm\psi_t^{-1}\varphi \int_{T^*X} \operatorname{Tr_s}\left[g|p|^2 r\left(-t\mathfrak{C}^{\mathcal{M},2}_{\phi_{\sqrt{t}b},\pm\mathcal{H}-\sqrt{t}b\omega^H}\right)\left(g^{-1}(x,p),(x,p)\right)\right] dv_{T^*X}. \tag{13.2.1}$$

Proposition 13.2.1. *Given $\beta > 0, N \in \mathbf{N}$, there exists $C_{\beta,N} > 0$ such that for $b \in]0,1], t \in]0,1]$,*

$$\left|\int_{\pi^{-1}(X\setminus\mathcal{U}_{\beta})} \operatorname{Tr_s}\left[g|p|^2 r\left(-t\mathfrak{C}^{\mathcal{M},2}_{\phi_{\sqrt{t}b},\pm\mathcal{H}-\sqrt{t}b\omega^H}\right)\left(g^{-1}(x,p),(x,p)\right)\right] dv_{T^*X}\right| \leq C_{\beta,N}t^N. \tag{13.2.2}$$

Proof. We use (13.1.4) for $v_{b\sqrt{t},b}$, (13.1.6) for $r\left(-t\mathfrak{C}^{2}_{\phi_{\sqrt{t}b},\pm\mathcal{H}-\sqrt{t}b\omega^H}\right)$, and also equation (17.22.6) in Theorem 17.22.2. Again we have used the arguments in section 3.7 to extend the results of chapter 17, which are valid for $\mathfrak{A}'^2_{\phi_b,\pm\mathcal{H}}$, to the curvature $\mathfrak{C}^{\mathcal{M},2}_{\phi_b,\pm\mathcal{H}-b\omega^H}$. □

Remark 13.2.2. By Proposition 13.2.1, the contribution of $\pi^{-1}(X\setminus\mathcal{U}_\beta)$ to the integral in the right-hand side of (13.2.1) is compatible with (13.1.13). Therefore the estimate (13.1.13) can be localized near $\pi^{-1}X_g$.

We use the notation of chapters 11 and 13. We start from equation (11.1.4) for $\underline{\widehat{\mathfrak{M}}}_{c,t}$. Set

$$\underline{\widehat{\mathcal{M}}}_{c,t} = K_b\underline{\widehat{\mathfrak{M}}}_{c,t}K_b^{-1}. \tag{13.2.3}$$

By (4.3.3), (11.1.1), (11.1.2), and (13.2.3), we get

$$v_{b\sqrt{t},t} = \pm\varphi(-1)^n \operatorname{Tr_s}\left[g\,|p|^2\,r\left(-\underline{\widehat{\mathcal{M}}}_{c,t}\right)\right]. \tag{13.2.4}$$

For $N\in\mathbf{N}$ large enough, let $K_{b,t,\lambda,N}((x,p),(x',p'))$ be the Schwartz kernel associated to the operator $\left(\lambda-\underline{\widehat{\mathcal{M}}}_{c,t}\right)^{-N}$. Similarly let $K_{0,t,\lambda,N}(x,x')$ be the Schwartz kernel associated to the operator $\left(\lambda-A^2_{\pm,t}\right)^{-N}$.

Definition 13.2.3. For $N\in\mathbf{N}$ large enough, set

$$\begin{aligned} v_{\sqrt{t}b,t,N}(x,\lambda) \\ &= \pm(-1)^n\varphi(N-1)!\int_{T^*X}\operatorname{Tr_s}\left[g\,|p|^2\,K_{b,t,\lambda,N}((x,p),(gx,gp))\right]dp. \end{aligned} \tag{13.2.5}$$

Similarly set

$$\begin{aligned} v_{0,t,N}(x,\lambda) &= (N-1)!\frac{n}{2}\operatorname{Tr_s}\left[gK_{0,t,\lambda,N}(x,gx)\right] \text{ for } c>0, \\ &= -(N-1)!\frac{(-1)^n n}{2}\operatorname{Tr_s}\left[gK_{0,t,\lambda,N}(x,gx)\right] \text{ if } c<0. \end{aligned} \tag{13.2.6}$$

By (1.7.5) and by (13.2.4), for N large enough,

$$\begin{aligned} v_{\sqrt{t}b,t} &= \int_X\left[\frac{1}{2i\pi}\int_\gamma r_{N-1}(-\lambda)\,v_{\sqrt{t}b,t,N}(x,\lambda)\,d\lambda\right]dx, \\ v_{0,t} &= \int_X\frac{1}{2i\pi}\left[\int_\gamma r_{N-1}(-\lambda)\,v_{0,t,N}(x,\lambda)\,d\lambda\right]dx. \end{aligned} \tag{13.2.7}$$

By using again equation (17.22.6) in Theorem 17.22.2, we find that given $\beta>0, N'\in\mathbf{N}$, there exists $C_{\beta,N,N'}>0$ such that for $b\in]0,1], t\in]0,1], \lambda\in\gamma$,

$$\left|\int_{\pi^{-1}(X\setminus\mathcal{U}_\beta)}v_{\sqrt{t}b,t,N}(x,\lambda)\,dx\right|\le C_{\beta,N,N'}t^{N'}. \tag{13.2.8}$$

By (4.7.1), (13.2.7), and (13.2.8), we find that to establish (13.1.13), we only need to show that for any $\lambda \in \gamma, x \in X_g, t \in]0,1], b \in]0,1], t \leq b^\beta$,

$$\left| \int_{\{y \in N_{X_g/X}, |y| \leq \beta\}} \left(v_{\sqrt{t}b,t,N}(x,y,\lambda) - v_{0,t,N}(x,y,\lambda) \right) k(x,y)\, dy \right| \\ \leq C_{\beta,N}(t^\alpha + b^\alpha). \quad (13.2.9)$$

For $x \in X_g, \lambda \in \gamma$, set

$$w_{\sqrt{t}b,t,N}(x,\lambda) = \int_{\{y \in N_{X_g/X}, |y| \leq \beta\}} v_{\sqrt{t}b,t,N}(x,y,\lambda)\, k(x,y)\, dy, \quad (13.2.10)$$

$$w_{0,t,N}(x,\lambda) = \int_{\{y \in N_{X_g/X}, |y| \leq \beta\}} v_{0,t,N}(x,y,\lambda)\, k(x,y)\, dy.$$

Then (13.2.9) is equivalent to

$$\left| w_{\sqrt{t}b,t,N}(x,\lambda) - w_{0,t,N}(x,\lambda) \right| \leq C_N(t^\alpha + b^\alpha), \quad (13.2.11)$$

in the range $t \leq b^\beta$.

13.3 A UNIFORM RESCALING ON THE CREATION ANNIHILATION OPERATORS

First we start giving a new approach to the results of chapter 11.

We use the same notation as in section 11.1. In particular we fix $x \in X_g$. We define $\underline{\widehat{\mathfrak{N}}}_{c,t}$ as in that section. Also we obtain the operator $\underline{\widehat{\mathcal{N}}}_{c,t}$ from $\underline{\widehat{\mathfrak{N}}}_{c,t}$ as in (13.2.3):

$$\underline{\widehat{\mathcal{N}}}_{c,t} = K_b \underline{\widehat{\mathfrak{N}}}_{c,t} K_b^{-1}. \quad (13.3.1)$$

Put

$$\underline{\widehat{\mathcal{O}}}_{c,t} = I_{\sqrt{t}} \underline{\widehat{\mathcal{N}}}_{c,t} I_{1/\sqrt{t}}. \quad (13.3.2)$$

As in sections 4.8 and 12.1, we introduce Grassmann variables $\mathfrak{e}_i, i_{\mathfrak{e}^i}, 1 \leq i \leq \ell$.

Definition 13.3.1. Let $\underline{\widehat{\mathcal{P}}}_{c,t}$ be the operator obtained from the operator $\underline{\widehat{\mathcal{O}}}_{c,t}$ by making the following replacements for $1 \leq i \leq \ell$:

- e^i is replaced by $e^i - \sqrt{t} i_{\widehat{e}_i}$.
- i_{e_i} is replaced by $-e^i/t + i_{e_i} + \left(\mathfrak{e}_i + i_{\mathfrak{e}^i} - i_{\widehat{e}_i}\right)/\sqrt{t}$.
- $\widehat{e}^i$ is replaced by $\widehat{e}^i - i_{\widehat{e}_i} + \mathfrak{e}_i + i_{\mathfrak{e}^i} + \sqrt{t} i_{e_i}$.
- $i_{\widehat{e}_i}$ is unchanged.

And for $\ell + 1 \leq i \leq n$:

- e^i is replaced by $\sqrt{t}\left(e^i - i_{\widehat{e}_i}\right)$.
- i_{e_i} is replaced by $i_{e_i - \widehat{e}_i}/\sqrt{t}$.
- $\widehat{e}^i$ is replaced by $\widehat{e}^i - i_{\widehat{e}_i} + e^i + i_{e_i}$.
- $i_{\widehat{e}_i}$ is unchanged.

One verifies easily that the above transformations are still compatible with the obvious commutation relations. Moreover, for $\ell + 1 \leq i, j \leq n$, $e^i i_{e_j} + \widehat{e}^i i_{\widehat{e}_j}$ is changed into $e^i i_{e_j} + \widehat{e}^i i_{\widehat{e}_j} + i_{e_j} i_{\widehat{e}_i} + i_{e_i} i_{\widehat{e}_j}$. If A is an $(n-\ell, n-\ell)$ antisymmetric matrix, its action on $\Lambda^{\cdot}\left(N^*_{X_g/X}\right)\widehat{\otimes}\Lambda^{\cdot}\left(N^*_{X_g/X}\right)$ is given by

$$-\sum_{\ell+1 \leq i,j \leq n} \langle Ae_i, e_j\rangle \left(e^i i_{e_j} + \widehat{e}^i i_{\widehat{e}_j}\right).$$

The above indicates that the action of A is unchanged when making the preceding transformations. More generally, this transformation commutes with the obvious action of $\mathrm{O}(n-\ell)$ on $\Lambda^{\cdot}\left(N^*_{X_g/X}\right)\widehat{\otimes}\Lambda^{\cdot}\left(N^*_{X_g/X}\right)$. In particular it commutes with the action of g.

As we did after Definition 11.1.1, we will now use the notation

$$\begin{aligned} \widehat{c}(e_i) &= \mathfrak{e}_i + i_{\mathfrak{e}^i} \text{ if } 1 \leq i \leq \ell, \\ &= e^i + i_{e_i} \text{ if } \ell+1 \leq i \leq n. \end{aligned} \tag{13.3.3}$$

Again we hope this notation does not cause extra confusion. Indeed for $1 \leq i \leq \ell, \widehat{c}(e_i)$ has no relation whatsoever with $e^i + i_{e_i}$, while it is still its former self for $\ell+1 \leq i \leq n$.

By taking (13.3.3) into account, we can then use almost the same conventions as in section 11.1 for the definition of $\widehat{\mathrm{Tr}_{\mathrm{s}}}$ and of $\widehat{\mathrm{Tr}_{\mathrm{s}}}^{\max}$. In this definition, only the indices $1 \leq i \leq \ell$ deserve a special treatment. We expand the kernel $r\left(-\widehat{\underline{\mathfrak{P}}}_{c,t}\right)\left(g^{-1}(y,p),(y,p)\right)$ as in (4.8.12) while using the transformations of Definition 13.3.1 instead of the transformations of Definitions 4.8.4 or 11.1.1. We treat the $\widehat{c}(e_i), 1 \leq i \leq \ell$ as a block, i.e., we forget about the expression $\widehat{c}(e_i) = \mathfrak{e}_i + i_{\mathfrak{e}^i}, 1 \leq i \leq \ell$. We reduce the corresponding expressions in normal form with respect to the $i_{e_i}, 1 \leq i \leq \ell$. We eliminate any term ultimately containing $i_{e_i}, 1 \leq i \leq \ell$, and we make the convention that

$$\widehat{\mathrm{Tr}_{\mathrm{s}}}\left[\prod_{i=1}^{\ell} e^i \widehat{c}(e_i)\right] = (-1)^\ell, \tag{13.3.4}$$

while the $\widehat{\mathrm{Tr}_{\mathrm{s}}}$ of any other monomial in the $e^i, \widehat{c}(e_i)$ will be zero. The other variables $\widehat{e}^i, i_{\widehat{e}_i}, 1 \leq i \leq \ell$, as well as all annihilation and creation operators for $\ell+1 \leq i \leq n$, are treated as standard operators, and they contribute to $\widehat{\mathrm{Tr}_{\mathrm{s}}}$ by their classical supertrace.

Proposition 13.3.2. *The following identity holds:*

$$t^{-\ell/2}\mathrm{Tr_s}\left[g\,|p|^2\,r\left(-\widehat{\underline{\mathcal{O}}}_{c,t}\right)\left(\left(g^{-1}(y,p)\right),(y,p)\right)\right]$$
$$=\widehat{\mathrm{Tr}}_{\mathrm{s}}\left[g\,|p|^2\,r\left(-\widehat{\underline{\mathcal{P}}}_{c,t}\right)\left(\left(g^{-1}(y,p)\right),(y,p)\right)\right]^{\max}. \quad (13.3.5)$$

Proof. The proof of our proposition is essentially the same as the proof of Proposition 11.1.2. Under the replacements of Definition 13.3.1, for $1\le i\le \ell$, $e^i i_{e_i}$ is replaced by $\left(e^i-\sqrt{t}\,i_{\widehat{e}_i}\right)\left(-e^i/t+i_{e_i}+\left(\widehat{c}\,(e_i)-i_{\widehat{e}_i}\right)/\sqrt{t}\right)$ and $\widehat{e}^i i_{\widehat{e}_i}$ is changed into $\left(c\,(\widehat{e}_i)+\widehat{c}\,(e_i)+\sqrt{t}\,i_{e_i}\right)i_{\widehat{e}_i}$, and for $\ell+1\le i\le n$, $e^i i_{e_i}$ is changed into $\left(e^i-i_{\widehat{e}_i}\right)i_{e_i-\widehat{e}_i}$, while $\widehat{e}^i i_{\widehat{e}_i}$ is changed into $\left(\widehat{e}^i+e^i+i_{e_i}\right)i_{\widehat{e}_i}$. In particular for $\ell+1\le i\le n$, $e^i i_{e_i}\widehat{e}^i i_{\widehat{e}_i}$ is changed into $e^i i_{e_i}\widehat{e}^i i_{\widehat{e}_i}+i_{e_i}i_{\widehat{e}_i}$. We deduce from these considerations that the supertrace of monomials in the $e^i, i_{e_i}, \widehat{e}^i, i_{\widehat{e}_i}, \ell+1\le i\le n$ acting on $\Lambda^{\cdot}\left(N^*_{X_g/X}\right)\widehat{\otimes}\Lambda^{\cdot}\left(N^*_{X_g/X}\right)$ is unchanged when making the above replacements. When combining the observation we made just before (13.3.3) with the above arguments, we get (13.3.5). The proof of our proposition is completed. □

13.4 THE LIMIT AS t → 0 OF THE RESCALED OPERATOR

Now we establish a more precise version of Theorem 11.1.3. In this version, the $\mathcal{O}$ are uniform in the considered range of parameters. Strictly speaking, this result is not needed in our proof of (8.4.7). However, the fact it is true is useful in understanding the proof.

Definition 13.4.1. Let $\widehat{\underline{\mathcal{P}}}_c$ be the operator

$$\widehat{\underline{\mathcal{P}}}_c=\frac{1}{4b^2}\left(-\Delta^V+|p|^2\pm\left(2i_{\widehat{e}_i}\widehat{e}^i-n\right)\right)\mp\frac{1}{2b}\left(\nabla_p+\left\langle i^*R^{TX}y,p\right\rangle\right)$$
$$+\frac{1}{4}\left\langle e_i,i^*R^{TX}e_j\right\rangle\left(\widehat{c}\,(e_i)+c\,(\widehat{e}_i)\right)\left(\widehat{c}\,(e_j)+c\,(\widehat{e}_j)\right)-\frac{1}{4}i^*\omega\left(\nabla^F,g^F\right)^2. \quad (13.4.1)$$

In the sequel, we will be still more precise in our treatment of the notation $\mathcal{O}$. Indeed we use the notation $\mathcal{O}(1)$ to indicate a function of y,t which is bounded together with its derivatives in the variable y. The same rule applies to all the other $\mathcal{O}$.

Theorem 13.4.2. *As* $t\to 0$,

$$\widehat{\underline{\mathcal{P}}}_{c,t}\to\widehat{\underline{\mathcal{P}}}_c. \quad (13.4.2)$$

More precisely,

$$\widehat{\underline{\mathcal{P}}}_{c,t}=\widehat{\underline{\mathcal{P}}}_c+\frac{\sqrt{t}}{b}\left(\mathcal{O}(1)\,\widehat{\nabla}+\mathcal{O}(y)\,\nabla_p+\mathcal{O}(1)\,p_ip_j\widehat{\nabla}_{\widehat{e}^k}+\mathcal{O}\left(|y|^2\right)p_i\right)$$
$$+\sqrt{t}\,\mathcal{O}\left(1+|y|+|p|^2\right). \quad (13.4.3)$$

Proof. The proof is essentially the same as the proof of Theorem 11.1.3. First note that equations (11.1.12) and (11.1.13) remain valid. Also for $1 \leq i \leq n$, $\widehat{e}^i - e^i/\sqrt{t} - \sqrt{t} i_{e_i}$ is changed into $\widehat{e}^i + i_{\widehat{e}_i}$, so that $i_{\widehat{e}_i}\left(\widehat{e}^i - e^i/\sqrt{t} - \sqrt{t} i_{e_i}\right)$ is changed into $i_{\widehat{e}_i}\widehat{e}^i$. This already indicates that even before making $t \to 0$, in the formula for $\underline{\widehat{\mathfrak{P}}}_{c,t}$ corresponding to (11.1.4), the first line is changed into

$$\frac{1}{4b^2}\left(-\Delta^V + |p|^2 \pm \left(2 i_{\widehat{e}_i}\widehat{e}^i - n\right)\right). \tag{13.4.4}$$

Also observe that for $1 \leq i \leq \ell$, $\widehat{e}^i$ is changed into $\widehat{c}(e_i) + c(\widehat{e}_i) + \sqrt{t} i_{e_i}$, and for $\ell + 1 \leq i \leq n$, into $\widehat{c}(e_i) + c(\widehat{e}_i)$. This accounts for the first term in the second line in (13.4.1).

As in the proof of Theorem 11.1.3, we must control the term

$$\mp \frac{\sqrt{t}}{2b} {}^1\nabla_{t, Y^{\mathcal{H}}}^{\Lambda^\cdot(T^*X)\widehat{\otimes}\Lambda^\cdot(T^*X)\widehat{\otimes}\Lambda^n(TX)\widehat{\otimes}F, u}.$$

We still use equations (11.1.16)-(11.1.18). This concludes the proof of equation (13.4.2). Proving (13.4.3) is just keeping track of the various terms in the right-hand side of (11.1.4) as $t \to 0$. □

Remark 13.4.3. It is very interesting to study directly the limit as $b \to 0$ of the operator $\underline{\widehat{\mathfrak{P}}}_c$. It has indeed the preferred matrix form, which was already considered in [BL91, sections 11-13] and in [B05], and which is used systematically in our treatment of the limit $b \to 0$ in chapter 17. From the methods of this chapter, we find that in the appropriate sense, as $b \to 0$, $\underline{\widehat{\mathcal{P}}}_c$ converges to the operator $\underline{\widehat{\mathcal{P}}}$ given by

$$\begin{aligned} \underline{\widehat{\mathcal{P}}} = &-\frac{1}{4}\left(\nabla_{e_i} + \left\langle i^* R^{TX} y, e_i\right\rangle\right)^2 \\ &+ \frac{1}{4}\left\langle e_i, i^* R^{TX} e_j\right\rangle \widehat{c}(e_i)\widehat{c}(e_j) - \frac{1}{4} i^*\omega\left(\nabla^F, g^F\right)^2. \end{aligned} \tag{13.4.5}$$

Recall that the superconnections A_+ and A_- were defined in section 2.6. As in section 1.7, $A_{+,t}$ and $A_{-,t}$ denote the corresponding superconnections which are associated to the metric g^{TX}/t. After the appropriate rescaling of $A^2_{\pm,2t}$, the operator $\underline{\widehat{\mathcal{Q}}}$ was obtained in [BLo95, proof of Theorem 3.16] as the limit of these operators as $t \to 0$. The operator $\underline{\widehat{\mathcal{Q}}}$ is given by the formula

$$\begin{aligned} \underline{\widehat{\mathcal{Q}}} = -\frac{1}{2}\left(\nabla_{e_i} + \frac{1}{2}\left\langle i^* R^{TX} y, e_i\right\rangle\right)^2 + \frac{1}{4}\left\langle e_i, i^* R^{TX} e_j\right\rangle \widehat{c}(e_i)\widehat{c}(e_j) \\ - \frac{1}{4} i^*\omega\left(\nabla^F, g^F\right)^2. \end{aligned} \tag{13.4.6}$$

Note that $\underline{\widehat{\mathcal{P}}}$ is obtained from $\underline{\widehat{\mathcal{Q}}}$ by conjugation by $I_{\sqrt{2}}$. When computing the local supertrace on M_g of $\exp\left(-\underline{\widehat{\mathcal{Q}}}\right)$, we get a correcting factor $1/2^{\ell/2}$. This is compensated by the fact that in [BLo95], up to a sign which is the same as here, $\prod_1^\ell \widehat{c}(e_i)$ contributes to the supertrace by the factor $2^{\ell/2}$. The conclusion is that the limit as $b \to 0$ of $\widehat{\underline{\mathcal{P}}}_{c,t}$ is precisely the operator which appears in the local index theorem for the elliptic superconnection.

In the sequel, we will also use the notation

$$\widehat{\underline{\mathcal{P}}}_{c,0} = \widehat{\underline{\mathcal{P}}}_{c}. \tag{13.4.7}$$

13.5 REPLACING X BY $\mathbf{T_xX}$

We fix $x \in X_g$. Set

$$P_{b,t} = 2b^2 \widehat{\underline{\mathcal{N}}}_{c,t}. \tag{13.5.1}$$

Now we use the notation of section 17.4. Using (11.1.4), (13.2.3), we find easily that the operator $P_{b,t}$ has the same structure as the operator P_h in chapter 17, with $h = \sqrt{t}b$. We will then use freely the notation and the estimates of chapter 17 applied to the operator $P_{b,t}$, while replacing h by $\sqrt{t}b$.

We define $P^0_{b,t}$ as in (17.6.1):

$$P^0_{b,t} = P_{b,t} + P_{\pm}. \tag{13.5.2}$$

Let $\eta : \mathbf{R}^n \to [0,1]$ be a smooth function. We assume that η is equal to 1 on the ball $\left\{y \in \mathbf{R}^n, \sup_{1 \le i \le n} \left|y^i\right| \le 1\right\}$, and that its support is included in $\left\{y \in \mathbf{R}^n, \sup_{1 \le i \le n} \left|y^i\right| \le 2\right\}$. Set

$$\sigma(y) = \sum_{m \in \mathbf{Z}^n} \eta(y - m). \tag{13.5.3}$$

For $m \in \mathbf{Z}^n$, set

$$\psi_m(y) = \frac{\eta(y-m)}{\sigma(y)}. \tag{13.5.4}$$

Then $\psi_m, m \in \mathbf{Z}^n$ is a partition of unity on $\mathbf{R}^n$. The function $\sum_{m \in \mathbf{Z}^n} \psi^2_m(y)$ is periodic with periods in $\mathbf{Z}^n$ and also positive, so it has a positive lower bound. It follows that if $u \in L^2(\mathbf{R}^n \times \mathbf{R}^n)$, the standard L^2 norm of u is equivalent to the norm $\left(\sum_{m \in \mathbf{Z}^n} \left|\psi_m(y)u\right|^2\right)^{1/2}$. Similar considerations apply to the other norms which are considered in chapter 17.

Now we will use the results of chapter 17. The operator R was defined in (17.5.17). Using inequality (17.5.22), we get, for $b \in]0,1], t \in]0,1], s \ge 0$,

$$\begin{aligned} \tau^{-4} \left|\psi_m U\right|^2_{\lambda,\mathrm{sc},s} + \left|\widehat{\nabla}\psi_m U\right|^2_{\lambda,\mathrm{sc},s} + \tau^{-3/2}\left|\psi_m U\right|^2_{\lambda,\mathrm{sc},s+1/4} \\ + \tau^{5/4}\left|\widehat{\nabla}\psi_m U\right|^2_{\lambda,\mathrm{sc},s+1/8} \le C_s |R\psi_m U|^2_{\lambda,\mathrm{sc},s}. \end{aligned} \tag{13.5.5}$$

Note that the constants in (13.5.5) are uniform in $m \in \mathbf{N}$, because for $y \in T_xX$ with $|y|$ large enough, our operator has constant coefficients in the variable y.

Using the fact that $[R,\psi_m(y)] = \mathcal{O}(\tau^{-1})$, from (13.5.5) we get

$$\tau^{-4}|U|^2_{\lambda,\mathrm{sc},s} + \left|\widehat{\nabla} U\right|^2_{\lambda,\mathrm{sc},s} + \tau^{-3/2}|U|^2_{\lambda,\mathrm{sc},s+1/4} + \tau^{5/4}\left|\widehat{\nabla} U\right|^2_{\lambda,\mathrm{sc},s+1/8} \leq C_s\left(|RU|^2_{\lambda,\mathrm{sc},s} + \tau^{-2}|U|^2_{\lambda,\mathrm{sc},s}\right). \quad (13.5.6)$$

The argument we used to prove that (17.5.21) implies (17.5.22) can be used to show that (17.5.22) still holds. Namely, we can as well replace ψ_m by 1 in (13.5.5). In particular the conclusion of Theorems 17.5.2, 17.6.1, and 17.6.3 are still valid.

Recall that the Sobolev spaces $\mathcal{H}^s$ were defined in Definition 15.3.1. Still observe that our base X being now $\mathbf{R}^n$, the embedding $\mathcal{H}^{s+1/4} \to \mathcal{H}^s$ is no longer compact.

As in Definition 17.6.2, put

$$S_{b,t,\lambda} = \left(P^0_{b,t} - \lambda\right)^{-1}. \quad (13.5.7)$$

The analysis for the parametrix of $S_{b,t,\lambda}$, which is done in sections 17.5-17.13, is still valid, since it is of a local nature.

Remark 13.5.1. Let $\Phi : \mathbf{R}^n \to \mathbf{R}$ be a smooth function which is uniformly bounded together with its derivatives. One such function is $\Phi(y) = c_0(1 + |y|^2)^{1/2}$. If $h = \sqrt{t}b$, by (11.1.4) and (13.2.3), we get

$$e^{\Phi(y)/h} P^0_{b,t} e^{-\Phi(y)/h} = P^0_{b,t} \mp \nabla_{Y^{\mathcal{H}}}\Phi. \quad (13.5.8)$$

The term $\nabla_{Y^{\mathcal{H}}}\Phi$ is linear in the variable p, and so it can be absorbed by the harmonic oscillator. In particular if $\sup_{y\in\mathbf{R}^n} \nabla\Phi(y)$ is small enough, we find that the analogue of the first equation in (17.6.14) still holds. Namely, given $a \in \mathbf{R}$, $s \in \mathbf{R}$, there exists $h_0 > 0, C_{a,s} > 0$ such that if $\mathrm{Re}\,\lambda \leq \lambda_1, b \in]0, h_0], t \in]0,1]$,

$$\left\| < p >^a e^{\Phi(y)/h} S_{b,t,\lambda} e^{-\Phi(y)/h} v \right\|_{\lambda,\mathrm{sc},s+1/4} \leq C_{a,s} \left\| < p >^a v \right\|_{\lambda,\mathrm{sc},s}. \quad (13.5.9)$$

Equation (13.5.9) shows that the resolvent $S_{h,\lambda}$ is local in y modulo exponentially small errors in the parameter h, and that if $|y - y'| \geq c_1 > 0$, the magnitude of the interaction between y and y' can be dominated uniformly by $e^{-c_0|y-y'|/h}$, for some $c_0 > 0$.

We shall denote by $\widehat{\underline{\mathcal{N}}}_{\infty,t}$ the limit as $c \to \pm\infty$ of the operator $\widehat{\underline{\mathcal{N}}}_{c,t}$. This limit is taken in the sense of section 17.21. As we will see in that section, $\widehat{\underline{\mathcal{N}}}_{\infty,t}$ is given by $A^2_{\pm,t,x}$, which is the curvature of the Levi-Civita superconnection over the total space of TX, which is associated to the metric $g^{T_xX}_x$ and to the local version of the horizontal subbundle that was described in section 4.8.

Let $K^1_{b,t,\lambda,N}((x,p),(x',p')), K^1_{0,t,\lambda,N}(x,x')$ be the Schwartz kernels which are associated to the operators $\left(\lambda - \widehat{\underline{\mathcal{N}}}_{c,t}\right)^{-N}, \left(\lambda - \widehat{\underline{\mathcal{N}}}_{\infty,t}\right)^{-N}$. If $x' \in T_xX$, we define $v^1_{\sqrt{t}b,t}(x',\lambda), v^1_{0,t}(x',\lambda)$ as in (13.2.5), (13.2.6), by making the obvious replacement. Similarly, if $x \in X_g$, we define $w^1_{\sqrt{t}b,t,N}(x,\lambda), w^1_{0,t,N}(x,\lambda)$ as in (13.2.10). Recall that $\beta > 0$ is fixed.

Proposition 13.5.2. *For $N \in \mathbf{N}^*$ large enough, there exist $\alpha > 0$, $C > 0$ such that for $b \in]0,1], t \in]0, b^\beta], \lambda \in \gamma, x \in X_g$,*

$$\left| w_{\sqrt{t}b,t,N}(x,\lambda) - w^1_{\sqrt{t}b,t,N}(x,\lambda) \right| \le Ct^\alpha, \tag{13.5.10}$$
$$\left| w_{0,t,N}(x,\lambda) - w^1_{0,t,N}(x,\lambda) \right| \le Ct^\alpha.$$

Proof. We use the notation in Theorem 17.22.2. Take $x \in X_g$. Consider a smooth section d whose support is included in a small compact neighborhood K of x. We may and we will assume that $\rho = 1$ on a neighborhood of K. Set

$$e = \left(\lambda - \underline{\widehat{\mathcal{M}}}_{c,t}\right)^{-N} d. \tag{13.5.11}$$

Let $\phi(x') : X \to [0,1]$ be a smooth function which is equal to 1 on K, whose support is included in $(\rho = 1)$. By Theorem 17.22.2 and by Remark 17.22.3, if $N' \in \mathbf{N}, M \in \mathbf{N}$,

$$\|(1-\phi)e\|_{t,M} \le C_{N',M} t^{N'} \|d\|_{t,-M}. \tag{13.5.12}$$

Let f be defined by the equation

$$f = d - \left(\lambda - \underline{\widehat{\mathcal{M}}}_{c,t}\right)^{N} \phi e. \tag{13.5.13}$$

Since the support of d is included in the support of ϕ, the same is true when replacing d by f. By (13.5.11), can rewrite (13.5.13) in the form

$$f = \left(\lambda - \underline{\widehat{\mathcal{M}}}_{c,t}\right)^{N} (1-\phi)e. \tag{13.5.14}$$

The coefficients of the $\underline{\widehat{\mathcal{M}}}_{c,t}$ are singular as $b \to 0$, with a singularity which is at most $1/b^2$. Since $t \le b^\beta$, by (13.5.12), (13.5.14), given $L \in \mathbf{N}, M \in \mathbf{N}$,

$$\|f\|_{t,M} \le C_{L,M} t^L \|d\|_{t,-M}. \tag{13.5.15}$$

We can consider $\phi e, d, f$ as being defined on T_xX. Since the support of ϕ is included in $\rho = 1$, by (13.5.13), we get

$$\left(\lambda - \underline{\widehat{\mathcal{N}}}_{c,t}\right)^{N} \phi e = d - f, \tag{13.5.16}$$

and so

$$\phi e = \left(\lambda - \underline{\widehat{\mathcal{N}}}_{c,t}\right)^{-N} (d-f). \tag{13.5.17}$$

Moreover, (13.5.17) can be written in the form

$$\phi\left(\lambda - \underline{\widehat{\mathcal{M}}}_{c,t}\right)^{-N} d - \left(\lambda - \underline{\widehat{\mathcal{N}}}_{c,t}\right)^{-N} d = -\left(\lambda - \underline{\widehat{\mathcal{N}}}_{c,t}\right)^{-N} f. \tag{13.5.18}$$

Assume first that the base S is reduced to a point. We claim that the obvious analogue of the first inequality in (17.21.24) (which is part of Theorem 17.21.3) holds when replacing L_c by $\underline{\widehat{\mathcal{N}}}_{c,1}$, so that given $\ell \in \mathbf{N}$, for $N \in \mathbf{N}^*$ large enough,

$$\left\| \left(\lambda - \underline{\widehat{\mathcal{N}}}_{c,1}\right)^{-N} \right\|_\ell \le C_N. \tag{13.5.19}$$

The difference with the proof in Theorem 17.21.3 is that $\mathbf{R}^n$ is noncompact. However, the techniques of the proof of this theorem can be adapted without any change. Since $\underline{\widehat{\mathcal{N}}}_{c,t} = t\underline{\widehat{\mathcal{N}}}_{c/t,1}$, for $N \in \mathbf{N}^*$ large enough, we get from (13.5.19),

$$\left\| \left(\lambda - \underline{\widehat{\mathcal{N}}}_{c,t}\right)^{-N} \right\|_{\ell} \leq C_N t^{-N}. \tag{13.5.20}$$

By (13.5.15), (13.5.18), (13.5.20), we deduce that if ψ is a smooth function whose support is a small neighborhood of x,

$$\left\| \psi(y) \left(K_{b,t,N} - K^1_{b,t,N}\right) \left((y,p),(y',p')\right) \psi(y') \right\|_L \leq C_{N'} t^{N'}. \tag{13.5.21}$$

Of course all the constants are uniform in $x \in X_g$. By (13.2.10) and (13.5.21), we get the first identity in (13.5.10). The same argument can be used for the second identity.

In the case where S is not reduced to a point, using (2.8.6), the above proof extends in full generality. □

13.6 A PROOF OF (13.2.11)

In this section, we establish (13.2.11), which will conclude the proof of (8.4.7). By (13.5.10), we find that to establish (13.2.11), it is enough to show that

$$\left| w^1_{\sqrt{t}b,t,N}(x,\lambda) - w^1_{0,t,N}(x,\lambda) \right| \leq C_N \left(t^\alpha + b^\alpha\right). \tag{13.6.1}$$

By the proof of Proposition 13.1.1, we know that

$$\left| w^1_{\sqrt{t}b,t,N}(x,\lambda) - w^1_{0,t,N}(x,\lambda) \right| \leq C_N t^{-u} b^v. \tag{13.6.2}$$

Recall that the operator $\underline{\widehat{\mathcal{P}}}_c$ was defined in Theorem 13.4.2.

Definition 13.6.1. For $N \in \mathbf{N}, x \in X_g$, let $\overline{K}_{b,0,\lambda,N}\left((y,p),(y',p')\right)$ be the Schwartz kernel associated to the operator $\left(\lambda - \underline{\widehat{\mathcal{P}}}_{c,0}\right)^{-N}$. For $N \in \mathbf{N}$ large enough, if $x \in X_g, y \in N_{X_g/X,x}$, set

$$\overline{v}_{b,N}(x,y,\lambda) = \pm(-1)^n \varphi (N-1)! \int_{T^*_x X} \widehat{\mathrm{Tr_s}}\left[g|p|^2 \overline{K}_{b,0,\lambda,N}\left((y,p),(gy,gp)\right)\right]^{\max} dp. \tag{13.6.3}$$

For $x \in X_g$, put

$$\overline{w}_{b,N}(x,\lambda) = \int_{N_{X_g/X,x}} \overline{v}_{b,N}(x,y,\lambda)\, dy. \tag{13.6.4}$$

Similarly let $\overline{K}_{\lambda,N}(y,y')$ be the Schwartz kernel associated to $\left(\lambda - \underline{\widehat{\mathcal{P}}}\right)^{-N}$. For $N \in \mathbf{N}$ large enough, set

$$\overline{v}_{0,N}(x,y,\lambda) = \pm\varphi(N-1)!\frac{n}{2}\widehat{\mathrm{Tr_s}}\left[g\overline{K}_{\lambda,N}(y,gy)\right]. \tag{13.6.5}$$

Put

$$\overline{w}_{0,N}(x,\lambda) = \int_{N_{X_g/X,x}} \overline{v}_{0,N}(x,y)\, dy. \tag{13.6.6}$$

Theorem 13.6.2. *For $N \in \mathbf{N}$ large enough, for $x \in X_g, y \in N_{X_g/X,x}, \lambda \in \gamma$, as $t \to 0$,*

$$t^{(n-\ell)/2} v^1_{\sqrt{t}b,t,N}\left(x, \sqrt{t}y\right) k\left(x, \sqrt{t}y\right) \to \overline{v}_{b,N}(x,y), \tag{13.6.7}$$
$$t^{(n-\ell)/2} v_{0,t,N}\left(x, \sqrt{t}y\right) k\left(x, \sqrt{t}y\right) \to \overline{v}_{0,N}(x,y).$$

Moreover, given $m \in \mathbf{N}$, there exist $C > 0, a > 0$ such that for $b \in]0,1], t \in]0,1], x \in X_g, y \in N_{X_g/X,x}, |y| \leq \beta/\sqrt{t}$,

$$\left|t^{(n-\ell)/2} v^1_{\sqrt{t}b,t,N}\left(x, \sqrt{t}y\right) k\left(x, \sqrt{t}y\right) - \overline{v}_{b,N}(x,y)\right| \leq C(1+|y|)^{-m} t^a, \tag{13.6.8}$$
$$\left|t^{(n-\ell)/2} v^1_{0,t,N}\left(x, \sqrt{t}y, \lambda\right) - \overline{v}_{0,N}(x,y)\right| \leq C(1+|y|)^{-m} t^a.$$

Remark 13.6.3. We will show how to derive (13.6.1) from Theorem 13.6.2. Indeed using (13.6.7), (13.6.8), we find that as $t \to 0$,

$$w^1_{\sqrt{t}b,t,N}(x,\lambda) \to \overline{w}_{b,N}(x,\lambda), \qquad w^1_{0,t,N}(x,\lambda) \to \overline{w}_{0,N}(x,\lambda). \tag{13.6.9}$$

More precisely, the same references show that

$$\left|w^1_{\sqrt{t}b,t,N}(x,\lambda) - \overline{w}_{b,N}(x,\lambda)\right| \leq Ct^\alpha, \tag{13.6.10}$$
$$\left|w^1_{0,t,N}(x,\lambda) - \overline{w}_{0,N}(x,\lambda)\right| \leq Ct^\alpha.$$

By (13.6.2) and (13.6.10), we get

$$\left|\overline{w}_{b,N}(x,\lambda) - \overline{w}_{0,N}(x,\lambda)\right| \leq C_N\left(t^{-u}b^{2v} + t^\alpha\right). \tag{13.6.11}$$

Since the left-hand side of (13.6.11) does not depend on t, we find that there exists $\alpha > 0$ (possibly different from the one in (13.6.11)) such that

$$\left|\overline{w}_{b,N}(x,\lambda) - \overline{w}_{0,N}(x,\lambda)\right| \leq C_N b^\alpha. \tag{13.6.12}$$

Incidentally note that (13.6.12) can be given a simple direct proof.

By (13.6.10) and (13.6.12), we get (13.6.1). This concludes the proof of (8.4.7).

So now we concentrate on the proof of Theorem 13.6.2.

13.7 A PROOF OF THEOREM 13.6.2

Note that the second sort of inequalities in (13.6.7), (13.6.8) refers to classical elliptic relative index theory for which such results should be well known. Note that using the arguments we gave in chapters 5 and 17, we find that as $b \to 0$,

$$v^1_{\sqrt{t}b,t,N}(x,y) \to v_{0,t,N}(x,y), \qquad \overline{v}_{b,0,N}(x,y) \to \overline{v}_{0,N}(x,y). \tag{13.7.1}$$

By (13.7.1), the second lines in (13.6.7), (13.6.8) follow from the first lines. So we concentrate now on the proof of the first identities in (13.6.7), (13.6.8).

We start from equation (11.1.4) for $\widehat{\underline{\mathfrak{M}}}_{c,t}$. In the sequel we will also use equation (11.1.4) for $\widehat{\underline{\mathfrak{N}}}_{c,t}$, which is an operator of the same type as $\widehat{\underline{\mathfrak{M}}}_{c,t}$. Theorem 13.4.2 gives us the asymptotic expansion of the operator $\widehat{\underline{\mathcal{P}}}_{c,t}$ as $t \to 0$.

Set

$$\alpha'_{\pm} = \frac{1}{2}\left(-\Delta^{V} + |p|^{2} \mp n\right) \pm i_{\widehat{e}_i}\widehat{e}^{i}. \tag{13.7.2}$$

The operator $\alpha'_{\pm}$ is self-adjoint and nonnegative. As explained in the proof of Theorem 13.4.2, the critical fact in $\widehat{\underline{\mathfrak{P}}}_{c,t}$ is that the first line in equation (11.1.4) for $\widehat{\underline{\mathfrak{M}}}_{c,t}$ contributes to $\widehat{\underline{\mathcal{P}}}_{c,t}$ by the operator $\alpha'_{\pm}/2b^2$ which does not depend on t. Moreover, even before taking the asymptotic expansion, the coefficient of $1/b$ in $\widehat{\underline{\mathcal{P}}}_{c,t}$ is an operator which maps $\ker \alpha'_{\pm}$ into its orthogonal. Indeed this component either depends linearly on p or contains one of the operators $\nabla_{\widehat{e}^i}$, or it contains an odd expression of the type $p_i p_j \nabla^{\widehat{e}^k}$.

In equation (13.4.1) for $\widehat{\underline{\mathcal{P}}}_{c}$, the term $\mp\frac{1}{2b}\left(\nabla_p + \left\langle i^* R^{TX} y, p\right\rangle\right)$ appears. This term raises an extra difficulty because it depends linearly in the variable y, and so it is not controlled any more by the estimates which are used in chapters 15 and 17. However, such a difficulty already appears in standard local index theory for elliptic Dirac operators. This should be clear by equation (13.4.5) for $\underline{\mathcal{P}}$, which now depends quadratically on y.

In the case of one single fiber, equation (4.8.4) makes clear that for $|y| \geq 4\eta_0$ large enough, the metric g^{T_xX} is flat on T_xX, and so the corresponding fiberwise curvature vanishes for $|y| \geq 4\eta_0$.

In the case of a family, the argument is subtler. Indeed the construction of $g_x^{T_xX}$ and of the new $T^H M_x$ given in section 4.7 also ensures that the full curvature tensor R^{TX} still vanishes for $|y| \geq 4\eta_0$.

To control the dependency in y, we will use the same method as the one which was developed in [BL91, subsection 11k)] in the context of elliptic local index theory. Indeed the above support conditions will permit us to introduce L^2 norms whose weight takes the degree into account.

Definition 13.7.1. For $0 \leq \sigma \leq m, 0 \leq \sigma' \leq \ell$, let $\mathbf{I}_x^{\sigma,\sigma'}$ be the vector space of L^2 sections of

$$\Lambda^{\sigma}\left(T^*_{\pi x}S\right)\widehat{\otimes}\Lambda^{\sigma'}\left(T^*_x X_g\right)\widehat{\otimes}\Lambda^{\cdot}\left(N^*_{X_g/X}\right)\widehat{\otimes}\Lambda^{\cdot}\left(T^*_x X\right)\widehat{\otimes}\widehat{c}\left(T_x X_g\right)\widehat{\otimes}F_x$$

over T_xX.

If $s \in \mathbf{I}_x^{\sigma,\sigma'}$, set

$$|s|^2_{0,t} = \int_{T_xX} |s|^2 \left(1 + \rho\left(\sqrt{t}y/2\right)|y|\right)^{2\left(m+\ell-\sigma-\sigma'\right)} dy. \tag{13.7.3}$$

Note that we have included $\widehat{c}\left(T_xX_g\right)$ in Definition 13.7.1 because of the inclusion of the extra $\widehat{c}\left(e_i\right) = \mathfrak{e}_i + i_{\mathfrak{e}^i}, 1 \leq i \leq \ell$ in the construction of $\widehat{\underline{\mathcal{P}}}_{c,t}$.

By (4.8.3), $\rho\left(\sqrt{t}y/2\right)$ is equal to 1 when $\sqrt{t}\,|y| \leq 4\eta_0$, and so it is equal to 1 on the support of $\rho\left(\sqrt{t}y\right)$. As in [BL91, Proposition 11.24], we find that for $1 \le i, j \le \ell$, the operators

$$1_{\sqrt{t}|y| \le 4\eta_0} |y| \, e^i e^j, \quad 1_{\sqrt{t}|y| \le 4\eta_0} |y| \sqrt{t} e^i, \qquad 1_{\sqrt{t}|y| \le 4\eta_0} |y| \, t^{3/2} i_{e_i} \qquad (13.7.4)$$

are uniformly bounded. In (13.7.4), one can replace any of the e^i by a $f^\alpha, 1 \le \alpha \le m$ and still get the same boundedness result.

In our context, over $T_x X \simeq \mathbf{R}^n$, one can then develop the same arguments as in chapters 15 and 17. Note of course that $T_x X$ is noncompact, whereas in these chapters, we assumed X to be compact. To compensate for noncompactness, in all the norms used in these chapters, we introduce the weights in (13.7.4) instead of the classical unweighted L^2 norms. Of course the analysis in the variable p remains unchanged. In particular we define the norm $\|A\|_\ell$ as in (17.21.22), by changing the L^2 norm as indicated above.

Definition 13.7.2. For $a \in \mathbf{R}, y, z \in \mathbf{R}^n$, set

$$m_{a,z}(y) = 1 + a^2 + |y - z|^2 . \qquad (13.7.5)$$

Put

$$\widehat{\underline{\mathcal{P}}}^{a,z}_{c,t} = m^{a/2}_{a,z} \widehat{\underline{\mathcal{P}}}_{c,t} m^{-a/2}_{a,z}. \qquad (13.7.6)$$

Now we establish an analogue of the first inequality in (17.21.24).

Proposition 13.7.3. *There exists $b_0 \in]0,1]$ such that given $\ell \in \mathbf{N}$, there exist $N \in \mathbf{N}, C_\ell > 0$ for which if $b \in]0, b_0], t \in [0,1], \lambda \in \gamma, x \in X_g, a \in \mathbf{R}, y, z \in \mathbf{R}^n$, then*

$$\left\| \left(\widehat{\underline{\mathcal{P}}}^{a,z}_{c,t} - \lambda \right)^{-N} \right\|_\ell \le C_\ell. \qquad (13.7.7)$$

Proof. For $a = 0$, the proof of (13.7.7) is the same as the proof of (17.21.24) in Theorem 17.21.3. Indeed the main point is contained Theorem 13.4.2 and its proof. As was observed after (13.7.2), the coefficient of $1/b^2$ in $\widehat{\underline{\mathcal{P}}}_{c,t}$ is the operator $\alpha'_\pm$, which does not depend on t. Combining this observation with the boundedness results in (13.7.4) leads easily to a proof of (13.7.7) when $a = 0$.

Clearly,

$$\widehat{\underline{\mathcal{P}}}^{a,z}_{c,t} = \widehat{\underline{\mathcal{P}}}_{c,t} \pm \frac{a}{2b m_{a,z}} \left\langle g^{TX,-1}_{\sqrt{t}y} (y - z), p \right\rangle . \qquad (13.7.8)$$

The functions $a(y-z)/m_{a,z}$ are uniformly bounded together with all their derivatives in the variable $y \in \mathbf{R}^n$. Also the term $\frac{a}{2m_{a,z}} \left\langle g^{TX,-1}_{\sqrt{t}y} (y - z), p \right\rangle$ maps $\ker \alpha_\pm$ in its orthogonal. Therefore the operator in (13.7.8) still has the preferred matrix structure which is needed in proving the results of chapter 17. The same methods lead to the estimate (13.7.7) also in the case of an arbitrary $a \in \mathbf{R}$. □

Proposition 13.7.4. *For $a' = 2$, for any $L \in \mathbf{N}$, there exist $C > 0, N \in \mathbf{N}$ for which for any $y, y', p, p' \in \mathbf{R}^n, k \in \mathbf{N}$ and all multiindices α such that*

$|\alpha| + k \leq L$, *for* $b \in]0, b_0], t \in [0,1], \lambda \in \gamma$, *then*

$$\left| < p >^k < p' >^k < y - y' >^k \partial^\alpha_{y,y',p,p'} \left(\left(\widehat{\underline{\mathcal{P}}}_{c,t} - \lambda \right)^{-N} - \left(\widehat{\underline{\mathcal{P}}}_{c,0} - \lambda \right)^{-N} \right) \right.$$
$$\left. ((y,p),(y',p')) \right| \leq Ct < y >^{a'} . \quad (13.7.9)$$

Proof. Recall that the operator $\widehat{\underline{\mathcal{P}}}^{a,z}_{c,t}$ was defined in (13.7.6). We claim that for $a' = 2$, given $\ell \in \mathbf{N}$, there exist $\ell' \in \mathbf{N}, C_\ell > 0$ for which for any $b \in]0, b_0], t \in [0,1], a \in \mathbf{R}, z \in \mathbf{R}^n$,

$$\begin{aligned}
&\left\| \left(\widehat{\underline{\mathcal{P}}}^{a,z}_{c,t} - \lambda \right)^{-1} u \right\|_\ell \leq C_\ell \|u\|_{\ell'} , \\
&\left\| m^{-a'/2}_{a',0} \left(\widehat{\underline{\mathcal{P}}}^{a,z}_{c,t} - \lambda \right)^{-1} m^{a'/2}_{a',0} u \right\|_\ell \leq C_\ell \|u\|_{\ell'} , \qquad (13.7.10) \\
&\left\| m^{-a'/2}_{a',0} \left(\left(\widehat{\underline{\mathcal{P}}}^{a,z}_{c,t} - \lambda \right)^{-1} - \left(\widehat{\underline{\mathcal{P}}}^{a,z}_{c,0} - \lambda \right)^{-1} \right) u \right\|_\ell \leq C_\ell \sqrt{t} \|u\|_{\ell'} .
\end{aligned}$$

Indeed for $\ell = 0$, the first two inequalities in (13.7.10) follow from the analogue of equation (17.21.23) in Theorem 17.21.3, and from the conjugation argument based on (13.7.8), which was used earlier.

Now we prove the first two equations in (13.7.10) for arbitrary $\ell \in \mathbf{N}$. Let $\mathbf{I}_x$ be the direct sum of the vector spaces considered in Definition 13.7.1. As in (17.2.6), we split $\mathbf{I}_x$ into

$$\mathbf{I}_x = \ker \alpha'_\pm \oplus \ker \alpha'^\perp_\pm . \qquad (13.7.11)$$

We will use the formal $(2,2)$ matrix expression for $\left(\widehat{\underline{\mathcal{P}}}_{c,t} - \lambda \right)^{-1}$, which is given in (17.2.12) and (17.21.2), replacing h by b. In particular, we use the notation in (17.16.1), i.e., we set

$$\Theta_{b,t,\lambda} = P^\perp_\pm \left(b^2 \widehat{\underline{\mathcal{P}}}_{c,t} - \lambda \right)^{-1} P^\perp_\pm . \qquad (13.7.12)$$

Also we define $T_{b,t,\lambda}$ as in (17.17.3). The other notation will be modified in the obvious way, by replacing the index h, λ by b, t, λ. Then we get a strict analogue of equation (17.21.2) for $\left(\widehat{\underline{\mathcal{P}}}_{c,t} - \lambda \right)^{-1}$. Note that these equations are also valid for $t = 0$.

Put

$$J_{b,t,\lambda} = i_\pm \left(T_{b,t,b^2\lambda} - \lambda \right)^{-1} P_\pm . \qquad (13.7.13)$$

As in (17.21.36), we get

$$\left(\widehat{\underline{\mathcal{P}}}_{c,t} - \lambda \right)^{-1} = J_{b,t,\lambda} + b R_{b,t,\lambda} . \qquad (13.7.14)$$

The semiclassical norms $\| \, \|_{b^2\lambda, \mathrm{sc}, s}$ are defined in equation (17.4.5), which is part of Definition 17.4.1. We make the obvious extension of these norms in our context. Similarly we define a family of operators $\mathcal{R}^\ell$ the way we do before (17.21.21), and we also obtain corresponding norms $\| \|_\ell$. To establish the first two identities in (13.7.10), we will use the commutation estimates contained in (17.21.46), (17.21.47), whose proof in the present context is strictly similar to the proof which is given in chapter 17. Also note that if $Q \in \mathcal{R}^\ell$,

$$\|Qu\|_{b^2\lambda,\mathrm{sc},M} \le C_\ell \|u\|_{\ell+2M}\,. \tag{13.7.15}$$

Finally, note that for $\ell \in \mathbf{N}$,

$$\|u\|_\ell = \sum_{Q\in\mathcal{R}^\ell} \|Qu\|_{b^2\lambda,\mathrm{sc},0}\,. \tag{13.7.16}$$

By using the analogue of the commutator estimates in (17.21.46), (17.21.47) and also (13.7.15), (13.7.16), we obtain the first two inequalities in (13.7.10) for arbitrary $\ell \in \mathbf{N}$.

Now we establish the third inequality in (13.7.10). Equation (13.4.3) gives the precise form of $\widehat{\mathcal{P}}_{c,t} - \widehat{\mathcal{P}}_{c,0}$. By conjugation by $m_{a,z}^{a/2}$, we obtain the corresponding asymptotics of $\widehat{\mathcal{P}}_{c,t}^{a,z} - \widehat{\mathcal{P}}_{c,0}^{a,z}$. The right-hand side of (13.4.3) is slightly modified by the conjugation. Indeed $\widehat{\mathcal{P}}_{c,0}$ is replaced by $\widehat{\mathcal{P}}_{c,0}^{a,z}$, and the term which appears as a factor of $\frac{\sqrt{t}}{b}$ contains an extra $\mathcal{O}(y)\,p$.

Put

$$\mathcal{D}_{c,t,\lambda}^{a,z} = \left(\widehat{\mathcal{P}}_{c,t}^{a,z} - \lambda\right)^{-1} - \left(\widehat{\mathcal{P}}_{c,0}^{a,z} - \lambda\right)^{-1}. \tag{13.7.17}$$

Then

$$\mathcal{D}_{c,t,\lambda}^{a,z} = -\left(\widehat{\mathcal{P}}_{c,t}^{a,z} - \lambda\right)^{-1}\left(\widehat{\mathcal{P}}_{c,t}^{a,z} - \widehat{\mathcal{P}}_{c,0}^{a,z}\right)\left(\widehat{\mathcal{P}}_{c,0}^{a,z} - \lambda\right)^{-1}. \tag{13.7.18}$$

To establish the third inequality in (13.7.10), note that the first two inequalities allow us to handle conjugation by $m_{a',0}^{-a'/2}$. Using the analogue of equation (13.4.3) for $\widehat{\mathcal{P}}_{c,t}^{a,z} - \widehat{\mathcal{P}}_{c,0}^{a,z}$, which was described above, and also the first inequality in (13.7.10), we see that to establish the third inequality in (13.7.10), only the term in the right-hand side of (13.4.3) which contains as a factor $\sqrt{t}/b$ is potentially troublesome. We will denote this term by $A_{b,t}^{a,z}$. It maps $\ker \alpha'_\pm$ into $\ker \alpha'^{\perp}_\pm$.

Now we write the analogue of (13.7.14) for $\left(\widehat{\mathcal{P}}_{c,t}^{a,z} - \lambda\right)^{-1}$, and we obtain

$$\left(\widehat{\mathcal{P}}_{c,t}^{a,z} - \lambda\right)^{-1} = J_{b,t,\lambda}^{a,z} + bR_{b,t,\lambda}^{a,z}. \tag{13.7.19}$$

Note that

$$J_{b,t,\lambda}^{a,z} A_{b,t}^{a,z} J_{b,0,\lambda}^{a,z} = 0. \tag{13.7.20}$$

Using (13.7.18)-(13.7.20), we find that in the right-hand side of (13.7.18), the singularity $1/b$ in A disappears.

Moreover, when taking $a' = 2$, by the analogue of (13.4.3), we get

$$m_{a',0}^{-a'/2}\left(\widehat{\mathcal{P}}_{c,t}^{a,z} - \widehat{\mathcal{P}}_{c,0}^{a,z}\right)$$
$$= \frac{\sqrt{t}}{b}\left(\mathcal{O}(1)\,\widehat{\nabla} + \mathcal{O}(1)\,\nabla_p + \mathcal{O}(1)\,p_i p_j \widehat{\nabla}_{\widehat{e}^k} + \mathcal{O}(1)\,p_i\right) + \sqrt{t}\mathcal{O}\left(1 + |p|^2\right). \tag{13.7.21}$$

Using the first two estimates in (13.7.10), and also (13.7.17)-(13.7.21), we obtain the third estimate in (13.7.10).

Note that in the estimates in (13.7.10), there is a loss of derivatives. We will now show that this loss can be compensated using an interpolation argument.

Take $a \in \mathbf{R}$. We claim that for $N \in \mathbf{N}$ large enough, the obvious analogue of (13.7.10) holds when replacing $\left(\widehat{\mathcal{P}}_{c,t} - \lambda\right)^{-1}$ by $\left(\widehat{\mathcal{P}}_{c,t} - \lambda\right)^{-N}$. Indeed this is obvious by iteration of the inequalities in (13.7.10).

For $N \in \mathbf{N}$, set

$$\mathcal{D}_{c,t,\lambda}^{a,z,N} = \left(\widehat{\mathcal{P}}_{c,t}^{a,z} - \lambda\right)^{-N} - \left(\widehat{\mathcal{P}}_{c,0}^{a,z} - \lambda\right)^{-N}. \tag{13.7.22}$$

Using (13.7.18), we have the obvious equality

$$\mathcal{D}_{c,t,\lambda}^{a,z,N} = \left(\widehat{\mathcal{P}}_{c,t}^{a,z} - \lambda\right)^{-(N-1)} \mathcal{D}_{c,t,\lambda}^{a,z}$$
$$+ \left(\widehat{\mathcal{P}}_{c,t}^{a,z} - \lambda\right)^{-(N-2)} \mathcal{D}_{c,t,\lambda}^{a,z}\left(\widehat{\mathcal{P}}_{c,0}^{a,z} - \lambda\right)^{-1} + \cdots \tag{13.7.23}$$

By proceeding as in Proposition 13.7.3, for $p \in \mathbf{N}$ large enough,

$$\left\| m_{a',0}^{-a'/2}\left(\widehat{\mathcal{P}}_{c,t}^{a,z} - \lambda\right)^{-p} m_{a',0}^{a'/2} \right\|_{\ell} \leq C_p. \tag{13.7.24}$$

Consider $\ell' \in \mathbf{N}$ which is associated to $\ell = 0$ in (13.7.10). Using (13.7.10), (13.7.24), for a given $\ell \geq \ell'$, for $n_+, n_- \in \mathbf{N}, n_+ + n_- = N, n_\pm \geq N_\ell \in \mathbf{N}$, we get inequalities with constants depending only on ℓ, n_+, n_-:

$$\left\| m_{a',0}^{-a'/2}\left(\widehat{\mathcal{P}}_{c,t}^{a,z} - \lambda\right)^{-n_+} \mathcal{D}_{c,t,\lambda}^{a,z}\left(\widehat{\mathcal{P}}_{c,0}^{a,z} - \lambda\right)^{-n_-} u \right\|_{\ell}$$
$$\leq C\left\| m_{a',0}^{-a'/2}\mathcal{D}_{c,t,\lambda}^{a,z}\left(\widehat{\mathcal{P}}_{c,0}^{a,z} - \lambda\right)^{-n_+} u \right\|_{-\ell}$$
$$\leq C\left\| m_{a',0}^{-a'/2}\mathcal{D}_{c,t,\lambda}^{a,z}\left(\widehat{\mathcal{P}}_{c,0}^{a,z} - \lambda\right)^{-n_-} u \right\|_{0}$$
$$\leq C\sqrt{t}\left\| \left(\widehat{\mathcal{P}}_{c,0}^{a,z} - \lambda\right)^{-n_-} u \right\|_{\ell'} \leq C\sqrt{t}\left\| \left(\widehat{\mathcal{P}}_{c,0}^{a,z} - \lambda\right)^{-n_-} u \right\|_{\ell} \leq C\sqrt{t}\,\|u\|_{-\ell}. \tag{13.7.25}$$

Moreover, for $n_+ \leq N_\ell, n_+ + n_- = N - 1$, using the second inequality in

(13.7.10), there is $\ell' \in \mathbf{N}$ depending only on ℓ such that

$$\left\| m_{a',0}^{-a'/2} \left(\widehat{\underline{\mathcal{P}}}_{c,t}^{a,z} - \lambda \right)^{-n_+} \mathcal{D}_{c,t,\lambda}^{a,z} \left(\widehat{\underline{\mathcal{P}}}_{c,0}^{a,z} - \lambda \right)^{-n_-} u \right\|_{\ell} \\ \le C_\ell \left\| m_{a',0}^{-a'/2} \mathcal{D}_{c,t,\lambda}^{a,z} \left(\widehat{\underline{\mathcal{P}}}_{c,0}^{a,z} - \lambda \right)^{-n_-} u \right\|_{\ell'} . \quad (13.7.26)$$

By the third inequality in (13.7.10), there is $\ell'' \in \mathbf{N}$ depending only on ℓ', and so depending only on ℓ, such that

$$\left\| \mathcal{D}_{c,t,\lambda}^{a,z} \left(\widehat{\underline{\mathcal{P}}}_{c,0}^{a,z} - \lambda \right)^{-n_-} u \right\|_{\ell'} \le C_\ell \sqrt{t} \left\| \left(\widehat{\underline{\mathcal{P}}}_{c,0}^{a,z} - \lambda \right)^{-n_-} u \right\|_{\ell''} . \quad (13.7.27)$$

Finally, since $n_- \ge N - 1 - n_\ell$, by (13.7.7) in Proposition 13.7.3, by taking $N \in \mathbf{N}$ large enough and still depending on ℓ, we get

$$\left\| \left(\widehat{\underline{\mathcal{P}}}_{c,0}^{a,z} - \lambda \right)^{-n_-} u \right\|_{\ell''} \le \|u\|_{-\ell} . \quad (13.7.28)$$

By (13.7.26)-(13.7.28), we obtain

$$\left\| m_{a',0}^{-a'/2} \left(\widehat{\underline{\mathcal{P}}}_{c,t}^{a,z} - \lambda \right)^{-n_+} \mathcal{D}_{c,t,\lambda}^{a,z} \left(\widehat{\underline{\mathcal{P}}}_{c,0}^{a,z} - \lambda \right)^{-n_-} u \right\|_{\ell} \le C\sqrt{t} \|u\|_{-\ell} . \quad (13.7.29)$$

If $n_- \le N_\ell$, using the nonconjugated form of the second inequality in (13.7.10) and the third inequality in the same equation, given $\ell \in \mathbf{N}$, there is $\ell' \in \mathbf{N}$ such that

$$\left\| \left(\widehat{\underline{\mathcal{P}}}_{c,t} - \lambda \right)^{-n_-} m_{a',0}^{-a'/2} \mathcal{D}_{c,t,\lambda}^{a,z} u \right\|_{\ell} \le C_\ell \sqrt{t} \|u\|_{\ell'} . \quad (13.7.30)$$

The dual equation to (13.7.30) gives

$$\left\| \mathcal{D}_{c,t,\lambda}^{a,z} m_{a',0}^{-a'/2} \left(\widehat{\underline{\mathcal{P}}}_{c,t} - \lambda \right)^{-n_-} u \right\|_{-\ell'} \le C_\ell \sqrt{t} \|u\|_{-\ell} . \quad (13.7.31)$$

Conjugating $\mathcal{D}_{c,t,\lambda}^{a,z}$ by $m_{a',0}^{-a'/2}$ does not modify the estimates, so that by (13.7.31) we get

$$\left\| m_{a',0}^{-a'/2} \mathcal{D}_{c,t,\lambda}^{a,z} \left(\widehat{\underline{\mathcal{P}}}_{c,t} - \lambda \right)^{-n_-} u \right\|_{-\ell'} \le C_\ell \sqrt{t} \|u\|_{-\ell} . \quad (13.7.32)$$

Take $n_+ \in \mathbf{N}$ so that $n_+ + n_- = N - 1$ as in each of the terms in the right-hand side of (13.7.23). Since $n_- \le N_\ell$, then $n_+ \ge N - 1 - N_\ell$. We can then take N large enough so that $n_+ \ge N_{\ell+\ell'}$. Using (13.7.32), we get

$$\left\| m_{a',0}^{-a'/2} \left(\widehat{\underline{\mathcal{P}}}_{c,t}^{a,z} - \lambda \right)^{-n_+} \mathcal{D}_{c,t,\lambda}^{a,z} \left(\widehat{\underline{\mathcal{P}}}_{c,0}^{a,z} - \lambda \right)^{-n_-} u \right\|_{\ell} \\ \le C_\ell \left\| m_{a',0}^{-a'/2} \mathcal{D}_{c,t,\lambda}^{a,z} \left(\widehat{\underline{\mathcal{P}}}_{c,0}^{a,z} - \lambda \right)^{-n_-} u \right\|_{-\ell'} \le C_\ell \sqrt{t} \|u\|_{-\ell} . \quad (13.7.33)$$

By (13.7.23), (13.7.25), (13.7.29), and (13.7.33), given $\ell \in \mathbf{N}$, for $N \in \mathbf{N}$ large enough,

$$\left\| m_{a',0}^{-a'/2} \mathcal{D}_{c,t,\lambda}^{a,z,N} u \right\|_{\ell} \le C_\ell \sqrt{t} \|u\|_{-\ell} . \quad (13.7.34)$$

By taking $\ell \in \mathbf{N}$ large enough in (13.7.34), we deduce from (13.7.34) that there exists $N \in \mathbf{N}$ such that under the conditions given in the statement of our proposition,

$$\left| < p >^k < p' >^k \frac{m_{a,z}^{a/2}(y)}{m_{a,z}^{a/2}(y')} \partial^{\alpha}_{y,y',p,p'} \left(\left(\widehat{\underline{\mathcal{P}}}_{c,t} - \lambda \right)^{-N} - \left(\widehat{\underline{\mathcal{P}}}_{c,0} - \lambda \right)^{-N} \right) \right.$$
$$\left. \left((y,p), (y',p') \right) \right| \leq C\sqrt{t} m_{a',0}^{a'/2}(y) . \quad (13.7.35)$$

Since the constant in the bound in (13.7.35) is independent of z, we are free to take $z = y'$. For a given value of $a \in \mathbf{N}$, (13.7.35) implies (13.7.9). The proof of our theorem is completed. □

Chapter Fourteen

The integration by parts formula

The purpose of this chapter is to apply the basic techniques of the Malliavin calculus to the hypoelliptic diffusion which is associated with the hypoelliptic Laplacian.

This chapter should be put in historical context. Malliavin invented his calculus in [M78], as a technique to derive integration by parts formulas with respect to the Brownian measure. He showed in particular that solutions of stochastic differential equations were accessible to his calculus. One main application was the proof by Malliavin of the regularity of the heat kernel associated to a second order hypoelliptic differential operator of the form considered by Hörmander [Hör67]. Malliavin's analysis was completed by Stroock [St81b, St81a], who proved smoothness of the heat kernel in great generality.

Another approach to the Malliavin calculus was given in [B81b] using the Girsanov transformation to obtain a direct proof of integration by parts, which itself is related to the Haussmann representation of Brownian martingales [Ha79].

The way the Malliavin calculus is applied to hypoelliptic diffusions is by showing that a map Φ from classical Wiener space to a smooth manifold X is a.s. nonsingular, and more precisely by obtaining proper estimates on the inverse of the Malliavin covariance matrix $\Phi'\Phi'^*$.

In [B84], a geometric form of integration by parts for an elliptic diffusion was given. In fact in the elliptic case, the transversality of the map Φ is obvious. In [B84], it was pointed out that the same situation could occur with mildly hypoelliptic diffusions. This will turn out to be also the case here, which explains why the sophistication in the arguments can be kept at a minimum. In addition, the objects which appear in the proof of the integration by parts formula are intimately related to the second order differential operators on the circle, which were considered in chapter 7.

The reader should be familiar with the theory of Brownian motion and stochastic integration.

This chapter is organized as follows. In section 14.1, we give the main arguments in [B84] from which one can derive the integration by parts formula for geometric elliptic diffusions, which are associated with the standard Laplacian of a Riemannian manifold X.

In section 14.2, the case of our hypoelliptic diffusion is considered. We establish a corresponding integration by parts formula.

In section 14.3, we briefly show how to derive estimates on the hypoelliptic heat kernel from the integration by parts formula. Although the proof of smoothness of the heat kernel is not developed in detail, it should be accessible to any reader with a reasonable knowledge of the subject.

In section 14.4, we give a path integral representation for the gradient of the logarithm of the heat kernel, similar to the one given in [B84] in the elliptic case.

14.1 THE CASE OF BROWNIAN MOTION

In this section, we recall the main results in [B84] on the integration by parts formula. We make the same assumptions as in section 1.2 and we use the corresponding notation. In particular S^X denotes the Ricci tensor of the Riemannian manifold X.

Let Δ^X be the Laplace-Beltrami operator acting on smooth real functions on X. Then Δ^X is the restriction to smooth functions of $-\square^X$.

Let $P \overset{\mathrm{O}(n)}{\longrightarrow}$ be the $\mathrm{O}(n)$ principal bundle of orthonormal frames in TX. Let (θ, ω) be the canonical 1-forms on P. The Maurer-Cartan equations on P can be written as

$$d\theta = -\omega \wedge \theta, \qquad d\omega = -\omega \wedge \omega + \Omega. \tag{14.1.1}$$

We will navigate freely between intrinsic tensorial objects on X and their equivariant representations with respect to the principal bundle P.

Take $x \in X$. Recall that T_xX is a Euclidean vector space. Let $s \in \mathbf{R}_+ \to w_s \in T_xX$ be a Brownian motion. Given the choice of an orthonormal basis in T_xX, it will often be convenient to consider $w_\cdot$ as a standard Brownian motion with values in $\mathbf{R}^n$. Let $s \in \mathbf{R}_+ \to x_s \in X$ be the curve in X, starting at x at time 0, whose development in T_xX is precisely w_s. Then $x_\cdot$ is a Brownian motion on X. Equivalently, if $\tau_s^0 \in \mathrm{Hom}\,(T_xX, T_{x_s}X)$ denotes parallel transport with respect to the Levi-Civita connection, then x is a solution of the stochastic differential equation in the sense of Stratonovitch,

$$\dot{x} = \tau_s^0 \dot{w}. \tag{14.1.2}$$

Note that equation (14.1.2) makes sense although $w_\cdot$ is a.s. nowhere differentiable. The theory of stochastic differential equations is precisely developed to make sense of an equation like (14.1.2).

Let P_x denote the probability law on $\mathcal{C}\,(\mathbf{R}_+, X)$ of the process $x_\cdot$.

Let h_s be a bounded adapted process with values in $\mathbf{R}^n$, and let A_s be a bounded adapted process taking values in antisymmetric (n, n) matrices. Now we describe the integration by parts formula of [B84, Theorem 2.2]. Let δw be the Itô differential of $w_\cdot$, as opposed to $\dot{w}ds$, the Stratonovitch differential of $w_\cdot$.

For $\ell \in \mathbf{R}$, consider the stochastic process,

$$w_t^\ell = \int_0^t e^{\ell A_s} \delta w_s + \int_0^t \ell h_s ds. \tag{14.1.3}$$

Then using the Girsanov formula as in [B84], we know that the probability law of $w^\ell_\cdot$ is equivalent to the law of $w_\cdot$, with an explicit density.

In (14.1.2), we replace $w_\cdot$ by $w^\ell_\cdot$, and we calculate the differential of $x_\cdot$ with respect to ℓ at $\ell = 0$. This computation is done as follows. Consider the stochastic differential equation on the processes (ϑ_s, ϖ_s) along the path x_s,

$$d\vartheta = \left(-\frac{1}{2}S^X\vartheta + h\right)ds + (\varpi + A)\,\delta w, \qquad d\varpi = -\Omega\,(\vartheta, \dot{w})\,ds, \qquad (14.1.4)$$
$$\vartheta(0) = 0, \qquad \varpi_0 = 0.$$

The integration by parts formula of [B84, Theorem 2.2] asserts in its simplest form that if $f : X \to \mathbf{R}$ is a smooth function, then

$$E^{P_x}\left[\langle f'(x_t), \vartheta_t\rangle\right] = E^{P_x}\left[f(x_t)\int_0^t \langle h_s, \delta w\rangle\right]. \qquad (14.1.5)$$

Consider the stochastic differential equation,

$$d\vartheta = \left(-\frac{1}{2}S^X\vartheta + h\right)ds, \qquad d\varpi = -\Omega\,(\vartheta, \dot{w})\,ds, \qquad (14.1.6)$$
$$\vartheta(0) = 0, \qquad \varpi_0 = 0.$$

Observe that (14.1.6) is a special case of (14.1.4), by simply taking $A = -\varpi$. Then the key observation in [B84] is that (14.1.5) still holds.

Let now $Y \in T_xX$. Let ϑ^Y be the solution of the differential equation along $x_\cdot$,

$$d\vartheta^Y = -\frac{1}{2}S^X\vartheta^Y ds, \qquad \vartheta^Y_0 = Y. \qquad (14.1.7)$$

Then by [B84, Theorem 2.14], we get

$$\nabla_Y E^{P_x}\left[f(x_t)\right] = E^{P_x}\left[\langle f'(x_t), \vartheta^Y_t\rangle\right]. \qquad (14.1.8)$$

Let $E_s : Y \to \vartheta^Y_s$ be the obvious linear map.

Let $s \in [0,1] \to \varphi_s \in \mathbf{R}$ be a smooth function such that $\varphi_0 = 0, \varphi_1 = 1$. Put

$$\overline{\vartheta}^Y_s = \varphi_{s/t}\vartheta^Y_s. \qquad (14.1.9)$$

Then $\overline{\vartheta}^Y_s$ is a solution of equation (14.1.6), with $h_s = \frac{1}{t}\varphi'_{s/t}\vartheta^Y_s$. By (14.1.5), (14.1.8), we get

$$\nabla_Y E^{P_x}\left[f(x_t)\right] = E^{P_x}\left[\int_0^t f(x_t)\int_0^t \left\langle \frac{1}{t}\varphi'_{s/t}\vartheta^Y_s, \delta w\right\rangle\right]. \qquad (14.1.10)$$

Let $p_t(x,y)$ be the smooth heat kernel associated to $e^{t\Delta^X/2}$. Given $y \in X$, let $P^t_{x,y}$ be the probability law of the Brownian bridge connecting x at time 0 and y at time t. The probability law $P^t_{x,y}$ is obtained via a regular disintegration of P_x with respect to $x_t = y$. By [B84, Theorem 2.15], under $P^t_{x,y}$, $w_\cdot$ and $x_\cdot$ are semimartingales. Using (14.1.10), we get, as in [B84],

$$\frac{\nabla p_t(x,y)}{p_t(x,y)} = E^{P^t_{x,y}}\left[\int_0^t \frac{1}{t}\varphi'_{s/t}\widetilde{E}_s\delta w\right]. \qquad (14.1.11)$$

Note that in [B84], the choice $\varphi_s = s$ was made. However, it should be pointed out that the choice of a function φ_s whose support is included in $[0,a]$ with $a<1$ is also very interesting, since estimates are much easier in the stochastic integral in the right-hand side of (14.1.11) if one stays away from $s=t$.

Now we follow [B84] to explain how (14.1.11) can be given a direct interpretation. The Weitzenböck formula in (1.2.13) asserts in particular that we have the identity of operators acting on smooth 1-forms,

$$\square^X = -\Delta^H + S^X. \tag{14.1.12}$$

By applying the de Rham operator d^X to the heat equation

$$\frac{\partial}{\partial s}p_s(x,y) - \frac{1}{2}\Delta^X p_s(x,y) = 0, \tag{14.1.13}$$

we get

$$\frac{\partial}{\partial s}\nabla p_s(x,y) + \frac{1}{2}\square^X \nabla p_s(x,y) = 0. \tag{14.1.14}$$

In (14.1.13), (14.1.14), the operators $\Delta^X, \square^X$ act on the variable x.

Now using the Feynman-Kac formula, one verifies easily that under P_x, for $s<t$, $\widetilde{E}_s \nabla p_{t-s}(x_s,y)$ is a martingale. Therefore, for $s<t$, $\widetilde{E}_s \frac{\nabla p_{t-s}}{p_{t-s}}(x_s,y)$ is a martingale with respect to $P^t_{x,y}$ for $s<t$. Finally, as explained in [B84, eq. (2.87)], under $P^t_{x,y}$, $\overline{w}_s = w_s - \int_0^s \frac{\nabla p_{t-u}}{p_{t-u}}(x_u,y)du$ is a Brownian martingale. Using the trivial

$$\int_0^t \frac{1}{t}\varphi'_{s/t}ds = 1, \tag{14.1.15}$$

we find that (14.1.11) is a consequence of the above considerations.

Remark 14.1.1. This is a simple remark for the experts in the Malliavin calculus. Observe that if k is a bounded adapted process, the solution of

$$\dot{\theta} = k, \qquad \theta_0 = 0 \tag{14.1.16}$$

is also a solution of (14.1.6) with $h = \frac{1}{2}S^X\theta + k$. If k is a constant k_0, then $\theta_t = tk_0$. In particular for $t>0$, the map $h \to \theta_t$ is surjective. This is another version of the invertibility of the Malliavin covariance matrix, which is obvious in this case. We will extend this remark to the case of our hypoelliptic operators after equation (14.3.7).

14.2 THE HYPOELLIPTIC DIFFUSION

Recall that $\mathcal{H} = |p|^2/2$. For $c \in \mathbf{R}^*, c = \pm 1/b^2, b>0$, we consider the second order operator on T^*X,

$$\mathcal{L}_c = \frac{1}{2}\Delta^V - c\nabla_{\widehat{p}} + c\nabla_{Y^{\mathcal{H}}}. \tag{14.2.1}$$

The operator $\frac{\partial}{\partial t} - \mathcal{L}_c$ is hypoelliptic by [Hör67].

We use the same notation as in section 4.7. Take $z = (x,p) \in T^*X$. Consider the stochastic differential equation for the process $z_s = (x_s, p_s) \in T^*X$,

$$
\begin{aligned}
\dot{x} &= cp, & \dot{p} &= -cp + \tau_s^0 \dot{w}, \\
x_0 &= x, & p_0 &= p.
\end{aligned}
\tag{14.2.2}
$$

We can rewrite (14.2.2) in the form

$$\ddot{x} = c\left(-\dot{x} + \tau_s^0 \dot{w}\right), \tag{14.2.3}$$

a second order differential equation already considered in [B05, eq. (0.11)]. Again (14.2.2), (14.2.3) have to be properly interpreted in the sense of Stratonovitch. Let P_z be the probability law on $\mathcal{C}\left(\mathbf{R}_+, T^*X\right)$ of the process $z_\cdot$. Incidentally note that formally, when making $c = \infty$, equation (10.2.20) restricts to the equation (14.1.2) $\dot{x} = \tau_s^0 \dot{w}$, and when making $c = 0$, (14.2.3) restricts to the equation of geodesics $\ddot{x} = 0$.

We will write a formula of integration by parts with respect to the new process $(x_\cdot, p_\cdot)$. As before, we replace $w_\cdot$ by $w_\cdot^\ell$ taken as in (14.1.3), and we calculate the differential of $(x_\cdot, p_\cdot)$ with respect to ℓ at $\ell = 0$. Set

$$J_s = \frac{\partial}{\partial \ell} x_s^\ell. \tag{14.2.4}$$

In the sequel, we use the notation $\dot{J}, \ddot{J}$ instead of $\frac{D}{Ds}J$, $\frac{D^2}{Ds^2}J$. Then by the first equation in (14.2.2),

$$\dot{J}_s = c\frac{Dp_s}{D\ell}. \tag{14.2.5}$$

By the second equation in (14.2.2), we obtain

$$
\begin{aligned}
&\ddot{J} + c\dot{J} + R^{TX}(J,\dot{x})\dot{x} = +c\left(h + (A + \varpi)\frac{\delta w}{ds}\right), \\
&\dot{\varpi} = R^{TX}(\dot{x}, J), \\
&J_0 = 0, \qquad \dot{J}_0 = 0, \qquad \varpi_0 = 0.
\end{aligned}
\tag{14.2.6}
$$

Let $g : T^*X \to \mathbf{R}$ be a smooth function on T^*X with compact support. Then using integration by parts on Wiener space as in [B84], we get

$$E^{P_{x,p}}\left[\left\langle dg(x_t, p_t), \left(J_t, \frac{\dot{J}_t}{c}\right)\right\rangle\right] = E^{P_{x,p}}\left[g(x_t, p_t)\int_0^t \langle h, \delta w\rangle\right]. \tag{14.2.7}$$

Consider now the differential equation

$$
\begin{aligned}
&\ddot{J} + c\dot{J} + R^{TX}(J,\dot{x})\dot{x} = ch, & \dot{\varpi} &= -R^{TX}(J,\dot{x}), \\
&J_0 = 0, \dot{J}_0 = 0, & \varpi_0 &= 0.
\end{aligned}
\tag{14.2.8}
$$

Then (14.2.8) is a special case of (14.2.7), by taking

$$A = -\varpi. \tag{14.2.9}$$

Note that taking into account what we said after (14.2.3), for $c = 0$, (14.2.8) is just the equation for Jacobi fields. Recovering equation (14.1.6) for $c = \pm\infty$ is subtler and will not be considered here. Then equation (14.2.7) still holds.

14.3 ESTIMATES ON THE HEAT KERNEL

Let h be an adapted process. Consider the differential equation

$$\ddot{J} + c\dot{J} + R^{TX}(J, \dot{x})\,\dot{x} = ch, \qquad (14.3.1)$$
$$J_0 = 0, \ \dot{J}_0 = 0.$$

Let $s \in [0,1] \to \psi_s \in \mathbf{R}$ be a smooth function such that

$$\psi_0 = \psi'_0 = 0, \qquad \psi_1 = 1, \psi'_1 = 0. \qquad (14.3.2)$$

A special case of such a function is just

$$\psi_s = s^2(3 - 2s). \qquad (14.3.3)$$

Set

$$\overline{\psi}_s = s^2(-1 + s). \qquad (14.3.4)$$

Then

$$\overline{\psi}_0 = \overline{\psi}'_0 = 0, \qquad \overline{\psi}_1 = 0, \overline{\psi}'_1 = 1. \qquad (14.3.5)$$

We claim that for $t > 0$, the map $h \to \left(J_t, \dot{J}_t\right)$ is surjective. Indeed given $A, B \in T_xX$, set

$$J_s = \psi_{s/t}A + t\overline{\psi}_{s/t}B. \qquad (14.3.6)$$

Then

$$\left(J_t, \dot{J}_t\right) = (A, B). \qquad (14.3.7)$$

Moreover, J_s is a solution of equation (14.3.1), with an h which is explicitly determined from the equation. Note that estimating h so as to prove the proper estimates on the objects appearing in the integration by parts formula is very easy in this case.

Again the expert will notice that what we just established is the invertibility of the Malliavin covariance matrix. We should stress that this invertibility is based on trivial considerations on differential equations, and does not rely on sophisticated probabilistic arguments. We are therefore very far from general hypoelliptic diffusions such as the ones considered by Malliavin [M78], and very close to the situation considered in [B84, subsection 1c)]. Indeed the same arguments as in [B84, Theorem 1.10] lead directly to the invertibility of the Malliavin covariance matrix, in the context of the deterministic Malliavin calculus of [B84].

Given $A, B \in T_xX$ taken as before, we can then use the integration by parts formula (14.2.7) with the above choice of h. Let Z be a smooth vector field on T^*X, which is bounded together with its derivatives. By proceeding as Malliavin in [M78] and in [B81b], which involves a slight extension of the above integral by parts formula, we ultimately obtain a formula for $E^{P_{x,p}}\left[Zg\left(x_{t,p_t}\right)\right]$ in terms of the expectation of a quantity where only

$g(x_t, p_t)$ appears. By iterating these computations, if $Z_1, \ldots, Z_m$ are smooth vector fields on T^*X, we obtain a corresponding formula for

$$E^{P_{x,p}}\left[Z_1 \ldots Z_m g(x_t, p_t)\right].$$

So we find that this expectation can be viewed as the integral of g with respect to some measure.

We can then deduce that there is given $z = (x, p) \in T^*X$, $\exp(t\mathcal{L}_c)$ is obtained via a smooth kernel $q_t(z, z')$ with respect to the symplectic volume dv_{T^*X}. The above considerations also lead to estimates on the heat kernel and its derivatives in z'. Joint smoothness in z, z' also follows from the previous considerations, as we will briefly show in the next section.

14.4 THE GRADIENT OF THE HEAT KERNEL

By (2.1.2), $TT^*X = \pi^*(TX \oplus T^*X)$. Also we identify T^*X to TX by the metric g^{TX}. Take $Z = (Y, Y') \in T_zT^*X$. Consider the differential equation

$$\ddot{J}^Z + c\dot{J}^Z + R^{TX}\left(J^Z, \dot{x}\right)\dot{x} = 0, \qquad (14.4.1)$$
$$J_0 = Y, \dot{J}_0 = Y'.$$

Then by proceeding as in [B84, Theorem 2.14] and using the same techniques as before, we get

$$\nabla_{(Y,Y'/c)} E^{P_{x,p}}\left[g(x_t, p_t)\right] = E^{P_z}\left[\left\langle g'(x_t, p_t), \left(J_t, \frac{\dot{J}_t}{c}\right)\right\rangle\right]. \qquad (14.4.2)$$

Set

$$\overline{J}^Z_s = \psi_{s/t} J^Z_s. \qquad (14.4.3)$$

Then $\overline{J}^Z_s$ is a solution of (14.2.8), with $h = h^Z$ given by

$$h^Z = \frac{\psi''_{s/t}}{ct^2} J^Z_s + \frac{2\psi'_{s/t}}{ct} \dot{J}^Z_s + \frac{\psi'_{s/t}}{t} J^Z_s. \qquad (14.4.4)$$

By (14.2.7) and (14.4.2), we get

$$\nabla_{(Y,Y'/c)} E^{P_{x,p}}\left[g(x_t, p_t)\right] = E^{P_{x,p}}\left[g(x_t, p_t) \int_0^t \left\langle h^Z, \delta w\right\rangle\right]. \qquad (14.4.5)$$

Recall that $q_t(z, z'), z, z' \in T^*X$ is the smooth heat kernel for $\exp(t\mathcal{L}_c)$ with respect to the symplectic volume dv_{T^*X} of T^*X. Observe that

$$q_t(z, z') = \int_{T^*X} q_{t/2}(z, z'') q_{t/2}(z'', z') dv_{T^*X}(z''). \qquad (14.4.6)$$

Now using the theorem on support of Stroock and Varadhan [StV72], it is not difficult to show that for any $s \in \mathbf{R}^*_+$, the support of the probability measure $q_s(z, z'') dv_{TX}(z'')$ is the full T^*X. Taking adjoints, the roles of z and z'' can be exchanged. From (14.4.6), we conclude that $q_t(z, z')$ is everywhere positive.

If $z, z' \in T^*X$, let $P^t_{z,z'}$ be the regular conditional law of the process $z.$ under P_z, conditioned on $z_t = z'$. In principle the disintegration of P^z with respect to z_t only exists z' a.e.. However, by proceeding as in [B84, chapter 2], this construction can be done smoothly for every $z' \in T^*X$, by using the h-process associated to the smooth kernel $q_t(z, z')$.

Let F_s, G_s be the linear maps $Z \in T_zT^*X \to J^Z_s \in T_{x_s}X, Z \in T_zT^*X \to \dot{J}^Z_s \in T_{x_s}X$. Their adjoints $\widetilde{F}_s, \widetilde{G}_s$ map $T_{x_s}X$ into T_zT^*X. By (14.4.4), (14.4.5), we get

$$\frac{\nabla_{(Y,Y'/c)}q_t(z,z')}{q_t(z,z')} = E^{P^t_{z,z'}}\left[\int_0^t \left\langle \frac{\psi''_{s/t}}{ct^2}J^Z_s + \frac{2\psi'_{s/t}}{ct}\dot{J}^Z_s + \frac{\psi'_{s/t}}{t}J^Z_s, \delta w_s\right\rangle\right]. \tag{14.4.7}$$

Equivalently,

$$\frac{\nabla_{(\cdot,\cdot/c)}q_t(z,z')}{q_t(z,z')} = E^{P^t_{z,z'}}\left[\int_0^t \left(\left(\frac{\psi''_{s/t}}{ct^2} + \frac{\psi'_{s/t}}{t}\right)\widetilde{F}_s + \frac{2\psi'_{s/t}}{ct}\widetilde{G}_s\right)\delta w_s\right]. \tag{14.4.8}$$

We will give an interpretation of (14.4.8) similar to the one in (14.1.12)-(14.1.15). Indeed we start from the analogue of equation (14.1.13),

$$\frac{\partial}{\partial s}q_s(z,z') - \mathcal{L}_c q_s(z,z') = 0. \tag{14.4.9}$$

In (14.4.9), the operator $\mathcal{L}_c$ acts on the variable z. Now we will use the results of chapter 2 with F the trivial line bundle. By applying the d^{T^*X} operator to (14.4.9), and using the fact that $-\mathcal{L}_c$ is the restriction to functions of the operator $2A^2_{\phi,\mathcal{H}^c}$, we get

$$\frac{\partial}{\partial s}\nabla q_s(z,z') + 2A^2_{\phi,\mathcal{H}^c}\nabla q_s(z,z') = 0. \tag{14.4.10}$$

Still the operator $A^2_{\phi,\mathcal{H}^c}$ acts on the variable z.

Using Itô's formula, we find easily that under $P^t_{z,z'}$, for $0 \le s < t$, the process $\overline{w}_s = w_s - \int_0^s \frac{\nabla^V q_{t-u}}{q_{t-u}}(z_u, z')\,du$ is a Brownian martingale. Now we use (2.2.3) and the Weitzenböck formula of Theorem 2.2.1 for $A^2_{\phi,\mathcal{H}^c}$. We trivialize $TX \simeq T^*X$ along $x.$ by parallel transport with respect to the Levi-Civita connection. Then by Itô's formula, we find easily that under P_z, for $s < t$, $\widetilde{F}_s\nabla^H q_{t-s}(z_s,z') + \frac{1}{c}\widetilde{G}_s\nabla^V q_{t-s}(z_s,z')$ is a martingale. It is equivalent to say that if $Z \in TT^*X$, under P_z, for $s < t$,

$$\left\langle J^Z_s, \nabla^H q_{t-s}(z_s,z')\right\rangle + \left\langle \frac{\dot{J}^Z}{c}, \nabla^V q_{t-s}(z_s z')\right\rangle \tag{14.4.11}$$

is a martingale. This martingale can be written explicitly as a stochastic integral with respect to the Brownian motion w. Note that this last fact will follow from equation (14.4.16).

Therefore under $P^t_{z,z'}$, for $s < t$,

$$\left\langle J^Z_s, \frac{\nabla^H q_{t-s}}{q_{t-s}}(z_s,z')\right\rangle + \left\langle \frac{\dot{J}^Z}{c}, \frac{\nabla^V q_{t-s}}{q_{t-s}}(z_s,z')\right\rangle \tag{14.4.12}$$

is a martingale.

Take $a \in]0,1[$. Let $\psi_s : [0,1] \to \mathbf{R}$ be a smooth function such that $\psi_0 = \psi'_0 = 0$, which is constant and equal to 1 on $[a,1]$. In particular it satisfies (14.3.2). By the above, it follows that

$$E^{P^t_{z,z'}}\left[\int_0^t \left\langle \frac{\psi''_{s/t}}{ct^2} J^Z_s + \frac{2\psi'_{s/t}}{ct} \dot{J}^Z_s + \frac{\psi'_{s/t}}{t} J^Z_s, \delta w_s \right\rangle\right] = \frac{1}{q_t(z,z')}$$

$$E^{P_z}\left[\int_0^t \left\langle \frac{\psi''_{s/t}}{ct^2} J^Z_s + \frac{2\psi'_{s/t}}{ct} \dot{J}^Z_s + \frac{\psi'_{s/t}}{t} J^Z_s, \nabla^V q_{t-s}(z_s,z') \right\rangle ds\right]. \quad (14.4.13)$$

Let T_c be the differential operator

$$T_c = \frac{1}{c}\left(\frac{D^2}{Ds^2} + c\frac{D}{Ds} + R^{TX}(\cdot,\dot{x})\dot{x}\right). \quad (14.4.14)$$

By construction,

$$\frac{\psi''_{s/t}}{ct^2} J^Z_s + \frac{2\psi'_{s/t}}{ct} \dot{J}^Z_s + \frac{\psi'_{s/t}}{t} J^Z_s = T_c\left(\psi_{s/t} J^Z_s\right). \quad (14.4.15)$$

Let D denote the covariant differential along the path $x.$ with respect to the Levi-Civita connection. By (2.2.3), (2.2.5), (14.4.10), and Itô's formula, we get, for $s < t$,

$$D\nabla^V q_{t-s}(z_s,z') = c\left(\nabla^V q_{t-s} - \nabla^H q_{t-s}\right)(z_s,z')\,ds + \nabla^V_{\delta w}\nabla^V q_{t-s}(z_s,z'),$$

$$(14.4.16)$$

$$D\nabla^H q_{t-s}(z_s,z') = cR^{TX}\left(\nabla^V q_{t-s}(z_s,z'),p\right)p\,ds + \nabla^V_{\delta w}\nabla^H q_{t-s}(z_s,z').$$

In the right-hand side of (14.4.13), we can replace $\int_0^t$ by $\int_0^{at}$. By (14.4.15) and using the Itô calculus, we get

$$\begin{aligned}\int_0^t &\left\langle \frac{\psi''_{s/t}}{ct^2} J^Z_s + \frac{2\psi'_{s/t}}{ct} \dot{J}^Z_s + \frac{\psi'_{s/t}}{t} J^Z_s, \nabla^V q_{t-s}(z_s,z') \right\rangle ds \\
&= -\int_0^{at} \left\langle \left(\frac{1}{c}\frac{D}{Ds} + 1\right)\psi_{s/t} J^Z_s, D\nabla^V q_{t-s}(z_s,z') \right\rangle \\
&\quad + \int_0^{at} \left\langle \psi_{s/t} J^Z_s, \frac{1}{c} R^{TX}\left(\nabla^V q_{t-s}(z_s,z'),\dot{x}\right)\dot{x} \right\rangle ds \\
&\quad + \left\langle \frac{1}{c}\dot{J}^Z_{at} + J^Z_{at}, \nabla^V q_{(1-a)t}(z_{at},z') \right\rangle. \end{aligned} \quad (14.4.17)$$

By (14.4.16), we get

$$\begin{aligned}\int_0^{at} &\left\langle \left(\frac{1}{c}\frac{D}{Ds} + 1\right)\psi_{s/t} J^Z_s, D\nabla^V q_{t-s}(z_s,z') \right\rangle \\
&= \int_0^{at} \left\langle \left(\frac{1}{c}\frac{D}{Ds} + 1\right)\psi_{s/t} J^Z_s, c\left(\nabla^V q_{t-s} - \nabla^H q_{t-s}\right)(z_s,z') \right\rangle ds \\
&\quad + \int_0^{at} \left\langle \left(\frac{1}{c}\frac{D}{Ds} + 1\right)\psi_{s/t} J^Z_s, \nabla^V_{\delta w}\nabla^V q_{t-s}(z_s,z') \right\rangle. \end{aligned} \quad (14.4.18)$$

Using (14.4.16) again, we obtain

$$\begin{aligned}\int_0^{at} \left\langle \left(\frac{1}{c}\frac{D}{Ds}+1\right)\psi_{s/t}J_s^Z, c\left(\nabla^V q_{t-s}-\nabla^H q_{t-s}\right)(z_s,z')\right\rangle ds \\ = \left\langle J_{at}^Z, \left(\nabla^V q_{(1-a)t}-\nabla^H q_{(1-a)t}\right)(z_{at},z')\right\rangle \\ + \int_0^{at}\left\langle \psi_{s/t}J_s^Z, cR^{TX}\left(\nabla^V q_{t-s}(z_s,z'),p\right)p\right\rangle ds \\ - \int_0^{at}\left\langle \psi_{s/t}J_s^Z, \nabla^V_{\delta w}\left(\nabla^V q_{t-s}-\nabla^H q_{t-s}\right)(z_s,z')\right\rangle. \quad (14.4.19)\end{aligned}$$

From now, we will use the notation $\simeq$ instead of $=$ every time an Itô integral containing δw is ignored. By (14.4.17)-(14.4.19), we get

$$\begin{aligned}\int_0^t \left\langle \frac{\psi''_{s/t}}{ct^2}J_s^Z + \frac{2\psi'_{s/t}}{ct}\dot{J}_s^Z + \frac{\psi'_{s/t}}{t}J_s^Z, \nabla^V q_{t-s}(z_s,z')\right\rangle ds \\ \simeq \left\langle J_{at}^Z, \nabla^H q_{(1-a)t}(z_{at},z')\right\rangle + \left\langle \frac{\dot{J}_{at}^Z}{c}, \nabla^V q_{(1-a)t}(z_{at},z')\right\rangle. \quad (14.4.20)\end{aligned}$$

As we saw before, the right-hand side of (14.4.20) is a martingale for $a<1$. Moreover, the Itô integrals which appear as the missing terms in (14.4.20) are trivially square integrable, and so they are martingales for $a<1$. In particular their expectation with respect to P_z vanishes.

From the above, we get

$$\begin{aligned}E^{P_z}\left[\int_0^t \left\langle \frac{\psi''_{s/t}}{ct^2}J_s^Z + \frac{2\psi'_{s/t}}{ct}\dot{J}_s^Z + \frac{\psi'_{s/t}}{t}J_s^Z, \nabla^V q_{t-s}(z_s,z')\right\rangle ds\right] \\ = \nabla_{(Y,Y'/c)}q_t(z,z'). \quad (14.4.21)\end{aligned}$$

Combining (14.4.13) and (14.4.21), we have given a direct proof of (14.4.7).

Chapter Fifteen

The hypoelliptic estimates

The purpose of this chapter is to prove the basic hypoelliptic estimates on the Laplacian $\mathfrak{A}'^2_{\phi_b,\pm\mathcal{H}}$. The proofs extend very easily to the curvature $\mathfrak{C}^{\mathcal{M},2}_{\phi_b,\pm\mathcal{H}-b\omega^H}$ of the superconnection $\mathfrak{C}^{\mathcal{M}}_{\phi_b,\pm\mathcal{H}-b\omega^H}$ defined in (2.4.15).

We consider $\mathfrak{A}'^2_{\phi_b,\pm\mathcal{H}}, \mathfrak{C}^{\mathcal{M},2}_{\phi_b,\pm\mathcal{H}-b\omega^H}$ instead of $\mathfrak{A}'^2_{\phi,\mathcal{H}^c}, \mathfrak{C}^2_{\phi,\mathcal{H}^c-\omega^H}$ because in chapter 17, we will establish estimates which are uniform in $b \in]0, b_0]$, where $b_0 > 0$ is a positive constant, for which the choice of $\mathfrak{A}'^2_{\phi_b,\pm\mathcal{H}}, \mathfrak{C}^{\mathcal{M},2}_{\phi_b,\pm\mathcal{H}-b\omega^H}$ is more natural. Part of the work which is needed in chapter 17 will then have been already done in the present chapter.

This chapter is organized as follows. In section 15.1, we recall simple properties on $\mathfrak{A}'^2_{\phi_b,\pm\mathcal{H}}$.

In section 15.2, we construct a Littlewood-Paley decomposition in the variable p. This way, we will be able to work instead on bounded balls or annuli of T^*X.

In section 15.3, we embed these bounded subsets in the projectivization of T^*X, so as to replace T^*X by a compact manifold.

In section 15.4, we prove the basic hypoelliptic estimates.

In section 15.5, we derive estimates for the resolvent of $\mathfrak{A}'^2_{\phi_b,\pm\mathcal{H}}$ on the real line.

In section 15.6, these results are extended to the full resolvent. In particular, we show in Theorem 15.7.1 that there exist $\lambda_0 > 0, c_0 > 0$ such that if

$$\lambda = -\lambda_0 + \sigma + i\tau,\ \sigma \leq c_0 \left|\tau\right|^{1/6},$$

then the resolvent $\left(\mathfrak{A}'^2_{\phi_b,\pm\mathcal{H}} - \lambda\right)^{-1}$ exists.

Finally in section 15.7, we show that for $N \in \mathbf{N}$ large enough, the operator $\left(\mathfrak{A}'^2_{\phi_b,\pm\mathcal{H}} - \lambda\right)^{-N}$ is trace class.

Throughout the chapter, we make the same assumptions as in sections 2.1-2.3 and we use the corresponding notation.

15.1 THE OPERATOR $\mathfrak{A}'^2_{\phi_b,\pm\mathcal{H}}$

We will often identify the fibers of TX and T^*X by the metric g^{TX}. Also $e_1, \ldots, e_n$ is an orthonormal basis of TX.

Recall that p denotes the canonical section of T^*X. The corresponding radial vector field along the fibers is denoted by $\widehat{p}$, so that

$$\widehat{p} = p_i \widehat{e}^i. \tag{15.1.1}$$

Put

$$L_c = 2\mathfrak{A}'^2_{\phi_b, \pm\mathcal{H}}. \tag{15.1.2}$$

As in (2.3.12), set

$$\begin{aligned}
\alpha_{\pm} &= \frac{1}{2}\left(-\Delta^V + |p|^2 \pm (2\widehat{e}_i i_{\widehat{e}^i} - n)\right), \\
\beta_{\pm} &= -\left(\pm \nabla^{\Lambda^{\cdot}(T^*T^*X)\widehat{\otimes}F,u}_{Y^{\mathcal{H}}} + \frac{1}{2}\omega\left(\nabla^F, g^F\right)(e_i)\nabla_{\widehat{e}^i}\right), \qquad (15.1.3)\\
\gamma_{\pm} &= -\frac{1}{4}\left\langle R^{TX}(e_i, e_j)e_k, e_\ell\right\rangle \left(e^i - \widehat{e}_i\right)\left(e^j - \widehat{e}_j\right) i_{e_k + \widehat{e}^k} i_{e_\ell + \widehat{e}^\ell} \\
&\quad - \left(\pm\left\langle R^{TX}(p, e_i)p, e_j\right\rangle + \frac{1}{2}\nabla^F_{e_i}\omega\left(\nabla^F, g^F\right)(e_j)\right)\left(e^i - \widehat{e}_i\right) i_{e_j + \widehat{e}^j}.
\end{aligned}$$

By (2.3.13), we get

$$L_c = \frac{\alpha_{\pm}}{b^2} + \frac{\beta_{\pm}}{b} + \gamma_{\pm}. \tag{15.1.4}$$

Incidentally, note that equation (2.6.2) gives an expansion of the same type when replacing $\mathfrak{A}'^2_{\phi_b,\pm\mathcal{H}}$ by $\mathfrak{C}^{\mathcal{M},2}_{\phi_b,\pm\mathcal{H}-b\omega^H}$. This is the key argument which allows us to extend without further mention our results on $\mathfrak{A}'^2_{\phi_b,\pm\mathcal{H}}$ to corresponding results for $\mathfrak{C}^{\mathcal{M},2}_{\phi_b,\pm\mathcal{H}-b\omega^H}$.

Let $T'X, T''X \subset TX \oplus T^*X$ be the subvector bundles of $TX \oplus T^*X$ which are spanned respectively by $U + \widehat{U}, U - \widehat{U}$, with $U \in TX$. Then

$$TX \oplus T^*X = T'X \oplus T''X, \tag{15.1.5}$$

and the splitting in the right-hand side of (15.1.5) is orthogonal.

From (15.1.5), we deduce that

$$\Lambda^{\cdot}(T^*X \oplus TX) = \Lambda^{\cdot}(T'X)\widehat{\otimes}\Lambda^{\cdot}(T''X). \tag{15.1.6}$$

Let N', N'' be the number operators of $\Lambda^{\cdot}(T'X), \Lambda^{\cdot}(T''X)$. Then the number operator N^{T^*X} of $\Lambda^{\cdot}(T^*X \oplus TX)$ splits naturally as

$$N^{T^*X} = N' + N''. \tag{15.1.7}$$

Observe that if $a > 0$,

$$\begin{aligned}
&a^{N''}\left\langle R^{TX}(p, e_i)p, e_j\right\rangle\left(e^i - \widehat{e}_i\right) i_{e_j + \widehat{e}^j} a^{-N''} \\
&\qquad = a\left\langle R^{TX}(p, e_i)p, e_j\right\rangle\left(e^i - \widehat{e}_i\right) i_{e_j + \widehat{e}^j}. \qquad (15.1.8)
\end{aligned}$$

By taking $a > 0$ small enough, we can make the right-hand side of (15.1.8) arbitrarily small. In particular given $b > 0$, we can choose $a > 0$ so that (15.1.8) is dominated by $\frac{|p|^2}{4b^2}$.

In the sequel, to make our notation simpler, we will not note the above conjugation, so that formally, some of our arguments are valid only for $b > 0$ small enough. However, we will indicate explicitly those points where a conjugation of the operator should be done so that our arguments remain valid even for large values of b.

15.2 A LITTLEWOOD-PALEY DECOMPOSITION

Recall that the Schwartz space $\mathcal{S}^{\cdot}(T^*X, \pi^*F)$ was defined in section 3.2. Let $\mathcal{S}'^{\cdot}(T^*X, \pi^*F^*)$ be the corresponding dual space of tempered currents. Let H be the Hilbert space of square-integrable sections of $\pi^*\left(\Lambda^{\cdot}(T^*X)\widehat{\otimes}F\right)$ on T^*X, let $\langle\ \rangle$ be the corresponding Hermitian product on H, and let $|\ |$ be the associated norm.

Let $\Delta^{H,u}$ be the obvious horizontal Laplacian of T^*X. Namely, if $e_1, \ldots, e_n$ is a locally defined orthonormal basis of TX, then

$$\Delta^{H,u} = \nabla_{e_i}^{\Lambda^{\cdot}(T^*X)\widehat{\otimes}F,u,2} - \nabla_{\nabla_{e_i}^{TX}e_i}^{\Lambda^{\cdot}(T^*X)\widehat{\otimes}F,u}. \tag{15.2.1}$$

Set

$$S = -\Delta^{H,u} - \Delta^V + |p|^2. \tag{15.2.2}$$

Then S is a self-adjoint positive operator. If $s \in \mathbf{R}$, $u \in \mathcal{S}^{\cdot}(T^*X, \pi^*F)$, set

$$|u|_s = \left|S^{s/2}u\right|_H. \tag{15.2.3}$$

Let H^s be the completion of $\mathcal{S}^{\cdot}(T^*X, \pi^*F)$ with respect to the norm $|\ |_s$. Then the H^s define a chain of Sobolev spaces. Put

$$H^\infty = \cap_{s\in\mathbf{R}}H^s. \tag{15.2.4}$$

Then $H^\infty = \mathcal{S}^{\cdot}(T^*X, \pi^*F)$.

We introduce a Littlewood-Paley decomposition in the radial variable $|p|$. Take $r_0 \in]1,2[$. Let $\phi(r)$ be a smooth function defined on $\mathbf{R}_+$ with values in $[0,1]$, which is decreasing and such that $\phi(r) = 1$ if $|r| \le \frac{1}{r_0}$ and $\phi(r) = 0$ if $|r| \ge 1$. Set

$$\chi(r) = \phi(r/2) - \phi(r). \tag{15.2.5}$$

Then χ takes its values in $[0,1]$ and its support is included in $[\frac{1}{r_0}, 2]$. For $j \in \mathbf{N}$, put

$$\chi_j(r) = \chi(2^{-j}r). \tag{15.2.6}$$

Then one has the obvious equality

$$\phi(r) + \sum_{j=0}^{\infty}\chi_j(r) = 1. \tag{15.2.7}$$

By (15.2.7),

$$\sum_{j=0}^{+\infty}\chi_j^2 \le 1. \tag{15.2.8}$$

Put

$$< p >= \left(1 + |p|^2\right)^{1/2}. \tag{15.2.9}$$

Let $u \in \mathcal{S}^{\cdot}(T^*X, \pi^*F)$. For $j \in \mathbf{N}$, set

$$\delta_j(u) = \chi_j(< p >)u. \tag{15.2.10}$$

Since $< p > \geq 1$, then $\phi(< p >) = 0$. By (15.2.7), (15.2.10), we get the Littlewood-Paley decomposition,

$$u = \sum_{j=0}^{\infty} \delta_j(u). \tag{15.2.11}$$

Set

$$\mathcal{B} = \left\{p \in T^*X, |p|^2 \leq 3\right\}. \tag{15.2.12}$$

The support of $\delta_0(u)$ is included in the ball $\mathcal{B}$. For $j \geq 1$ the support of the $\delta_j(u)$ is included in the annulus C_j given by

$$C_j = \{p, < p > \in [2^j/r_0 \ , \ 2^{j+1}]\}, \tag{15.2.13}$$

Observe that

$$C_j \cap C_{j+2} = \emptyset. \tag{15.2.14}$$

By (15.2.8), (15.2.11), and (15.2.14), we get

$$\sum_{j=0}^{+\infty} |\delta_j(u)|_H^2 \leq |u|_H^2 \leq 3 \sum_{j=0}^{+\infty} |\delta_j(u)|_H^2 . \tag{15.2.15}$$

Definition 15.2.1. For $u \in \mathcal{S}^{\cdot}(T^*X, \pi^*F)$, set

$$U_j(x,p) = \delta_j(u)(x, 2^j p). \tag{15.2.16}$$

Let $\mathcal{R}$ be the annulus:

$$\mathcal{R} = \{p, |p|^2 \in [\frac{1}{r_0^2} - \frac{1}{4}, 4]. \tag{15.2.17}$$

For any $j \in \mathbf{N}$, $U_j \in \mathcal{S}^{\cdot}(T^*X, \pi^*F)$. Moreover, the support of U_0 is included in the ball $\mathcal{B}$, and for $j \geq 1$, the support of the U_j is included in $\mathcal{R}$. We recover u from U by the formula

$$u(x,p) = \sum_{j=0}^{\infty} U_j(x, 2^{-j}p). \tag{15.2.18}$$

Put

$$\mathcal{B}_0 = \{p \in T^*X\}, |p|^2 \leq 5\}. \tag{15.2.19}$$

For any $j \in \mathbf{N}$, the support of the U_j is included in $\mathcal{B}_0$.

15.3 PROJECTIVIZATION OF T*X AND SOBOLEV SPACES

Let $\mathbf{P}(T^*X \oplus \mathbf{R})$ be the real projectivization of $T^*X \oplus \mathbf{R}$, and let Y be the total space of $\mathbf{P}(T^*X \oplus \mathbf{R})$. Then T^*X embeds as an open dense subset of $\mathbf{P}(T^*X \oplus \mathbf{R})$ by the map $p \in T^*X \to (p, 1) \in \mathbf{P}(T^*X \oplus \mathbf{R})$. The complement of T^*X in $\mathbf{P}(T^*X \oplus \mathbf{R})$ is just $\mathbf{P}(T^*X)$, whose total space is denoted by Z. We still denote by $\pi : Y \to X$ the obvious projection.

Recall that the metric g^{TT^*X} on TT^*X was defined after equation (2.1.23). Let $g^{T\mathbf{P}(T^*X\oplus\mathbf{R})}$ be a metric on $T\mathbf{P}(T^*X\oplus\mathbf{R})$ which restricts to g^{TT^*X} on the ball $2\mathcal{B}_0$. We denote by $dv_{\mathbf{P}(T^*X\oplus\mathbf{R})}$ the corresponding volume form.

Let $\mathcal{S}$ be the space of smooth sections of $\pi^*\left(\Lambda^{\cdot}(T^*X)\widehat{\otimes}\Lambda^{\cdot}(TX)\widehat{\otimes}F\right)$ on Y which vanish near Z. Note that $\mathcal{S}$ can be identified with the vector space of elements of $\mathcal{S}^{\cdot}(T^*X,\pi^*F)$ with compact support. Let H be the corresponding Hilbert space of L^2 sections of $\pi^*\left(\Lambda^{\cdot}(T^*X)\widehat{\otimes}\Lambda^{\cdot}(TX)\widehat{\otimes}F\right)$ on Y. Note that H embeds continuously into H. Let $|\,|$ denote the obvious norm on H.

Using the unitary connection $\nabla^{\Lambda^{\cdot}(T^*T^*X)\widehat{\otimes}F,u}$, we can define a natural self-adjoint Laplacian Δ^Y acting on $\mathcal{S}$. This is a second order self-adjoint elliptic differential operator. The precise choice of Δ^Y is irrelevant. Set

$$\mathbb{S}=-\Delta^Y+1. \tag{15.3.1}$$

Then $\mathbb{S}$ is a self-adjoint positive operator.

Definition 15.3.1. For $j\in\mathbf{N}$, set

$$\Lambda_j=\left(\mathbb{S}+2^{4j}\right)^{1/2}. \tag{15.3.2}$$

Given $s\in\mathbf{R},U\in\mathcal{S}$, we define the Sobolev norm $|U|_{j,s}$ by the formula

$$|U|_{j,s}=2^{jn/2}\left|\Lambda_j^sU\right|_{\mathsf{H}}. \tag{15.3.3}$$

If $u\in\mathcal{S}$, if the U_j are defined as in (15.2.16), put

$$\|u\|_s^2=\sum_{j=0}^{\infty}|U_j|_{j,s}^2. \tag{15.3.4}$$

We denote by $\mathcal{H}^s$ the completion of $\mathcal{S}$ with respect to the norm $\|\,\|_s$. We will use the notation $\mathcal{H}=\mathcal{H}^0$.

Remark 15.3.2. By (15.2.15), $\mathcal{H}=H$, and the associated norm on $\mathcal{H}$ is equivalent to the norm of H. Moreover, $u\in\mathcal{H}^1$ if and only if $u,\nabla_{e_i}u,<p>\nabla_{\widehat{e}^i}u,<p>^2u\in\mathcal{H}$. When u is restricted to have fixed compact support in T^*X, the $\mathcal{H}^s$ are the usual Sobolev spaces. Also for $s'>s$, the embedding of $\mathcal{H}^{s'}$ in $\mathcal{H}^s$ is compact.

We claim that if $s\in\mathbf{R}$, $\mathcal{H}^s\subset H^s$, and the corresponding embedding is continuous. Indeed note that by (15.2.18), if $s\in\mathbf{R},u\in\mathcal{S}$,

$$|u|_s\le C_s\|u\|_s. \tag{15.3.5}$$

Also using basic properties of the harmonic oscillator, we know that given $m\in\mathbf{N}$, for $s\in\mathbf{R}$ large enough, for $|\alpha|\le m$, the map $u\in H^s\to<p>^m\widehat{\nabla}^\alpha u\in H$ is uniformly bounded. So we find easily that given $s\in\mathbf{R}$, for $s'\ge s$ large enough, $H^{s'}\subset\mathcal{H}^s$, and the corresponding embedding is continuous.

Set

$$\mathcal{H}^\infty=\cap_{s\in\mathbf{R}}\mathcal{H}^s. \tag{15.3.6}$$

Using (15.2.4) and the above results, we get

$$\mathcal{H}^\infty=H^\infty=\mathcal{S}^{\cdot}(T^*X,\pi^*F). \tag{15.3.7}$$

15.4 THE HYPOELLIPTIC ESTIMATES

In the sequel, $\widehat{\nabla} U$ denotes the differential of U in the directions along the fibers T^*X or $\mathbf{P}(T^*X \oplus \mathbf{R})$, and ∇U denotes the differential of U in horizontal directions.

Recall that for $a \in \mathbf{R}, u \in \mathcal{S}^{\cdot}(T^*X, \pi^*F)$, then $K_a u(x,p) = u(x, ap)$.

Definition 15.4.1. For $\tau > 0$, set

$$L_{c,\tau} = K_\tau^{-1} L_c K_\tau. \tag{15.4.1}$$

Put

$$\begin{aligned} \alpha_{\pm,\tau} &= \frac{1}{2}\left(-\tau^2 \Delta^V + \tau^{-2}|p|^2 \pm (2\widehat{e}_i i_{\widehat{e}^i} \mp n)\right), \\ \beta_{\pm,\tau} &= -\left(\pm \tau^{-1} \nabla_{Y^{\mathcal{H}}}^{\Lambda^{\cdot}(T^*T^*X)\widehat{\otimes} F,u} + \frac{1}{2}\tau\omega\left(\nabla^F, g^F\right)(e_i)\nabla_{\widehat{e}^i}\right), \qquad (15.4.2) \\ \gamma_{\pm,\tau} &= -\frac{1}{4}\left\langle R^{TX}(e_i,e_j)e_k, e_\ell\right\rangle \left(e^i - \widehat{e}_i\right)\left(e^j - \widehat{e}_j\right) i_{e_k + \widehat{e}^k} i_{e_\ell + \widehat{e}^\ell} \\ &\quad - \left(\pm\tau^{-2}\left\langle R^{TX}(p,e_i)p, e_j\right\rangle + \frac{1}{2}\nabla_{e_i}^F \omega\left(\nabla^F, g^F\right)(e_j)\right)\left(e^i - \widehat{e}_i\right) i_{e_j + \widehat{e}^j}. \end{aligned}$$

By (15.1.3), (15.1.4), (15.4.1), we get

$$L_{c,\tau} = \frac{\alpha_{\pm,\tau}}{b^2} + \frac{\beta_{\pm,\tau}}{b} + \gamma_{\pm,\tau}. \tag{15.4.3}$$

Also observe that with respect to the standard Hermitian product on H, $\beta_\pm$ is skew-adjoint.

Let $L'_{c,\tau}, L''_{c,\tau}$ be the self-adjoint and skew-adjoint parts of $L_{c,\tau}$ with respect to the standard L^2 Hermitian product, so that

$$L_{c,\tau} = L'_{c,\tau} + L''_{c,\tau}. \tag{15.4.4}$$

From (15.4.2), one can deduce obvious formulas for $L'_{c,\tau}, L''_{c,\tau}$.

Given $\lambda_0 > 0$, set

$$P_c = L_c + \lambda_0, \qquad P_{c,\tau} = L_{c,\tau} + \lambda_0. \tag{15.4.5}$$

Let $P'_{c,\tau}, P''_{c,\tau}$ be the self-adjoint and skew-adjoint parts of $P_{c,\tau}$.

In the sequel $|\,|$ denotes the norm in H or in $\mathcal{H}$. Since the support of the sections which we will consider is included in $\mathcal{B}_0$, the restriction of these norms to such sections are equivalent. Also $\langle\,\rangle$ denotes the standard Hermitian product on H.

Theorem 15.4.2. *If $\lambda_0 > 0$ is large enough, for any $s \in \mathbf{R}$, there exists $C_s > 0$ such that for any $j \in \mathbf{N}$, for any $U \in \mathcal{S}^{\cdot}(T^*X, \pi^*F)$ whose support is included in the ball $\mathcal{B}$ for $j = 0$ and in the annulus $\mathcal{R}$ for $j \geq 1$, then*

$$2^{4j}|U|_{j,s}^2 + \left|\widehat{\nabla} U\right|_{j,s}^2 + 2^{3j/2}|U|_{j,s+1/4}^2 + 2^{-5j/4}\left|\widehat{\nabla} U\right|_{j,s+1/8}^2 \leq C_s \left|P_{c,2^{-j}} U\right|_{j,s}^2. \tag{15.4.6}$$

Proof. The proof of our theorem is organized as follows.

- First we establish elementary coercitivity estimates.
- Then we prove three important lemmas. In particular Lemma 15.4.4 is a crude hypoellipticity estimate, where the Hörmander property of our second order operator plays a key role.
- The combination of the three lemmas leads to a proof of Theorem 15.4.2.

Actually, we follow closely the proof by Hörmander [Hör85, chapter 22, pp. 353-359] of the hypoelliptic estimates, with special attention to the dependence on the scaling parameter $\tau = 2^{-j}$. A proof of a similar result which involves different norms was given by Helffer-Nier [HeN05] and Hérau-Nier [HN04] when establishing the hypoellipticity of the Fokker-Planck operator.

Note that we can ignore the normalizing constant $2^{jn/2}$ in the right-hand side of (15.3.3). For simplicity, we will not write the index j explicitly in the Sobolev norms. Also we will take $\tau \in]0,1]$.

We will denote by C positive constants not depending on τ, s, and by C_s positive constants not depending on τ and still depending on s. These constants may still depend on c. Also they may vary from line to line.

We claim that if $b > 0$ is small enough but fixed, for $\lambda_0 = \lambda_0 (b) > 0$ large enough, there exists $C > 0$ such that for any $\tau \in]0,1]$ and $U \in \mathcal{S}^{\cdot} (T^*X, \pi^*F)$ with support in the annulus $\mathcal{R}$, then

$$\left|\widehat{\nabla} U\right|^2 + \tau^{-4} |U|^2 \leq C \left\langle P'_{c,\tau} U, \tau^{-2} U\right\rangle. \tag{15.4.7}$$

Indeed recall that by (15.4.2), only $\alpha_{\pm,\tau}, \gamma_{\pm,\tau}$ contribute to $P'_{c,\tau}$. Also note that given $b_0 > 0, b_1, b_2 \in \mathbf{R}$, for $\lambda_0 > 0$ large enough, the polynomial

$$Q(\tau) = (\lambda_0 + b_2)\tau^2 + b_1\tau + b_0$$

has a positive lower bound when $\tau \in]0,1]$. We use also the fact that on $\mathcal{R}$, $|p|^2$ has a positive lower bound. For $b > 0$ small enough, the quadratic term in the variable p in $\gamma_{\pm,\tau}$ is small with respect to $\tau^{-2}|p|^2/2$. Then (15.4.7) follows from (15.4.2). Of course if $\tau = 1$, the estimate (15.4.7) still holds for $\lambda_0 > 0$ large enough when the support of U is included in the ball $\mathcal{B}$.

For $b > 0$ arbitrary, as explained in (15.1.8), we should conjugate our operator by $a^{N''}$ with a small enough, so that the operator obtained from P_c by this conjugation is such that (15.4.7) still holds. We will continue the proof in the case where $b > 0$ is small, the proof for b large being the one suggested here.

Note that

$$\left|\left\langle P'_{c,\tau} U, \tau^{-2} U\right\rangle\right| \leq \left|\left\langle P_{c,\tau} U, \tau^{-2} U\right\rangle\right|. \tag{15.4.8}$$

Using (15.4.7), (15.4.8), we find that under the same assumptions as before,

$$\left|\widehat{\nabla} U\right|^2 + \tau^{-4} |U|^2 \leq C \left|P_{c,\tau} U\right|^2. \tag{15.4.9}$$

For $1 \le i \le n$, we still denote $\widehat{e}^i$ a vector field along the fibers $\mathbf{P}(T^*X \oplus \mathbf{R})$ of Y which coincides with our given $\widehat{e}^i$ on $\mathcal{B}_0$.

We will use classical pseudodifferential operators with weight Λ on Y. A symbol of degree d is a smooth function $a(y,\zeta,\tau)$ on $T^*Y \times]0,1]$ with values in

$$\pi^*\mathrm{End}\left(\Lambda^{\cdot}(T^*X)\widehat{\otimes}\Lambda^{\cdot}(TX)\widehat{\otimes}F\right),$$

such that for any α, β, there exist $C_{\alpha,\beta} > 0$ such that

$$\left|\partial_y^\alpha \partial_\zeta^\beta a(y,\zeta,\tau)\right| \le C_{\alpha,\beta}(\tau^{-4} + |\zeta|^2)^{\frac{d-|\beta|}{2}}. \tag{15.4.10}$$

We denote by S^d the set of symbols of degree d.

To keep in line with the notation in (15.3.2), set

$$\Lambda = \left(\mathbb{S} + \tau^{-4}\right)^{1/2}. \tag{15.4.11}$$

If $U \in \mathcal{S}$, set

$$|U|_s = |\Lambda^s U|. \tag{15.4.12}$$

By definition in our context, a smoothing operator on Y is a family of operators $B(\tau), \tau \in]0,1]$ such that for $s,t \in \mathbf{R}$, there exist $C_{s,t}$ such that

$$|B(\tau)U|_s \le C_{s,t}|U|_t. \tag{15.4.13}$$

We use a local coordinate system on X, and we trivialize the above vector bundles in this coordinate system. This induces in turn a local coordinate system on Y.

Let $\widehat{U}$ denote the Fourier transform of $U(y)$ in the variable $y \in \mathbf{R}^{2n}$, i.e.,

$$\widehat{U}(\zeta) = \int_{\mathbf{R}^{2n}} e^{-i\langle y,\zeta\rangle} U(y)\,dy. \tag{15.4.14}$$

To a symbol a, given the above coordinate system and trivialization, we associate an operator $\mathrm{Op}(a)$ by the formula

$$\mathrm{Op}(a)(y, D_y, \tau)U(y) = (2\pi)^{-2n}\int_{\mathbf{R}^{2n}} e^{i\langle y,\zeta\rangle} a(y,\zeta,\tau)\widehat{U}(\zeta)\,d\zeta. \tag{15.4.15}$$

Note that in (15.4.15), $D_y = -i\partial_y$.

Let $\mathcal{E}^d$ be the corresponding set of pseudodifferential operators $A(\tau)$ of degree d on $\mathcal{P}(T^*X \oplus \mathbf{R})$. Then $A(\tau) \in \mathcal{E}^d$ if and only if for any small compact $K \subset \mathcal{P}(T^*X \oplus \mathbf{R})$, for any cutoff function $\theta(y)$ with support close enough to K, there exist $a \in S^d$, a cutoff function $\theta'(y)$ with support close enough to K and a smoothing operator $B(\tau)$ such that

$$A(\tau)\theta = \theta' \mathrm{Op}(a)\theta + B(\tau). \tag{15.4.16}$$

For $A \in \mathcal{E}^d$, the class $\sigma(A)$ of A in $\mathcal{E}^d/\mathcal{E}^{d-1}$ will be called the principal symbol of A.

If $E_d \in \mathcal{E}^d, E_{d'} \in \mathcal{E}^{d'}$, then $E_d E_{d'} \in \mathcal{E}^{d+d'}$, $\sigma(E_d E_{d'}) = \sigma(E_d)\sigma(E_{d'})$. If a, a' are symbols, let $[a,a']$ denote their pointwise commutator, and let

$\{a, a'\}$ denote their Poisson bracket. If $E_d = \mathrm{Op}(a), E_{d'} = \mathrm{Op}(a')$, then $[E_d, E_{d'}] - \mathrm{Op}([a, a'] - i\{a, a'\}) \in \mathcal{E}^{d-2}$.

When acting on $\mathcal{H}$, operators in $\mathcal{E}^0$ act as a family of bounded operators, which is uniformly bounded in the parameter $\tau \in]0, 1]$. Also observe that $\Lambda, \nabla_{e_i}, \nabla_{\widehat{e}^j} \in \mathcal{E}^1$.

Let $\psi(r) : \mathbf{R}_+ \to [0, 1]$ be a smooth function which is equal to 1 when $r^2 \leq 5$, and to 0 for $r^2 \geq 6$. Set

$$\theta_0(p) = \psi(|p|). \tag{15.4.17}$$

Then θ_0 is equal to 1 on $\mathcal{B}_0$.

Consider the operator R given by

$$R = \theta_0 P_{c,\tau} \theta_0. \tag{15.4.18}$$

If the support of U is included in $\mathcal{B}_0$,

$$P_{c,\tau} U = RU. \tag{15.4.19}$$

Let R', R'' be the self-adjoint and skew-adjoint components of R, so that

$$R = R' + R''. \tag{15.4.20}$$

Clearly,

$$R' = \theta_0 P'_{c,\tau} \theta_0, \qquad R'' = \theta_0 P''_{c,\tau} \theta_0. \tag{15.4.21}$$

Of course $P'_{c,\tau}, P''_{c,\tau}$ can be obtained via (15.4.2)-(15.4.5). Using these explicit formulas, and also the fact that $\tau^{-2} \in \mathcal{E}^1$, we get

$$[R, \mathcal{E}_d] \in \tau \mathcal{E}^d \widehat{\nabla} + \tau^{-2} \mathcal{E}^d. \tag{15.4.22}$$

Moreover, if $E \in \mathcal{E}^d$ is such that $\sigma(E)$ is scalar, then we have the stronger

$$[R, E] \in \tau^2 \mathcal{E}^d \widehat{\nabla} + \tau^{-1} \mathcal{E}^d. \tag{15.4.23}$$

The difference between (15.4.22) and (15.4.23) comes from the fact that the term $-\frac{1}{2}\tau\omega(\nabla^F, g^F)(e_i)\nabla_{\widehat{e}^i}$ in $\beta_{\pm,\tau}$ was the only term contributing to $\tau \mathcal{E}^d \widehat{\nabla}$ in (15.4.22), and that this contribution disappears when commuting with an operator with scalar principal symbol and also because $\tau^{-2}\mathcal{E}^d$ can be replaced by $\tau^{-1}\mathcal{E}^d$. Equation (15.4.23) will be used in particular in (15.4.56).

We take $\lambda_0 > 0$ large enough so that (15.4.7)-(15.4.9) hold.

Lemma 15.4.3. *If $\lambda_0 > 0$ is large enough, there exists $C > 0$ such that for any $U \in \mathcal{S}^{\cdot}(T^*X, \pi^*F)$ with support included in the ball $\mathcal{B}$ if $j = 0$, and in the annulus $\mathcal{R}$ for $j \geq 1$, then*

$$|R''U|^2_{-1/2} \leq C\tau |RU|^2. \tag{15.4.24}$$

Proof. By (15.4.9), we get

$$\left|\widehat{\nabla} U\right| + \tau^{-2}|U| \leq C|RU|. \tag{15.4.25}$$

Using (15.4.2)-(15.4.5), we find that $\tau R'' \in \mathcal{E}^1$. In fact multiplication by τ kills the term factor τ^{-1} which appears in the skew-adjoint part of $\beta_{\pm,\tau}$. Set

$$E_0 = \Lambda^{-1}\tau R''. \tag{15.4.26}$$

Then $E_0 \in \mathcal{E}^0$. Also

$$|R''U|_{-1/2}^2 = \tau^{-1} \langle R''U, E_0 U \rangle . \tag{15.4.27}$$

Moreover,

$$\tau^{-1} \langle R''U, E_0 U \rangle = \tau^{-1} \langle RU, E_0 U \rangle - \tau^{-1} \langle R'U, E_0 U \rangle . \tag{15.4.28}$$

Also since $E_0 \in \mathcal{E}^0$, it acts as a uniformly bounded operator on $\mathcal{H}$. By (15.4.25), we get

$$\tau^{-1} |\langle RU, E_0 U \rangle| \le \tau |RU|^2 . \tag{15.4.29}$$

For λ_0 large enough, R' is self-adjoint and nonnegative, and so by Cauchy-Schwarz, we get

$$2\tau^{-1} |\langle R'U, E_0 U \rangle| \le \tau^{-1} |\langle R'U, U \rangle| + \tau^{-1} |\langle R' E_0 U, E_0 U \rangle| . \tag{15.4.30}$$

By (15.4.25), we have

$$\tau^{-1} |\langle R'U, U \rangle| = \tau^{-1} |\mathrm{Re} \langle RU, U \rangle| \le C\tau |RU|^2 . \tag{15.4.31}$$

Finally, we write

$$\tau^{-1} \langle R' E_0 U, E_0 U \rangle = \tau^{-1} \mathrm{Re} \langle E_0 RU, E_0 U \rangle + \tau^{-1} \mathrm{Re} \langle [R, E_0] U, E_0 U \rangle . \tag{15.4.32}$$

By (15.4.22) and (15.4.25), we obtain

$$|[R, E_0] U| \le C \left(\tau \left| \widehat{\nabla} U \right| + \tau^{-2} |U| \right) \le C |RU| . \tag{15.4.33}$$

Our lemma now follows from (15.4.27)-(15.4.33). □

Now we prove a result which plays a crucial role in the proof of the hypoellipticity of our operator.

Lemma 15.4.4. *There exists $C > 0$ such that for any $U \in \mathcal{S}^{\cdot} (T^*X, \pi^* F)$ with support included in the ball $\mathcal{B}$ if $j = 0$, and in the annulus $\mathcal{R}$ if $j \ge 1$, then*

$$|U|_{1/4} \le C\tau^{3/4} |RU|. \tag{15.4.34}$$

Proof. We have to prove that there exists $C > 0$ such that if j, U are taken as indicated, and if $\tau = 2^{-j}$, then

$$\left|\tau^{-2} U\right|_{-3/4} + |\nabla U|_{-3/4} + \left|\widehat{\nabla} U\right|_{-3/4} \le C\tau^{3/4} |RU| . \tag{15.4.35}$$

Note that $\Lambda^{-3/4} \le \tau^{3/2}$, and so

$$|U|_{-3/4} \le C\tau^{3/2} |U| . \tag{15.4.36}$$

By (15.4.9), (15.4.25), and (15.4.36), we obtain

$$\left|\tau^{-2} U\right|_{-3/4} + \left|\widehat{\nabla} U\right|_{-3/4} \le C\tau^{3/2} |RU| . \tag{15.4.37}$$

Let e be a smooth section of TX over X. Observe the critical fact (which guarantees hypoellipticity by Hörmander [Hör67]) that

$$\left[\nabla_{\widehat{e}}, \nabla_{Y^{\mathcal{H}}}^{\Lambda^{\cdot}(T^*T^*X)\widehat{\otimes}F,u}\right] = \nabla_{e}^{\Lambda^{\cdot}(T^*T^*X)\widehat{\otimes}F,u} - \nabla_{\widehat{\nabla_{Y^{\mathcal{H}}}^{TX} e}}. \tag{15.4.38}$$

Observe that $\nabla_{Y^{\mathcal{H}}}^{TX} e$ depends linearly in p, and so $\nabla_{\widehat{\nabla_{Y^{\mathcal{H}}}^{TX} e}}$ is a vertical differentiation operator of type $p\partial_p$. The scaling which we used before ensures that such an operator is of the type $\widehat{\nabla}$ which is controlled by the estimate (15.4.38). Using (15.4.37), (15.4.38), we find that to establish (15.4.35), we only need to prove that

$$\left|\left[\nabla_{\widehat{e}}, \nabla_{Y^{\mathcal{H}}}^{\Lambda^{\cdot}(T^*T^*X)\widehat{\otimes}F,u}\right] U\right|_{-3/4} \leq C\tau^{3/4} \left|RU\right|. \tag{15.4.39}$$

Let S be the differential operator

$$S = \theta_0 \nabla_{Y^{\mathcal{H}}}^{\Lambda^{\cdot}(T^*T^*X)\widehat{\otimes}F,u} \theta_0. \tag{15.4.40}$$

Because of the support condition on U, we may replace $\nabla_{Y^{\mathcal{H}}}^{\Lambda^{\cdot}(T^*T^*X)\widehat{\otimes}F,u}$ in (15.4.39) by S. This will be done repeatedly in the sequel. Also in the sequel we will often use the notation $\widehat{\nabla}$ instead of $\nabla_{\widehat{e}}$.

Set

$$E_{-1/2} = \Lambda^{-3/2}[\widehat{\nabla}, S]. \tag{15.4.41}$$

Then $E_{-1/2} \in \mathcal{E}^{-1/2}$. Moreover,

$$\left|\left[\widehat{\nabla}, S\right] U\right|_{-3/4}^2 = \left\langle \left[\widehat{\nabla}, S\right] U, E_{-1/2} U\right\rangle. \tag{15.4.42}$$

Also observe that since $\nabla_{\widehat{e}}$ and $\nabla_{Y^{\mathcal{H}}}^{\Lambda^{\cdot}(T^*T^*X)\widehat{\otimes}F,u}$ are skew-adjoint operators with respect to the standard L^2 Hermitian product $\langle\,\rangle$, then

$$\begin{aligned} \left\langle \widehat{\nabla} SU, E_{-1/2}U\right\rangle &= -\left\langle E^*_{-1/2}SU, \widehat{\nabla} U\right\rangle - \left\langle SU, [\widehat{\nabla}, E_{-1/2}]U\right\rangle, \\ \left\langle S\widehat{\nabla} U, E_{-1/2}U\right\rangle &= -\left\langle \widehat{\nabla} U, E_{-1/2}SU\right\rangle - \left\langle \widehat{\nabla} U, [S, E_{-1/2}]U\right\rangle. \end{aligned} \tag{15.4.43}$$

By (15.4.2), and using obvious notation, we get

$$S = \mp \tau R'' + \tau^2 \widehat{\nabla} + \tau^{-1}. \tag{15.4.44}$$

By (15.4.11), $\Lambda^{-1/2} \leq \tau$, and so

$$\left|\tau^{-1} U\right|_{-1/2} \leq \left|U\right|. \tag{15.4.45}$$

By (15.4.9), by (15.4.24) in Lemma 15.4.3, and by (15.4.44), (15.4.45), we get

$$\left|SU\right|_{-1/2} \leq C\tau^{3/2} \left|RU\right|. \tag{15.4.46}$$

Therefore,

$$\left|E^*_{-1/2}SU\right| \leq C\tau^{3/2} \left|RU\right|. \tag{15.4.47}$$

Using (15.4.9) and (15.4.47), we get

$$\left|\left\langle E^*_{-1/2}SU, \widehat{\nabla}U\right\rangle\right| \le C\tau^{3/2}\left|RU\right|^2. \tag{15.4.48}$$

Note that the operators $[\nabla_{\widehat{e}^i}, E_{-1/2}]$ and $[S, E_{-1/2}]$ lie in $\mathcal{E}^{-1/2}$. Using (15.4.9) and (15.4.46), we obtain

$$\left|\left\langle SU, \left[\widehat{\nabla}, E_{-1/2}\right]U\right\rangle\right| \le C\tau^{3/2}\left|RU\right|\left|U\right| \le C\tau^{7/2}\left|RU\right|^2. \tag{15.4.49}$$

Still using (15.4.9) and the obvious analogue of (15.4.47), we can control the first term in the right-hand side of the second line in (15.4.43) as in (15.4.48). By proceeding as before,

$$\left|\left\langle \widehat{\nabla}U, \left[S, E_{-1/2}\right]U\right\rangle\right| \le C\left|RU\right|\left|U\right| \le C\tau^2\left|RU\right|^2. \tag{15.4.50}$$

By (15.4.43)-(15.4.50), we obtain the required estimate in (15.4.39). The proof of our lemma is completed. □

By (15.4.7), if $\delta > 0$ and if $U \in \mathcal{S}^{\cdot}(T^*X, \pi^*F)$ verifies the same support conditions as before,

$$\begin{aligned}\left|\widehat{\nabla}U\right|^2 &\le C\mathrm{Re}\left\langle \Lambda^{-1/8}RU, \Lambda^{1/8}\tau^{-2}U\right\rangle \\ &\le C\left(\delta\left|\Lambda^{-1/8}RU\right|^2 + \frac{\tau^{-4}}{\delta}\left|\Lambda^{1/8}U\right|^2\right).\end{aligned} \tag{15.4.51}$$

Let $\theta_1(p)$ be a smooth function of $|p|$ with values in $[0,1]$, which has compact support and which is constructed in the same way as the function θ_0. We assume that θ_1 is equal to 1 on a neighborhood of the annulus $\mathcal{R}$, and that it vanishes near $p = 0$. Finally, if $j \ge 1$, set $\theta_j = \theta_1$. On the sequel, for $j \in \mathbf{N}$, we use the notation $\theta = \theta_j$.

By Lemma 15.4.4, for any $s \in \mathbf{R}$, we get

$$|\Lambda^{1/4}\theta\Lambda^s U| \le C\tau^{3/4}|R\theta\Lambda^s U|. \tag{15.4.52}$$

Moreover, we have the commutator identities

$$\begin{aligned}\Lambda^{1/4}\theta\Lambda^s &= \Lambda^{s+1/4}\theta + \Lambda^{1/4}\left[\theta, \Lambda^s\right], \\ R\theta\Lambda^s &= \Lambda^s R\theta + R[\theta, \Lambda^s] + [R, \Lambda^s]\theta.\end{aligned} \tag{15.4.53}$$

Since $\theta = 1$ on the support of U, the essential support of $[\theta, \Lambda^s]$ does not intersect the support of U, so for any s, σ, there exist $C_{s,\sigma} > 0$ such that

$$\left|[\theta, \Lambda^s]U\right|_\sigma \le C_{s,\sigma}\left|U\right|_s. \tag{15.4.54}$$

Also $\theta U = U$, and for $\alpha \ge 0$, if $V \in \mathcal{H}$,

$$|V|_{-\alpha} \le \tau^{2\alpha}\left|V\right|. \tag{15.4.55}$$

Moreover, by (15.4.23),

$$[R, \Lambda^s] \in \tau^2\mathcal{E}^s\widehat{\nabla} + \tau^{-1}\mathcal{E}^s. \tag{15.4.56}$$

Using (15.4.52)-(15.4.56), we find that

$$|U|_{s+1/4}^2 \leq \tau^{3/2}\left(C\,|RU|_s^2 + C_s\tau^4\left|\widehat{\nabla}U\right|_s^2 + C_s\tau^{-2}\,|U|_s^2\right). \tag{15.4.57}$$

If we apply (15.4.51) to $\theta\Lambda^{s+1/8}U$, by using the same commutation arguments as before, we get

$$\left|\widehat{\nabla}U\right|_{s+1/8}^2 \leq \delta\left(C\,|RU|_s^2 + C_s\tau^4\left|\widehat{\nabla}U\right|_s^2 + C_s\tau^{-2}\,|U|_s^2\right) + \frac{\tau^{-4}}{\delta}\left(C\,|U|_{s+1/4}^2 + C_s\tau^4\,|U|_s^2\right) + C_s\,|U|_{s+1/8}^2. \tag{15.4.58}$$

Now we scale (15.4.57) by the factor $\tau^{-3/2}$, and (15.4.58) by the factor $1/\delta$. Also for $A > 0$, we take $\delta^{-1} = A\tau^{5/4}$. Adding up these two inequalities, we obtain

$$\tau^{-3/2}\,|U|_{s+1/4}^2 + A\tau^{5/4}\left|\widehat{\nabla}U\right|_{s+1/8}^2 \leq C\,|RU|_s^2 + C_s\tau^4\left|\widehat{\nabla}U\right|_s^2 + C_s\tau^{-2}\,|U|_s^2 + CA^2\tau^{-3/2}\,|U|_{s+1/4}^2 + C_s\left(A^2\tau^{5/2}\,|U|_s^2 + A\tau^{5/4}\,|U|_{s+1/8}^2\right). \tag{15.4.59}$$

In (15.4.59), we fix A so that $CA^2 \leq 1/2$. Using the trivial fact that $\tau^{5/2}\,|U|_s^2 \leq \tau^{5/4}\,|U|_{s+1/8}^2$, we obtain that

$$\tau^{-3/2}\,|U|_{s+1/4}^2 + \tau^{5/4}\left|\widehat{\nabla}U\right|_{s+1/8}^2 \leq C\,|RU|_s^2 + C_s\tau^4\left|\widehat{\nabla}U\right|_s^2 + C_s\tau^{-2}\,|U|_s^2 + C_s\tau^{5/4}\,|U|_{s+1/8}^2. \tag{15.4.60}$$

By equation (15.4.7) applied to $\theta\Lambda^s U$, and still using (15.4.56), we obtain

$$\left|\widehat{\nabla}U\right|_s^2 + \tau^{-4}\,|U|_s^2 \leq C\,|RU|_s^2 + C_s\tau^4\left|\widehat{\nabla}U\right|_s^2 + C_s\tau^{-2}\,|U|_s^2. \tag{15.4.61}$$

By adding (15.4.60) and (15.4.61), we get

$$\left|\widehat{\nabla}U\right|_s^2 + \tau^{-4}\,|U|_s^2 + \tau^{-3/2}\,|U|_{s+1/4}^2 + \tau^{5/4}\left|\widehat{\nabla}U\right|_{s+1/8}^2 \leq C\,|RU|_s^2 + C_s\tau^4\left|\widehat{\nabla}U\right|_s^2 + C_s\tau^{-2}\,|U|_s^2 + C_s\tau^{5/4}\,|U|_{s+1/8}^2. \tag{15.4.62}$$

We will complete the proof of Theorem 15.4.2 by a contradiction argument. For $s \in \mathbf{R}$, let W^s be the Hilbert space

$$W^s = \left\{u \in \mathcal{H}^s, \widehat{\nabla}u \in \mathcal{H}^{s-1/8}\right\}. \tag{15.4.63}$$

Suppose that for some value of s, equation (15.4.6) does not hold. Then there exist $\tau_k = 2^{-j_k}, U_k$ such that the left-hand side of (15.4.6) , in which we make $j = j_k, U = U_k$, is equal to 1, and moreover $\lim_{k\to+\infty}\left|P_{c,\tau_k}U_k\right|_s = 0$. By (15.4.62), the sequence j_k remains bounded, so we may suppose that j_k remains constant and equal to j.

So now $\tau = 2^{-j}$ is fixed. The support of the U_k is included in $\mathcal{B}_0$, and moreover the U_k form a bounded sequence in the Hilbert space $W^{s+1/4}$. By Sobolev embedding, we can as well assume that U_k converges in $W^{s+1/8}$ to U_∞. We claim that $U_\infty \neq 0$. Indeed if $U_\infty = 0$, as $k \to +\infty$, the right-hand side of the inequality (15.4.62) would tend to 0, while the $\underline{\lim}$ of the left-hand side would be positive.

By (15.4.62), for any $U \in \mathcal{S}^{\cdot}(T^*X, \pi^*F)$ with support included in $\mathcal{B}_0$,

$$|U|^2_{\sigma+1/4} + \left|\widehat{\nabla} U\right|^2_{\sigma+1/8} \leq C_\sigma \left(|RU|^2_\sigma + |U|^2_{\sigma+1/8} + \left|\widehat{\nabla} U\right|^2_\sigma \right). \tag{15.4.64}$$

By iterating (15.4.64), we find that for $N \in \mathbf{N}$ large enough,

$$|U|^2_{\sigma+1/4} + \left|\widehat{\nabla} U\right|^2_{\sigma+1/8} \leq C_{\sigma,N} \left(|RU|^2_\sigma + |U|^2_{\sigma-N} \right). \tag{15.4.65}$$

Recall that $\tau = 2^{-j}$ is fixed.

Lemma 15.4.5. *For any $\sigma \in \mathbf{R}$, and any $U \in \mathcal{S}'^{\cdot}(T^*X, \pi^*F^*)$ whose support is included in the ball $\mathcal{B}_0$, if $RU \in \mathcal{H}^\sigma$, then $U \in \mathcal{H}^{\sigma+1/4}, \widehat{\nabla} U \in \mathcal{H}^{\sigma+1/8}$, and inequality* (15.4.65) *holds for U.*

Proof. This is a standard consequence of (15.4.65). In fact let Ψ_m be a sequence of scalar regularizing operators in $\mathcal{E}^\infty$ which is bounded in $\mathcal{E}^0$ and which converges to the identity as $m \to +\infty$. We may and we will assume that support of the Ψ_m is as close as necessary to the diagonal. Note that there are uniformly bounded $A_m, B_m \in \mathcal{E}^0$ such that

$$[R, \Psi_m] = A_m \widehat{\nabla} + B_m. \tag{15.4.66}$$

The $V_m = \Psi_m U$ still have compact support, this support being included in an open ball $\mathcal{B}'_0$ which is slightly bigger as $\mathcal{B}_0$. Of course, inequality (15.4.65) is still valid for $U \in \mathcal{S}^{\cdot}(T^*X, \pi^*F)$ with support included in $\mathcal{B}'_0$. Using (15.4.65) with U replaced by $V_m = \Psi_m U$, we find that given t, N, there exists $C_{t,N} > 0$ such that for any $m \in \mathbf{N}$,

$$|V_m|^2_{t+1/4} + \left|\widehat{\nabla} V_m\right|^2_{t+1/8} \leq C_{t,N} \left(|RU|^2_t + |U|^2_{t-N} + \left|A_m \widehat{\nabla} U\right|^2_t + |B_m U|^2_t \right). \tag{15.4.67}$$

Also there exists $t \in \mathbf{R}$ such that $U \in \mathcal{H}^{t+1/4}, \widehat{\nabla} U \in \mathcal{H}^{t+1/8}$. Using the fact that $RU \in \mathcal{H}^\sigma$, we conclude that the set of $t \in \mathbf{R}$ which are taken as before contains σ. The proof of our lemma is completed. □

As we saw before, $\lim_{k \to +\infty} \left|P_{c,\tau} U_k\right|_s = 0$. Since the sequence U_k converges to U_∞ in $W^{s+1/8}$, we conclude that $P_{c,\tau} U_\infty = 0$, which implies $RU_\infty = 0$. By Lemma 15.4.5, we conclude that U_∞ is smooth. By (15.4.9), we find that $U_\infty = 0$. This gives us the required contradiction. The proof of Theorem 15.4.2 is completed. □

15.5 THE RESOLVENT ON THE REAL LINE

For $s \in \mathbf{R}$, let $D_s(L_c)$ be the domain of L_c, i.e.,

$$D_s(L_c) = \{u \in \mathcal{H}^s, L_c u \in \mathcal{H}^s\}. \tag{15.5.1}$$

We can also define $D_s(P_c)$, and

$$D_s(P_c) = D_s(L_c). \tag{15.5.2}$$

Theorem 15.5.1. *Take $\lambda_0 > 0$ large enough. If $s \in \mathbf{R}, u \in \mathcal{S}'^{\cdot}(T^*X, \pi^*F^*)$, if $P_c u \in \mathcal{H}^s$, then $u \in \mathcal{H}^{s+1/4}$, and there exists $C_s > 0$ such that for any u taken as before,*

$$\|u\|_{s+1/4} \leq C_s \|P_c u\|_s. \tag{15.5.3}$$

Moreover, $\mathcal{S}^{\cdot}(T^*X, \pi^*F)$ is dense in $D_s(P_c)$, i.e., for any $u \in D_s(P_c)$, there exists a sequence $u_k \in \mathcal{S}^{\cdot}(T^*X, \pi^*F)$ such that

$$\lim_{k\to+\infty} \left(\|u - u_k\|_s + \|P_c(u - u_k)\|_s\right) = 0. \tag{15.5.4}$$

Proof. We use the notation of sections 15.2-15.4. Take $u \in \mathcal{S}'^{\cdot}(T^*X, \pi^*F^*)$, and set $v = P_c u$. For $j \in \mathbf{N}$, we define the corresponding U_j, V_j as in (15.2.10), (15.2.16). In the sequel we take $\tau = 2^{-j}$, and we drop the index j in U_j, V_j.

Set

$$W = K_{\tau^{-1}}\left[P_c, \chi(\tau < p >)\right] u. \tag{15.5.5}$$

A straightforward computation shows that

$$P_{c,\tau} U = V + W. \tag{15.5.6}$$

Since the Hamiltonian vector field $Y^{\mathcal{H}}$ preserves $\mathcal{H} = |p|^2/2$, then

$$\left[\nabla^{\Lambda^{\cdot}(T^*T^*X)\widehat{\otimes}F,u}_{Y^{\mathcal{H}}}, \chi(\tau < p >)\right] = 0. \tag{15.5.7}$$

Now we use (15.1.3), (15.1.5) and we get

$$\begin{aligned}\left[P_c, \chi(\tau < p >)\right] = \frac{1}{2b^2}\left(-\Delta^V \chi(\tau < p >) - 2\left(\nabla_{\widehat{e}^i}\chi(\tau < p >)\right)\nabla_{\widehat{e}^i}\right)\\ - \frac{1}{2b}\omega\left(\nabla^F, g^F\right)(e_i)\nabla_{\widehat{e}^i}\chi(\tau < p >).\end{aligned} \tag{15.5.8}$$

By (15.5.8), we get

$$\begin{aligned}K_{\tau^{-1}}\left[P_c, \chi(\tau < p >)\right]K_\tau = \frac{1}{2b^2}\Big(-K_{\tau^{-1}}\left(\Delta^V\chi(\tau < p >)\right)\\ - 2K_{\tau^{-1}}\left(\nabla_{\widehat{e}^i}\chi(\tau < p >)\right)\tau\nabla_{\widehat{e}^i}\Big)\\ - \frac{1}{2b}\omega\left(\nabla^F, g^F\right)(e_i)K_{\tau^{-1}}\nabla_{\widehat{e}^i}\chi(\tau < p >).\end{aligned} \tag{15.5.9}$$

We write (15.2.18) in the form

$$u = \sum K_{\tau'} U'. \tag{15.5.10}$$

In (15.5.10), the sum is made over the $\tau' = 2^{-j}, j \in \mathbf{N}$, and the corresponding U' is just U_j. By (15.5.5), (15.5.9), (15.5.10), we get

$$W = \left(\frac{1}{2b^2} \left(-K_{\tau^{-1}} \left(\Delta^V \chi \left(\tau < p > \right) \right) - 2K_{\tau^{-1}} \left(\nabla_{\widehat{e}^i} \chi \left(\tau < p > \right) \right) \tau \nabla_{\widehat{e}^i} \right) \right.$$
$$\left. - \frac{1}{2b} \omega \left(\nabla^F, g^F \right) (e_i) K_{\tau^{-1}} \nabla_{\widehat{e}^i} \chi \left(\tau < p > \right) \right) \sum K_{\tau'/\tau} U'. \tag{15.5.11}$$

The support of $K_{\tau^{-1}} \chi (\tau < p >)$ is included in $\left\{ |p|^2 \in \left[\frac{1}{r_0^2} - \tau^2, 4 - \tau^2 \right] \right\}$. The support of U' is included in $\left\{ |p|^2 \in \left[\frac{1}{r_0^2} - \tau'^2, 4 - \tau'^2 \right] \right\}$, and so the support of $K_{\tau'/\tau} U'$ is included in $\left\{ |p|^2 \in \left[\frac{\tau^2}{\tau'^2} \frac{1}{r_0^2} - \tau^2, 4 \frac{\tau^2}{\tau'^2} - \tau^2 \right] \right\}$. We then find that only the τ' which are such that $1/2r_0 \le \tau'/\tau \le 2r_0$ contribute to the sum in the right-hand side of (15.5.11). Since $r_0 \in]1, 2[$, and also $\tau = 2^{-j}, \tau' = 2^{-j'}$, then the nonzero terms in (15.5.11) are such that $0 \le |j - j'| \le 1$, i.e., there are at most three nonzero terms.

By (15.5.11), if $u \in \mathcal{S}^{\cdot} \left(T^* X, \pi^* F \right), t \in \mathbf{R}$, then

$$|W_j|_t \le C_t 2^{-j} \sum_{|j'-j| \le 1} |U_{j'}|_t + C_t 2^{-2j} \sum_{|j'-j| \le 1} \left| \widehat{\nabla} U_{j'} \right|_t. \tag{15.5.12}$$

We use (15.5.5), (15.5.6), (15.5.11), and (15.5.12). Assume that $t \in \mathbf{R}$ is such that $u, \widehat{\nabla} u, Pu \in \mathcal{H}^t$. Then

$$\left| P_{c,2^{-j}} U_j \right|_t \le C |V_j|_t + C_t 2^{-j} \sum_{|j'-j| \le 1} |U_{j'}|_t + C_t 2^{-2j} \sum_{|j'-j| \le 1} \left| \widehat{\nabla} U_{j'} \right|_t. \tag{15.5.13}$$

By inequality (15.4.6) in Theorem 15.4.2, we get

$$2^{3j/4} |U_j|_{t+1/4} + 2^{-5j/8} \left| \widehat{\nabla} U_j \right|_{t+1/8} \le C_t \left| P_{c,2^{-j}} U_j \right|_t. \tag{15.5.14}$$

By (15.5.13), (15.5.14), we obtain

$$2^{3j/4} |U_j|_{t+1/4} + 2^{-5j/8} \left| \widehat{\nabla} U_j \right|_{t+1/8}$$
$$\le C |V_j|_t + C_t 2^{-j} \sum_{|j'-j| \le 1} |U_{j'}|_t + C_t 2^{-2j} \sum_{|j'-j| \le 1} \left| \widehat{\nabla} U_{j'} \right|_t. \tag{15.5.15}$$

Set

$$\beta_{j,t} = 2^{3j/4} |U_j|_{t+1/4} + 2^{-5j/8} \left| \widehat{\nabla} U_j \right|_{t+1/8}. \tag{15.5.16}$$

By (15.5.15), we get

$$\beta_{j,t} \le C\,|V_j|_t + C_t 2^{-7j/4} \sum_{|j'-j|\le 1} \beta_{j',t-1/4} + C_t 2^{-11j/8} \sum_{|j'-j|\le 1} \beta_{j',t-1/8}. \tag{15.5.17}$$

Since $\Lambda^{-1/8} \le \tau^{1/4}$, and $11/8 < 2$, from (15.5.17), we get

$$\beta_{j,t} \le C\,|V_j|_t + C_t 2^{-11j/8} \sum_{|j'-j|\le 1} \beta_{j',t-1/8}. \tag{15.5.18}$$

By (15.5.18), we obtain

$$\beta_{j,t}^2 \le C\,|V_j|_t^2 + C_t 2^{-11j/4} \sum_{|j'-j|\le 1} \beta_{j',t-1/8}^2. \tag{15.5.19}$$

Since $u \in \mathcal{S}'^{\cdot}(T^*X, \pi^*F^*)$, for $s' \in \mathbf{R}$ small enough, $u \in H^{s'}$. By using the same arguments as in Remark 15.3.2, we find that for $t \in \mathbf{R}$ small enough, $\beta_{j,t} \in \ell_2$. In the statement of our theorem, we made the assumption that if $v = P_c u$, then $v \in \mathcal{H}^s$, which is just the fact that $|V_j|_s \in \ell_2$. By (15.5.19), it should be clear that $\beta_{j,s} \in \ell_2$. In particular, $u \in \mathcal{H}^{s+1/4}$.

Let $\mathcal{W}^s$ be the vector space of the $u \in \mathcal{S}'^{\cdot}(T^*X, \pi^*F^*)$ which are such that $\beta_{j,s} \in \ell_2$. We equip $\mathcal{W}^s$ with the norm

$$\|u\|_{\mathcal{W}^s} = |\beta_{j,s}|_{\ell_2}. \tag{15.5.20}$$

By (15.5.19), we get

$$\|u\|_{\mathcal{W}^s} \le C_s \left(\|P_c u\|_s + \|u\|_{\mathcal{W}^{s-1/8}}\right). \tag{15.5.21}$$

Using a contradiction argument similar to the one we used in the proof of Theorem 15.4.2 after equation (15.4.62), we find that

$$\|u\|_{\mathcal{W}^s} \le C_s \|P_c u\|_s, \tag{15.5.22}$$

from which (15.5.3) follows.

Finally the fact that $\mathcal{S}^{\cdot}(T^*X, \pi^*F)$ is dense in $D_s(P_c)$ can be established by an argument similar to the one we used in the proof of Lemma 15.4.5. The proof of our theorem is completed. □

15.6 THE RESOLVENT ON C

Recall that for $s \in \mathbf{R}$, $D_s(L_c)$ was defined in (15.5.1). For $s = 0$, we use instead the notation $D(L_c)$.

Theorem 15.6.1. *There exist $\lambda_0 > 0, C_0 > 0$ such that for any $u \in D(L_c), \lambda \in \mathbf{C}, \mathrm{Re}(\lambda) \le -\lambda_0$,*

$$|\lambda|^{1/6} \|u\|_0 + \|u\|_{1/4} \le C_0 \|(L_c - \lambda) u\|_0. \tag{15.6.1}$$

There exists $C_1 > 0$ such that for $\sigma, \tau \in \mathbf{R}, \sigma \le C_1 |\tau|^{1/6}$, if $\lambda = -\lambda_0 + \sigma + i\tau$, then

$$(1 + |\lambda|)^{1/6} \|u\|_0 \le C_1 \|(L_c - \lambda) u\|_0. \tag{15.6.2}$$

Proof. Note that identity (15.6.2) follows from (15.6.1). Indeed in (15.6.1), we take $\lambda = -\lambda_0 + i\tau, v = (L_c - (\lambda + \sigma))\, u$, so that $(L_c - \lambda)\, u = v + \sigma u$, and (15.6.1) shows that

$$|\lambda|^{1/6} \|u\|_0 \leq C_0 \left(\|v\|_0 + |\sigma| \, \|u\|_0\right), \tag{15.6.3}$$

from which (15.6.2) follows.

To prove (15.6.1), we take λ_0 as in section 15.4. As before, $P_c = L_c + \lambda_0$. If $\lambda \in \mathbf{C}$, set

$$P_{c,\lambda} = L_c - \lambda. \tag{15.6.4}$$

If $\alpha \in \mathbf{R}_+, \beta \in \mathbf{R}$, let $\lambda \in \mathbf{C}$ be given by

$$\lambda = -\lambda_0 - \alpha + i\beta, \tag{15.6.5}$$

so that

$$P_{c,\lambda} = P_c + \alpha - i\beta. \tag{15.6.6}$$

Now we follow closely the proof of Theorems 15.4.2 and 15.5.1, but we modify the weight Λ_j in (15.3.2) by putting instead

$$\Lambda_{\lambda,j} = \left(\mathbb{S} + 2^{4j} + 2^{-2j}|\lambda|^2\right)^{1/2}. \tag{15.6.7}$$

For $u \in \mathcal{S}$, equations (15.3.3), (15.3.4) are now replaced by

$$|U|_{\lambda,j,s} = 2^{jn/2} \left|\Lambda^s_{\lambda,j} U\right|_{\mathcal{H}}, \qquad \|u\|^2_{\lambda,s} = \sum_{j=0}^{\infty} |U_j|^2_{\lambda,j,s}. \tag{15.6.8}$$

Clearly, there is $C > 0$ such that for $t > 0, a > 0$,

$$t^2 + \frac{a}{t} \geq C a^{2/3}. \tag{15.6.9}$$

From (15.6.9), we get

$$2^{4j} + 2^{-2j} |\lambda|^2 \geq C \, |\lambda|^{4/3}. \tag{15.6.10}$$

Using (15.6.10), we find that (15.6.1) will follow from the estimate

$$\|u\|_{\lambda,1/4} \leq C \, \|P_{c,\lambda} u\|_0. \tag{15.6.11}$$

Note that this estimate is the obvious extension of equation (15.5.3) in Theorem 15.5.1. So we concentrate on the proof of (15.6.11), by following the same strategy as in our proof of Theorem 15.5.1.

We define $P_{c,\lambda,\tau}$ from $P_{c,\lambda}$ as in (15.4.1), so that by (15.6.6),

$$P_{c,\lambda,\tau} = P_{c,\tau} + \alpha - i\beta. \tag{15.6.12}$$

First, we will show that Theorem 15.4.2 is still valid for $\lambda_0 > 0$ large enough, when using instead the norms in (15.6.8) and replacing $P_{c,\tau}$ by $P_{c,\lambda,\tau}$, with constants not depending on λ, so that equation (15.4.6) is replaced by

$$2^{4j} |U|^2_{\lambda,j,s} + \left|\widehat{\nabla} U\right|^2_{\lambda,j,s} + 2^{3j/2} |U|^2_{\lambda,j,s+1/4} + 2^{-5j/4} \left|\widehat{\nabla} U\right|^2_{\lambda,j,s+1/8} \leq C_s \left|P_{c,\lambda,2^{-j}} U\right|^2_{\lambda,j,s}. \tag{15.6.13}$$

Instead of (15.4.8), (15.4.9), for $\lambda_0 > 0$ large enough, if the support of $U \in \mathcal{S}^{\cdot}(T^*X, \pi^*F)$ is included in the annulus $\mathcal{R}$, we have

$$\left|\widehat{\nabla} U\right|^2 + \tau^{-4}|U|^2 + \alpha\tau^{-2}|U|^2 \leq C\left\langle P'_{c,\tau,\lambda}U, \tau^{-2}U\right\rangle, \tag{15.6.14}$$
$$\left|\widehat{\nabla} U\right|^2 + \tau^{-4}|U|^2 + \alpha\tau^{-2}|U|^2 \leq C|P_{c,\tau,\lambda}U|^2.$$

We will modify our construction of pseudodifferential operators on the total space Y of $\mathbf{P}(T^*X \oplus \mathbf{R})$. In inequality (15.4.10), we replace the weight $\tau^{-4}+\zeta^2$ by the weight $\tau^{-4}+\tau^2|\lambda|^2+|\zeta|^2$. We denote by S^d_λ the corresponding set of symbols, and by $\mathcal{E}^d_\lambda$ the associated set of pseudodifferential operators. Instead of (15.4.11), we now set

$$\Lambda = \left(\mathbb{S} + \tau^{-4} + \tau^2|\lambda|^2\right)^{1/2}. \tag{15.6.15}$$

We take θ_0 as in (15.4.18), and we define R_λ by the formula

$$R_\lambda = \theta_0 P_{c,\tau,\lambda}\theta_0. \tag{15.6.16}$$

The obvious analogues of equation (15.4.22) and (15.4.23) still hold.

The same arguments as before lead to the analogue of Lemma 15.4.3. We claim indeed that as in the proof of Lemma 15.4.3, $\tau R''_\lambda \in \mathcal{E}^1$. This is because the contribution of λ to $\tau R''_\lambda$ is just $\tau \mathrm{Im}\,\lambda$, and our choice of the weight Λ in (15.6.15) has been made so that $\tau|\lambda| \in \mathcal{E}^1$. The analogue of (15.4.24) is then verified, i.e.,

$$|R''_\lambda U|^2_{\lambda,-1/2} \leq C\tau|R_\lambda U|^2. \tag{15.6.17}$$

We claim that the analogue of Lemma 15.4.4 also holds. Indeed by proceeding as in the proof of (15.4.39) in Lemma 15.4.4, we get

$$\left|\left[\widehat{\nabla}, \nabla^{\Lambda^{\cdot}(T^*T^*X\widehat{\otimes}F),u}_{Y^{\mathcal{H}}}\right]U\right|_{\lambda,-3/4} \leq C\tau^{3/4}|R_\lambda U|. \tag{15.6.18}$$

From (15.6.18), we obtain

$$\left|\left[\widehat{\nabla}, \tau R''_\lambda\right]U\right|_{\lambda,-3/4} \leq C\tau^{3/4}|R_\lambda U|. \tag{15.6.19}$$

Clearly,

$$\Lambda^{-3/4} \leq (|\lambda|\tau)^{-1/2}\tau^{1/2} = |\lambda|^{-1/2}. \tag{15.6.20}$$

By the second inequality in (15.6.14) and by (15.6.20), using the fact that $0 \leq \alpha \leq |\lambda|$, we get

$$\tau|\alpha||U|_{\lambda,-3/4} \leq C\tau^2|R_\lambda U|. \tag{15.6.21}$$

By (15.6.12),

$$i\beta = -R''_\lambda + R''. \tag{15.6.22}$$

Now since $\Lambda^{-1/4} \leq \tau^{1/2}$, by (15.6.17), we get

$$\tau|R''_\lambda U|_{\lambda,-3/4} \leq C\tau^2|R_\lambda U|. \tag{15.6.23}$$

Moreover, by (15.4.2), (15.4.3), (15.6.14), and (15.6.18),

$$\tau \left|R''U\right|_{\lambda,-3/4} \leq C\tau^{3/4} \left|R_\lambda U\right|. \tag{15.6.24}$$

By (15.6.22)-(15.6.24), we obtain

$$\tau|\beta| \left|U\right|_{\lambda,-3/4} \leq C\tau^{3/4} \left|R_\lambda U\right|. \tag{15.6.25}$$

By (15.6.21), (15.6.25), we get

$$\tau \left|\lambda\right| \left|U\right|_{\lambda,-3/4} \leq C\tau^{3/4} \left|R_\lambda U\right|. \tag{15.6.26}$$

The estimate (15.6.26) is precisely the one which permits us to complete the proof of the analogue of Lemma 15.4.4, and so to get

$$\left|U\right|_{\lambda,1/4} \leq C\tau^{3/4} \left|R_\lambda U\right|. \tag{15.6.27}$$

We can then obtain the analogue of (15.4.62), i.e.,

$$\begin{aligned} &\left|\widehat{\nabla} U\right|^2_{\lambda,s} + \tau^{-4}\left|U\right|^2_{\lambda,s} + \tau^{-3/2}\left|U\right|^2_{\lambda,s+1/4} + \tau^{5/4}\left|\widehat{\nabla} U\right|^2_{\lambda,s+1/8} \\ &\leq C\left|R_\lambda U\right|^2_{\lambda,s} + C_s\tau^4\left|\widehat{\nabla} U\right|^2_{\lambda,s} + C_s\tau^{-2}\left|U\right|^2_{\lambda,s} + C_s\tau^{5/4}\left|U\right|^2_{\lambda,s+1/8}. \end{aligned} \tag{15.6.28}$$

To deduce (15.6.13) from (15.6.28), we use the same argument used after (15.4.62), with associated sequences $\tau_k = 2^{-j_k}, U_k, \lambda_k$. Using (15.6.28), we may and we will assume that $j_k = j$, so that $\tau_k = \tau$ remains constant. Also we claim that we may as well assume that $|\lambda_k|$ remains bounded. Indeed for $\sigma \in \mathbf{R}, \epsilon > 0$,

$$\left|U\right|_{\lambda,\sigma-\epsilon} \leq \left(\tau\left|\lambda\right|\right)^{-\epsilon}\left|U\right|_{\lambda,\sigma}. \tag{15.6.29}$$

By (15.6.29), we deduce that if $|\lambda_k| \to +\infty$, for k large enough, we get (15.6.13), with the given j, which is impossible. Therefore we have completed the proof of (15.6.13).

We deduce (15.6.11) from (15.6.13) exactly as in the proof of Theorem 15.5.1. In fact we obtain the obvious analogue of equation (15.5.21). To derive the analogue of (15.5.22), we still use a contradiction argument. Namely, we assume there exists u_k, λ_k, with $\left\|P_{c,\lambda_k} u_k\right\|_s \to 0$, while $\|u_k\|_{\mathcal{W}^s} = 1$. First we claim that $|\lambda_k|$ remains uniformly bounded. Indeed if $|\lambda_k| \to +\infty$, by still using (15.6.29), we find that the analogue of (15.5.21) would contradict (15.5.22). So we may as well assume that $\lambda_k \to \lambda$. It is now easy to proceed, and so we obtain a proof of the analogue of (15.5.22), i.e., we have completed the proof of (15.6.11).

This concludes the proof of Theorem 15.6.1. □

15.7 TRACE CLASS PROPERTIES OF THE RESOLVENT

If $A \in \mathcal{L}(\mathcal{H})$, let $\|A\|$ be the norm of A. Let $\mathcal{L}_1(A)$ be the vector space of trace class operators acting on $\mathcal{H}$. If $A \in \mathcal{L}_1(\mathcal{H})$, set

$$\|A\|_1 = \operatorname{Tr}\left[\left(A^*A\right)^{1/2}\right]. \tag{15.7.1}$$

Then $\| \|_1$ is a norm on $\mathcal{L}_1(\mathcal{H})$.

If A is a possibly unbounded operator acting on $\mathcal{H}$ with compact resolvent, let $\mathrm{Sp}\, A \subset \mathbf{C}$ be the spectrum of A. By definition, if $\lambda \in \mathrm{Sp} A$, the characteristic subspace of $\mathcal{H}$ associated to λ is the image of the spectral projection operator attached to λ.

In what follows, the Hilbert space H is equipped with its canonical Hermitian product. The formal adjoint L_c^* of L_c is taken with respect to this formal adjoint.

Theorem 15.7.1. *The adjoint of the unbounded operator L_c acting on H is the formal adjoint L_c^* of L_c acting on $\mathcal{S}^{\cdot}(T^*X, \pi^*F)$, with domain $D(L_c^*) = \{u \in H, L_c^* u \in H\}$.*

There exist $\lambda_0 > 0, c_0 > 0, C > 0$ such that if $\mathcal{U} \subset \mathbf{C}$ is given by

$$\mathcal{U} = \{\lambda = -\lambda_0 + \sigma + i\tau,\ \ \sigma, \tau \in \mathbf{R}, \sigma \le c_0 |\tau|^{1/6}\}, \tag{15.7.2}$$

if $\lambda \in \mathcal{U}$, the resolvent $(L_c - \lambda)^{-1}$ exists, and moreover

$$\left\|(L_c - \lambda)^{-1}\right\| \le \frac{C}{(1 + |\lambda|)^{1/6}}. \tag{15.7.3}$$

There exists $C > 0$ such that if $\lambda \in \mathbf{R}, \lambda \le -\lambda_0$, then

$$\left\|(L_c - \lambda)^{-1}\right\| \le C (1 + |\lambda|)^{-1}. \tag{15.7.4}$$

If $\lambda \notin \mathrm{Sp}\, L_c$, then $(L_c - \lambda)^{-1}$ is a compact operator acting on $\mathcal{H}$, and for $N \in \mathbf{N}, N > 12n$, $(L_c - \lambda)^{-N}$ is trace class. There exists $C' > 0$ such that if $\lambda \in \mathcal{U}$,

$$\left\|(L_c - \lambda)^{-N}\right\|_1 \le C' (1 + |\lambda|)^N. \tag{15.7.5}$$

If $\lambda \notin \mathrm{Sp}\, L_c$, for any $s \in \mathbf{R}$, the resolvent $(L_c - \lambda)^{-1}$ maps $\mathcal{H}^s$ into $\mathcal{H}^{s+1/4}$. In particular $(L_c - \lambda)^{-1}$ maps $\mathcal{S}^{\cdot}(T^*X, \pi^*F)$ into itself. Given $s \in \mathbf{R}$, there exists $C_s > 0$ such that for $\lambda \in \mathcal{U}, u \in \mathcal{S}^{\cdot}(T^*X, \pi^*F)$,

$$\left\|(L_c - \lambda)^{-1} u\right\|_{s+1/4} \le C_s (1 + |\lambda|)^{4|s|+1} \|u\|_s. \tag{15.7.6}$$

The spectrum $\mathrm{Sp}\, L_c$ of L_c is discrete. If $\lambda \in \mathrm{Sp}\, L_c$, let V_λ be the characteristic subspace of L_c associated to λ. Then V_λ is a finite dimensional subspace of $\mathcal{S}^{\cdot}(T^*X, \pi^*F)$.

Proof. As we saw in Theorem 15.5.1, $\mathcal{S}^{\cdot}(T^*X, \pi^*F)$ is dense in the domain $D(L_c)$, which implies the first part of our theorem.

By (15.6.2), we find that by an adequate choice of $\lambda_0 > 0, c_0 > 0$, if $\lambda \in \mathcal{U}$, the range of $L_c - \lambda$ is closed, and moreover $L_c - \lambda$ is injective. The operator L_c^* being of the same type as L_c, the same results also hold for $L_c^* - \overline{\lambda}$. In particular the range of $L_c - \lambda$ is dense. Since it is closed, $L_c - \lambda$ is surjective. By still using (15.6.2), if $\lambda \in \mathcal{U}$, the operator $(L_c - \lambda)^{-1}$ is well defined, and we get the uniform bound (15.7.3).

Now we establish (15.7.4). Indeed by (15.6.14), if $\lambda \in \mathbf{R}, \lambda \le -\lambda_0$, then

$$|\lambda + \lambda_0| \, \tau^{-2} \, |U|^2 \le C \left\langle \left(L'_{c,\tau} - \lambda\right) U, \tau^{-2} U \right\rangle . \tag{15.7.7}$$

By the obvious analogue of (15.4.8) and by (15.7.7), if $\lambda \in \mathbf{R}, \lambda \le -\lambda_0$, we get

$$|\lambda + \lambda_0| \, |U| \le C \left| \left(L_{c,\tau} - \lambda\right) U \right| . \tag{15.7.8}$$

By (15.7.8), we obtain

$$|\lambda + \lambda_0| \, \|u\|_0 \le C \left\| \left(L_c - \lambda\right) u \right\|_0 . \tag{15.7.9}$$

By (15.7.9), we get (15.7.4).

If $\lambda \notin \operatorname{Sp} L_c$,

$$\left(L_c - \lambda\right)^{-1} - \left(L_c + \lambda_0\right)^{-1} = \left(\lambda_0 + \lambda\right) \left(L_c + \lambda_0\right)^{-1} \left(L_c - \lambda\right)^{-1} . \tag{15.7.10}$$

Since the embedding of $\mathcal{H}^{1/4}$ into $\mathcal{H}$ is compact, by equation (15.5.3) in Theorem 15.5.1, $\left(L_c + \lambda_0\right)^{-1}$ is a compact operator. By (15.7.10), $\left(L_c - \lambda\right)^{-1}$ is also compact.

Recall that the self-adjoint operator S was defined in (15.2.2). It is easy to see that as $t \to 0$,

$$\operatorname{Tr}\left[e^{-tS}\right] = \frac{C}{t^{3n/2}} + o\left(t^{-3n/2}\right) . \tag{15.7.11}$$

By (15.7.11), we find that for $s \in \mathbf{R}, s > 3n$, $S^{-s/2}$ is trace class. In particular, for $s > 3n$, the embedding $H^s \to H$ is trace class. By Remark 15.3.2, $\mathcal{H}^s$ embeds continuously in H^s. It follows that for $s > 3n$, the embedding $\mathcal{H}^s \to H$ is trace class.

By Theorem 15.5.1, $\left(L_c + \lambda_0\right)^{-1}$ maps $\mathcal{H}^s$ into $\mathcal{H}^{s+1/4}$. Therefore, if $N \in \mathbf{N}, N > 12n$, $\left(L_c + \lambda_0\right)^{-N}$ is trace class.

If $\lambda \notin \operatorname{Sp} L_c, N \in \mathbf{N}$, we have the obvious extension of (15.7.10),

$$\left(L_c - \lambda\right)^{-N} = \left(L_c + \lambda_0\right)^{-N} \left[\sum_{j=1}^{N} C_N^j \left(\lambda + \lambda_0\right)^j \left(L_c - \lambda\right)^{-j} \right] . \tag{15.7.12}$$

By (15.7.12), we conclude that if $\lambda \notin \operatorname{Sp} L_c, N \in \mathbf{N}, N > 12n$, $\left(L_c - \lambda\right)^{-N}$ is trace class. Moreover, if $\lambda \in \mathcal{U}$, using the uniform bound (15.7.3) for $\left\| \left(L_c - \lambda\right)^{-1} \right\|$ and also (15.7.12), we get the estimate (15.7.5).

Take λ_0 as in Theorem 15.5.1. By (15.5.3),

$$\|u\|_{s+1/4} \le C_s \left\| \left(L_c - \lambda\right) u \right\|_s + |\lambda + \lambda_0| \, \|u\|_s . \tag{15.7.13}$$

Also since the $\mathcal{H}^s$ form a chain of Sobolev spaces, for any $s \ge 0, A > 0$,

$$\|u\|_s \le A^{-1/4} \|u\|_{s+1/4} + A^s \|u\|_0 . \tag{15.7.14}$$

By taking $A = 2^4 |\lambda + \lambda_0|^4$, we deduce from (15.7.13), (15.7.14) that for $s \ge 0$,

$$\|u\|_{s+1/4} \le C_s \left(\left\| \left(L_c - \lambda\right) u \right\|_s + |\lambda + \lambda_0|^{4s+1} \|u\|_0 \right) . \tag{15.7.15}$$

By (15.7.15), we find that for $s \geq 0$, if $\lambda \notin \operatorname{Sp} L_c$, $(L_c - \lambda)^{-1}$ maps $\mathcal{H}^s$ into $\mathcal{H}^{s+1/4}$. By duality and interpolation, this result extends to $s \in \mathbf{R}$. Using (15.3.7), we find that if $\lambda \notin \operatorname{Sp} L_c$, $(L_c - \lambda)^{-1}$ maps $\mathcal{S}^{\cdot}(T^*X, \pi^*F)$ into itself. Using (15.7.3) and (15.7.15), we get (15.7.6) for $s \geq 0$. The case of a general $s \in \mathbf{R}$ follows by duality and interpolation.

As we saw before, if $\lambda \notin \operatorname{Sp} L_c$, $(L_c - \lambda)^{-1}$ is compact. Therefore $\operatorname{Sp} L_c$ is discrete. If $\lambda \in \operatorname{Spec} L_c$, by Riesz theory, the characteristic subspace V_λ is finite dimensional. Indeed let $\delta_\lambda \in \mathbf{C}$ be a small circle of center λ such that λ is the only element of $\operatorname{Sp} L_c$ contained in the closed disk bounded by δ_λ. Then the circle δ_λ is a compact subset of the resolvent set. In particular for $\mu \in \delta_\lambda$, the operators $(L_c - \mu)^{-1} \in \operatorname{End} \mathcal{H}$ are uniformly bounded. Set

$$\mathfrak{P}_\lambda = \frac{1}{2i\pi} \int_{\delta_\lambda} (\mu - L_c)^{-1} \, d\mu, \tag{15.7.16}$$
$$\mathfrak{Q}_\lambda = 1 - \mathfrak{P}_\lambda.$$

By [Y68, Theorem VIII.8.1], $\mathfrak{P}_\lambda$ is a projector. Since the operators which appear in the integral are bounded and compact, the projector $\mathfrak{P}_\lambda$ is compact. Therefore, its image $E_{c,\lambda}$ is finite dimensional. By integrating by parts as many times as necessary and using the fact that the $(L_c - \mu)^{-1}$ map $\mathcal{H}^s$ into $\mathcal{H}^{s+1/4}$, we find that $E_{c,\lambda} \subset \mathcal{S}^{\cdot}(T^*X, \pi^*F)$. Finally, $\mathfrak{Q}_\lambda$ projects on a complementary vector space $F_{c,\lambda}$.

Clearly L_c acts on $E_{c,\lambda}$, and its only eigenvalue on this vector space is λ. Moreover, L_c also preserves $F_{c,\lambda}$, and λ does not lie in the spectrum of the restriction. By using Jordan's theory, we find that in fact $E_{c,\lambda}$ is just the characteristic subspace of L_c.

The proof of our theorem is completed. □

Chapter Sixteen

Harmonic oscillator and the J_0 function

The purpose of this chapter is to introduce the basic tools which are needed in the proof of the convergence of the resolvent of the operator $2\mathfrak{A}'^2_{\phi_b,\pm\mathcal{H}}$ to the resolvent of $\square^X/2$ as $b \to 0$.

Here we essentially consider the case where X is a flat torus, or even $\mathbf{R}^n$, and we give an explicit formula for the resolvent for $b=1$. In chapter 17, this will be used when studying the semiclassical symbol of the resolvent of $2\mathfrak{A}'^2_{\phi_b,\pm\mathcal{H}}$ and establishing its main properties.

This chapter is organized as follows. In section 16.1, we introduce the formalism of the bosonic annihilation and creation operators, and we construct the Bargman kernel, which passes from the classical form of the harmonic oscillator $\frac{1}{2}\left(-\Delta^V + |p|^2 - n\right)$ to its spectral decomposition, where the harmonic oscillator is just the bosonic number operator $\mathcal{N}$.

In section 16.2, we introduce an operator $B(\xi)$.

In section 16.3, we compute the spectrum of $B(i\xi)$, and we give a formula for the resolvent $(B(i\xi)-\lambda)^{-1}$. We introduce functions $J_k(y,\lambda)$, which give certain matrix coefficients of the resolvent.

In section 16.4, we give special attention to the function $J_0(y,\lambda)$, which is the matrix coefficient of the resolvent with respect to the ground state. The precise analysis of the function $J_0(y,\lambda)$ plays a critical role in our study in chapter 17 of the behavior of the resolvent of $2\mathfrak{A}'^2_{\phi_b,\pm\mathcal{H}}$ as $b \to 0$.

Finally, in section 16.5, we study instead the resolvent of $B(i\xi)+P$, where P is an orthogonal projection operator. The point of adding P is that $B(i\xi)+P$ is invertible for any ξ. The corresponding procedure will also play an important role in chapter 17.

16.1 FOCK SPACES AND THE BARGMAN TRANSFORM

Let V be a real finite dimensional Euclidean vector space of dimension n, and let V^* be its dual. Whenever necessary, we identify V and V^* by the scalar product.

Let $e_1,\ldots,e_n$ be an orthonormal basis of V, and let $e^1,\ldots,e^n$ be the corresponding dual basis of V^*.

Let $B(V^*)$ be the Heisenberg algebra associated to V^*. This algebra is generated by $1, a(U), a^*(V), U, V \in V^*$, with the commutation relations

$$a(U)\,a^*(V) - a^*(V)\,a(U) = \langle U, V\rangle . \tag{16.1.1}$$

In (16.1.1), $a(U), a^*(V)$ should be considered as even, so that we can write this identity in the form

$$[a(U), a^*(V)] = \langle U, V \rangle . \tag{16.1.2}$$

The associated bosonic number operator $\mathcal{N}$ is given by

$$\mathcal{N} = a^*(e^i)\, a(e^i) . \tag{16.1.3}$$

The Heisenberg representation of $B(V^*)$ on the L^2 space of V^* with respect to the Gaussian measure $e^{-|p|^2/2}dp/(2\pi)^{n/2}$ is given by

$$a(U) = \nabla_U, \quad a^*(U) = -\nabla_U + \langle V, p \rangle , \quad \mathcal{N} = -\Delta^V + \nabla_{\widehat{p}}. \tag{16.1.4}$$

Under the isometry of $L^2\left(e^{-|p|^2/2}dp/(2\pi)^{n/2}\right)$ to the standard L^2 space given by $f \to e^{-|p|^2/2} f\left(\sqrt{2}p\right)/\pi^{n/4}$, the operators in (16.1.2), (16.1.3) become

$$a(U) = \frac{1}{\sqrt{2}}\left(\nabla_U + \langle U, p \rangle\right), \quad a^*(V) = \frac{1}{\sqrt{2}}\left(-\nabla_V + \langle V, p \rangle\right), \tag{16.1.5}$$
$$\mathcal{N} = \frac{1}{2}\left(-\Delta^V + |p|^2 - n\right).$$

If $z \in V^* \otimes_{\mathbf{R}} \mathbf{C}$, let $|z|^2 \in \mathbf{C}$ denote the complexification of the square of the Euclidean norm, and let $\|z\|^2 \in \mathbf{R}_+$ be the square of the Hermitian norm of z. Let $\mathcal{H}_z$ be the Hilbert space of holomorphic functions of $z = q + ir \in V^* \otimes_{\mathbf{R}} \mathbf{C}$ which are square integrable with respect to $\exp\left(-\|z\|^2\right) dqdr$. Let $P(p)$ be the standard heat kernel associated to $e^{\Delta^V/2}$ on V^*,

$$P(p) = \frac{1}{(2\pi)^{n/2}} e^{-|p|^2/2}. \tag{16.1.6}$$

The map $u \in L^2\left(e^{-|p|^2/2}dp/(2\pi)^{n/2}\right) \to P * u(z)/\pi^{n/2} \in \mathcal{H}_z$ is obviously an isometry.

When identifying $L^2\left(e^{-|p|^2/2}dp/(2\pi)^{n/2}\right)$ to the standard L^2 as above, the corresponding isometry $L^2 \to \mathcal{H}_z$ is given by the action of the Bargman kernel B given by

$$u \to Bu(z) = \frac{1}{\pi^{3n/4}} \int_{V^*} \exp\left(-|p|^2/2 - |z|^2/2 + \sqrt{2}\langle p, z\rangle\right) u(p)\, dp. \tag{16.1.7}$$

The inverse B^{-1} is given by the formula

$$B^{-1} f(p) = \frac{1}{\left(2\sqrt{\pi}\right)^{n/2}} \int_{z=a+ib} [e^{|p|^2/2 - \sqrt{2}\langle z, p\rangle + |z|^2/2} f(z)\, db. \tag{16.1.8}$$

To verify (16.1.8), it is enough to take $f(z) = e^{\langle \alpha, z\rangle}$, in which case explicit computations lead easily to (16.1.8).

When acting on $\mathcal{H}_z$, under the above isomorphism, we now have

$$a(U) = \nabla_U, \quad a^*(V) = \langle V, z \rangle , \quad \mathcal{N} = z^i \frac{\partial}{\partial z^i}. \tag{16.1.9}$$

Note here that B maps $\frac{1}{\pi^{n/4}}e^{-|p|^2/2}$ into $\pi^{-n/2}$. This corresponds to the canonical identification of $\ker \mathcal{N}$ under the various representations of the Heisenberg algebra. Also note that $\mathcal{H}_z$ is usually called the Heisenberg space representation of the bosonic algebra.

Recall that the Hermite polynomials on $\mathbf{R}^n$ are defined by

$$e^{\langle \alpha,p\rangle - \frac{1}{2}|p|^2} = \sum_{k\in\mathbf{N}^n} \alpha^k H_k(p). \tag{16.1.10}$$

In $\mathcal{H}_z$, the expansion (16.1.10) corresponds to the expansion

$$e^{\langle \alpha,z\rangle} = \sum_{k\in\mathbf{N}^n} \alpha^k \frac{z^k}{k!}. \tag{16.1.11}$$

The $H_k(p)$ are mutually orthogonal in $L^2\left(e^{-|p|^2/2}\right)dp/(2\pi)^{n/2}$, and the $z^k/k!$ are mutually orthogonal in $\mathcal{H}_z$.

In the sequel, we will use the classical notation

$$: e^{\langle \alpha,p\rangle} := e^{\langle \alpha,p\rangle - \frac{1}{2}|\alpha|^2}. \tag{16.1.12}$$

The notation :: is for normal ordering. The notation reconciles the Gaussian and holomorphic representation of the Heisenberg algebra.

Proposition 16.1.1. *For $\xi \in \mathbf{R}^n$, in $L^2\left(e^{-|p|^2/2}dp/(2\pi)^{n/2}\right)$, we have the identity*

$$e^{(a(\xi)-a^*(\xi))} : e^{\langle \alpha,p\rangle} := e^{\langle \alpha,\xi\rangle - |\xi|^2/2} : e^{\langle \alpha-\xi,p\rangle} : . \tag{16.1.13}$$

Proof. By (16.1.2), $[a(\xi,), a^*(\xi)] = |\xi|^2$, and so

$$e^{(a(\xi)-a^*(\xi))} = e^{-|\xi|^2/2}e^{-a^*(\xi)}e^{a(\xi)}. \tag{16.1.14}$$

By (16.1.9), we get

$$e^{a(\xi)} : e^{\langle \alpha,p\rangle} := e^{\langle \alpha,\xi\rangle} : e^{\langle \alpha,p\rangle} :, \quad e^{-a^*(\xi)} : e^{\langle \alpha,p\rangle} :=: e^{\langle \alpha-\xi,p\rangle} : . \tag{16.1.15}$$

Then (16.1.13) follows from (16.1.14) and (16.1.15). □

16.2 THE OPERATOR $\mathbf{B}(\xi)$

Definition 16.2.1. Set

$$B(\xi) = \frac{1}{2}\left(-\Delta^V + |p|^2 - n\right) - \langle p,\xi\rangle. \tag{16.2.1}$$

Using the original representation of the operators $a, a^*, \mathcal{N}$ as in (16.1.5), which corresponds to original version of our operator, we can write $B(\xi)$ in the form

$$B(\xi) = \mathcal{N} - \frac{1}{\sqrt{2}}(a(\xi) + a^*(\xi)). \tag{16.2.2}$$

We rewrite (16.2.1) in the form

$$B(\xi) = \frac{1}{2}\left(-\Delta^V + |p-\xi|^2\right) - \frac{1}{2}|\xi|^2. \tag{16.2.3}$$

By (16.2.1), we get

$$e^{\nabla_\xi} B(\xi) e^{-\nabla_\xi} = \mathcal{N} - \frac{1}{2}|\xi|^2. \tag{16.2.4}$$

By (16.1.5), we can rewrite (16.2.4) as

$$e^{\frac{1}{\sqrt{2}}(a(\xi)-a^*(\xi))} B(\xi) e^{-\frac{1}{\sqrt{2}}(a(\xi)-a^*(\xi))} = \mathcal{N} - \frac{1}{2}|\xi|^2. \tag{16.2.5}$$

Proposition 16.2.2. *For $t > 0, \alpha, \xi \in V$,*

$$e^{-tB(\xi)} : e^{\langle \alpha, p\rangle} := \exp\left(\frac{t}{2}|\xi|^2 + (1-e^{-t})\left(\left\langle \alpha, \frac{\xi}{\sqrt{2}}\right\rangle - \frac{|\xi|^2}{2}\right)\right)$$
$$: \exp\left(\left\langle e^{-t}\left(\alpha - \frac{\xi}{\sqrt{2}}\right) + \frac{\xi}{\sqrt{2}}, p\right\rangle\right) :. \tag{16.2.6}$$

Proof. By (16.2.4),

$$e^{-tB(\xi)} = e^{t\frac{|\xi|^2}{2}} e^{-\frac{1}{\sqrt{2}}(a(\xi)-a^*(\xi))} e^{-t\mathcal{N}} e^{\frac{1}{\sqrt{2}}(a(\xi)-a^*(\xi))}. \tag{16.2.7}$$

Moreover, since $\mathcal{N}$ is the number operator, if $\beta \in \mathbf{R}^n$,

$$e^{-t\mathcal{N}} : e^{\langle \beta, p\rangle} :=: e^{\langle e^{-t}\beta, p\rangle} :. \tag{16.2.8}$$

Using (16.1.13), (16.2.7), and (16.2.8), we get (16.2.6). □

Scalar products are now taken in $L^2\left(e^{-|p|^2/2} dp/(2\pi)^{n/2}\right)$.

Proposition 16.2.3. *The following identity holds:*

$$\left\langle e^{-tB(\xi)} : e^{\langle \alpha, p\rangle} :, : e^{\langle \beta, p\rangle} : \right\rangle = \exp\left(\left(e^{-t} - 1 + t\right)\frac{|\xi|^2}{2}\right)$$
$$\exp\left(\left(1-e^{-t}\right)\left\langle \alpha + \beta, \frac{\xi}{\sqrt{2}}\right\rangle + e^{-t}\langle \alpha, \beta\rangle\right). \tag{16.2.9}$$

In particular,

$$\left\langle e^{-tB(\xi)}1, 1\right\rangle = \exp\left(\left(e^{-t} - 1 + t\right)\frac{|\xi|^2}{2}\right) \tag{16.2.10}$$

Proof. To get (16.2.9), we use Proposition 16.2.2 and the fact that

$$\left\langle : e^{\langle \alpha, p\rangle} :, : e^{\langle \beta, p\rangle} : \right\rangle = e^{\langle \alpha, \beta\rangle}. \tag{16.2.11}$$

By making $\alpha = 0, \beta = 0$, we get (16.2.10). □

Remark 16.2.4. The operator $B(\xi)$ depends linearly on $\xi \in V$. It still makes sense as an operator when $\xi \in V \otimes_{\mathbf{R}} \mathbf{C}$. By analyticity, we find that the identities in Propositions 16.2.2 and 16.2.3 extend to $\xi \in V \otimes_{\mathbf{R}} \mathbf{C}$. In the sequel we will use these identities when replacing $\xi \in \mathbf{C}$ by $i\xi, \xi \in V$. Note that the self-adjoint part of the operator $B(i\xi)$ is nonnegative. One then finds easily that if $\xi \neq 0$, the operator $B(i\xi)$ is invertible. This is reflected by the fact that as $t \to +\infty$, (16.2.6) and (16.2.9) converge to 0 at the rate $e^{-t|\xi|^2/2}$.

16.3 THE SPECTRUM OF B (iξ)

Note that if $\xi \in V$,

$$B(i\xi) = \frac{1}{2}\left(-\Delta^V + |p|^2 - n\right) - i\langle p, \xi\rangle. \tag{16.3.1}$$

Now we consider the holomorphic representation of the Heisenberg algebra, so that $\mathcal{N}$ is given by (16.1.9). In particular, if $\xi \in \mathbf{R}^n$,

$$B(i\xi) = z^i \frac{\partial}{\partial z^i} - \frac{i}{\sqrt{2}}\left(\langle \xi, z\rangle + \nabla_\xi\right). \tag{16.3.2}$$

Note that $B(i\xi)$ acts as an unbounded operator on $L^2(V^*)$. Also $B(i\xi)$ is a compact perturbation of the harmonic oscillator $B(0) = \mathcal{N}$. Therefore it has compact resolvent. In particular $B(i\xi)$ has discrete spectrum and finite dimensional characteristic subspaces. Let $\operatorname{Sp} B(i\xi)$ be the spectrum of $B(i\xi)$.

For $\mu \in \mathbf{C} \setminus (-\mathbf{N})$, let D_μ be the inverse of the operator $\mathcal{N} + \mu$ acting on holomorphic functions on $\mathbf{C}^n$. Then D_μ is the holomorphic extension in the variable $\mu \in \mathbf{C}$ of the operator defined for $\operatorname{Re}\mu > 0$ by the formula

$$D_\mu f(z) = \int_0^1 t^{\mu-1} f(tz)dt. \tag{16.3.3}$$

Equivalently, if t_+^μ is the holomorphic extension of the distribution $1_{t\geq 0}t^\mu$, we have

$$D_\mu(f)(z) = \int^1 (t_+)^{\mu-1} f(tz)dt. \tag{16.3.4}$$

Proposition 16.3.1. *Let $v : V \otimes_{\mathbf{R}} \mathbf{C} \to \mathbf{C}$ be a holomorphic function. If $\lambda \in \mathbf{C}, \lambda \notin \frac{|\xi|^2}{2} + \mathbf{N}$, the equation*

$$(B(i\xi) - \lambda)\, u = v \tag{16.3.5}$$

has a unique holomorphic solution u, which is given by the formula

$$u(z) = \int^1 t_+^{|\xi|^2/2-\lambda-1} v\left(t\left(z - \frac{i\xi}{\sqrt{2}}\right) + i\frac{\xi}{\sqrt{2}}\right) \exp\left((1-t)\left(\left\langle \frac{i\xi}{\sqrt{2}}, z\right\rangle + \frac{|\xi|^2}{2}\right)\right) dt. \tag{16.3.6}$$

Moreover,

$$\operatorname{Sp} B(i\xi) = \frac{|\xi|^2}{2} + \mathbf{N}. \tag{16.3.7}$$

Proof. By (16.3.2), we get

$$B(i\xi) = \left(z^i - \frac{i}{\sqrt{2}}\xi^i\right)\left(\frac{\partial}{\partial z^i} - \frac{i}{\sqrt{2}}\xi^i\right) + \frac{|\xi|^2}{2}. \tag{16.3.8}$$

By (16.3.8), we get

$$e^{i\nabla_\xi/\sqrt{2}}e^{-i\langle\xi/\sqrt{2},z\rangle}B(i\xi)\,e^{i\langle\xi/\sqrt{2},z\rangle}e^{-i\nabla_\xi/\sqrt{2}}=z^i\frac{\partial}{\partial z^i}+\frac{|\xi|^2}{2}. \tag{16.3.9}$$

Incidentally, note that in view of (16.1.14), (16.3.9) is just a form of (16.2.5). By (16.3.4), (16.3.9), we get (16.3.6).

Now we establish (16.3.7). Indeed if $\lambda\in\operatorname{Sp}B(i\xi)$, then $\ker(B(i\xi)-\lambda)$ is not reduced to 0. By the above it follows that $\lambda\in\frac{1}{2}|\xi|^2+\mathbf{N}$. Conversely by (16.2.4), we get

$$B(i\xi)=e^{-i\nabla_\xi}\left(\mathcal{N}+\frac{1}{2}|\xi|^2\right)e^{i\nabla_\xi}. \tag{16.3.10}$$

Moreover, the spectrum of the self-adjoint $\mathcal{N}+\frac{|\xi|^2}{2}$ is just $\frac{|\xi|^2}{2}+\mathbf{N}$, and the corresponding eigenvectors are the product of $\exp\left(-|p|^2/2\right)$ by the Hermite polynomials. The operator $e^{-i\nabla_\xi}$ acts naturally on such eigenvectors. We still obtain in this way an eigenvector of $B(i\xi)$. Therefore $\frac{|\xi|^2}{2}+\mathbf{N}$ is included in $\operatorname{Sp}B(i\xi)$. □

Remark 16.3.2. Note that for $\lambda\in\mathbf{C},\operatorname{Re}\lambda<0$,

$$(B(i\xi)-\lambda)^{-1}=\int_0^{+\infty}e^{t(\lambda-B(i\xi))}dt. \tag{16.3.11}$$

Using (16.2.6) and (16.3.11), we recover (16.3.6).

By (16.3.6), we find that if $\lambda\notin\frac{|\xi|^2}{2}+\mathbf{N}$,

$$(B(i\xi)-\lambda)^{-1}1=\int^1 t_+^{|\xi|^2/2-\lambda-1}\exp\left((1-t)\left(\left\langle\frac{i\xi}{\sqrt{2}},z\right\rangle+\frac{|\xi|^2}{2}\right)\right)dt. \tag{16.3.12}$$

Definition 16.3.3. If $k\in\mathbf{N},y\in\mathbf{R}$, set

$$J_k(y,\lambda)=\int^1(t_+)^{y^2-\lambda-1}e^{(1-t)y^2}(1-t)^k\,dt. \tag{16.3.13}$$

By expanding (16.3.12) in the variable z, we get

$$(B(i\xi)-\lambda)^{-1}1=\sum_{k=0}^{+\infty}\frac{1}{k!}J_k\left(|\xi|/\sqrt{2},\lambda\right)\left(i\left\langle\frac{\xi}{\sqrt{2}},z\right\rangle\right)^k. \tag{16.3.14}$$

Note that

$$J_0(y,\lambda)=\int^1(t_+)^{y^2-\lambda-1}e^{(1-t)y^2}dt. \tag{16.3.15}$$

For $\operatorname{Re}\lambda<y^2$, we can rewrite (16.3.15) in the form

$$J_0(y,\lambda)=\int_0^{+\infty}e^{\lambda t}e^{\left(1-t-e^{-t}\right)y^2}dt. \tag{16.3.16}$$

Writing the function $J_0(y,\lambda)$ in the form (16.3.16) is natural in view of (16.2.10) and (16.3.11).

16.4 THE FUNCTION $\mathbf{J_0}(\mathbf{y}, \lambda)$

Let $\delta = (\delta_0, \delta_1, \delta_2)$ with $\delta_0 \in \mathbf{R}, \delta_1 > 0, \delta_2 > 0$. Set

$$\mathcal{W}_\delta = \left\{\lambda \in \mathbf{C}, \operatorname{Re}\lambda \leq \delta_0 + \delta_1 \left|\operatorname{Im}\lambda\right|^{\delta_2}\right\}. \tag{16.4.1}$$

Set

$$u = y^2 - \lambda. \tag{16.4.2}$$

Theorem 16.4.1. *If $y \in \mathbf{R}$, the function $\lambda \in \mathbf{C} \to J_0(y, \lambda) \in \mathbf{C}$ is meromorphic, with simple poles at $\lambda \in y^2 + \mathbf{N}$. Moreover,*

$$J_0(y, \lambda) = \sum_{k \geq 0} \frac{y^{2k}}{u(u+1)...(u+k)}, \tag{16.4.3}$$

and the series in (16.4.3) converges uniformly on the compact subsets of the domain of definition of $J_0(y, \lambda)$.

There exist $\delta = (\delta_0, \delta_1, \delta_2)$ with $\delta_0 \in]0, 1[, \delta_1 > 0, \delta_2 = 1$ such that

- *If $k \in \mathbf{N}$, there exists $C_k > 0$ such that if $(y, \lambda) \in \mathbf{R} \times \mathcal{W}_\delta, |y| + |\lambda| \geq 1$, then*

$$\left|\partial_y^k J_0(y, \lambda)\right| \leq C_k \left(1 + |y| + |\lambda|\right)^{-1-k}. \tag{16.4.4}$$

- *If $(y, \lambda) \in \mathbf{R} \times \mathcal{W}_\delta$, then $J_0(y, \lambda) \neq 0, -1$, and there exists $C_0 > 0$ such that if $(y, \lambda) \in \mathbf{R} \times \mathcal{W}_\delta$, then*

$$\left|J_0(y, \lambda)\right| \geq C_0 \left(1 + |y| + |\lambda|\right)^{-1}. \tag{16.4.5}$$

- *There is $C > 0$ such that if $\lambda \in \mathcal{W}_\delta, \lambda \neq y^2$, then*

$$\left|J_0(y, \lambda) - \left(y^2 - \lambda\right)^{-1}\right| \leq C \frac{y^2}{|y^2 - \lambda| \left(1 + |y| + |\lambda|\right)}. \tag{16.4.6}$$

Proof. If $\operatorname{Re}\lambda < y^2$, then $\operatorname{Re} u > 0$. By (16.3.15), we get

$$J_0(y, \lambda) = \sum_{k=0}^{+\infty} \int_0^1 t^{u-1} (1-t)^k \, dt \frac{y^{2k}}{k!}. \tag{16.4.7}$$

If $a > 0, b > 0$,

$$\int_0^1 t^{a-1} (1-t)^{b-1} \, dt = \frac{\Gamma(a)\Gamma(b)}{\Gamma(a+b)}. \tag{16.4.8}$$

By (16.4.7), (16.4.8), we obtain

$$J_0(y, \lambda) = \sum_{k=0}^{+\infty} \frac{\Gamma(u)}{\Gamma(u+k+1)} y^{2k}. \tag{16.4.9}$$

Note that

$$\Gamma(u+1) = u\Gamma(u). \tag{16.4.10}$$

From (16.4.9), (16.4.10), we get (16.4.3). Since both sides of (16.4.3) are meromorphic in λ, the equality extends to its obvious domain of definition.

Set

$$\psi(t) = \log(t) + 1 - t. \tag{16.4.11}$$

Note that for $x \in]0,1]$,

$$\psi(1-x) \simeq -\frac{x^2}{2}, \qquad \psi(1-x) \leq -\frac{x^2}{2}. \tag{16.4.12}$$

To prove (16.4.4), first observe that if $(y,\lambda) \in \mathbf{R} \times \mathcal{W}_\delta$, if $0 < \delta_0 < 1$, the only possible pole of $J_0(y,\lambda)$ as a function of λ is $\lambda = y^2$. Therefore if $\delta_0 \in]0,1[$ is small enough, if $\lambda \in \mathcal{W}_\delta, |y| + |\lambda| \geq 1$, $J_0(y,\lambda)$ has no pole. To establish the estimate (16.4.4), we can as well take $C > 0$ as large as necessary, and assume that $|y| + |\lambda| \geq C$.

If $|y| + |\lambda| \geq C$ and $|\lambda|$ is small, then $y^2 \gg |\lambda|$, and so by (16.3.15), (16.4.11),

$$J_0(y,\lambda) = \int_0^1 e^{y^2\psi(t)} \frac{dt}{t^{\lambda+1}}. \tag{16.4.13}$$

Using (16.4.12), (16.4.13), we get (16.4.4) in the above range.

Take $\phi_0 \in]0,\pi/2[$. Consider the domain of $D \in \mathbf{C}$ which is limited by the half lines $\operatorname{Arg} z = \pm\left(\frac{\pi}{2} - \phi_0\right)$ and which contains the line $-\mathbf{R}_+$. In the sequel, we fix $\varepsilon \in]0,1/2]$ as small as necessary, and we also assume that $\lambda \in D$, with $|\lambda| \geq \varepsilon > 0$.

Observe that there exists $C > 0$ such that if $\lambda \in D, \operatorname{Re}\lambda \geq 0$, then

$$|\operatorname{Im}\lambda| \geq C|\lambda|. \tag{16.4.14}$$

We claim that there exists $C' > 0$ such that if $\lambda \in D, x \geq 0$,

$$|\lambda - x| \geq C'|\lambda|. \tag{16.4.15}$$

Indeed this is clear if $\operatorname{Re}\lambda \leq 0$, and for $\operatorname{Re}\lambda \geq 0$, this follows from (16.4.14).

By (16.4.15), we find that

$$|u(u+1)\dots(u+k)| \geq C'^{k+1}|\lambda|^{k+1}. \tag{16.4.16}$$

By (16.4.3) and (16.4.16), we see that if $c > 0$ is small enough and $y^2 \leq c|\lambda|$, then (16.4.4) holds. Observe that the various constants above remain uniform as long as ϕ_0 remains away from 0.

Now we fix $c > 0$ as above. We can choose ϕ_0 small enough so that if $\lambda \in D, y^2 \geq c|\lambda|$, then

$$\operatorname{Re}\lambda \leq y^2/2. \tag{16.4.17}$$

Note that if $|\lambda| + |y| \geq 1$ and $y^2 \geq c|\lambda|$, then $y^2/c + |y| \geq 1$, which implies that there is $\alpha > 0$ such that $|y| \geq \alpha$. By (16.4.17), if $\lambda \in D$,

$$y^2 - \operatorname{Re}\lambda \geq \frac{y^2}{2} \geq \frac{\alpha^2}{2}. \tag{16.4.18}$$

Put

$$\Phi(t) = \left(y^2 - \lambda\right)\log(t) + y^2(1-t). \tag{16.4.19}$$

Note that

$$\Phi(t) = y^2\psi(t) - \lambda \log(t). \tag{16.4.20}$$

For $\rho = \pm\frac{1}{2}$, set

$$\theta(s) = \rho(1-s). \tag{16.4.21}$$

Consider the complex path γ parametrized by $s \in [0,1]$,

$$t_s = se^{i\theta(s)}. \tag{16.4.22}$$

Then

$$J_0(y,\lambda) = \int_\gamma e^{\Phi(t)} \frac{dt}{t}. \tag{16.4.23}$$

Note that by (16.4.18), the integral in (16.4.23) is indeed convergent.

Over the path γ, we have the obvious

$$\begin{aligned}\operatorname{Re}\Phi(t) = \left(y^2 - \operatorname{Re}\lambda\right)\psi(s) + sy^2\left(1-\cos\left(\rho(1-s)\right)\right) \\ + (1-s)\left(\rho \operatorname{Im}\lambda + \operatorname{Re}\lambda\right).\end{aligned} \tag{16.4.24}$$

We claim that if ϕ_0 is small enough, there is $C > 0$ such that given $\lambda \in D$, we can choose $\rho = \pm\frac{1}{2}$ so that

$$\operatorname{Re}\Phi(t_s) \leq C\left(y^2\psi(s) - (1-s)|\lambda|\right). \tag{16.4.25}$$

Observe that

$$\psi(s) \leq -\frac{1}{2}(1-s)^2, \qquad (1-\cos(\rho(1-s))) \leq \frac{1}{2}\rho^2(1-s)^2. \tag{16.4.26}$$

By (16.4.18), (16.4.26), if $\rho = \pm\frac{1}{2}$,

$$\left(y^2 - \operatorname{Re}\lambda\right)\psi(s) + sy^2\left(1-\cos\left(\rho(1-s)\right)\right) \leq \frac{1}{4}y^2\psi(s). \tag{16.4.27}$$

Moreover, given $\eta > 0$, by taking ϕ_0 small enough, if $\lambda \in D, \operatorname{Re}\lambda \geq 0$, then

$$\operatorname{Re}\lambda \leq \eta\,|\operatorname{Im}\lambda|. \tag{16.4.28}$$

We take $\eta = \frac{1}{4}, \rho = -\frac{1}{2}\operatorname{sgn}(\operatorname{Im}\lambda)$. Using (16.4.14) and (16.4.28), we find that if $\lambda \in D$, if $\operatorname{Re}\lambda \geq 0$

$$\rho\operatorname{Im}\lambda + \operatorname{Re}\lambda \leq -\frac{1}{4}|\operatorname{Im}\lambda| \leq -C|\lambda|. \tag{16.4.29}$$

Of course, if $\operatorname{Re}\lambda \leq 0$, we still have the inequality

$$\rho\operatorname{Im}\lambda + \operatorname{Re}\lambda \leq -\frac{1}{2}|\lambda|. \tag{16.4.30}$$

By (16.4.24), (16.4.27), (16.4.29), (16.4.30), we get (16.4.25).

Since $y^2 \geq c|\lambda|$, using (16.4.25) and also the fact that $|\lambda| \geq \epsilon$, we find that the contribution of the $s \in [0,1]$ which are away from 1 in the integral in the right-hand side of (16.4.23) is compatible with (16.4.4).

Using (16.4.25), the first equation in (16.4.26), and the fact that if $s \in \mathbf{R}$, then $s^2 \geq s - 1$, we get

$$\left| \int_{1/2}^{1} e^{\Phi(t_s)} \frac{dt_s}{t_s} \right| \leq C \int_0^{+\infty} \exp\left(-C'\left(1 + |\lambda| / |y|\right) s\right) \frac{ds}{|y|}. \tag{16.4.31}$$

By (16.4.31) and again taking into account the fact that $|\lambda| \geq \epsilon$, we finally obtain (16.4.4) for $k = 0$. Higher derivatives in the variable y can be handled in the same way. Therefore we get (16.4.4) in full generality.

We claim that if $C > 0$ is large enough, if $|y| + |\lambda| \geq C$, then (16.4.5) holds. First if $|y| + |\lambda|$ is large and $|\lambda|$ is small, we use again the fact that $y^2 \gg |\lambda|$, so that by using (16.4.12) and (16.4.13), (16.4.5) holds in this range.

Then we consider the case where $\lambda \in D$. By (16.4.3),

$$J_0(y, \lambda) = \frac{1}{u} \sum_{k \geq 0} \frac{y^{2k}}{(u+1) \dots (u+k)}, \tag{16.4.32}$$

so that using a bound similar to (16.4.16), if $|y|^2 \leq c|\lambda|$, we get from (16.4.32),

$$|J_0(y, \lambda)| \geq \frac{C}{|u|}, \tag{16.4.33}$$

from which (16.4.5) still follows.

Now we may as well assume that $\lambda \in D, |\lambda| \geq \epsilon, y^2 \geq c|\lambda|$. By using again (16.4.25) and the first equation in (16.4.26), it is clear that in (16.4.23), the obvious bounds on the integral away from $s = 1$ are compatible with (16.4.5) with $|y| + |\lambda|$ large, as long as we show that the integral near $s = 1$ verifies the corresponding bound.

We will show that the inequality in (16.4.31) can be turned into an equivalence. The sum of the first two terms in (16.4.24) vanishes to order 2 at $s = 1$, while the third term is negative and controlled by (16.4.29). By making the change of variable $|y|(1 - s) = s'$ and using integration by parts, we get

$$|J_0(y, \lambda)| \geq \frac{C}{|y| + |\lambda|}, \tag{16.4.34}$$

which is exactly what we need.

Now we show that with the adequate choice of δ, if $(y, \lambda) \in \mathbf{R} \times \mathcal{W}_\delta$, then $J_0(y, \lambda) \neq 0, -1$. Using (16.4.4), (16.4.5), what is left to prove is that if $\phi_0 > 0$ is small enough, if $y \in \mathbf{R}, \lambda \in D, |y| + |\lambda| \leq C$, then $J_0(y, \lambda)$ is never equal to 0 or -1. Since the considered domain is compact, we may as well assume that $\operatorname{Re} \lambda \leq 0$.

Here we take $V = \mathbf{R}$. Set

$$f = \left(B(i\sqrt{2}y) - \lambda\right)^{-1} 1. \tag{16.4.35}$$

By (16.3.14),

$$\left\langle f, \left(B(i\sqrt{2}y) - \lambda\right) f \right\rangle = J_0(y, \lambda). \tag{16.4.36}$$

By (16.1.9), (16.3.2), (16.4.36), we get

$$\langle \mathcal{N} f, f\rangle - \operatorname{Re}\lambda \left|f\right|^2 = \operatorname{Re} J_0\left(y, \lambda\right). \tag{16.4.37}$$

By (16.4.37), if $\operatorname{Re}\lambda \leq 0$, we obtain

$$\operatorname{Re} J_0\left(y, \lambda\right) \geq 0, \tag{16.4.38}$$

the equality in (16.4.38) being possible only if $y = 0, \operatorname{Re}\lambda = 0$, in which case $J_0\left(y, \lambda\right) = -1/\lambda$. In any case, the values $0, -1$ are excluded. This concludes the proof of (16.4.5).

Now we establish (16.4.6). First note that if $y \in \mathbf{R}$, for $\lambda \in \mathcal{W}_\delta$, the only possible pole for the function $J_0\left(y, \lambda\right)$ is $\lambda = y^2$. Also note that by (16.4.3),

$$J_0\left(y, \lambda\right) = \frac{1}{y^2 - \lambda} + \frac{y^2}{y^2 - \lambda} J_0\left(y, \lambda - 1\right). \tag{16.4.39}$$

Moreover, if $\lambda \in \mathcal{W}_\delta, \lambda - 1 \in \mathcal{W}_\delta$ and $|\lambda - 1|$ has a positive lower bound on $\mathcal{W}_\delta$. If $y \in \mathbf{R}$, the poles of $J\left(y, \cdot - 1\right)$ are given by $y^2 + 1 + \mathbf{N}$, and this set does not intersect $\mathcal{W}_\delta$. By (16.4.4), if $y \in \mathbf{R}, \lambda \in \mathcal{W}_\delta$, then

$$\left|J_0\left(y, \lambda - 1\right)\right| \leq C\left(1 + |y| + |\lambda|\right)^{-1}. \tag{16.4.40}$$

By (16.4.39), (16.4.40), we get (16.4.6). The proof of our theorem is completed. □

Let $S \subset \mathbf{R}_+$ be a nonempty closed set. If $\lambda \in \mathbf{C}$, put

$$\begin{aligned} r_1\left(\lambda\right) &= d\left(\lambda, S\right), \\ \rho\left(\lambda\right) &= 1 + \frac{1}{|\lambda|} \text{ if } \operatorname{Re}\lambda \leq 0, \\ &= 1 + \frac{1}{|\lambda|} + \frac{1 + \operatorname{Re}\lambda}{r_1\left(\lambda\right) + \left|\operatorname{Im}\lambda\right|} \text{ if } \operatorname{Re}\lambda > 0. \end{aligned} \tag{16.4.41}$$

Observe that if $\lambda \in S$, then $\rho\left(\lambda\right) = +\infty$.

We take $\delta = \left(\delta_0, \delta_1, \delta_2\right)$ as in Theorem 16.4.1. If $h \in]0, 1], \lambda \in \mathcal{W}_\delta$, then $h^2\lambda \in \mathcal{W}_\delta$.

Given $\delta_2' \in]0, 1[$, by taking $\delta_0' \in]0, 1[, \delta_1' > 0$ small enough, we have the inclusion

$$\mathcal{W}_{\delta'} \subset \mathcal{W}_\delta. \tag{16.4.42}$$

It follows that if $h \in]0, 1], \lambda \in \mathcal{W}_{\delta'}$, then $h^2\lambda \in \mathcal{W}_\delta$.

Proposition 16.4.2. *Given $r > 0$, there exist $C > 0, C_r > 0$ such that if $h \in]0, 1]$, if $\lambda \in \mathbf{C}$ is such that $h^2\lambda \in \mathcal{W}_\delta$, and if*

$$r \operatorname{Re}\lambda + 1 \leq \left|\operatorname{Im}\lambda\right|, \tag{16.4.43}$$

then

$$\begin{aligned} \left|h^2 J_0\left(hy, h^2\lambda\right)\right| &\leq \frac{Ch + C_r\left(1 + |\lambda|\right)^{-1/2}}{1 + |\lambda|^{1/2} + |y|}, \\ \left|h^2 J_0\left(hy, h^2\lambda\right)\right| &\leq C\frac{h|y| + 1}{|y|^2 + 1}. \end{aligned} \tag{16.4.44}$$

There exist $C > 0, h_0 \in]0,1]$ *such that for* $h \in]0, h_0], y^2 \in S, h^2\lambda \in \mathcal{W}_\delta$, *then*

$$\left|h^2 J_0\left(hy, h^2\lambda\right)\right| \leq C\rho\left(\lambda\right)\frac{1+h^2\left|\lambda\right|+h\left|y\right|}{1+\left|\lambda\right|+y^2}. \tag{16.4.45}$$

Proof. By equation (16.4.4), if $h \in]0,1], h^2\lambda \in \mathcal{W}_\delta, h\left|y\right| + h^2\left|\lambda\right| \geq 1$,

$$\left|h^2 J_0\left(hy, h^2\lambda\right)\right| \leq C\left(1+\left|y\right|/h+\left|\lambda\right|\right)^{-1}. \tag{16.4.46}$$

Also,

$$\frac{1+\left|\lambda\right|^{1/2}+\left|y\right|}{1+\left|y\right|/h+\left|\lambda\right|} \leq C\left(h+\left(1+\left|\lambda\right|\right)^{-1/2}\right), \tag{16.4.47}$$

$$\frac{h}{\left|y\right|+h} \leq \frac{h\left|y\right|+1}{\left|y\right|^2+h\left|y\right|+1} \leq \frac{h\left|y\right|+1}{\left|y\right|^2+1}.$$

By (16.4.46), (16.4.47), we get (16.4.44) when $h\left|y\right| + h^2\left|\lambda\right| \geq 1$.

If $r = \frac{1+c_r}{1-c_r}, -1 < c_r < 1$, if $\lambda = a + ib, a > 0$ is such that (16.4.43) holds, if $y \in \mathbf{R}$,

$$\left|y^2-\lambda\right| \geq y^2 - a + \left|b\right| \geq y^2 + c_r\left(\left|a\right|+\left|b\right|\right)+1-c_r, \tag{16.4.48}$$

By (16.4.48), we get

$$\left|y^2-\lambda\right| \geq C_r\left(1+y^2+\left|\lambda\right|\right). \tag{16.4.49}$$

Moreover, there is $d_r > 0$ such that if $\lambda = a + ib$ verifies (16.4.43), then $\left|\lambda\right| \geq d_r$. In particular if $a < 0$, we still have

$$\left|y^2-\lambda\right| = y^2+\left|\lambda\right| \geq C_r\left(1+y^2+\left|\lambda\right|\right). \tag{16.4.50}$$

For $k \in \mathbf{N}^*$, replacing y^2 by $y^2 + k/h^2$ in (16.4.49), (16.4.50), if λ verifies (16.4.43), we get

$$\left|h^2\left(y^2-\lambda\right)+k\right| \geq C_r\left(h^2\left(1+y^2+\left|\lambda\right|\right)+k\right) \geq C_r k. \tag{16.4.51}$$

By (16.4.3), (16.4.49)-(16.4.51), we conclude that if $h \in]0,1], \lambda \in \mathbf{C}, \left|hy\right| + h^2\left|\lambda\right| \leq 1$, if (16.4.43) holds,

$$h^2\left|J_0\left(hy, h^2\lambda\right)\right| \leq C_r\left(1+\left|y\right|^2+\left|\lambda\right|\right)^{-1}. \tag{16.4.52}$$

Also observe that

$$\frac{1+\left|\lambda\right|^{1/2}+\left|y\right|}{1+\left|y\right|^2+\left|\lambda\right|} \leq C\left(1+\left|\lambda\right|\right)^{-1/2}. \tag{16.4.53}$$

By (16.4.52), (16.4.53), we still get (16.4.44).

Now we will establish (16.4.45). First assume that $h\left|y\right| + h^2\left|\lambda\right| \geq 1$. By (16.4.4) in Theorem 16.4.1, we get

$$\left|h^2 J_0\left(hy, h^2\lambda\right)\right| \leq \frac{Ch^2}{1+h\left|y\right|+h^2\left|\lambda\right|}. \tag{16.4.54}$$

My multiplying the numerator and the denominator in the right-hand side of (16.4.54) by $1/h^2 + \left|y\right|/h + \left|\lambda\right|$, we get an estimate like (16.4.45), in which

$\rho(\lambda)$ is replaced by 1. Since $\rho(\lambda) \geq 1$, we have established (16.4.45) in this case.

Assume now that $h|y| + h^2|\lambda| \leq 1$ and that $\lambda \in \mathcal{W}_{\delta'}$, so that $h^2\lambda \in \mathcal{W}_\delta$. First suppose that $\operatorname{Re}\lambda \leq 0$. Then for $k \in \mathbf{N}$,

$$\left|h^2\left(y^2-\lambda\right)+k\right| \geq k. \tag{16.4.55}$$

Using (16.4.3) and (16.4.55), we obtain

$$\left|h^2 J_0\left(hy, h^2\lambda\right)\right| \leq \frac{C}{y^2+|\lambda|} \leq C\frac{1+1/|\lambda|}{1+|\lambda|+y^2}, \tag{16.4.56}$$

which fits with (16.4.45).

Now assume that $\operatorname{Re}\lambda > 0$. First we will show that there exists $h_0 \in]0,1], C > 0$ such that if $h \in]0, h_0], \lambda \in \mathcal{W}_{\delta'}$,

$$\left|h^2 J_0\left(hy, h^2\lambda\right)\right| \leq \frac{C}{|y^2-\lambda|}. \tag{16.4.57}$$

Indeed for $k \in \mathbf{N}^*$, if $\lambda = a + ib, a > 0$,

$$\left|h^2\left(y^2-\lambda\right)+k\right| \geq k + h^2\left(y^2 - a + |b|\right). \tag{16.4.58}$$

If $|b| \geq a$, a lower bound for (16.4.58) is still k. If $|b| \leq a$, since $\lambda \in \mathcal{W}_{\delta'}$, and $\delta'_2 < 1$, we see that such λ vary in a compact set. From (16.4.58), we find that for $h \in]0,1]$ small enough, a lower bound for (16.4.58) is $k/2$. Therefore we get (16.4.57). So to establish (16.4.45) also in this case, what remains to prove is that

$$\frac{1+|\lambda|+y^2}{|y^2-\lambda|} \leq C\rho(\lambda). \tag{16.4.59}$$

First we consider the case where $y^2 \geq a$. Since $y^2 \in S$, by construction

$$y^2 - a + |b| \geq r_1(\lambda). \tag{16.4.60}$$

In this case, the left-hand side of (16.4.59) is just $\frac{y^2+a+|b|+1}{y^2-a+|b|}$. This is a decreasing function of y^2, so that its maximum on the considered domain of variations is at $y^2 = a$. The value at the maximum is just $\frac{2a+|b|+1}{|b|}$. If $|b| \geq r_1(\lambda)$, this is dominated by $1+2(2a+1)/(|b|+r_1(\lambda))$. This bound is compatible with (16.4.59). If $|b| < r_1(\lambda)$, by (16.4.60), the minimum value of y^2 is just $a + r_1(\lambda) - |b|$. The maximum value of the considered function is attained at the minimum value of y^2, and the value of the maximum is now $\frac{(2a+r_1(\lambda)+1)}{r_1(\lambda)}$. Again this is compatible with the bound (16.4.59).

Now suppose that $y^2 < a$. Since $y^2 \in S$, instead of (16.4.60), we now have

$$a - y^2 + |b| \geq r_1(\lambda). \tag{16.4.61}$$

The left-hand side of (16.4.59) is $\frac{y^2+a+|b|+1}{-y^2+a+|b|}$. This is an increasing function of y^2, so that the maximum value of this function on its domain of variation is still $\frac{2a+|b|+1}{|b|}$. If $|b| < r_1(\lambda)$, the domain of variation of y^2 has now the upper bound $a + |b| - r_1(\lambda)$. At this point the value of the considered function is given by $\frac{2a+2|b|-r_1(\lambda)+1}{r_1(\lambda)} \leq \frac{2a+r_1(\lambda)+1}{r_1(\lambda)}$. The proof continues as before. We have completed the proof of our proposition. □

Now we extend Theorem 16.4.1 to arbitrary J_k. We make the convention that if $k \in \mathbf{Z}, k < 0$, set $J_k = 0$.

Theorem 16.4.3. *For any $k \in \mathbf{N}$, if $y \in \mathbf{R}$, the function $\lambda \in \mathbf{C} \to J_k(y,\lambda) \in \mathbf{C}$ is meromorphic, with simple poles at $\lambda \in y^2 + \mathbf{N}$. Moreover,*

$$J_k(y,\lambda) = \sum_{k' \geq 0} \frac{(k'+1)\ldots(k'+k)}{u(u+1)\ldots(u+k'+k)} y^{2k'}, \tag{16.4.62}$$

and the series in (16.4.62) converges uniformly on the compact subsets of the domain of definition of $J_k(y,\lambda)$. Also for $k \in \mathbf{N}$,

$$k(J_k - J_{k-1}) + y^2 J_{k+1} - \lambda J_k = \delta_{k,0}. \tag{16.4.63}$$

There exists $\delta = (\delta_0, \delta_1, \delta_2)$ with $\delta_0 \in]0,1[, \delta_1 > 0, \delta_2 = 1$ such that for $k, k' \in \mathbf{N}$, there exists $C_{k,k'} > 0$ such that if $(y,\lambda) \in \mathbf{R} \times \mathcal{W}_\delta, |y| + |\lambda| \geq 1$, then

$$\left|\partial_y^k J_{k'}(y,\lambda)\right| \leq C_{k,k'}(1 + |y| + |\lambda|)^{-1-k-k'}. \tag{16.4.64}$$

Proof. To establish (16.4.62), we use equation (16.3.13) for J_k, and we proceed as in (16.4.7)-(16.4.10). Also we use (16.3.2) with $n = 1$ and the fact that by (16.3.14),

$$\left(B\left(i\sqrt{2}y\right) - \lambda\right) \sum_{k=0}^{+\infty} \frac{1}{k!} J_k(y,\lambda)(iyz)^k = 1, \tag{16.4.65}$$

and we get (16.4.63).

Using (16.3.13) and (16.4.62), and proceeding as in the proof of Theorem 16.4.1, we get (16.4.64). The proof of our theorem is completed. □

Now we assume that $V = \mathbf{R}^n$. If $\lambda \notin \frac{|\xi|^2}{2} + \mathbf{N}$, if $\beta \in \mathbf{N}^n$, set

$$(B(i\xi) - \lambda)^{-1} z^\beta = \sum_{\alpha \in \mathbf{N}^n} \psi_\alpha^\beta z^\alpha. \tag{16.4.66}$$

Comparing (16.3.14) and (16.4.66), we get

$$\sum_{\alpha \in \mathbf{N}^n} \psi_\alpha^0 z^\alpha = \sum_{k=0}^{+\infty} \frac{J_k}{k!} \left(|\xi|/\sqrt{2}, \lambda\right) \left(i\left\langle \frac{\xi}{\sqrt{2}}, z\right\rangle\right)^k. \tag{16.4.67}$$

By (16.4.67), we deduce in particular that

$$\psi_0^0 = J_0\left(|\xi|/\sqrt{2}, \lambda\right). \tag{16.4.68}$$

Proposition 16.4.4. *For any $\beta \in \mathbf{N}^n$,*

$$\sum_{\alpha \in \mathbf{N}^n} \psi_\alpha^\beta z^\alpha = \sum_{\substack{k \in \mathbf{N} \\ \beta_1, \beta_2 \in \mathbf{N}^n \\ \beta_1 + \beta_2 = \beta}} J_{k+|\beta_1|}\left(|\xi|/\sqrt{2}, \lambda\right)$$
$$\frac{(\beta_1 + \beta_2)!}{k!\beta_1!\beta_2!} \left\langle i\xi/\sqrt{2}, z\right\rangle^k \left(-z + i\xi/\sqrt{2}\right)^{\beta_1} z^{\beta_2}. \tag{16.4.69}$$

Proof. By equation (16.3.6) in Proposition 16.3.1, if $z' \in \mathbf{C}^n$,

$$(B(i\xi)-\lambda)^{-1} e^{\langle z',z\rangle} = \int^1 t_+^{|\xi|^2/2-\lambda-1} e^{(1-t)|\xi|^2/2}$$
$$\exp\left((1-t)\left(i\left\langle \xi/\sqrt{2}, z\right\rangle + \left\langle z', -z+i\xi/\sqrt{2}\right\rangle\right)\right) dt e^{\langle z',z\rangle}. \quad (16.4.70)$$

Moreover, by (16.4.66),

$$(B(i\xi)-\lambda)^{-1} e^{\langle z',z\rangle} = \sum_{\alpha,\beta\in\mathbf{N}^n} \frac{1}{\beta!}\psi_\alpha^\beta z^\alpha z'^\beta. \quad (16.4.71)$$

We can now expand the term $\exp(((1-t)\ldots)\ldots)$ in the second line in right-hand side of (16.4.70) using (16.3.13). Comparing with (16.4.71), we get (16.4.69). □

16.5 THE RESOLVENT OF $\mathbf{B}(\mathbf{i}\xi)+\mathbf{P}$

Let P be the orthogonal projection operator on $\ker \mathcal{N}$. Note that in L^2, $\ker\mathcal{N}$ is spanned by $e^{-|p|^2/2}$, and in $\mathcal{H}_z$ it is spanned by the function 1. Consider the equation

$$(B(i\xi)+P-\lambda)^{-1} z^\beta = \sum_{\alpha\in\mathbf{N}^n} a_\alpha^\beta z^\alpha. \quad (16.5.1)$$

By (16.4.66), (16.5.1), we get

$$a_\alpha^\beta = \psi_\alpha^\beta - \psi_\alpha^0 a_0^\beta. \quad (16.5.2)$$

In particular, by Theorem 16.4.1, by (16.4.68) and (16.5.2), if $\lambda \in \mathcal{W}_\delta$, then $\psi_0^0 = J_0\left(|\xi|/\sqrt{2},\lambda\right) \neq -1$, and moreover,

$$a_\alpha^\beta = \psi_\alpha^\beta - \frac{\psi_\alpha^0\psi_0^\beta}{1+\psi_0^0}. \quad (16.5.3)$$

A symbol $a(\xi,\lambda)$ of degree d is a smooth function of ξ with values in $\mathbf{C}$, which is holomorphic in the parameter $\lambda\in\mathcal{W}_\delta$, such that if $\gamma\in\mathbf{N}^n$, there exists $C_\gamma>0$ such that

$$\left|\partial_\xi^\gamma a(\xi,\lambda)\right| \le C_\gamma (1+|\lambda|+|\xi|)^{d-|\gamma|}. \quad (16.5.4)$$

We denote by $\mathbb{S}_\delta^d$ the corresponding class of symbols.

Proposition 16.5.1. *There exists $\delta=(\delta_0,\delta_1,\delta_2)$, with $\delta_0\in]0,1[, \delta_1>0, \delta_2=1$ such that if $\lambda\in\mathcal{W}_\delta$, the resolvent $(B(i\xi)+P-\lambda)^{-1}$ exists, and further if $\alpha,\beta\in\mathbf{N}^n$, $a_\alpha^\beta\in\mathbb{S}_\delta^{-1}$.*

Proof. We take δ as in Theorem 16.4.1. Since $B(i\xi)-\lambda$ has compact resolvent, $B(i\xi)+P$ also has compact resolvent. Assume that a nonzero $f\in\mathcal{H}_z$ is such that

$$(B(i\xi)+P-\lambda)f=0. \quad (16.5.5)$$

Assume first that $\lambda \in \mathcal{W}_\delta, \lambda \neq \frac{|\xi|^2}{2}$. By Proposition 16.3.1, $\lambda \notin \operatorname{Sp} B(i\xi)$, and so

$$f = -(B(i\xi) - \lambda)^{-1} Pf. \tag{16.5.6}$$

Since $f \neq 0$, then $Pf \neq 0$. By (16.3.14), (16.5.6), we obtain

$$Pf = -J_0\left(|\xi|/\sqrt{2}, \lambda\right) Pf. \tag{16.5.7}$$

By Theorem 16.4.1, $J_0\left(y, |\xi|/\sqrt{2}\right) \neq -1$, which contradicts (16.5.7).

If $\lambda = \frac{|\xi|^2}{2}, \lambda \in \mathcal{W}_\delta$, then $\lambda \leq \delta_0$. Moreover,

$$\operatorname{Re}\langle (B(i\xi) + P - \lambda) f, f\rangle \geq (1-\lambda)|f|^2 \geq (1-\delta_0)|f|^2, \tag{16.5.8}$$

which contradicts (16.5.5). We have thus proved that $\mathcal{W}_\delta$ is included in the resolvent set of $B(i\xi) + P$.

We use the estimates in equation (16.4.64) in Theorem 16.4.3, and also equation (16.4.69) to express the ψ_α^β as a finite linear combination of products of monomials in the components of ξ by the J_k. Ultimately we find that for $(y, \lambda) \in \mathbf{R} \times \mathcal{W}_\delta, |y| \geq 1$, the ψ_α^β verify the estimates in (16.5.4) with $d = -1$. Also recall that by Theorem 16.4.1, $J_0(y, \lambda)$ verifies the estimates in (16.4.4), and also that for $(y, \lambda) \in \mathbf{R} \times \mathcal{W}_\delta$, $J_0(y, \lambda) \neq -1$. Using (16.4.68) and (16.5.3), we conclude that $a_\alpha^\beta \in \mathbb{S}_\delta^{-1}$. The proof of our proposition is completed. □

Set

$$P^\perp = 1 - P. \tag{16.5.9}$$

We take δ as in Proposition 16.5.1.

Proposition 16.5.2. *If $\lambda \in \mathcal{W}_\delta$, then $P^\perp (B(i\xi) - \lambda) P^\perp$ is invertible.*

Proof. By Proposition 16.5.1, if $\lambda \in \mathcal{W}_\delta$, $B(i\xi) + P - \lambda$ is invertible. As we explain in section 17.1, the invertibility of $P^\perp (B(i\xi) - \lambda) P^\perp$ is equivalent to the invertibility of $P(B(i\xi) + P - \lambda)^{-1} P$. Also $\ker \mathcal{N}$ is 1-dimensional. By (16.5.1),

$$P(B(i\xi) + P - \lambda)^{-1} P = a_0^0. \tag{16.5.10}$$

By (16.4.68) and (16.5.3),

$$a_0^0 = \frac{J_0}{1 + J_0}\left(|\xi|/\sqrt{2}, \lambda\right). \tag{16.5.11}$$

Note that (16.5.11) is an equality of holomorphic functions, and so it is also valid at the poles of $J_0\left(|\xi|/\sqrt{2}, \lambda\right)$.

By Theorem 16.4.1 and by (16.5.11), we find that if $\lambda \in \mathcal{W}_\delta$, then $a_0^0 \neq 0$. The proof of our theorem is completed. □

Definition 16.5.3. If $\lambda \in \mathcal{W}_\delta$, set

$$\mathfrak{T}_{\xi,\lambda} = \frac{1}{2} P a(\xi) \left(P^\perp (B(i\xi) - \lambda) P^\perp\right)^{-1} a^*(\xi) P. \tag{16.5.12}$$

Then $\mathfrak{T}_{\xi,\lambda} \in \operatorname{End} \ker \mathcal{N}$. Since $\ker \mathcal{N}$ is 1-dimensional, $\mathfrak{T}_{\xi,\lambda} \in \mathbf{C}$.

By Theorem 16.4.1, we know that if $\lambda \in \mathcal{W}_\delta$, then $J_0(y,\lambda) \neq 0$. Note that if $h \in]0,1]$, then $h\mathcal{W}_\delta \subset \mathcal{W}_\delta$.

Proposition 16.5.4. *If $\lambda \in \mathcal{W}_\delta$,*

$$\mathfrak{T}_{\xi,\lambda} - \lambda = J_0^{-1}\left(|\xi|/\sqrt{2},\lambda\right), \qquad \mathfrak{T}_{\xi,\lambda} = \frac{|\xi|^2}{2}\frac{J_1}{J_0}\left(|\xi|/\sqrt{2},\lambda\right). \tag{16.5.13}$$

Also if $h \in]0,1]$, $\frac{1}{h^2}\mathfrak{T}_{h\xi,h^2\lambda}$ extends continuously at $h=0$. More precisely,

$$\frac{1}{h^2}\mathfrak{T}_{h\xi,h^2\lambda}|_{h=0} = \frac{1}{2}|\xi|^2. \tag{16.5.14}$$

Finally,

$$\frac{J_1}{J_0}(y,\lambda) \in \mathbb{S}_\delta^{-1}. \tag{16.5.15}$$

Proof. By Proposition 16.3.1, we find that if $\lambda \in \mathcal{W}_\delta, \lambda \neq |\xi|^2/2$, $B(i\xi) - \lambda$ is invertible. Moreover, by (16.3.14),

$$P(B(i\xi)-\lambda)^{-1}P = J_0\left(|\xi|/\sqrt{2},\lambda\right). \tag{16.5.16}$$

By what we just saw, (16.5.16) is invertible. By the argument we give after (17.1.5), $\mathfrak{T}_{\xi,\lambda} - \lambda$ is invertible, and moreover,

$$\mathfrak{T}_{\xi,\lambda} - \lambda = \left(P(B(i\xi)-\lambda)^{-1}P\right)^{-1}. \tag{16.5.17}$$

By (16.5.16), (16.5.17), we get the first equation in (16.5.13) when $\lambda \neq |\xi|^2/2$. Also both sides extend to $\lambda = |\xi|^2/2$ by continuity, so that the first equation in (16.5.13) still holds there. The second equation follows from (16.4.63) for $k=0$.

By (16.5.13),

$$\frac{1}{h^2}\mathfrak{T}_{h\xi,h^2\lambda} - \lambda = \left(h^2 J_0\right)^{-1}\left(h|\xi|/\sqrt{2},h^2\lambda\right). \tag{16.5.18}$$

Using (16.4.3) and (16.5.18), we get (16.5.14). We can derive (16.5.14) directly from (16.5.12). In fact we get

$$\frac{1}{h^2}\mathfrak{T}_{h\xi,h^2\lambda}|_{h=0} = \frac{1}{2}Pa(\xi)\left(P^\perp \mathcal{N} P^\perp\right)^{-1}a^*(\xi)P. \tag{16.5.19}$$

By (16.5.19), we get

$$\frac{1}{h^2}\mathfrak{T}_{h\xi,h^2\lambda}|_{h=0} = \frac{1}{2}Pa(\xi)a^*(\xi)P = \frac{1}{2}|\xi|^2. \tag{16.5.20}$$

Now we establish (16.5.15). Using equation (16.4.5) in Theorem 16.4.1 and Theorem 16.4.3, we find that for $|y|+|\lambda| \geq 1$, $\frac{J_1}{J_0}$ verifies the proper estimates. Moreover, $\lambda = y^2$ is the only possible pole of J_1 in $\mathbf{R} \times \mathcal{W}_\delta$ and this pole is simple. Finally, $\lambda = y^2$ is a simple pole of J_0, and by (16.4.3),

$$\mathrm{Res}_{\lambda=y^2} J_0(y,\lambda) = -e^{y^2}, \tag{16.5.21}$$

which does not vanish. Therefore $\frac{J_1}{J_0}(y,\lambda)$ is holomorphic in λ when $(y,\lambda) \in \mathbf{R} \times \mathcal{W}_\delta$. This concludes the proof of (16.5.15). The proof of our proposition is completed. □

By (16.5.13), we find that if $h \in]0,1], \lambda \in \mathcal{W}_\delta, \lambda \neq |\xi|^2/2$,

$$\left(\frac{1}{h^2}\mathfrak{T}_{h\xi,h^2\lambda} - \lambda\right)^{-1} = h^2 J_0\left(h|\xi|/\sqrt{2},h^2\lambda\right). \tag{16.5.22}$$

Chapter Seventeen

The limit of $\mathfrak{A}'^2_{\phi_b,\pm\mathcal{H}}$ as $b \to 0$

The purpose of this chapter is to study the asymptotics of the hypoelliptic Laplacian $L_c = 2\mathfrak{A}'^2_{\phi_b,\pm\mathcal{H}}$ as $b \to 0$. Our main result is that, as anticipated in [B05], it converges in the proper sense to the standard Laplacian $\square^X/2$. As in chapter 15, we only consider the case of one single fiber, the more general case of the hypoelliptic curvature of a family does not introduce any significant new difficulty.

Since this chapter is analytically quite involved, we will try to describe its organization in painstaking detail. The operator L_c is of order 1 in the horizontal directions, while $\square^X/2$ is of order 2. The crucial algebraic link between these two operators was given in [B05, Theorem 3.14], and stated as the second identity in Theorem 2.3.2. The underlying motivation for the computations in [B05] is the evaluation of the resolvent $(L_c - \lambda)^{-1}$ as a $(2,2)$ matrix with respect to the splitting $H = \ker \alpha_\pm \oplus \ker \alpha_\pm^\perp$. A formal algebraic formula for this resolvent is given in (17.2.12), which is based on a trivial computation on matrices. At least at a formal level, it is clear that when $b \to 0$, $(L_c - \lambda)^{-1} \to i_\pm \left(\square^X/2 - \lambda\right)^{-1} P_\pm$. The main point of the present chapter is to justify this formal argument, and also to provide the proper functional analytic framework so that the convergence takes place in suitable Sobolev-like spaces. We will describe the relevant formulas in more detail.

We will set here $b = h$, so as to underline that h is a semiclassical parameter. Put

$$\begin{aligned} P_h &= h^2 L_c, \\ \Theta_{h,\lambda} &= P_\pm^\perp \left(P_h - \lambda\right) P_\pm^\perp, \\ T_{h,\lambda} &= P_\pm \gamma_\pm P_\pm - P_\pm \left(\beta_\pm + h\gamma_\pm\right) \Theta_{h,\lambda}^{-1} \left(\beta_\pm + h\gamma_\pm\right) P_\pm. \end{aligned} \tag{17.0.1}$$

In (17.2.12), (17.21.2), using the fact that $\beta_\pm$ maps $\ker \alpha_\pm$ into $\ker \alpha_\pm^\perp$, we obtain the formal equality of operators acting on $\ker \alpha_\pm$,

$$P_\pm \left(L_c - \lambda\right)^{-1} i_\pm = \left(T_{h,h^2\lambda} - \lambda\right)^{-1}. \tag{17.0.2}$$

Formal considerations show that as $h \to 0$,

$$\Theta_{h,h^2\lambda} \to P_\pm^\perp \alpha_\pm P_\pm^\perp. \tag{17.0.3}$$

Let $\alpha_\pm^{-1}$ be the inverse of the restriction of $\alpha_\pm$ to $\ker \alpha_\pm^\perp$. By (17.0.1), (17.0.3), we find that as $h \to 0$,

$$T_{h,h^2\lambda} \to P_\pm \left(\gamma_\pm - \beta_\pm \alpha_\pm^{-1} \beta_\pm\right) P_\pm. \tag{17.0.4}$$

The crucial formula in [B05, Theorem 3.14], which is given in Theorem 2.3.2, asserts precisely that

$$P_{\pm}\left(\gamma_{\pm} - \beta_{\pm}\alpha_{\pm}^{-1}\beta_{\pm}\right)P_{\pm} = \frac{\square^X}{2}. \tag{17.0.5}$$

Equations (17.0.3)-(17.0.5) provide one of the main arguments in favor of the fact that as $h \to 0$, $(L_c - \lambda)^{-1} \to i_{\pm}\left(\square^X/2 - \lambda\right)^{-1} P_{\pm}$.

As $h \to 0$, P_h behaves like a semiclassical operator in the x variable, since differentiation in x is multiplied by h. It is then natural to use a semiclassical pseudodifferential calculus to handle the convergence in (17.0.3).

For $h > 0$ small enough, we will show in Theorem 17.17.4 that $T_{h,h^2\lambda}$ is a classical pseudodifferential operator of order 1 acting on $\Omega^{\cdot}(X,F)$ or on $\Omega^{\cdot}(X, F \otimes o(TX))$. The convergence of operators will be analyzed as a convergence of a family of pseudodifferential operators of order 1 to a differential operator of order 2. Of course, these are standard pseudodifferential operators. These simple considerations indicate that while the convergence (17.0.3) involves semiclassical pseudodifferential operators, the convergence (17.0.4) will be obtained via classical pseudodifferential calculus.

From the above we find that we should combine at the same time a semiclassical pseudodifferential calculus with an ordinary pseudodifferential calculus. Corresponding to these two calculi, there will be two kinds of norms, semiclassical norms to handle the convergence in (17.0.3) and classical ones to handle the convergence in (17.0.4). The issue gets even more involved when one has to get a precise view of the behavior of the heat kernel of L_c as $t \to 0$, which is uniform as $h \to 0$. One of the important points which is established at the end of this chapter is precisely such a uniformity result. Incidentally, observe that since no wave equation is associated to our operators, finite propagation speed methods cannot be used to establish localization of the heat kernel, not to speak of uniform localization.

It is probably easier to describe briefly the tormented convergence which takes place from a dynamical perspective. Indeed as explained in detail in [B06], the stochastic process $(x_{\cdot}, p_{\cdot})$ which is associated to the scalar part of L_c is a Langevin process, such that $\dot{x} = p/h$. In particular $x_{\cdot}$ has C^1 trajectories. When $h \to 0$, the component x converges to a standard Brownian motion on X, so that in distribution sense, $\dot{x}$ converges to the time derivative of Brownian motion, whose trajectories are nowhere differentiable. These convergences were handled in a related context by Stroock and Varadhan [StV72]. What is being done here is the functional analytic counterpart to the convergence of the dynamics.

Let us now describe in more detail the semiclassical aspects of the analysis. Studying the operator $\Theta_{h,\lambda}$ is made easier by replacing the operator P_h by the operator P_h^0, given by

$$P_h^0 = P_h + P_{\pm}. \tag{17.0.6}$$

While $\Theta_{h,\lambda}$ is unchanged by this transformation, $\alpha_{\pm}$ is replaced by $\alpha_{\pm} + P_{\pm}$, which is invertible. Let $S_{h,\lambda}$ be the resolvent for P_h^0, i.e.,

$$S_{h,\lambda} = \left(P_h^0 - \lambda\right)^{-1}. \tag{17.0.7}$$

A simple formula given in Theorem 17.16.2 expresses $\Theta_{h,\lambda}^{-1}$ in terms of $S_{h,\lambda}$. The convergence of the resolvent $(L_c - \lambda)^{-1}$ is obtained in particular by studying the behavior of $S_{h,\lambda}$ as $h \to 0$. As should be clear from the previous considerations, P_h^0 is a differential operator which is semiclassical in the x variable. However, because of lack of uniformity as $|p| \to +\infty$, for a given $h > 0$, $S_{h,\lambda}$ does not lie in a proper algebra of operators. Actually we deal with classes of operators which are semiclassical pseudodifferential operators in the x variable and ordinary operators in the p variable. These classes are described in detail in section 17.8. In fact if (x, ξ) are the canonical variables for the ordinary pseudodifferential calculus on X, in our classes of operators, there are increasing powers of $< p >$ which appear when considering the x, ξ differentials of the corresponding symbols, so that ultimately, even though our operators can be composed, their composition does not lie in the class. We are even led to consider classes of operators which are not invariant under change of coordinates to better describe the resolvent $S_{h,\lambda}$.

The proper use of pseudodifferential calculus allows us to prove in Theorem 17.10.1 that in the proper sense, for h small enough, $S_{h,\lambda}$ is an operator of order $-2/3$, which improves on the hypoelliptic estimates which were obtained in Theorems 15.5.1 and 15.6.1, where the gain of regularity was only $1/4$. Moreover, as explained before, we show in Theorem 17.15.3 that $P_{\pm} S_{h,\lambda} i_{\pm}$ is a semiclassical pseudodifferential operator of order -1 on X. These refined estimates are needed in the proof of the proper convergence of the resolvent $(L_c - \lambda)^{-1}$as $h \to 0$.

Now we describe the organization of this chapter in more detail.

Section 17.1 is devoted to elementary computations on $(2,2)$ matrices.

In section 17.2, we apply formally these computations to the evaluation of $(L_c - \lambda)^{-1}$ as a $(2,2)$ matrix.

Sections 17.3-17.15 are devoted to the semiclassical analysis of the operator L_c.

In section 17.3, we introduce the semiclassical Poisson bracket on smooth functions on the total space of $\mathbf{P}(T^*X \oplus \mathbf{R})$, in which only the base coordinate is rescaled by the factor h. When $h \to 0$, this semiclassical Poisson bracket converges to the fiberwise Poisson bracket along the fibers T^*X.

In section 17.4, we introduce semiclassical Sobolev norms. Their construction is adapted from section 15.3, where the case of a fixed h was considered.

In section 17.5, uniform hypoelliptic estimates on the operator P_h are established with respect to the semiclassical Sobolev norms, which extend the corresponding estimates in section 15.4. To establish these estimates, the proper algebra of semiclassical pseudodifferential operators on $\mathbf{P}(T^*X \oplus \mathbf{R})$ are introduced.

In section 17.6, corresponding estimates are established for P_h^0, and the associated resolvent $S_{h,\lambda}$ is considered for $\lambda \in \mathbf{R}$. The resolvent is shown to be an operator of order $-1/4$.

In section 17.7, the resolvent is extended to $\lambda \in \mathbf{C}$. In particular it is shown that a domain to the left of a curve with a cusp is included in the resolvent set.

In sections 17.8-17.11, we develop pseudodifferential operator techniques to improve on these estimates.

In section 17.8, classes of semiclassical symbols $\mathcal{S}^{d,k}_{\rho,\delta,c}$ on X with values in operators along the fiber T^*X are introduced. Let $\mathcal{P}^{d,k}_{\rho,\delta,c}$ be the corresponding classes of operators. The (d,ρ,δ) refer to the corresponding parameters in the classical Hörmander's classes $S^d_{\rho,\delta}$. The parameters $k, c = (c_0, c_1)$ refer to the growth as $|p| \to +\infty$ of these symbols.

In section 17.9, when $\lambda \in \mathbf{R}$, with $\operatorname{Re}\lambda$ bounded above, the full semiclassical symbol $Q^0_h(x,\xi) - \lambda$ of $P_h - \lambda$ is considered, as well as its inverse $e_{0,h,\lambda}(x,\xi)$. It is shown to lie in one of the above classes, with $d = -2/3$.

In section 17.10, under the same conditions on λ, a parametrix is obtained for $S_{h,\lambda}$, whose principal symbol is shown to be $e_{0,h,\lambda}(x,\xi)$. In particular, in a given coordinate chart, we show that it lies in one of the above classes with $d = -2/3$, which improves on the above $-1/4$.

In section 17.11, we show that the parametrix is local over X in the proper sense as $h \to 0$.

In sections 17.12 and 17.13, we show that $P_\pm S_{h,\lambda}$ and $S_{h,\lambda}P_\pm$ lie in a better class of operators, whose order is $-5/6$.

In section 17.14, the above results are extended to the set of $\lambda \in \mathbf{C}$ to the left of a cusp-shaped curve.

In section 17.15, we prove that $P_\pm S_{h,\lambda} i_\pm$ is a semiclassical elliptic pseudodifferential operator over X of order -1, whose principal symbol can be easily expressed in terms of the function $J_0(y,\lambda)$, whose properties were studied in chapter 16.

In section 17.16, we study the analytic properties of the operator $\Theta_{h,\lambda}$ introduced in (17.0.1) by expressing it in terms of $S_{h,\lambda}$.

In section 17.17, using the results which were obtained on $\Theta_{h,\lambda}$, we study the properties of the operator $T_{h,\lambda}$ defined in (17.0.7). In particular we obtain in Theorem 17.17.4 a key formula for $T_{h,\lambda}$ which shows that $T_{h,\lambda}$ is a pseudodifferential operator on X of order -1. The principal symbol of $T_{h,\lambda}$ is expressed in terms of the functions J_0, J_1, J_2 of chapter 16.

Sections 17.18-17.20 are devoted to the study of the asymptotics of $T_{h,h^2\lambda}$ as $h \to 0$. Part of the difficulty lies in the fact that we have to describe precisely in what sense this family of operators of order 1 converges to the operator $\square^X/2$ which is of order 2.

In section 17.18, we introduce the operator $(J_1/J_0)\left(hD^X/\sqrt{2},\lambda\right)$, which we will use to approximate $T_{h,h^2\lambda}$.

In section 17.19 we express $T_{h,h^2\lambda} - \lambda$ in terms of an operator $U_{h,h^2\lambda}$, whose asymptotics as $h \to 0$ is studied in detail. The refined properties of the function J_0, J_1, J_2 which were established in chapter 16 play a key role in our estimates.

In section 17.20, the asymptotics as $h \to 0$ of the operator $\left(T_{h,h^2\lambda} - \lambda\right)^{-1}$ is obtained.

In section 17.21, we obtain uniform estimates on the resolvent $(L_c - \lambda)^{-1}$

as $h \to 0$, using at the same time ordinary and semiclassical norms, and we study its convergence as $h \to 0$ at the level of the corresponding kernels.

Finally, in section 17.22, we obtain corresponding results on the resolvent $\left(\epsilon^2 L_c - \lambda\right)^{-1}$ which are uniform as $\epsilon \in]0, 1]$.

We use the same notation as in chapter 15. In particular H still denotes the Hilbert space of square integrable sections of $\pi^* \left(\Lambda^{\cdot} (T^*X) \widehat{\otimes} \Lambda^{\cdot} (TX) \widehat{\otimes} \pi^* F\right)$ on T^*X, and $| \, |$ is the corresponding L^2 norm on H.

17.1 PRELIMINARIES IN LINEAR ALGEBRA

Let $E = E_0 \oplus E_1$ be a $\mathbf{Z}_2$-graded vector space. Let $u \in \operatorname{End}(E)$. We write u in matrix form with respect to the splitting of E as

$$u = \begin{bmatrix} A & B \\ C & D \end{bmatrix}. \tag{17.1.1}$$

Assume that u is invertible. We will give a matrix expression for the inverse u^{-1} of u under the assumption that D is invertible. When writing this matrix expression, we will assume implicitly that other matrix expressions are invertible as well. These implicit assumptions will be obvious in the formula anyway.

Set

$$H = A - BD^{-1}C. \tag{17.1.2}$$

We have the following easy formula:

$$u^{-1} = \begin{bmatrix} H^{-1} & -H^{-1}BD^{-1} \\ -D^{-1}CH^{-1} & D^{-1} + D^{-1}CH^{-1}BD^{-1} \end{bmatrix}. \tag{17.1.3}$$

Let $P, P^{\perp}$ be the projectors on E_0, E_1. We extend operators acting on E^0 or E^1 by the 0 operator on E^1 or E^0. From (17.1.3), we get

$$D^{-1} = u^{-1} - u^{-1}\left(Pu^{-1}P\right)^{-1} u^{-1}. \tag{17.1.4}$$

In fact if D is invertible, the invertibility of u is equivalent to the invertibility of $A - BD^{-1}C$. In the finite dimensional case, this is obvious by the formula

$$\det u = \det(D) \det\left(A - BD^{-1}C\right). \tag{17.1.5}$$

If u is invertible, the invertibility of D is equivalent to the invertibility $Pu^{-1}P$.

17.2 A MATRIX EXPRESSION FOR THE RESOLVENT

Recall that $b \in \mathbf{R}_+^*$ and that $c = \pm 1/b^2$. We will consider again the operator $L_c = 2\mathfrak{A}'^2_{\phi_b, \pm\mathcal{H}}$ as in equation (15.1.2). The operators $\alpha_\pm, \beta_\pm, \gamma_\pm$ were defined in (2.3.12) and in (15.1.3). By (15.1.4),

$$L_c = \frac{\alpha_\pm}{b^2} + \frac{\beta_\pm}{b} + \gamma_\pm. \tag{17.2.1}$$

Here we take $b_0 \in]0,1]$ small enough, and we assume that $b \in]0,b_0]$. In the sequel, we will set

$$h = b. \tag{17.2.2}$$

This is because h is a standard notation for a semiclassical parameter.

Let P_h be the operator

$$P_h = \alpha_\pm + h\beta_\pm + h^2\gamma_\pm, \tag{17.2.3}$$

so that

$$L_c = \frac{P_h}{h^2}. \tag{17.2.4}$$

Note that P_h is different from P_c in (15.4.5).

Let $\ker \alpha_\pm^\perp$ be the orthogonal subspace to $\ker \alpha_\pm$ in H with respect to the standard Hermitian product of H, so that with the notation in (2.3.14),

$$\ker\, \alpha_\pm^\perp = \operatorname{Im} \alpha_\pm. \tag{17.2.5}$$

Then we have the splitting

$$H = \ker \alpha_\pm \oplus \ker \alpha_\pm^\perp, \tag{17.2.6}$$

which is just the one in (2.3.14).

As was observed in section 2.3, $\beta_\pm$ maps $\ker \alpha_\pm$ into $\ker \alpha_\pm^\perp$. In the sequel, we will write the considered operators in matrix form with respect to the above splitting. So we get

$$\alpha_\pm = \begin{pmatrix} 0 & 0 \\ 0 & \alpha \end{pmatrix}, \qquad \beta_\pm = \begin{pmatrix} 0 & \beta_2 \\ \beta_3 & \beta_4 \end{pmatrix}, \qquad \gamma_\pm = \begin{pmatrix} \gamma_1 & \gamma_2 \\ \gamma_3 & \gamma_4 \end{pmatrix} \tag{17.2.7}$$

Set

$$\begin{aligned} L_1 &= \gamma_1, & L_2 &= \beta_2 + h\gamma_2, \\ L_3 &= \beta_3 + h\gamma_3, & L_4 &= \alpha + h\beta_4 + h^2\gamma_4. \end{aligned} \tag{17.2.8}$$

By (17.2.1)-(17.2.8), we obtain

$$L_c = \begin{pmatrix} L_1 & \frac{L_2}{h} \\ \frac{L_3}{h} & \frac{L_4}{h^2} \end{pmatrix}. \tag{17.2.9}$$

Let $\lambda \in \mathbf{C}$ be such that $L_c - \lambda$ is invertible. Set

$$H = L_1 - \lambda - L_2 \left(L_4 - h^2\lambda\right)^{-1} L_3. \tag{17.2.10}$$

There is no risk of confusion between the operator H and the Hilbert space H. Put

$$D_4 = L_4 - h^2\lambda. \tag{17.2.11}$$

By (17.1.3), at least formally, we can write $(L_c - \lambda)^{-1}$ in matrix form as

$$(L_c - \lambda)^{-1} = \begin{bmatrix} H^{-1} & -hH^{-1}L_2D_4^{-1} \\ -hD_4^{-1}L_3H^{-1} & h^2D_4^{-1} + h^2D_4^{-1}L_3H^{-1}L_2D_4^{-1} \end{bmatrix}. \tag{17.2.12}$$

The remainder of this chapter is devoted to the analysis of equation (17.2.12).

As we already saw in section 2.3 after equation 2.3.13, the fiberwise kernel of the operator $\alpha_\pm$ restricted to fiberwise forms is 1-dimensional. If $F = \mathbf{R}$, the kernel of α_+ is spanned by the Gaussian $\exp\left(-|p|^2/2\right)$ and the kernel of α_- restricted to fiberwise forms is spanned by $\exp\left(-|p|^2\right)\eta$, where η is a fiberwise volume form.

Let $P_\pm$ be the fiberwise orthogonal projector from H on $\ker \alpha_\pm$. Set

$$P_\pm^\perp = 1 - P_\pm. \tag{17.2.13}$$

If $u \in \Lambda^\cdot(T^*X)\widehat{\otimes}\Lambda^\cdot(TX)\widehat{\otimes}F$, let Q_+u (resp. Q_-u) be the orthogonal projection of u on $\Lambda^\cdot(T^*X)\widehat{\otimes}F$ (resp. on $\Lambda^\cdot(T^*X)\widehat{\otimes}\Lambda^n(TX)\widehat{\otimes}F$). Then one has the obvious formula

$$P_\pm u = \pi^{-n/2}e^{-|p|^2/2}\int_{T^*X} e^{-|q|^2/2}Q_\pm u(q)\,dv_{T^*X}(q). \tag{17.2.14}$$

17.3 THE SEMICLASSICAL POISSON BRACKET

We use the notation of section 15.3. Recall that Y is the total space of $\mathbf{P}(T^*X \oplus \mathbf{R})$. Let $i : T^VY \to TY$ be the vector subbundle TY which consists of tangent vectors to the fibers $\mathbf{P}(T^*X \oplus \mathbf{R})$, and let $T^{V*}Y$ be its dual. We have the obvious exact sequences which are dual to each other,

$$\begin{aligned} &0 \to T^VY \overset{i}{\to} TY \to \pi^*TX \to 0, \\ &0 \to \pi^*T^*X \to T^*Y \overset{i^*}{\to} T^{V*}Y \to 0. \end{aligned} \tag{17.3.1}$$

Given $h > 0$, we denote by $T^*Y_{\mathrm{sc}} \subset T^*Y \oplus T^{V*}Y$ the graph of the morphism i^*/h. Then T^*Y_{sc} is a vector bundle on $Y\times]0,1]$. The Grassmann graph construction asserts that T^*Y_{sc} extends to a smooth vector bundle over $Y \times [0,1]$. In particular,

$$T^*Y_{\mathrm{sc}}|_{Y\times\{0\}} = \pi^*T^*X \oplus T^{V*}Y. \tag{17.3.2}$$

Denote by $T^HY \subset TY$ a horizontal vector bundle on Y, so that $TY = T^HY \oplus T^VY$. The corresponding dual splitting is $T^*Y = \pi^*T^*X \oplus T^{V*}Y$. A smooth trivialization of T^*Y_{sc} is given by

$$(\alpha,\beta) \in \pi^*T^*X \oplus T^{V*}Y \to (\pi^*\alpha + h\beta, i^*\beta) \in T^*Y_{\mathrm{sc}}. \tag{17.3.3}$$

Also observe that over $Y\times]0,1]$, the map $\gamma \in T^*Y \to (h\gamma, i^*\gamma) \in T^*Y_{\mathrm{sc}}$ identifies the two vector bundles.

We have obvious morphisms

$$T^{V*}Y \overset{k}{\longleftarrow} T^*Y_{\mathrm{sc}} \overset{j}{\longrightarrow} T^*Y. \tag{17.3.4}$$

Then over $]0,1]$, j identifies T^*Y_{sc} and T^*Y.

Recall that T^*Y is a symplectic manifold. Let ω^{T^*Y} be the corresponding symplectic form. The symplectic form ω^{T^*Y} on T^*Y pulls back to a closed

2-form $j^*\omega^{T^*Y}$ on the manifold T^*Y_{sc}. The restriction of the form $j^*\omega^{T^*Y}$ to the h-fibers for $h \in]0,1]$ is a symplectic form.

Let $\{\}^{T^*Y} \in \Lambda^2(TT^*Y)$ be the Poisson bracket on T^*Y, i.e., the 2-form on T^*T^*Y which is dual to the symplectic form ω^{T^*Y}. For $h \in]0,1]$, set

$$\{\}^{T^*Y_{\mathrm{sc}}} = h^2 j^* \{\}^{T^*Y}. \tag{17.3.5}$$

One verifies easily that $\{\}^{T^*Y_{\mathrm{sc}}}$ extends smoothly at $h = 0$.

Let $\{\}^{T^{V*}Y} \in \Lambda^2\left(T^V T^{V*}Y\right)$ be the Poisson bracket along the fibers of $T^{V*}Y$. By (17.3.2), $\{\}^{T^{V*}Y}$ can be considered a Poisson bracket on $T^*Y_{\mathrm{sc}}|_{h=0}$. Then one verifies easily that

$$\{\}^{T^*Y_{\mathrm{sc}}}|_{h=0} = \{\}^{T^{V*}Y}. \tag{17.3.6}$$

Let $f, g : T^*Y_{\mathrm{sc}} \to \mathbf{R}$ be two smooth functions. Note here that these two functions also depend implicitly on h. Set

$$\{f,g\}_{\mathrm{sc}} = \{df, dg\}^{T^*Y_{\mathrm{sc}}}. \tag{17.3.7}$$

Equation (17.3.7) defines the semiclassical Poisson bracket. By (17.3.6), (17.3.7), we get

$$\{f,g\}_{\mathrm{sc}|h=0} = \{f,g\}^{T^{V*}Y}_{|h=0}. \tag{17.3.8}$$

17.4 THE SEMICLASSICAL SOBOLEV SPACES

We use the Littlewood-Paley decomposition of elements of $\mathcal{S}^{\cdot}(T^*X, \pi^*F)$ similar to the one in section 15.2. As in (15.2.16), (15.2.18), we now have

$$U_j(x,p) = \delta_j(u)\left(x, 2^j p\right), \qquad u(x,p) = \sum_{j=0}^{\infty} U_j(x, 2^{-j}p). \tag{17.4.1}$$

We will now define semiclassical Sobolev norms with small parameter h, while keeping track of the explicit dependence on the spectral parameter $\lambda \in \mathbf{C}$.

Recall that in section 15.3, we introduced the vector space $\mathcal{S}$, the Laplacian Δ^Y acting on $\mathcal{S}$, and the operator $\mathbb{S}$ in (15.3.1). Let $\Delta^{Y,V}$ be the fiberwise Laplacian along the fibers $\mathbf{P}(T^*X \oplus \mathbf{R})$.

Definition 17.4.1. Let $\mathbb{S}_{\mathrm{sc}}$ be the second order self-adjoint positive operator acting on $\mathcal{S}$,

$$\mathbb{S}_{\mathrm{sc}} = h^2\mathbb{S} - \Delta^{Y,V} + 1. \tag{17.4.2}$$

For $j \in \mathbf{N}, \lambda \in \mathbf{C}$, set

$$\Lambda_{\lambda,\mathrm{sc},j} = \left(\mathbb{S}_{\mathrm{sc}} + 2^{4j} + 2^{-2j}|\lambda|^2\right)^{1/2}. \tag{17.4.3}$$

This weight is closely related to the one in (15.6.7).

For $s \in \mathbf{R}, U \in \mathcal{S}$, let $|U|_{\lambda,\mathrm{sc},j,s}$ be the semiclassical Sobolev norm of U given by

$$|U|_{\lambda,\mathrm{sc},j,s} = 2^{jn/2} \left|\Lambda^s_{\lambda,\mathrm{sc},j} U\right|. \tag{17.4.4}$$

If $u(x,p) \in \mathcal{S}$, set

$$\|u\|^2_{\lambda,\mathrm{sc},s} = \sum_{j=0}^{\infty} |U_j|^2_{\lambda,\mathrm{sc},j,s}, \qquad \|u\|^2_{\mathrm{sc},s} = \|u\|^2_{0,\mathrm{sc},s}. \tag{17.4.5}$$

Remark 17.4.2. For given $h > 0, \lambda \in \mathbf{C}$, the completion of $\mathcal{S}$ with respect to the norm $\| \; \|_{\lambda,\mathrm{sc},s}$ is the vector space $\mathcal{H}^s$ defined in Definition 15.3.1, but for a given $\lambda \in \mathbf{C}$, the norms $\| \; \|_{\lambda,\mathrm{sc},s}$ and $\| \; \|_s$ are not uniformly equivalent as $h \to 0$, except when $s = 0$, where both are equivalent to the usual L^2 norm on H.

Notice that $\|u\|^2_{\lambda,\mathrm{sc},1}$ is equivalent to

$$\left|\left(<p>^2 + \frac{|\lambda|}{<p>}\right) u\right|^2 + h^2 |\nabla u|^2 + \left|<p> \widehat{\nabla} u\right|^2,$$

and this uniformly for $h > 0, \lambda \in \mathbf{C}$.

17.5 UNIFORM HYPOELLIPTIC ESTIMATES FOR $\mathbf{P_h}$

Lemma 17.5.1. *Recall that $P_\pm$ is given by (17.2.14). For any $s \in \mathbf{R}$, $P_\pm$ maps $\mathcal{H}^s$ into itself. Given $s \in \mathbf{R}$, there exist $C_s > 0$ such that for any $h \in]0,1], \lambda \in \mathbf{C}, u \in \mathcal{S}$,*

$$\|P_\pm u\|_{\lambda,\mathrm{sc},s} \leq C_s \|u\|_{\lambda,\mathrm{sc},s}. \tag{17.5.1}$$

Proof. This is an obvious consequence of (17.2.14). □

Now we establish an obvious uniform analogue of Theorems 15.5.1 and 15.6.1.

Theorem 17.5.2. *There exist $h_0 > 0, \lambda_0 > 0$ such that if $h \in]0,h_0], \lambda \in \mathbf{C}, \operatorname{Re}\lambda \leq -\lambda_0, s \in \mathbf{R}, u \in \mathcal{S}'^{\cdot}(T^*X, \pi^*F^*)$, if $(P_h - \lambda) u \in \mathcal{H}^s$, then $u \in \mathcal{H}^{s+1/4}$. Moreover, there exist constants $C > 0, C_s > 0$ such that if h, λ, s, u are taken as before,*

$$\|u\|_{\mathrm{sc},s+1/4} \leq C_s \|(P_h + \lambda_0)u\|_{\mathrm{sc},s}, \tag{17.5.2}$$
$$|\lambda|^{1/6} \|u\|_{\mathrm{sc},0} + \|u\|_{\mathrm{sc},1/4} \leq C \|(P_h - \lambda) u\|_{\mathrm{sc},0}.$$

Proof. Comparing with Theorems 15.5.1 and 15.6.1, the main point here is to check that the constants C, C_s in (17.5.2) are uniform in $h \in]0,h_0]$.

We will closely follow the proofs of Theorems 15.4.2, 15.5.1, and 15.6.1. If $\lambda \in \mathbf{C}$, put

$$P_{h,\lambda} = P_h - \lambda. \tag{17.5.3}$$

As in (15.6.5), if $\lambda_0 > 0, \alpha > 0, \beta \in \mathbf{R}$, set

$$\lambda = -\lambda_0 - \alpha + i\beta. \tag{17.5.4}$$

In the sequel, λ_0 will be precisely determined.

For $\tau \in]0,1]$, set

$$P_{h,\lambda,\tau} = K_{\tau^{-1}} P_{h,\lambda} K_\tau. \tag{17.5.5}$$

We define $\alpha_{\pm,\tau}, \beta_{\pm,\tau}, \gamma_{\pm,\tau}$ as in (15.4.2). Then $\alpha_{\pm,\tau}$ is self-adjoint with respect to the standard Hermitian product on H, and $\beta_{\pm,\tau}$ is skew-adjoint. By (17.2.3), (17.5.4), (17.5.5), we get

$$P_{h,\lambda,\tau} = \alpha_{\pm,\tau} + \lambda_0 + \alpha - i\beta + h\beta_{\pm,\tau} + h^2\gamma_{\pm,\tau}. \tag{17.5.6}$$

Let $Q'_{h,\lambda,\tau}, Q''_{h,\lambda,\tau}$ be the self-adjoint and skew-adjoint parts of $P_{h,\lambda,\tau}$, so that

$$P_{h,\lambda,\tau} = Q'_{h,\lambda,\tau} + Q''_{h,\lambda,\tau}. \tag{17.5.7}$$

Note that $\alpha_{\pm,\tau}$ appears only in $Q'_{h,\lambda,\tau}$, and $\beta_{\pm,\tau}$ appears only in $Q''_{h,\lambda,\tau}$.

To keep in line with (17.4.3), set

$$\Lambda_{\lambda,\mathrm{sc}} = \left(\mathbb{S}_{\mathrm{sc}} + \tau^{-4} + \tau^2 |\lambda|^2\right)^{1/2}. \tag{17.5.8}$$

As in chapter 15, we will make $\tau = 2^{-j}, j \in \mathbf{N}$.

Take $U \in \mathcal{S}^{\cdot}(T^*X, \pi^*F)$ with support in the annulus $\mathcal{R}$. By proceeding as in (15.6.14), we find for $h_0 > 0$ small enough, if $\lambda_0 > 0$ is large enough, there exists $C > 0$ such that for $h \in]0, h_0]$ and λ taken as in (17.5.4),

$$\begin{aligned} &\left|\widehat{\nabla} U\right|^2 + \tau^{-4}|U|^2 + \alpha\tau^{-2}|U|^2 \le C\left\langle Q'_{h,\lambda,\tau}U, \tau^{-2}U\right\rangle, \\ &\left|\widehat{\nabla} U\right|^2 + \tau^{-4}|U|^2 + \alpha\tau^{-2}|U|^2 \le C\left|P_{h,\lambda,\tau}U\right|^2. \end{aligned} \tag{17.5.9}$$

Also (17.5.9) still holds when $\tau = 1$, and the support of U is included in $\mathcal{B}$. From now on $h_0 > 0, \lambda_0 > 0$ will be chosen so that (17.5.9) holds.

We will use semiclassical pseudodifferential operators on the total space Y of $\mathbf{P}(T^*X \oplus \mathbf{R})$ with weight Λ. In particular, as in the proof of Theorem 15.6.1, we incorporate the spectral parameter λ in the weight. Recall that π denotes the projection $T^*X \to X$ or $Y \to X$.

If $U \in \mathcal{S}^{\cdot}(T^*X, \pi^*F)$, and if U has compact support, if $s \in \mathbf{R}$, set

$$|U|_{\lambda,\mathrm{sc},s} = \left|\Lambda^s_{\lambda,\mathrm{sc}}U\right|. \tag{17.5.10}$$

In a local coordinate system on X, and using the appropriate trivialization, the map $\pi : Y \to X$ is written as $y = (x,p) \to x$ with (x,p) varying in a compact subset of $\mathbf{R}^{2n}$. In the given coordinate system, then $T^*Y \simeq \mathbf{R}^{2n}$. In the above coordinate system, $\zeta = (\xi, \eta) \in T^*Y_{\mathrm{sc}}$ corresponds to $\frac{\xi}{h} + \eta \in T^*Y \simeq \mathbf{R}^{2n}$.

By definition, a symbol of degree d is a smooth function $a(y,\zeta,h,\tau,\lambda)$ defined on $T^*Y_{\mathrm{sc}} \times]0,1] \times \mathbf{C}$ with values in $\mathrm{End}\left(\Lambda^{\cdot}(T^*X)\widehat{\otimes}\Lambda^{\cdot}(TX)\widehat{\otimes}F\right)$, which is such that for any multiindices α, β, there exist $C_{\alpha,\beta} > 0$ for which if $\mathrm{Re}\lambda \le -\lambda_0$, and all other variables vary in their natural domain of definition,

$$\left|\partial^\alpha_y \partial^\beta_\zeta a(y,\zeta,h,\tau,\lambda)\right| \le C_{\alpha,\beta}\left(\tau^{-4} + \tau^2|\lambda|^2 + |\zeta|^2\right)^{\frac{d-|\beta|}{2}}. \tag{17.5.11}$$

We denote by S^d the set of symbols of degree d. When $a(y,\zeta,h,\tau,\lambda)$ is defined also for $h=0$, then (17.5.11) should also be valid for $h=0$.

A smoothing operator on Y is a family of operators $B(h,\tau,\lambda)$, where h,τ,λ are taken as before, such that for any $s,t\in\mathbf{R}$, there exist $C_{s,t}>0$ with

$$\left|B(h,\tau,\lambda)U\right|_{\lambda,\mathrm{sc},s}\le C_{s,t}\left|U\right|_{\lambda,\mathrm{sc},t}. \tag{17.5.12}$$

We quantify a symbol a into an operator $A=\mathrm{Op}(a)$ by the formula

$$A(x,p,hD_x,D_p,h,\tau,\lambda)U(x,p)$$
$$=(2\pi)^{-2n}h^{-n}\int_{\mathbf{R}^{2n}}e^{\frac{i}{h}\langle x,\xi\rangle+i\langle p,\eta\rangle}a(x,p,\xi,\eta,h,\tau,\lambda)\widehat{U}\left(\frac{\xi}{h},\eta\right)d\xi d\eta, \tag{17.5.13}$$

where $\widehat{U}(\xi,\eta)$ is the Fourier transform of U in the variables x,p.

Let $\mathcal{E}^d$ be the associated set of pseudodifferential operators of degree d on Y. Then $A\in\mathcal{E}^d$ if, for any small compact subset $K\subset Y$, for any cutoff function $\theta(y)$ with support included in a small neighborhood of K, there exists a cutoff function θ' equal to 1 near the support of θ, such that in the appropriate coordinate system, there exists $a\in S^d$ and B smoothing such that

$$A(\tau)\theta=\theta'\mathrm{Op}(a)\theta+B(h,\tau,\lambda). \tag{17.5.14}$$

For $A\in\mathcal{E}^d$, the principal symbol $\sigma(A)$ of A is the class of a in the quotient space S^d/S^{d-1}.

If $E_d\in\mathcal{E}^d, E_{d'}\in\mathcal{E}^{d'}$, then $E_dE_{d'}\in\mathcal{E}^{d+d'}$, $\sigma(E_dE_{d'})=\sigma(E_d)\sigma(E_{d'})$. Moreover, if $E_d=\mathrm{Op}(e), E_{d'}=\mathrm{Op}(e')$, then

$$[E_d,E_{d'}]-\mathrm{Op}\left([e,e']+\frac{1}{i}\{e,e'\}_{\mathrm{sc}}\right)\in\mathcal{E}^{d+d'-2}. \tag{17.5.15}$$

Elements of $\mathcal{E}^0$ are uniformly bounded operators on $\mathcal{H}$, and moreover $\Lambda_{\lambda,\mathrm{sc}},h\nabla_{e_i},\nabla_{\widehat{e}^i},\tau^{-2}\in\mathcal{E}^1$.

As should be clear from (17.5.13), the above class of semiclassical operators is naturally associated to what is known in the literature as the adiabatic limit [BeB94, MaMe90, Wi85]. Indeed consider the projection $\pi:Y\to X$. The adiabatic limit refers to a situation where the metric on Y is of the form $g^{TY}+\pi^*g^{TX}/h^2$. Let g^{T^VY} be the metric induced by g^{TY} on T^VY, and let $g^{T^{V*}Y}$ be the dual metric on $T^{V*}Y$. As $h\to0$, the principal symbol of the Laplacian Δ_h^Y is such that

$$\sigma\left(-\Delta_h^Y\right)=\left|i^*\xi\right|^2_{g^{TY}}+\mathcal{O}\left(h^2\right)\left|\xi\right|^2_{g^{T^*Y}} \tag{17.5.16}$$

Let $\theta_0(p)$ be a smooth radial cutoff function with values in $[0,1]$, which is equal to 1 near the ball $\mathcal{B}_0$, and which vanishes for $|p|^2\ge6$. Set

$$R=\theta_0P_{h,\lambda,\tau}\theta_0. \tag{17.5.17}$$

Let $R=R'+R''$ be the decomposition of R into its self-adjoint and skew-adjoint parts. The explicit forms of these operators can be obtained via equations (15.4.2) and (17.5.6). Then $\alpha_{\pm,\tau}$ does not appear in R'', and $\beta_{\pm,\tau}$ does not appear in R'.

Note that

$$\theta_0 h \nabla_{Y^{\mathcal{H}}}^{\Lambda^{\cdot}(T^*T^*X)\widehat{\otimes}F,u} \theta_0 \in \mathcal{E}^1. \tag{17.5.18}$$

Using the explicit formula for R, we get

$$\begin{aligned}
&[R, E_d] \in \tau \mathcal{E}^d \widehat{\nabla} + \tau^{-2} \mathcal{E}^d, \\
&[R, E_d] \in \tau^2 \mathcal{E}^d \widehat{\nabla} + \tau^{-1} \mathcal{E}^d \quad \text{if } \sigma(E_d) \quad \text{is scalar}, \\
&\tau R'' \in \mathcal{E}^1.
\end{aligned} \tag{17.5.19}$$

As in the proof of Lemmas 15.4.3 and 15.4.4, for any $U \in \mathcal{S}^{\cdot}(T^*X, \pi^*F)$ with support in the ball $\mathcal{B}$ if $j = 0$, and in the annulus $\mathcal{R}$ otherwise,

$$|R''U|_{\lambda,\mathrm{sc},-1/2} \leq C\tau^{1/2} |RU|_{\lambda,\mathrm{sc},0}, \qquad |U|_{\lambda,\mathrm{sc},1/4} \leq C\tau^{3/4} |RU|_{\lambda,\mathrm{sc},0}. \tag{17.5.20}$$

As in the proof of Theorem 15.4.2, from (17.5.20), we derive the following analogue of (15.4.62):

$$\begin{aligned}
&\left|\widehat{\nabla} U\right|^2_{\lambda,\mathrm{sc},s} + \tau^{-4} |U|^2_{\lambda,\mathrm{sc},s} + \tau^{-3/2} |U|^2_{\lambda,\mathrm{sc},s+1/4} + \tau^{5/4} \left|\widehat{\nabla} U\right|^2_{\lambda,\mathrm{sc},s+1/8} \\
&\leq C|RU|^2_{\lambda,\mathrm{sc},s} + C_s\tau^4 \left|\widehat{\nabla} U\right|^2_{\lambda,\mathrm{sc},s} + C_s\tau^{-2} |U|^2_{\lambda,\mathrm{sc},s} + C_s\tau^{5/4} |U|^2_{\lambda,\mathrm{sc},s+1/8}.
\end{aligned} \tag{17.5.21}$$

Note here that it is essential that the constant C in the second line of (17.5.21) does not depend on s.

We shall now deduce from (17.5.21) conclusions similar to the ones we obtained in equation (15.4.6) in Theorem 15.4.2 for nonnegative values of s. Namely, we will show that for $s \geq 0$, there exists $C_s > 0$ such that for any $\tau = 2^{-j}, h \in]0, h_0], \mathrm{Re}\lambda \leq -\lambda_0, U \in \mathcal{S}$, with the same support conditions as before,

$$\begin{aligned}
\tau^{-4} |U|^2_{\lambda,\mathrm{sc},s} + \left|\widehat{\nabla} U\right|^2_{\lambda,\mathrm{sc},s} &+ \tau^{-3/2} |U|^2_{\lambda,\mathrm{sc},s+1/4} \\
&+ \tau^{5/4} \left|\widehat{\nabla} U\right|^2_{\lambda,\mathrm{sc},s+1/8} \leq C_s |RU|^2_{\lambda,\mathrm{sc},s}.
\end{aligned} \tag{17.5.22}$$

To establish (17.5.22), we use our usual contradiction argument. Suppose that for some $s \geq 0$, equation (17.5.22) does not hold. Then there exist sequences $\tau_k = 2^{-j_k}, h_k, \lambda_k = -\lambda_0 + \alpha_k + i\beta_k, U_k$ such that the left-hand side of (17.5.22) is equal to 1 and $\lim_{k\to+\infty} |R(h_k, \tau_k, \lambda_k) U_k|_{\lambda_k,\mathrm{sc},s} = 0$. By (17.5.21), the sequence j_k is necessarily bounded, so we may suppose that $j_k = j$ is constant, i.e., $\tau = 2^{-j_k}$ remains constant.

By (17.5.8), for $\mu \geq 0$, if $\mu = \mu' + \mu''$, with $\mu' \geq 0, \mu'' \geq 0$, then

$$\Lambda^{\mu} \geq \tau^{-2\mu' + \mu''} |\lambda|^{\mu''}. \tag{17.5.23}$$

By (17.5.23), we get in particular

$$\Lambda^{\mu} \geq |\lambda|^{2\mu/3}. \tag{17.5.24}$$

By (17.5.24), we get

$$|V|_{\lambda,\mathrm{sc},s} \le C_\mu |\lambda|^{-2\mu/3} |V|_{\lambda,\mathrm{sc},s+\mu}. \tag{17.5.25}$$

Then if $|\lambda_k| \to +\infty$, it would follow from (17.5.21) and (17.5.25) that (17.5.22) would hold.

Therefore, we can also assume that the sequence λ_k converges to $\lambda \in \mathbf{C}$, so that ultimately we can forget about the dependence on λ.

Set

$$|||U|||^2_{\mathrm{sc},s} = |U|^2_{\lambda,\mathrm{sc},s+1/8} + \left|\widehat{\nabla} U\right|^2_{\lambda,\mathrm{sc},s}. \tag{17.5.26}$$

The norm $|||U|||_{\mathrm{sc},s}$ still depends on the parameter h. By (17.5.21), we get

$$|||U_k|||^2_{\mathrm{sc},s+1/8} \le C_s \left(|RU_k|^2_{\lambda,\mathrm{sc},s} + |||U_k|||^2_{\mathrm{sc},s}\right). \tag{17.5.27}$$

For $s \in \mathbf{R}, \varepsilon > 0$, there exists $C_{s,\varepsilon} > 0$ such that

$$|||U|||_{\mathrm{sc},s} \le \varepsilon |||U|||_{\mathrm{sc},s+1/8} + C_{s,\varepsilon} |U|. \tag{17.5.28}$$

By (17.5.9), (17.5.27), (17.5.28), for $s \ge 0$, we obtain

$$|||U_k|||_{\mathrm{sc},s+1/8} \le C_s |RU_k|_{\lambda,\mathrm{sc},s}. \tag{17.5.29}$$

Now we made the assumption that as $k \to +\infty$, the right-hand side of (17.5.29) tends to 0. Therefore the left-hand side of (17.5.29) also tends to 0, which contradicts the fact that the left-hand side of (17.5.22) is equal to 1.

Now we follow the proof of Theorem 15.5.1. Let $v = (P_h - \lambda)\,u$, and let U_j, V_j be associated to u, v by (15.2.16). To make our notation simpler, we will write $|V_j|_{\lambda,\mathrm{sc},t}$ instead of $|V_j|_{\lambda,\mathrm{sc},j,t}$ as in (17.4.4). As in (15.5.13), for $s \in \mathbf{R}$, we get

$$\begin{aligned}\left|P_{h,\lambda,2^{-j}} U_j\right|_{\lambda,\mathrm{sc},s} \le C |V_j|_{\lambda,\mathrm{sc},s} &+ C_s 2^{-j} \sum_{|j'-j|\le 1} |U_{j'}|_{\lambda,\mathrm{sc},s} \\ &+ C_s 2^{-2j} \sum_{|j'-j|\le 1} \left|\widehat{\nabla} U_j\right|_{\lambda,\mathrm{sc},s}.\end{aligned} \tag{17.5.30}$$

As in (15.5.16), set

$$\beta_{j,s} = 2^{3j/4} |U_j|_{\lambda,\mathrm{sc},s+1/4} + 2^{-5j/8} \left|\widehat{\nabla} U_j\right|_{\lambda,\mathrm{sc},s+1/8}. \tag{17.5.31}$$

As in (15.5.17), (15.5.18), we deduce from (17.5.30) that

$$\beta_{j,s} \le C |V_j|_{\lambda,\mathrm{sc},s} + C_s 2^{-11j/8} \sum_{|j'-j|\le 1} \beta_{j',s-1/8}. \tag{17.5.32}$$

We define the auxiliary norm $|||u|||_{\lambda,\mathrm{sc},s}$ by the formula

$$|||u|||_{\lambda,\mathrm{sc},s} = |\beta_{j,s}|_{\ell^2}. \tag{17.5.33}$$

When $s = 0$, $||u||_{\lambda,\mathrm{sc},s}$ does not depend on λ. With the conventions in (17.4.5), this is just $\|u\|_{\mathrm{sc},0}$, which is equivalent to the usual L^2 norm.

By (17.5.32), for $s \in \mathbf{R}$, we get

$$|||u|||_{\lambda,\mathrm{sc},s} \leq C_s \left(||v||_{\lambda,\mathrm{sc},s} + |||u|||_{\lambda,\mathrm{sc},s-1/8}\right). \tag{17.5.34}$$

Given $\varepsilon > 0, s \in \mathbf{R}$, there exists $C_{\varepsilon,s} > 0$ such that

$$|||u|||_{\lambda,\mathrm{sc},s-1/8} \leq \varepsilon|||u|||_{\lambda,\mathrm{sc},s} + C_{\varepsilon,s} \left\|u\right\|_{\mathrm{sc},0}. \tag{17.5.35}$$

Moreover, by (17.5.9),

$$\left\|u\right\|_{\mathrm{sc},0} \leq C \left\|v\right\|_{\mathrm{sc},0}. \tag{17.5.36}$$

Finally, for $s \geq 0$,

$$\left\|v\right\|_{\mathrm{sc},0} \leq \left\|v\right\|_{\lambda,\mathrm{sc},s}. \tag{17.5.37}$$

By (17.5.34)-(17.5.37), for $s \geq 0$, we obtain

$$|||u|||_{\lambda,\mathrm{sc},s} \leq C_s||v||_{\lambda,\mathrm{sc},s}. \tag{17.5.38}$$

Now we use again equation (15.6.9), which asserts that

$$\tau^2|\lambda|^2 + \tau^{-4} \geq C|\lambda|^{4/3}. \tag{17.5.39}$$

By proceeding as in the proof of equation (15.6.1) in Theorem 15.6.1, that is, using (17.5.38) with $s = 0$ (instead of (15.6.11)) and (17.5.39), we get the second inequality in (17.5.2). For $s \geq 0$, we obtain the first inequality in (17.5.2) from (17.5.38) with $\lambda = -\lambda_0$. In particular, since the adjoint P_h^* of P_h has the same structure as P_h, we find that for $h \in]0, h_0], s \geq 0$, $P_h + \lambda_0$ is one to one from $\{u \in \mathcal{H}^s, P_h u \in \mathcal{H}^s\}$ into $\mathcal{H}^s$, and that there exist $C_s > 0$ such that

$$\left\|(P_h + \lambda_0)^{-1} v\right\|_{\mathrm{sc},s} \leq C_s \left\|v\right\|_{\mathrm{sc},s}. \tag{17.5.40}$$

Also (17.5.40) still holds for P_h^*, and so by duality, we get (17.5.40) for any $s \in \mathbf{R}$.

By (17.5.21), for any $s \in \mathbf{R}$, by the same argument of contradiction as the one we used to derive (17.5.22) from (17.5.21),

$$|||u|||_{\lambda,\mathrm{sc},s} \leq C_s \left(||v||_{\lambda,\mathrm{sc},s} + \left\|u\right\|_{\lambda,\mathrm{sc},s}\right). \tag{17.5.41}$$

Using (17.5.41) with $\lambda = \lambda_0$ and (17.5.40), we find that the first equation in (17.5.2) still holds for $s < 0$. The proof of our theorem is completed. □

17.6 THE OPERATOR $\mathbf{P}_\mathbf{h}^0$ AND ITS RESOLVENT $\mathbf{S}_{\mathbf{h},\lambda}$ FOR $\lambda \in \mathbf{R}$

Set

$$P_h^0 = P_h + P_\pm. \tag{17.6.1}$$

By (17.2.3), (17.6.1), we obtain

$$P_h^0 = \alpha_\pm + h\beta_\pm + h^2\gamma_\pm + P_\pm. \tag{17.6.2}$$

Now we will extend Theorem 17.5.2 to the operator P_h^0.

Theorem 17.6.1. *There exist $h_0 > 0, \lambda_1 > 0$ such that for $h \in]0, h_0], \lambda \in \mathbf{C}, \mathrm{Re}\,\lambda \leq \lambda_1, s \in \mathbf{R}, u \in \mathcal{S}'^{\cdot}(T^*X, \pi^*F^*)$, if $\left(P_h^0 - \lambda\right) u \in \mathcal{H}^s$, then $u \in \mathcal{H}^{s+1/4}$. Moreover, there exist constants $C > 0, C_s > 0$ such that*

$$\|u\|_{\mathrm{sc},s+1/4} \leq C_s \| \left(P_h^0 - \lambda_1\right) u \|_{\mathrm{sc},s}, \tag{17.6.3}$$
$$|\lambda|^{1/6} \|u\|_{\mathrm{sc},0} + \|u\|_{\mathrm{sc},1/4} \leq C \| \left(P_h^0 - \lambda\right) u \|_{\mathrm{sc},0}.$$

Proof. By (15.1.3), (17.6.2), we get

$$\mathrm{Re}\left\langle \left(P_h^0 - \lambda\right) u, u\right\rangle = \mathrm{Re}\left\langle (\alpha_\pm + P_\pm) u, u\right\rangle - \mathrm{Re}\,\lambda \left|u\right|^2 + h^2 \mathrm{Re}\left\langle \gamma_\pm u, u\right\rangle. \tag{17.6.4}$$

By (15.1.3),

$$|\gamma_\pm| \leq C\left(|p|^2 + 1\right). \tag{17.6.5}$$

By (15.1.3), (17.6.4), (17.6.5), if $h_0 > 0, \lambda_1 > 0$ are small enough, for $h \in]0, h_0], \lambda \in \mathbf{C}, \mathrm{Re}\,\lambda \leq \lambda_1$,

$$\|u\|_{\mathrm{sc},0} \leq C \| \left(P_h^0 - \lambda\right) u \|_{\mathrm{sc},0}. \tag{17.6.6}$$

By Lemma 17.5.1 and by Theorem 17.5.2, for $s \in \mathbf{R}$, we obtain

$$\|u\|_{\mathrm{sc},s+1/4} \leq C_s \left(\| \left(P_h^0 - \lambda_1\right) u \|_{\mathrm{sc},s} + \|u\|_{\mathrm{sc},s}\right), \tag{17.6.7}$$
$$|\lambda|^{1/6} \|u\|_{\mathrm{sc},0} + \|u\|_{\mathrm{sc},1/4} \leq C\left(\| \left(P_h^0 - \lambda\right) u \|_{\mathrm{sc},0} + \|u\|_{\mathrm{sc},0}\right).$$

Also given $\epsilon > 0$, there exists $C_{s,\varepsilon} > 0$ such that

$$\|u\|_{\mathrm{sc},s} \leq \varepsilon \|u\|_{\mathrm{sc},s+1/4} + C_{s,\epsilon} \|u\|_{\mathrm{sc},0}. \tag{17.6.8}$$

Using (17.6.6)-(17.6.8), we get the second inequality in (17.6.3), and also the first one when $s \geq 0$. In particular, as in the proof of Theorem 17.5.2, $P_h^0 - \lambda_1$ is one to one from $\{u \in \mathcal{H}^s, P_h^0 u \in \mathcal{H}^s\}$ into $\mathcal{H}^s$, and also if $s \geq 0$, there exists $C_s > 0$ such that

$$\| \left(P_h^0 - \lambda_1\right)^{-1} v \|_{\mathrm{sc},s} \leq C_s \|v\|_{\mathrm{sc},s}. \tag{17.6.9}$$

Also the formal adjoint Q_h^{0*} of P_h^0 has the same properties as P_h^0, and so it verifies similar estimates. By duality, we find that (17.6.9) holds for any $s \in \mathbf{R}$, which together with (17.6.7) implies the first inequality in (17.6.3) for any $s \in \mathbf{R}$. The proof of our theorem is complete. □

Definition 17.6.2. For $\lambda \in \mathbf{C}, \mathrm{Re}\,\lambda \leq \lambda_1$, set

$$S_{h,\lambda} = \left(P_h^0 - \lambda\right)^{-1}. \tag{17.6.10}$$

Let $\varphi = (\varphi_1(x), ..., \varphi_N(x))$ be a family of smooth real functions on X. Let $\mathrm{Ad}_\varphi^N S$ be the iterated commutator,

$$\mathrm{Ad}_\varphi^N S_{h,\lambda} = [\varphi_N, ...[\varphi_2, [\varphi_1, S_{h,\lambda}]]...]. \tag{17.6.11}$$

Clearly,

$$[\varphi_1, S_{h,\lambda}] = -S_{h,\lambda}\left[\varphi_1, P_h^0\right] S_{h,\lambda} = \mp h S_{h,\lambda} \left(\nabla_{Y^{\mathcal{H}}} \varphi_1\right) S_{h,\lambda}. \tag{17.6.12}$$

By (17.6.11), (17.6.12), we get

$$\mathrm{Ad}_\varphi^N S_{h,\lambda} = (\mp h)^N \sum_{\sigma \in S_N} S_{h,\lambda} \left(\nabla_{Y^{\mathcal{H}}} \varphi_{\sigma_1}\right) S_{h,\lambda} \dots S_{h,\lambda} \left(\nabla_{Y^{\mathcal{H}}} \varphi_{\sigma_N}\right) S_{h,\lambda}. \tag{17.6.13}$$

Theorem 17.6.3. *If $N \in \mathbf{N}, a \in \mathbf{R}, s \in \mathbf{R}$, there exist $C_{a,s} > 0, C_s > 0$ such that for $\lambda \in \mathbf{C}, \operatorname{Re}\lambda \le \lambda_1, h \in]0, h_0]$, the operator $< p >^a S_{h,\lambda} < p >^{-a}$ is bounded from $\mathcal{H}^s$ to $\mathcal{H}^{s+1/4}$, the operator $\left(P_h^0 - \lambda\right) \operatorname{Ad}_\varphi^N \left(S_{h,\lambda}\right)$ is bounded from $\mathcal{H}^s$ to $\mathcal{H}^{s+N/5}$, and moreover if $v \in \mathcal{S}^{\cdot}\left(T^*X, \pi^*F\right)$,*

$$|| < p >^a S_{h,\lambda} v||_{\lambda,\mathrm{sc},s+1/4} \le C_{a,s}|| < p >^a v||_{\lambda,\mathrm{sc},s}, \quad (17.6.14)$$
$$|| \left(P_h^0 - \lambda\right) \operatorname{Ad}_\varphi^N S_{h,\lambda} v||_{\lambda,\mathrm{sc},s+N/5} \le C_s h^N ||v||_{\lambda,\mathrm{sc},s}.$$

Proof. We establish the first inequality in (17.6.14). Put $u = S_{h,\lambda} v$. Then

$$v = \left(P_h - (\lambda - \lambda_0 - \lambda_1)\right) u + \left(P_\pm - \lambda_0 - \lambda_1\right) u. \quad (17.6.15)$$

By (17.5.1), we get

$$\left\| \left(P_\pm - \lambda_0 - \lambda_1\right) u \right\|_{\lambda,\mathrm{sc},s} \le C \left\| u \right\|_{\lambda,\mathrm{sc},s}. \quad (17.6.16)$$

With the notation of the proof of Theorem 17.5.2, set

$$\gamma_{j,s} = 2^{2j} \left| U_j \right|_{\lambda,\mathrm{sc},s} + 2^{3j/4} \left| U_j \right|_{\lambda,\mathrm{sc},s+1/4} + 2^{-5j/8} \left| \widehat{\nabla} U_j \right|_{\lambda,\mathrm{sc},s+1/8}. \quad (17.6.17)$$

Put

$$||||u||||_{\lambda,\mathrm{sc},s} = \left|\gamma_{j,s}\right|_{\ell^2} \quad (17.6.18)$$

Recall that the norm $|||u|||_{\lambda,\mathrm{sc},s}$ was defined in (17.5.33). We have the trivial inequalities

$$\left\| u \right\|_{\lambda,\mathrm{sc},s+1/4} \le |||u|||_{\lambda,\mathrm{sc},s} \le ||||u||||_{\lambda,\mathrm{sc},s}. \quad (17.6.19)$$

Observe that under the given conditions on $\lambda \in \mathbf{C}$, $\lambda - \lambda_0 - \lambda_1$ verifies the conditions given in Theorem 17.5.2 for λ. By (17.5.22), (17.5.30), (17.6.15), (17.6.16), and the first inequality in (17.6.19), we get for $s \ge 0$,

$$||||u||||_{\lambda,\mathrm{sc},s} \le C_s \left(||v||_{\lambda,\mathrm{sc},s} + |||u|||_{\lambda,\mathrm{sc},s-1/8} \right). \quad (17.6.20)$$

Using (17.5.35) and (17.6.20), we get

$$||||u||||_{\lambda,\mathrm{sc},s} \le C_s \left(\left\| v \right\|_{\lambda,\mathrm{sc},s} + \left\| u \right\|_{\mathrm{sc},0} \right). \quad (17.6.21)$$

Moreover, by (17.6.3),

$$\left\| u \right\|_{\mathrm{sc},0} \le C \left\| v \right\|_{\mathrm{sc},0}. \quad (17.6.22)$$

Also for $s \ge 0$,

$$\left\| v \right\|_{\mathrm{sc},0} \le \left\| v \right\|_{\lambda,\mathrm{sc},s}. \quad (17.6.23)$$

By (17.6.21)-(17.6.23), we find that for $s \ge 0$,

$$||||u||||_{\lambda,\mathrm{sc},s} \le C_s \left\| v \right\|_{\lambda,\mathrm{sc},s}. \quad (17.6.24)$$

By (17.6.19) and (17.6.24), for $s \ge 0$, we get

$$\left\| u \right\|_{\lambda,\mathrm{sc},s+1/4} \le C_s \left\| v \right\|_{\lambda,\mathrm{sc},s}, \quad (17.6.25)$$

which is just the first inequality in (17.6.14) with $a = 0$.

To obtain this equation for $s \geq 0$ and arbitrary $a \in \mathbf{R}$, we just use the same proof as before, replacing $\gamma_{\lambda,\mathrm{sc},t,j}$ by $2^{ja}\gamma_{\lambda,\mathrm{sc},t,j}$.

For $1 \leq i \leq N$, put

$$w = -\left(\nabla_{Y^{\mathcal{H}}}\varphi\right) u. \tag{17.6.26}$$

The index i has not been written in w, φ to avoid complicating the notation. Recall that w grows linearly in p. We get

$$|W_j|_{\lambda,\mathrm{sc},s} \leq C_s 2^j |U_j|_{\lambda,\mathrm{sc},s}. \tag{17.6.27}$$

Clearly there is $C > 0$ such that for $t > 0, b > 0$,

$$bt^{1/4} + \frac{1}{t} \geq Cb^{4/5}. \tag{17.6.28}$$

By (17.6.28), we get

$$\Lambda^{1/5} \leq C\left(\tau^{1/4}\Lambda^{1/4} + \tau^{-1}\right). \tag{17.6.29}$$

By (17.6.27), (17.6.29), we obtain

$$\begin{aligned} |W_j|_{\lambda,\mathrm{sc},s+1/5} &\leq 2^j |W_j|_{\lambda,\mathrm{sc},s} + 2^{-j/4} |W_j|_{\lambda,\mathrm{sc},s+1/4} \\ &\leq C_s\left(2^{2j} |U_j|_{\lambda,\mathrm{sc},s} + 2^{3j/4} |U|_{\lambda,\mathrm{sc},s+1/4}\right). \end{aligned} \tag{17.6.30}$$

By comparing (17.6.17) with the right-hand side of (17.6.30), and using (17.6.24), for $s \geq 0$, we get

$$\|w\|_{\lambda,\mathrm{sc},s+1/5} \leq C_s \|v\|_{\lambda,\mathrm{sc},s}. \tag{17.6.31}$$

Using the first equation in (17.6.14) and (17.6.13), (17.6.31), we obtain the second equation in (17.6.14) for $s \geq 0$.

We get analogous estimates when replacing P_h^0 by its formal adjoint P_h^{0*}. We find that for $s \geq 0, a \in \mathbf{R}$,

$$\begin{aligned} \left\|<p>^a S_{h,\lambda}^* v\right\|_{\lambda,\mathrm{sc},s+1/4} &\leq C \left\|<p>^a v\right\|_{\lambda,\mathrm{sc},s}, \\ \left\|S_{h,\lambda}^* \nabla_{Y^{\mathcal{H}}}\varphi v\right\|_{\lambda,\mathrm{sc},s+1/5} &\leq C \|v\|_{\lambda,\mathrm{sc},s}. \end{aligned} \tag{17.6.32}$$

By duality, we conclude that the estimates in (17.6.14) still hold for $s \leq -1/4$. By interpolation, we find that they hold for arbitrary $s \in \mathbf{R}$. The proof of our theorem is completed. □

Lemma 17.6.4. *Let K be a compact subset of X, and $\varphi(x) \in C_0^\infty(X \setminus K)$. Given $s, t \in \mathbf{R}, N \in \mathbf{N}$, there exists $C_{s,t,N} > 0$ such that for $h \in]0, h_0], \lambda \in \mathbf{C}, \operatorname{Re}\lambda \leq \lambda_1$, if $v \in \mathcal{H}^s$, if the support of $v \in \mathcal{H}^s$ is included in $\pi^{-1}(K)$, then*

$$\|\varphi S_{h,\lambda} v\|_{\lambda,\mathrm{sc},t} \leq C_{s,t,N} h^N \|v\|_{\lambda,\mathrm{sc},s}. \tag{17.6.33}$$

Proof. Let $\psi \in C_0^\infty(X \setminus K)$, which is equal to 1 on the support of φ, so that $\varphi = \varphi\psi, \psi v = 0$. Then

$$\varphi S_{h,\lambda} v = \varphi \mathrm{Ad}_\psi^N S_{h,\lambda} v = \varphi S_{h,\lambda}\left(P_h^0 - \lambda\right) \mathrm{Ad}_\psi^N S_{h,\lambda} v. \tag{17.6.34}$$

Using (17.6.14) in Theorem 17.6.3 and (17.6.34), we get (17.6.33). □

17.7 THE RESOLVENT $\mathbf{S}_{\mathbf{h},\lambda}$ FOR $\lambda \in \mathbf{C}$

Definition 17.7.1. For $\lambda_1 > 0, c_0 > 0$, set

$$\mathcal{V} = \left\{ \lambda \in \mathbf{C}, \lambda = \mu + \nu, \operatorname{Re} \mu \leq \lambda_1, \nu \in \mathbf{R}, |\nu| \leq c_0 |\mu|^{1/6} \right\}. \tag{17.7.1}$$

The definition of $\mathcal{V}$ should be compared with the definition of $\mathcal{U}$ given in (15.7.2) with λ_0, c_0 replaced by $-\lambda'_0, c'_0$, which we repeat. Namely,

$$\mathcal{U} = \left\{ \lambda = \lambda'_0 + \sigma + i\tau, \sigma, \tau \in \mathbf{R}, \sigma \leq c'_0 |\tau|^{1/6} \right\}. \tag{17.7.2}$$

Note that if $\lambda \in \mathcal{U}$,

$$\lambda = \lambda'_0 + \sigma - c'_0 |\tau|^{1/6} + i\tau + c'_0 |\tau|^{1/6}, \tag{17.7.3}$$

so that if $\lambda_1 = \lambda'_0, c_0 = c'_0$, then $\mathcal{U} \subset \mathcal{V}$.

Conversely assume that $\lambda \in \mathcal{V}$. There are $a \geq 0, \tau \in \mathbf{R}$ such that

$$\mu = \lambda_1 - a + i\tau, \tag{17.7.4}$$

and so if $\sigma = \nu - a$,

$$\lambda = \lambda_1 + \sigma + i\tau. \tag{17.7.5}$$

If $\sigma \geq 0$, then $a \leq \nu$, so that

$$a \leq c_0 |\mu|^{1/6}. \tag{17.7.6}$$

By (17.7.4), (17.7.6), we get

$$a \leq C \left(1 + |\tau|^{1/6}\right), \tag{17.7.7}$$

with C depending only on c_0, λ_1. By (17.7.4), (17.7.7),

$$|\mu| \leq C' (1 + |\tau|). \tag{17.7.8}$$

From (17.7.8), we deduce that since $\lambda \in \mathcal{V}$,

$$\sigma \leq C'' \left(1 + |\tau|^{1/6}\right). \tag{17.7.9}$$

By (17.7.5), we get

$$\lambda = \lambda_1 + C'' + (\sigma - C'') + i\tau. \tag{17.7.10}$$

By (17.7.10), we find that if $\lambda'_0 = \lambda_1 + C'', c'_0 = C''$, then $\lambda \in \mathcal{U}$. Finally, note that if $\sigma < 0$, then $\lambda \in \mathcal{U}$.

Theorem 17.7.2. *There exists $\lambda_1 > 0,\ c_0 > 0$ such that for $h \in]0, h_0]$, $S_{h,\lambda}$ extends as a holomorphic function of $\lambda \in \mathcal{V}$, and moreover the conclusions of Theorem 17.6.3 and Lemma 17.6.4 remain valid when $\lambda \in \mathcal{V}$.*

Proof. We take $\lambda_0 > 0$ as in Theorem 17.5.2, and λ_1 as in Theorem 17.6.3

If $\lambda = \mu + \nu \in \mathcal{V}$ is taken as in (17.7.1), for $s \in \mathbf{R}$, the norms $\| \|_{\lambda,\mathrm{sc},s}$ and $\| \|_{\mu,\mathrm{sc},s}$ are uniformly equivalent. Also note that if $\left(P_h^0 - \lambda\right) u = v$, then

$$\left(P_h^0 - \mu\right) u = v + \nu u. \tag{17.7.11}$$

By (17.6.14) in Theorem 17.6.3, we deduce from (17.7.11) that given $s \in \mathbf{R}$, there is $C_s > 0$ such that

$$\|u\|_{\mu,\mathrm{sc},s+1/4} \leq C_s \left(\|v\|_{\mu,\mathrm{sc},s} + |\nu| \, \|u\|_{\mu,\mathrm{sc},s} \right). \tag{17.7.12}$$

Moreover, by (17.5.24),

$$\|u\|_{\mu,\mathrm{sc},s} \leq |\mu|^{-1/6} \|u\|_{\mu,\mathrm{sc},s+1/4} . \tag{17.7.13}$$

By (17.7.12), (17.7.13) we conclude that given $s \in \mathbf{R}$, if $c_0 > 0$ is small enough and $\lambda \in \mathcal{V}$, then

$$\|u\|_{\mu,\mathrm{sc},s+1/4} \leq C'_s \|v\|_{\mu,\mathrm{sc},s} . \tag{17.7.14}$$

Now we will show that for $c_0 > 0$ small enough, (17.7.14) is valid for any $s \in \mathbf{R}$. Given $\lambda \in \mathbf{C}$, we still define R_λ as in (17.5.17), that is

$$R_\lambda = \theta_0 P_{h,\lambda,\tau} \theta_0. \tag{17.7.15}$$

First we concentrate on the proof that for $c_0 > 0$ small enough, for $s \in \mathbf{R}$, the analogue of (17.5.22) holds for any $\lambda \in \mathcal{V}$, with R replaced by R_λ, the constant $C_s > 0$ depending only on s.

Indeed if $\lambda = \mu + \nu \in \mathcal{V}$, set

$$\mu' = \mu - \lambda_0 - \lambda_1, \tag{17.7.16}$$

so that

$$\operatorname{Re} \mu' \leq -\lambda_0. \tag{17.7.17}$$

By (17.7.16), (17.7.17), we get

$$|\mu| \leq C \, |\mu'| . \tag{17.7.18}$$

If U verifies the same support conditions as in (17.5.22), then $\theta_0^2 U = U$, and so

$$R_{\mu'} U = R_\lambda U + (\lambda_0 + \lambda_1 + \nu) \, U. \tag{17.7.19}$$

Therefore

$$|R_{\mu'} U|_{\mu',\mathrm{sc},s} \leq |R_\lambda U|_{\mu',\mathrm{sc},s} + (\lambda_0 + \lambda_1 + |\nu|) \, |U|_{\mu',\mathrm{sc},s} . \tag{17.7.20}$$

Using (17.5.24), (17.7.18), and (17.7.1), we get

$$|\nu| \, |U|_{\mu',\mathrm{sc},s} \leq C \frac{|\nu|}{|\mu|^{1/6}} |U|_{\mu',\mathrm{sc},s+1/4} \leq C c_0 \, |U|_{\mu',\mathrm{sc},s+1/4} . \tag{17.7.21}$$

By (17.7.20), (17.7.21), we obtain

$$|R_{\mu'} U|_{\mu',\mathrm{sc},s} \leq |R_\lambda U|_{\mu',\mathrm{sc},s} + (\lambda_0 + \lambda_1) \, |U|_{\mu',\mathrm{sc},s} + C c_0 \, |U|_{\mu',\mathrm{sc},s+1/4} . \tag{17.7.22}$$

Now because of (17.7.17), we can use the inequality (17.5.21), in which λ is replaced by μ', and R by $R_{\mu'}$. By combining this inequality with (17.7.22), we get

$$\begin{aligned}
&\left| \widehat{\nabla} U \right|^2_{\mu',\mathrm{sc},s} + \tau^{-4} \, |U|^2_{\mu',\mathrm{sc},s} + \tau^{-3/2} \, |U|^2_{\mu',\mathrm{sc},s+1/4} + \tau^{5/4} \left| \widehat{\nabla} U \right|^2_{\mu',\mathrm{sc},s+1/8} \\
&\leq C |R_\lambda U|^2_{\mu',\mathrm{sc},s} + C_s \tau^4 \left| \widehat{\nabla} U \right|^2_{\mu',\mathrm{sc},s} + C_s \tau^{-2} \, |U|^2_{\mu',\mathrm{sc},s} + C_s \tau^{5/4} \, |U|^2_{\mu',\mathrm{sc},s+1/8} \\
&\qquad + C \, (\lambda_0 + \lambda_1) \, |U|_{\mu',\mathrm{sc},s} + C c_0 \, |U|_{\mu',\mathrm{sc},s+1/4} .
\end{aligned} \tag{17.7.23}$$

Using (17.7.23) and also the fact that the constant C does not depend on s, we find by taking $c_0 > 0$ small enough and independent of s, a strict analogue of equation (17.5.21) with respect to the norms indexed by μ' and with R replaced by R_λ.

Now given $s \in \mathbf{R}$, the norms indexed by λ or by μ' are uniformly equivalent. This means that given $s \in \mathbf{R}$, we derive a strict analogue of (17.5.22) for $\lambda \in \mathcal{V}$, in which the original norms indexed by λ are considered, and $R = R_\lambda$. This proves the claim we made after (17.7.15).

It is now easy to continue our proof along the lines of the proof of Theorem 17.5.2, so as to obtain the analogue of this theorem. The proof of the analogue of Lemma 17.6.4 is now strictly similar to the proof of the lemma itself. The proof of our theorem is completed. □

17.8 A TRIVIALIZATION OVER X AND THE SYMBOLS $\mathcal{S}^{\mathbf{d},\mathbf{k}}_{\rho,\delta,\mathbf{c}}$

Let $x_0 \in X$ and let $x = \left(x^1, \dots, x^n\right)$ be the geodesic coordinate system centered at x_0 on a small open neighborhood U of x_0. We trivialize TX on U by parallel transport with respect to the connection ∇^{TX} along geodesics centered at x_0, and we trivialize F by parallel transport with respect to the connection ∇^F along these geodesics. Therefore $\Lambda^{\cdot}(T^*X) \widehat{\otimes} \Lambda^{\cdot}(TX) \widehat{\otimes} F$ has been identified to $\left(\Lambda^{\cdot}(T^*X) \widehat{\otimes} \Lambda^{\cdot}(TX) \widehat{\otimes} F\right)_{x_0}$ by parallel transport along geodesics centered at x_0 with respect to the connection $\nabla^{\Lambda^{\cdot}(T^*T^*X)\widehat{\otimes} F}$.

Set

$$V = \left(\Lambda^{\cdot}(T^*X) \widehat{\otimes} \Lambda^{\cdot}(TX) \widehat{\otimes} F\right)_{x_0}. \tag{17.8.1}$$

Note that over U, T^*X has been identified to $T^*X_{x_0}$ by the above metric preserving trivialization. In particular

$$T^*X|_U \simeq U \times T^*_{x_0}X. \tag{17.8.2}$$

Let d be the obvious trivial connection on $T^*_{x_0}X$. Then the Levi-Civita connection ∇^{T^*X} is given by

$$\nabla^{T^*X} = d + \Gamma^{T^*X}, \tag{17.8.3}$$

so that Γ^{T^*X} is a 1-form valued in antisymmetric elements of $\operatorname{End}\left(T^*_{x_0}X\right)$.

If $x \in U$, let $\sigma_x : T^*_{x_0}X \to T^*_xX$ be the identification which is obtained by parallel transport with respect to ∇^{T^*X} along the geodesic connecting x_0 and x. Then σ_x is an isometry. The canonical 1-form θ on T^*X is given by $\sigma_x p$. In our coordinate system, the symplectic form ω is given by

$$\omega = d\sigma_x p. \tag{17.8.4}$$

Let dx be the Euclidean volume form on $T_{x_0}X$, let dp be the Euclidean volume form on $T^*_{x_0}X$. Let $k(x)$ be the smooth positive function on U such that in the given coordinates,

$$dv_X(x) = k(x)\, dx. \tag{17.8.5}$$

By the above, the symplectic volume form on T^*X is given over U by $dv_X(x)\,dp$.

In the above coordinate system, the vector field $Y^{\mathcal{H}}$ is given by

$$Y^{\mathcal{H}}(x,p) = \left(g^{TX}\right)^{-1} \sigma p - \Gamma^{T^*X}\left(\left(g^{TX}\right)^{-1} \sigma p\right) p. \tag{17.8.6}$$

The right-hand side of (17.8.6) gives the canonical splitting of $Y^{\mathcal{H}}$ as the sum of a horizontal and a vertical vector field in the (x,p) coordinates. Note that since Γ^{T^*X} takes its values in antisymmetric matrices, both components of $Y^{\mathcal{H}}$ in (17.8.6) preserve the function $|p|^2$. This is compatible with the fact that the Hamiltonian vector field $Y^{\mathcal{H}}$ preserves $\mathcal{H} = |p|^2/2$.

We denote by $\mathcal{S}$ the Schwartz space of functions of $p \in T^*_{x_0}X$ with values in V.

Recall that $\lambda_1 > 0$ was obtained in Theorem 17.6.1. In the sequel, we still denote by λ_1 a positive real number, possibly smaller than the one found in Theorem 17.6.1.

Clearly, in our given coordinate system, $T^*U \simeq U \times \mathbf{R}^n$. For technical reasons, in the sequel, we have to distinguish $T^*_{x_0}X$ from $\mathbf{R}^n$. Indeed here p varies in $T^*_{x_0}X$, while $\xi \in: R^n$.

For $p, \xi \in T^*_{x_0}X$, set

$$\begin{aligned} A(\xi,p,\lambda) &= <p>^2 + \frac{|2\lambda_1 - \lambda|}{<p>} + <\xi>, \\ B(\xi,p,\lambda) &= \left(\frac{|2\lambda_1 - \lambda|}{<p>} + <\xi>\right)^{2/3}. \end{aligned} \tag{17.8.7}$$

To keep in line with the notation already used, if $\tau \in]0,1], p, \xi \in T^*_{x_0}X$, set

$$\begin{aligned} A_\tau(\xi,\lambda) &= \tau^{-2} + \tau|2\lambda_1 - \lambda| + <\xi>, \\ B_\tau(\xi,\lambda) &= \left(\tau|2\lambda_1 - \lambda| + <\xi>\right)^{2/3}. \end{aligned} \tag{17.8.8}$$

It will be convenient to choose an isometric identification of $T^*_{x_0}X$ with $\mathbf{R}^n$.

For $u \in \mathcal{S}$, we consider again the Littlewood-Paley decomposition of U as in (17.4.1). In particular, we still have the identity

$$u(p) = \sum_{j=0}^{+\infty} U_j\left(2^{-j}p\right). \tag{17.8.9}$$

Let Δ^V be the Laplacian on $T^*_{x_0}X$. Let J_τ be the positive self-adjoint operator

$$J_\tau(\xi,\lambda) = \left(-\Delta^V + A^2_\tau(\xi,\lambda)\right)^{1/2}. \tag{17.8.10}$$

For $s \in \mathbf{R}, u \in \mathcal{S}$, set

$$|U|_{p,s,j} = 2^{jn/2}|J^s_{2^{-j}}U|_{L^2}, \qquad \|u\|^2_{p,s} = \sum_{j=0}^{\infty} |U_j|^2_{p,s,j}. \tag{17.8.11}$$

Given $s \in \mathbf{R}$, the norms in (17.8.11) depend on ξ, and for a given ξ, they are mutually equivalent. We denote by $\mathcal{H}_{p,s}$ the completion of $\mathcal{S}$ for the norm $\| \|_{p,s}$. Note that $\mathcal{H}_{p,s}$ does not depend on the choice of ξ.

In the sequel we will write that two norms are related by the equivalence sign $\simeq$ if they are uniformly equivalent with respect to the given family of parameters.

Observe that for $k \in \mathbf{N}$,

$$2^{-jn/2} |U_j|_{p,k,j} \simeq \sum_{|\beta|+l \leq k} A^l_{2^{-j}} \left|\partial^\beta_p U_j\right|_{L^2}. \tag{17.8.12}$$

Let $\theta_j(\xi, p), j \in \mathbf{N}$ be a family of smooth radial functions of $p \in \mathbf{R}^n$ depending on the parameter $\xi \in \mathbf{R}^n$ for which, given any multiindex β, there exists $C_\beta > 0$ such that if $j \in \mathbf{N}, p \in \mathcal{B}_0$,

$$|\partial^\beta_p \theta_j| \leq C_\beta A^{|\beta|}_{2^{-j}}. \tag{17.8.13}$$

By (17.8.12) and (17.8.13), if $s \in \mathbf{N}$,

$$|\theta_j U_j|_{p,s,j} \leq C_s |U_j|_{p,s,j}, \tag{17.8.14}$$

with a constant C_s depending only on the C_β for $|\beta| \leq s$. By duality and interpolation, we find that (17.8.14) is still valid for any $s \in \mathbf{R}$.

Let $m(\xi, r, h, \lambda)$ be a positive function, which is smooth in $r \in [1, +\infty[$, with parameters ξ, h, λ, such that for any $k \in \mathbf{N}$, there exist $C_k > 0$ for which

$$\left|\frac{\partial^k}{\partial r^k} m\right| \leq C_k m. \tag{17.8.15}$$

For $\tau \in]0, 1]$, set

$$m_\tau(\xi, h, \lambda) = m\left(\xi, \tau^{-1}, h, \lambda\right). \tag{17.8.16}$$

We assume that there exists $C > 0$ such that if $p \in \mathcal{B}$ for $j = 0$, or for $p \in \mathcal{R}$ for $j \geq 1$,

$$\frac{1}{C} m_{2^{-j}}(\xi, h, \lambda) \leq m\left(\xi, < 2^j p >, h, \lambda\right) \leq C m_{2^{-j}}(\xi, h, \lambda). \tag{17.8.17}$$

If m_1, m_2 verify (17.8.15) and (17.8.17), if $a \in \mathbf{R}$, the functions $m_1 + m_2$, $m_1 m_2, m_1^a$ verify the same equations.

Set

$$\theta_j = \frac{m\left(\xi, < 2^j p >, h, \lambda\right)}{m_{2^{-j}}(\xi, h, \lambda)}. \tag{17.8.18}$$

Observe that $2^{2j} \leq A_{2^{-j}}$. Using the weaker $2^j \leq A_{2^{-j}}$ and also (17.8.15), (17.8.17), we find that θ_j verifies (17.8.13).

If $u \in \mathcal{S}$, we can define the function $m(\xi, < p >, h, \lambda) u(p)$, which we denote mu for simplicity.

Using (17.8.9), we get

$$mu = \sum_{j=0}^{\infty} m(\xi, < p >, h, \lambda) U_j\left(2^{-j} p\right). \tag{17.8.19}$$

By (17.8.14), (17.8.15), and using the fact that as we just saw, $1/m$ verifies the same assumptions as m, there exists $C_s > 0$ such that

$$\frac{1}{C_s} m_{2^{-j}} \left|U_j\right|_{p,s,j} \leq \left|m\left(\xi, < 2^j p >, h, \lambda\right) U_j\right|_{p,s,j} \leq C_s m_{2^{-j}} \left|U_j\right|_{p,s,j}. \tag{17.8.20}$$

By (17.8.17), there exists $C(\xi, h, \lambda) > 0, M \in \mathbf{N}$ such that

$$\left|m\left(\xi, r, h, \lambda\right)\right| \leq C\left(\xi, h, \lambda\right)\left(1+r\right)^M. \tag{17.8.21}$$

By (17.8.15), similar inequalities hold for the $\frac{\partial^k}{\partial r^k} m$. It follows that for given h, λ, ξ, then $mu \in \mathcal{S}$. From (17.8.20), we deduce that for any $\xi, h \in]0, h_0], \lambda \in \mathbf{C}, u \in \mathcal{S}$,

$$\frac{1}{C_s} \sum_{j=0}^{\infty} m^2_{2^{-j}} \left|U_j\right|^2_{p,s,j} \leq \left\|mu\right\|^2_{p,s} \leq C_s \sum_{j=0}^{\infty} m^2_{2^{-j}} \left|U_j\right|^2_{p,s,j}. \tag{17.8.22}$$

Note that by the above, (17.8.22) contains only finite expressions.

One verifies easily that the functions $r, A_{r^{-1}}(\xi, \lambda), B_{r^{-1}}(\xi, \lambda)$ verify the conditions in (17.8.15), (17.8.17). Therefore if $N, a, b \in \mathbf{R}$, any monomial $r^N A^a_{r^{-1}} B^b_{r^{-1}}$ also verifies these conditions. Using (17.8.22), given $a, b, N \in \mathbf{R}$, we get the equivalence of norms,

$$\left\|< p >^N A^a B^b u\right\|^2_{p,s} \simeq \sum_{j=0}^{\infty} 2^{2jN} A^{2a}_{2^{-j}} B^{2b}_{2^{-j}} \left|U_j\right|^2_{p,s,j}. \tag{17.8.23}$$

Now we will introduce a class of pseudodifferential operators on U with values in operators acting on $\mathcal{S}$. Our classes extend the classes $S^m_{\rho,\delta}$ defined in [Hör85, section 18.1].

Definition 17.8.1. Let $e(x, \xi, h, \lambda)$ be a function which is smooth in $x \in \mathbf{R}^n, \xi \in \mathbf{R}^n$, holomorphic in $\lambda \in \mathbf{C}, \operatorname{Re}\lambda \leq \lambda_1$, which depends also on $h \in]0, h_0]$, and which takes its values in the set of linear continuous operators acting on $\mathcal{S}$.

Let $0 \leq \delta < \rho \leq 1 \in \mathbf{R}$, $c = (c_0, c_1) \in \mathbf{R}^2$, and $d, k \in \mathbf{R}$. The function e will be said to be a symbol in the class $\mathcal{S}^{d,k}_{\rho,\delta,c}$ if for any $s \in \mathbf{R}, N \in \mathbf{N}$, for any multiindices α, β, there exists $C_{s,N,\alpha,\beta} > 0$ such that if $u \in \mathcal{S}$,

$$\begin{aligned} &\left\|< p >^N \partial^\alpha_x \partial^\beta_\xi e\left(x, \xi, h, \lambda\right) u\right\|_{p,s} \\ &\qquad \leq C_{s,N,\alpha,\beta} \left\|< p >^{N+k+c_0|\alpha|+c_1|\beta|} u\right\|_{p,s+d+\delta|\alpha|-\rho|\beta|}. \end{aligned} \tag{17.8.24}$$

Here, if e, e' are symbols taken as above, we denote by ee' the pointwise product of e and e'. This is *not* the product of the symbols in any class of pseudodifferential operators. If $e \in \mathcal{S}^{d,k}_{\rho,\delta,c}$, $e' \in \mathcal{S}^{d',k'}_{\rho,\delta,c}$, then $ee' \in \mathcal{S}^{d+d',k+k'}_{\rho,\delta,c}$. Moreover, if e is taken as before, $\partial^\alpha_x \partial^\beta_\xi e \in \mathcal{S}^{d+\delta|\alpha|-\rho|\beta|,k+c_0|\alpha|+c_1|\beta|}_{\rho,\delta,c}$. If $c = (c_0, c_1)$ is such that $c_0 \geq 0$, any smooth function $\varphi(x)$ with bounded derivatives of any order lies in $\mathcal{S}^{0,0}_{\rho,\delta,c}$. Also note that tautologically, $\partial_x, \partial_\xi, < p >$ commute, so that the order of the operators in (17.8.24) is irrelevant.

If $e \in \mathcal{S}^{d,k}_{\rho,\delta,c}$, the L^2-adjoint $e^{*,p}$ of e in the variable p also lies in $\mathcal{S}^{d,k}_{\rho,\delta,c}$.

If the support of $u(x,\cdot) \in \mathcal{S}$ is compact, the partial Fourier transform $\widehat{u}(\xi,p)$ of $u(x,p)$ in the variable x is given by the obvious formula

$$\widehat{u}(\xi,p) = \int_{\mathbf{R}^n} e^{-i\langle x,\xi\rangle} u(x,p)\, dx. \tag{17.8.25}$$

Definition 17.8.2. Let $\mathcal{P}^{d,k}_{\rho,\delta,c}$ be the set of semiclassical pseudodifferential operators $E(x, hD_x, h, \lambda)$ with values in operators acting on the space $\mathcal{S}$, such that if $u \in \mathcal{S}$, and if the support of $u(x,\cdot)$ is compact, there is $e(x,\xi,h,\lambda) \in \mathcal{S}^{d,k}_{\rho,\delta,c}$ such that

$$E(x, hD_x, h, \lambda)\, u(x,p) = (2\pi h)^{-n} \int_{\mathbf{R}^n} e^{\frac{i}{h}\langle x,\xi\rangle} \left(e(x,\xi,h,\lambda)\,\widehat{u}\right)\left(\frac{\xi}{h}, p\right) d\xi. \tag{17.8.26}$$

Let $E_1(x, hD_x, h, \lambda) \in \mathcal{P}^{d_1,k_1}_{\rho,\delta,c}$ and $E_2(x, hD_x, h, \lambda) \in \mathcal{P}^{d_2,k_2}_{\rho,\delta,c}$ be compactly supported in x. Then one has the classical formula for E_1E_2,

$$E_1E_2 = E_3, \tag{17.8.27}$$

with E_3 associated to e_3 given by the following formula [Hör85, chapter 18]:

$$\begin{aligned} e_3(x,\xi,h,\lambda) = {} & \sum_{|\beta| \le M} \frac{(h/i)^{|\beta|}}{\beta!} \partial_\xi^\beta e_1 \partial_x^\beta e_2 \\ & + (h/i)^{M+1}(M+1) \sum_{|\beta|=M+1} \frac{R_{M,\beta}}{\beta!}, \end{aligned} \tag{17.8.28}$$

$$\begin{aligned} R_{M,\beta} = {} & (2\pi)^{-n} \int_0^1 (1-t)^M dt \int_{\mathbf{R}^n \times \mathbf{R}^n} e^{-iu\theta} \partial_\xi^\beta e_1(x, \xi + th\theta) \\ & \times \partial_x^\beta e_2(x+u,\xi)\, du d\theta = (2\pi)^{-n} \int_0^1 (1-t)^M dt \int_{\mathbf{R}^n \times \mathbf{R}^n} e^{-iu\theta} \frac{(1-\Delta_u)^L}{(1+|\theta|^2)^L} \\ & \times \partial_\xi^\beta e_1(x,\xi+th\theta)\, \partial_x^\beta e_2(x+u,\xi)\, du d\theta. \end{aligned}$$

In general, the symbol e_3 does not lie in any of the above classes, because of a lack of control in the powers of p. The purpose of many of the manipulations which follow is to circumvent this difficulty. Equivalently, the classes $\mathcal{P}^{\cdot,\cdot}_{\rho,\delta,c}$ do not form an algebra under composition, except when $c = (0,0)$.

Moreover, if $E(x, hD_x, h, \lambda) \in \mathcal{P}^{d,k}_{\rho,\delta,c}$ is compactly supported in x, its adjoint E^* is associated to the symbol e^* given by

$$e^*(x,\xi,h,\lambda) = \sum_{|\beta|\le M} \frac{(h/i)^{|\beta|}}{\beta!} \partial_x^\beta \partial_\xi^\beta e^{*,p} + (h/i)^{M+1}(M+1) \sum_{|\beta|=M+1} \frac{R_{*,M,\beta}}{\beta!}, \tag{17.8.29}$$

$$R_{*,M,\beta} = (2\pi)^{-n} \int_0^1 (1-t)^M dt \int e^{-iu\theta} \partial_x^\beta \partial_\xi^\beta e^{*,p}(x+u, \xi+th\theta)\, du d\theta.$$

From (17.8.28), (17.8.29), one can get estimates on $R_{M,\beta}, R_{*,M,\beta}$.

Remark 17.8.3. Observe that (17.8.24) indicates that the operators considered above induce a loss in the control at infinity in the variable p. Note that the condition $\rho > \delta$ is essential in guaranteeing that the operators considered above can be composed.

In the sequel we will not insist on the fact that the classes of symbols we consider be invariant under change of coordinates, even though the operators we will consider are globally defined on T^*X, i.e., are themselves invariant under these changes.

Let $\psi : \mathbf{R}^n \to \mathbf{R}^n$ be a diffeomorphism. The action of ψ on smooth real functions $f(\xi)$ is given by $f(\xi) \to f\left(\widetilde{\psi}'^{-1}(x)\,\xi\right)$. More generally ψ also acts on $\mathcal{S} \otimes V$ by an action K_ψ, which incorporates the action of ψ on $\Lambda^{\cdot}(T^*X)\,\widehat{\otimes}\Lambda^{\cdot}(TX)\,\widehat{\otimes}F$. Given our choice of trivialization, the actions of ψ on the ξ and p variables do not coincide.

Let $E(x, hD_x, h, \lambda)$ be a properly supported operator whose symbol is denoted $e(x, \xi, h, \lambda)$. Under ψ, $e(x, \xi, h, \lambda)$ is changed into $e_\phi(y, \eta, h, \lambda)$, with

$$e_\psi(\psi(x), \eta, h, \lambda) = e^{-i\langle\psi(x),\eta\rangle/h} K_\psi e(x, hD_x, h, \lambda)\, e^{i\langle\psi(x),\eta\rangle/h} K_\psi^{-1}. \tag{17.8.30}$$

From these considerations, we see that operators of the type $\xi\partial_\xi$ have to satisfy estimates which are compatible with the corresponding estimate involving ∂_x. The fact that the estimates with $\xi\partial_\xi$ should be compatible with those with ∂_x leads to the inequalities

$$c_0 \geq c_1, \qquad \rho + \delta \geq 1. \tag{17.8.31}$$

Note that the second condition in (17.8.31) is exactly the one which appears in [Hör85, section 18.1, p. 94] for the pseudodifferential operators associated to symbols in $S^m_{\rho,\delta}$.

In the sequel, we use the notation

$$\begin{aligned} \mathcal{S}^d_{\rho,\delta} &= \mathcal{S}^{d,0}_{\rho,\delta,(0,0)}, & \mathcal{P}^d_{\rho,\delta} &= \mathcal{S}^{d,0}_{\rho,\delta,(0,0)}, \\ \mathcal{S}^{d,k} &= \mathcal{S}^{d,k}_{2/3,1/3,(2,2)}, & \mathcal{P}^{d,k} &= \mathcal{S}^{d,k}_{2/3,1/3,(2,2)}. \end{aligned} \tag{17.8.32}$$

Lemma 17.8.4. *Let $K \subset \mathbf{R}^n$ be compact. Let $E(x, hD_x, h, \lambda) \in \mathcal{P}^{d,k}_{\rho,\delta,c}$ be such that its symbol $e(x, \xi, h, \lambda)$ vanishes for $x \notin K$. For $s \in \mathbf{R}, N \in \mathbf{N}$, there exist $M_s > 0, C_{s,N} > 0$ such that for $u(x, \cdot)$ with support included in K,*

$$\left\| <p>^N Eu \right\|_{\lambda,\mathrm{sc},s} \leq C_{s,N} \left\| <p>^{N+M_s} u \right\|_{\lambda,\mathrm{sc},s+d}. \tag{17.8.33}$$

Proof. Peetre's inequality asserts that if $a \in \mathbf{R}^*_+, \xi, \eta \in \mathbf{R}^n$,

$$\left(a + |\xi + \eta|^2\right) \leq \frac{2}{a}\left(a + |\xi|^2\right)\left(a + |\eta|^2\right). \tag{17.8.34}$$

In particular if $a \geq 1$,

$$\left(a + |\xi + \eta|^2\right) \leq 2\left(1 + |\eta|^2\right)\left(a + |\xi|^2\right). \tag{17.8.35}$$

By (17.8.35), we get

$$\frac{1}{C_s} < \eta >^{-|s|} \|u\|_{p,s,\xi} \leq \|u\|_{p,s,\xi+\eta} \leq C_s < \eta >^{|s|} \|u\|_{p,s,\xi} \,. \tag{17.8.36}$$

By (17.8.28), (17.8.29) and (17.8.36), we find that if $E_1 \in \mathcal{P}^{d_1}_{\rho,\delta}$ and $E_2 \in \mathcal{P}^{d_2}_{\rho,\delta}$ are compactly supported in the variable x, then $E_1E_2 \in \mathcal{P}^{d_1+d_2}_{\rho,\delta}$, and also that if E_1^* is the adjoint of E_1, then $E_1^* \in \mathcal{P}^{d_1}_{\rho,\delta}$. In this argument we use explicitly the fact that $\delta < \rho$. Note that stability under composition comes from the fact that, as explained in Remark 17.8.3, when $c = (0,0)$, there is no loss in powers of p in the definition of the composition in (17.8.27), (17.8.29).

By using the same arguments as in the proof in [Hör85, Theorem 18.1.11] of the continuity of the pseudodifferential operators of order 0 acting on L^2, we find that given $d \in \mathbf{R}$, $s \in \mathbf{R}$, there exists $L_s \in \mathbf{N}$ such that if the symbol $e(x,\xi,h,\lambda)$ of E verifies the estimates in (17.8.24) with $k = 0, c = (0,0)$ when $|\alpha| + |\beta| \leq L_s$, for any $N \in \mathbf{N}$, there is $C_{s,N} > 0$ such that

$$|| < p >^N Eu||_{\lambda,\mathrm{sc},s} \leq C_{s,N}|| < p >^N u||_{\lambda,\mathrm{sc},s+d}. \tag{17.8.37}$$

To complete the proof, we observe that given $s \in \mathbf{R}$ and $E(x, hD_x, h, \lambda) \in \mathcal{P}^{d,k}_{\rho,\delta,c}$, there exists $M \in \mathbf{N}$ such that $e(x,\xi,h,\lambda) < p >^{-M}$ verifies precisely the above assumptions. This completes the proof of (17.8.33). □

17.9 THE SYMBOL $\mathbf{Q^0_h(x,\xi) - \lambda}$ AND ITS INVERSE $\mathbf{e_{0,h,\lambda}(x,\xi)}$

In this section, we take $x_0 \in X$ as in section 17.8, and we use the corresponding notation. In particular, we use the trivialization of the various vector bundles which was considered there. So we may as well consider U as an open set in $\mathbf{R}^n$.

The operator P^0_h is considered as an operator acting on smooth sections of V over $\pi^{-1}U$. It will be more convenient to view P^0_h as acting on smooth sections of $\mathcal{S} \otimes V$ over U.

Take $(x,\xi) \in T^*U \simeq U \times \mathbf{R}^n$. In our semiclassical setting, we define the semiclassical symbol $Q_h(x,\xi)$ of P_h to be given by

$$Q_h(x,\xi) = e^{-i\langle x,\xi\rangle/h} P_h e^{i\langle x,\xi\rangle/h}. \tag{17.9.1}$$

We define the symbol $Q^0_h(x,\xi)$ of P^0_h by a similar formula.

Recall that P_h is given by (17.2.3). By using (17.8.6), we get

$$Q_h(x,\xi) = \alpha_\pm \mp \left(i \left\langle \left(g^{TX}\right)^{-1} \sigma p, \xi \right\rangle - h \left\langle \Gamma^{TX} \left(\left(g^{TX}\right)^{-1} \sigma p \right) e_i, e_j \right\rangle \right.$$
$$\left. \left(e^i i_{e_j} + \widehat{e}_i i_{\widehat{e}^j}\right) \right) \pm h \widehat{\nabla}_{\Gamma^{T^*X}\left((g^{TX})^{-1}\sigma p\right)p} - \frac{h}{2} \omega\left(\nabla^F, g^F\right)(e_i) \nabla_{\widehat{e}^i} + h^2 \gamma_\pm. \tag{17.9.2}$$

Of course all the matrix operators in (17.9.2) are evaluated in the considered trivializations. Moreover,

$$Q_h^0(x,\xi) = Q_h(x,\xi) + P_{\pm}. \tag{17.9.3}$$

It is fundamental to observe that in our trivialization, the operators $\alpha_{\pm}$ and $P_{\pm}$ are constant, i.e., they do not depend on $x \in U$. Also as we saw after (17.8.6),

$$\widehat{\nabla}_{\Gamma^{T^*X}((g^{TX})^{-1}\sigma p)p} |p|^2 = 0. \tag{17.9.4}$$

Let L^2 be the vector space of square integrable sections of V in the variable $p \in \mathbf{R}^n$. If $u \in L^2$, we denote by $\|u\|$ the norm of u in L^2. We still denote by $\widehat{\nabla}$ differentiation along the vertical $\mathbf{R}^n$, and by Δ^V the corresponding Laplacian in the variable p.

Given x, ξ, we consider $Q_h^0(x,\xi)$ as an unbounded operator acting on L^2 with domain

$$D_h^0(x,\xi) = \{u \in L^2, Q_h^0(x,\xi)\,u \in L^2\}. \tag{17.9.5}$$

The graph of $Q_h^0(x,\xi)$ in $L^2 \times L^2$ is closed. We equip $D_h^0(x,\xi)$ with the norm of the graph induced by $L^2 \times L^2$.

Recall that the functions $A(\xi,p,\lambda), B(\xi,p,\lambda)$ were defined in (17.8.7). First we state a fundamental theorem which gives L^2-estimates for $Q_h^0(x,\xi)$.

In the sequel, we use the notation

$$\lambda' = 2\lambda_1 - \lambda, \tag{17.9.6}$$

so that if $\operatorname{Re}\lambda \le \lambda_1$, then $\operatorname{Re}\lambda' \ge \lambda_1$.

Let $U' \subset U$ be an open set such that $\overline{U}' \subset U$. In what follows, $\| \ \|$ denotes the standard L^2 norm.

Theorem 17.9.1. *Given $\lambda_1 \in]0,1/2[$, there exist $h_0 \in]0,1]$ such that for $h \in]0,h_0], (x,\xi) \in T^*U', \lambda \in \mathbf{C}, \operatorname{Re}\lambda \le \lambda_1$, then $Q_h^0(x,\xi) - \lambda$ is one to one from $D_h^0(x,\xi)$ into L^2, and moreover $C_0^\infty(\mathbf{R}^n, V)$ is dense in $D_h^0(x,\xi)$.*

For any $a, b \in \mathbf{R}$, there exist $C_a > 0, C_b > 0$ such that if $\lambda \in \mathbf{C}$ is taken as before, if $u, v \in L^2$, and $\left(Q_h^0(x,\xi) - \lambda\right) u = v$, then

$$\begin{aligned} &\left\| <p>^a \widehat{\nabla} u \right\| + \left\| <p>^{a+1} u \right\| \le C_a \left\| <p>^{a-1} v \right\|, \\ &\left\| <p>^b \left(|\Delta^V u| + B(\xi,p,\lambda)\,|u|\right) \right\| \le C_b \left\| <p>^b (1 + h <p>) v \right\|. \end{aligned} \tag{17.9.7}$$

Proof. Set

$$L = \alpha_{\pm} + P_{\pm} - 2\lambda_1. \tag{17.9.8}$$

Observe that the operator L is self-adjoint and positive for $\lambda_1 < 1/2$.

In the sequel we fix λ_1 such that $0 < \lambda_1 < 1/2$. Then there exists $C > 0$ such that

$$\langle Lu, u \rangle \ge C \int_{\mathbf{R}^n} \left(|\widehat{\nabla} u|^2 + <p>^2 |u|^2 \right) dp. \tag{17.9.9}$$

Recall that λ' was defined in (17.9.6). Let $\theta(p)$ be a radial smooth compactly supported function, and let $u \in D_h^0(x,\xi)$. Since $Q_h(x,\xi)$ is a second order elliptic operator, then $u \in H^2_{\text{loc}}$. Using the specific form of $\beta_\pm, \gamma_\pm$ in (15.1.3) and also (17.9.2), we find that there exists $C > 0$ such that

$$\operatorname{Re}\left\langle \left(Q_h^0(x,\xi) - \lambda\right) u, \theta^2 u\right\rangle \geq \operatorname{Re}\left\langle (L+\lambda') u, \theta^2 u\right\rangle \\ - Ch \int_{\mathbf{R}^n} \left(< p >^2 \left|\widehat{\nabla}\theta^2\right| |u|^2 + \theta^2 \left(< p > |u|^2 + \left|\widehat{\nabla} u\right| |u|\right)\right) dp \\ - Ch^2 \int_{\mathbf{R}^n} < p >^2 \theta^2 |u|^2 \, dp. \quad (17.9.10)$$

Moreover, by (17.9.9), we get

$$\operatorname{Re}\left\langle (L+\lambda') u, \theta^2 u\right\rangle = \operatorname{Re}\langle (L+\lambda')\theta u, \theta u\rangle + \operatorname{Re}\langle [\theta, L] u, \theta u\rangle \\ \geq C \int_{\mathbf{R}^n} \theta^2 \left(\left|\widehat{\nabla} u\right|^2 + < p >^2 |u|^2\right) dp + \operatorname{Re}\lambda' \int_{\mathbf{R}^n} \theta^2 |u|^2 \, dp \\ - C' \int_{\mathbf{R}^n} \left(\left|\widehat{\nabla}\theta\right|^2 + \left|\theta\Delta^V \theta\right|\right) |u|^2 \, dp - \|\theta u\| \, \|[P_\pm, \theta] u\| . \quad (17.9.11)$$

Using the fact that if $\operatorname{Re}\lambda \leq \lambda_1$, then $\operatorname{Re}\lambda' \geq \lambda_1$, by (17.9.10), (17.9.11), for $h_0 > 0$ small enough and $h \in]0, h_0]$, we get

$$\operatorname{Re}\left\langle \left(Q_h^0(x,\xi) - \lambda\right) u, \theta^2 u\right\rangle \geq C \int_{\mathbf{R}^n} \theta^2 \left(|\widehat{\nabla} u|^2 + < p >^2 |u|^2\right) dp \\ + \lambda_1 \int_{\mathbf{R}^n} \theta^2 |u|^2 \, dp - C' \Bigg(h \int_{\mathbf{R}^n} < p >^2 |\widehat{\nabla}\theta^2| |u|^2 \, dp \\ + \int_{\mathbf{R}^n} \left(\left|\widehat{\nabla}\theta\right|^2 + \left|\theta\Delta^V\theta\right|\right) |u|^2 dp \Bigg) - \|\theta u\| \, \|[P_\pm, \theta] u\| . \quad (17.9.12)$$

Take $\theta(p) = \gamma(< p >)\rho(\varepsilon p)$, where γ is a power of $< p >$, and ρ is a smooth compactly supported radial function which is equal to 1 near $p = 0$. We choose first $\deg(\gamma) = -1/2$. As $\varepsilon \to 0$, the negative terms in the right-hand side of (17.9.12) are easily controlled. By making $\varepsilon \to 0$ in (17.9.12), we find that $< p >^{1/2} u \in L^2$. The same argument with different choices of $\deg\gamma \in [0,1]$ show that if $u \in D_h^0(x,\xi)$, then $< p >^2 u \in L^2$, $< p > \widehat{\nabla} u \in L^2$.

In particular, when taking $\theta = 1$ in (17.9.12) and using Cauchy-Schwarz, we obtain

$$\left\|\widehat{\nabla} u\right\| + \|< p > u\| \leq C \left\|\left(Q_h^0(x,\xi) - \lambda\right) u\right\| . \quad (17.9.13)$$

By taking $\deg\gamma = 1$ and using (17.9.13) to control $\|\theta u\| \, \|[P_\pm, \theta] u\|$, we get

$$\left\|< p > \widehat{\nabla} u\right\| + \|< p >^2 u\| \leq C \left\|\left(Q_h^0(x,\xi) - \lambda\right) u\right\| . \quad (17.9.14)$$

If $u \in D_h^0(x,\xi)$, if $u_\varepsilon = \rho(\varepsilon p) u$, then $u_\varepsilon \in D_h^0(x,\xi)$, and from (17.9.14), we find that as $\varepsilon \to 0$,

$$\left\| \left[Q_h^0(x,\xi), \rho(\varepsilon p)\right] u\right\| \to 0. \quad (17.9.15)$$

Therefore, as $\varepsilon \to 0$,

$$\|u_\varepsilon - u\| + \left\|Q_h^0(x,\xi)(u_\varepsilon - u)\right\| = 0. \tag{17.9.16}$$

By (17.9.16), we deduce easily that $C_0^\infty(\mathbf{R}^n, V)$ is dense in $D_h^0(x,\xi)$. In particular the adjoint $Q_h^{0*}(x,\xi)$ of $Q_h^0(x,\xi)$ is equal to the formal adjoint of $Q_h^0(x,\xi)$ with domain $D_h^{0*}(x,\xi) = \{u \in L^2, Q_h^{0*}(x,\xi)\, u \in L^2\}$. Observe that $Q_h^0(x,\xi)$ and $Q_h^{0*}(x,\xi)$ have the same structure. By (17.9.14), both are injective with closed range, so that $Q_h^0(x,\xi)$ is one to one from $D_h^0(x,\xi)$ into L^2.

Using now (17.9.12) with a symbol $\theta = <p>^a$, and noting that for any $b, c \in \mathbf{R}$, $<p>^c P_\pm <p>^b$ is a bounded operator, we get the first part of (17.9.7).

Now we establish the second part of (17.9.7). Assume that $u \in D_h^0(x,\xi)$, and set $\left(Q_h^0(x,\xi) - \lambda\right) u = v$. Put

$$v' = \left(-\frac{1}{2}\Delta^V \mp i\left\langle \left(g_x^{TX}\right)^{-1} \sigma_x p, \xi\right\rangle + \lambda'\right) u. \tag{17.9.17}$$

Note that by (15.1.3) and (17.9.2), we can give an explicit formula for v' in terms of u and v. By using the first inequality in (17.9.7) with $a = 0, 1, 2$, we get

$$\|v'\| \le C\left(\|v\| + h\,\|<p>v\|\right). \tag{17.9.18}$$

Incidentally observe that

$$\|(1 + h<p>)\, v\| \le \|v\| + h\,\|<p>v\| \le \sqrt{2}\left(\|(1+h<p>)\, v\|\right). \tag{17.9.19}$$

By (17.9.18), we find that to establish the second part of (17.9.7) with $b = 0$, we only need to establish the inequalities

$$\left\|\Delta^V u\right\| + |\xi|^{2/3}\|u\| \le C\,\|v'\|, \quad \left\|\left(\frac{|\lambda'|}{<p>}\right)^{2/3} u\right\| \le C\,\|(1+h<p>)\, v\|. \tag{17.9.20}$$

First we assume that $\xi = 0$. Observe that since $\operatorname{Re}\lambda' \ge \lambda_1 > 0$, the operator $\frac{-\Delta^V}{-\Delta^V + 2\lambda'}$ acts as a bounded operator on L^2. Therefore the first inequality in (17.9.20) follows from (17.9.17). Using (17.9.17) again and the inequality we just proved, we get

$$|\lambda'|\,\|u\| \le C\,\|v'\|, \tag{17.9.21}$$

which combined with (17.9.18), (17.9.19) also leads to a proof of the second inequality in (17.9.20).

Now we prove (17.9.20) with $\xi \neq 0$. Set

$$k(x) = \pm\widetilde{\sigma}_x\left(g_x^{TX}\right)^{-1}\xi. \tag{17.9.22}$$

By (17.9.2), the ξ dependent part of $Q_h^0(x,\xi)$ is given by $\mp i\langle k, p\rangle$. By making an x dependent rotation of the coordinates p, we may and we will assume that $k = \pm(|k|, 0, \ldots, 0)$. Recall that for $a \in \mathbf{R}$, $K_a u(p) = u(ap)$. Set

$$u' = K_{|k|^{-1/3}} u, \qquad w = |k|^{-2/3} K_{|k|^{-1/3}} v', \qquad \mu = |k|^{-2/3}\lambda'. \tag{17.9.23}$$

Then equation (17.9.17) can be written in the form

$$w = \left(-\frac{1}{2}\Delta^V + \mu - ip_1\right) u'. \tag{17.9.24}$$

Then the first inequality in (17.9.20) is equivalent to the inequality

$$\left\|\Delta^V u'\right\| + \|u'\| \leq C \|w\|. \tag{17.9.25}$$

We concentrate on the proof of (17.9.25).

We denote by $\widehat{u}'(\eta)$ the Fourier transform of u' in the variable p. Then equation (17.9.24) is equivalent to

$$\widehat{w} = \left(\frac{\partial}{\partial \eta^1} + \mu + \frac{1}{2}|\eta|^2\right) \widehat{u}'. \tag{17.9.26}$$

Now we write $\eta = (\eta^1, \eta')$, with $\eta' \in \mathbf{R}^{n-1}$. Since $\widehat{u}'$ is a tempered distribution, by (17.9.26), we can express it in the form

$$\widehat{u}'(\eta^1, \eta') = \int_0^{+\infty} e^{-(\mu+|\eta|^2/2)s + \eta^1 s^2/2 - s^3/6} \widehat{w}(\eta^1 - s, \eta')\, ds. \tag{17.9.27}$$

For $s \geq 0$,

$$\eta^{1,2} s - \eta^1 s^2 \geq -s^3/4, \tag{17.9.28}$$

and so

$$|\eta|^2 s - \eta^1 s^2 + s^3/3 \geq s|\eta'|^2 + s^3/12. \tag{17.9.29}$$

In the sequel, we write

$$\mu = \mu_r + i\mu_i,\ \mu_r \in \mathbf{R}_+, \mu_i \in \mathbf{R}. \tag{17.9.30}$$

Clearly,

$$\int_0^{+\infty} e^{-(\mu_r + |\eta'|^2/2)s - s^3/24}\, ds \leq C \frac{1}{1 + \mu_r + |\eta'|^2/2}. \tag{17.9.31}$$

Estimating (17.9.27) by a convolution and using (17.9.29) and (17.9.31), we obtain

$$\|\widehat{u}'(., \eta')\|_{L^2(\eta^1)} \leq \frac{C}{1 + \mu_r + |\eta'|^2} \|\widehat{w}(., \eta')\|_{L^2(\eta^1)}. \tag{17.9.32}$$

Set $p = (p_1, p')$. Let $\Delta^V_{p'}$ be the Laplacian in the variable p'. By (17.9.32), we deduce that

$$\left\|\Delta^V_{p'} u'\right\| + (1 + \mu_r)\|u'\| \leq C \|w\|. \tag{17.9.33}$$

Remember that our goal is to establish (17.9.25). In view of (17.9.33) we are now reduced to a 1-dimensional problem on the variable p_1. Set

$$L = -\frac{1}{2}\frac{\partial^2}{\partial p_1^2} - i(p_1 - \mu_i), \qquad w' = w + \left(\frac{1}{2}\Delta^V_{p'} - \mu_r\right) u'. \tag{17.9.34}$$

We rewrite (17.9.24) as an equation of Airy type,

$$Lu' = w'. \tag{17.9.35}$$

Clearly,

$$\left|\operatorname{Re}\left\langle Lu',u'\right\rangle\right| \leq \left\|u'\right\| \left\|w'\right\|. \tag{17.9.36}$$

From (17.9.33), (17.9.34), (17.9.36), we get

$$\left\|\frac{\partial u'}{\partial p_1}\right\| \leq C \left\|w\right\|. \tag{17.9.37}$$

Moreover,

$$\left\|Lu'\right\|^2 = \left\|\frac{1}{2}\frac{\partial^2 u'}{\partial p_1^2}\right\|^2 + \left\|\left(p_1-\mu_1\right)u'\right\|^2 + i\left\langle\frac{\partial u'}{\partial p_1}, u'\right\rangle. \tag{17.9.38}$$

By (17.9.33), (17.9.35), (17.9.37), (17.9.38), we get

$$\left\|\frac{1}{2}\frac{\partial^2 u'}{\partial p_1^2}\right\| + \left\|\left(p_1-\mu_1\right)u'\right\| \leq C \left\|w\right\|. \tag{17.9.39}$$

By (17.9.33), (17.9.39), we find that (17.9.25) holds.

By (17.9.18), (17.9.19), (17.9.23), (17.9.33) and (17.9.39), we get

$$\left\|\Delta^V u\right\| + \left(|k|^{2/3} + \operatorname{Re}\lambda'\right)\left\|u\right\| + \left\|\left|\langle k,p\rangle - \operatorname{Im}\lambda'\right|u\right\| \leq C\left\|\left(1+h<p>\right)v\right\|. \tag{17.9.40}$$

Also observe that

$$|k|^{2/3} + \operatorname{Re}\lambda' + \left|\langle k,p\rangle - \operatorname{Im}\lambda'\right| \geq C'\left(\frac{|\lambda'|}{<p>}\right)^{2/3}. \tag{17.9.41}$$

Indeed (17.9.41) is trivially true if $|k|\,|p| \leq \frac{1}{2}\left|\operatorname{Im}\lambda'\right|$. If not, then $|k| \geq C\frac{\operatorname{Im}\lambda'}{<p>}$, and (17.9.41) also holds. By (17.8.7), (17.9.40) and (17.9.41), we get the second identity in (17.9.20). Thus we have established the second identity in (17.9.7) when $b=0$.

Now we establish the second identity in (17.9.7) for arbitrary $b \in \mathbf{R}$. We have the obvious identity

$$\left(Q_h^0(x,\xi)-\lambda\right)<p>^b u = <p>^b v + \left[Q_h^0(x,\xi), <p>^b\right]u. \tag{17.9.42}$$

Using the considerations we made after (17.8.6) and also (17.9.2), (17.9.3), we get

$$\begin{aligned}\left[Q_h^0(x,\xi), <p>^b\right] = -\frac{1}{2}\left[\Delta^V + h\omega\left(\nabla^F, g^F\right)(e_i)\nabla_{\widehat{e}^i}, <p>^b\right]\\ + \left[P_\pm, <p>^b\right].\end{aligned} \tag{17.9.43}$$

Using now the first inequality in (17.9.7) and the second inequality in (17.9.7) with $b=0$ applied to $<p>^b u$, we obtain this second inequality for arbitrary b. The proof of our theorem is completed. □

By Theorem 17.9.1, we know that $Q_h^0(x,\xi)-\lambda$ is one to one from $D(x,\xi)$ into L^2.

Remark 17.9.2. The estimates in the second line of (17.9.7) are much more precise than those we would get using the classical hypoelliptic estimates of chapter 15. Indeed, because $B(\xi, p, \lambda) \geq |\xi|^{2/3}$, we gain $2/3$ derivatives in the variable x when passing from v to u, instead of the classical gain of $1/4$ which was obtained in Theorem 17.5.2. Also a simple scaling argument shows that $2/3$ is indeed optimal. Still the factor $h < p >$ in the right-hand side of the second line of (17.9.7) is certainly not optimal. In the right-hand side of equation (17.9.2), the term $\pm h\widehat{\nabla}_{\Gamma^{T^*X}((g^{TX})^{-1}\sigma p)p}$ is responsible for the appearance of $h < p >$.

Definition 17.9.3. Let $e_{0,h,\lambda}(x,\xi) : L^2 \to D^0_h(x,\xi)$ be the inverse of $Q^0_h(x,\xi) - \lambda$.

If $\rho \in \mathbf{R}$, set

$$c_1(\rho) = \frac{9}{2}\rho - 1. \tag{17.9.44}$$

We take $\lambda_1 \in]0, 1/2[$.

Theorem 17.9.4. *Take $\rho \in]1/3, 2/3]$ and assume that $h \in]0, h_0], \operatorname{Re}\lambda \leq \lambda_1$. For any $v \in \mathcal{S}$, the function $(x,\xi) \to u(x,\xi) = e_{0,h,\lambda}(x,\xi)v$ is smooth as a function of (x,ξ) with values in $\mathcal{S} \otimes V$. The symbol $e_{0,h,\lambda}(x,\xi)$ lies in $\mathcal{S}^{-2/3,1}_{\rho,1/3,(2,c_1(\rho))}$. In particular, $e_{0,h,\lambda}(x,\xi) \in \mathcal{S}^{-2/3,1}$.*

Proof. Note that for $\rho = 2/3$, then $c_1(\rho) = 2$, so that $\mathcal{S}^{-2/3,1}_{\rho,1/3,(2,c_1(\rho))} = \mathcal{S}^{-2/3,1}$. To establish our theorem, we just have to prove that $e_{0,h,\lambda}(x,\xi) \subset \mathcal{S}^{-2/3,1}_{\rho,1/3,(2,c_1(\rho))}$. Comparing with (17.8.24), we have to show that for any multiindices α, β, for $s \in \mathbf{R}, N \in \mathbf{N}$, there exists $C_{s,N,\alpha,\beta} > 0$ such that

$$\begin{aligned} &\left\| < p >^N \partial^\alpha_x \partial^\beta_\xi u \right\|_{p,s} \\ &\qquad \leq C_{s,N,\alpha,\beta} \left\| < p >^{N+1+2|\alpha|+c_1(\rho)|\beta|} v \right\|_{p,s-2/3+|\alpha|/3-\rho|\beta|}. \end{aligned} \tag{17.9.45}$$

Set

$$\begin{aligned} M_{s,N}(u) = &\left\| < p >^N \left(< p >^2 + \frac{B(\xi,p,\lambda)}{1 + h < p >} \right) u \right\|_{p,s} \\ &+ \left\| < p >^{N+1} \widehat{\nabla} u \right\|_{p,s} + \left\| < p >^N \frac{\Delta^V u}{1 + h < p >} \right\|_{p,s}. \end{aligned} \tag{17.9.46}$$

For $h \in]0, h_0]$, we have the obvious

$$\left\| < p >^N u \right\|_{p,s} \leq C M_{s-2/3,N+1}(u), \tag{17.9.47}$$

and so (17.9.45) will be a consequence of the estimate

$$M_{s,N}\left(\partial^\alpha_x \partial^\beta_\xi u\right) \leq C_{s,N,\alpha,\beta} \left\| < p >^{N+2|\alpha|+c_1(\rho)|\beta|} v \right\|_{p,s+|\alpha|/3-\rho|\beta|}. \tag{17.9.48}$$

To establish (17.9.48), we will argue by induction on $k = |\alpha| + |\beta|$.

We will obtain the case where $k = 0$ as a consequence of Theorem 17.9.1. Let U_j, V_j be associated to u, v as in (17.4.1). Set

$$D_{s,j}(U_j) = 2^{2j}|U_j|_{p,s,j} + |\widehat{\nabla} U_j|_{p,s,j} + \frac{2^{-2j}|\Delta^V U_j|_{p,s,j} + B_{2^{-j}}|U_j|_{p,s,j}}{1+h2^j}. \tag{17.9.49}$$

The definition of $D_{s,j}(U_j)$ should be compared with the one of $M_{s,N}(u)$ in (17.9.46), at least when $N = 0$. The scaling by K_τ accounts for the apparent discrepancy.

For $\tau \in]0,1]$, set

$$Q_{h,\tau}(x,\xi) = K_{\tau^{-1}} Q_h(x,\xi) K_\tau, \qquad Q^0_{h,\tau}(x,\xi) = K_{\tau^{-1}} Q^0_h(x,\xi) K_\tau. \tag{17.9.50}$$

We define $P_{\pm,\tau}$ by a similar formula. Then

$$Q^0_{h,\tau}(x,\xi) = Q_{h,\tau}(x,\xi) + P_{\pm,\tau}. \tag{17.9.51}$$

By (17.2.14), we get

$$P_{\pm,\tau} u = \pi^{-n/2}\tau^{-n} e^{-|p|^2/2\tau^2} \int_{\mathbf{R}^n} e^{-|q|^2/2\tau^2} Q_\pm u(q)\, dq. \tag{17.9.52}$$

By (17.9.2), we obtain

$$\begin{aligned} Q^0_{h,\tau}(x,\xi) = \alpha_{\pm,\tau} \mp \tau^{-1}\Big(i\left\langle \left(g^{TX}\right)^{-1}\sigma p, \xi\right\rangle - \\ h\left\langle \Gamma^{TX}\left(\left(g^{TX}\right)^{-1}\sigma p\right) e_i, e_j\right\rangle \left(e^i i_{e_j} + \widehat{e}_i i_{\widehat{e}^j}\right)\Big) \\ \pm h\tau^{-1}\widehat{\nabla}_{\Gamma^{T^*X}\left(\left(g^{TX}\right)^{-1}\sigma p\right)p} - \frac{h\tau}{2}\omega\left(\nabla^F, g^F\right)(e_i)\nabla_{\widehat{e}^i} + h^2\gamma_{\pm,\tau} + P_{\pm,\tau}. \end{aligned} \tag{17.9.53}$$

The following lemma plays a crucial role in the proof of (17.9.48) in the case $k = 0$.

Lemma 17.9.5. *For any $s \in \mathbf{R}$, there exists $C_s > 0$ such that for $u \in \mathcal{S}, j \in$* **N**,

$$\begin{aligned} \left|\widehat{\nabla} U_j\right|_{p,s,j} + 2^{2j}\left|U_j\right|_{p,s,j} \le C_s \left\|K_{2^{-j}}\left(Q^0_{h,2^{-j}}(x,\xi) - \lambda\right) U_j\right\|_{p,s}, \\ 2^{-2j}\left|\Delta^V U_j\right|_{p,s,j} + B_{2^{-j}}\left|U_j\right|_{p,s,j} \le C_s(1 + h/\tau) \\ \left\|K_{2^{-j}}\left(Q^0_{h,2^{-j}}(x,\xi) - \lambda\right) U_j\right\|_{p,s}. \end{aligned} \tag{17.9.54}$$

For any $s \in \mathbf{R}$, there exists a rapidly decreasing function Γ_s for which if $u, v \in \mathcal{S}$ are such that $\left(Q^0_h(x,\xi) - \lambda\right) u = v$, then

$$D_{s,j}(U_j) \le \sum_{k=0}^{+\infty} \Gamma_s\left(2^{|j-k|}\right) |V_k|_{p,s,k}. \tag{17.9.55}$$

Proof. We still use the notation in (17.8.8)-(17.8.11). Also we will often write $Q_h, Q_h^0, Q_{h,\tau}, Q_{h,\tau}^0$ instead of $Q_h(x,\xi), Q_h^0(x,\xi), Q_{h,\tau}(x,\xi), Q_{h,\tau}^0(x,\xi)$. Set

$$|U|_s = \tau^{-n/2}|J_\tau^s U|_{L^2}, \tag{17.9.56}$$

so that if $\tau = 2^{-j}$, then $|U|_{p,s,j} = |U|_s$.

Let $\mathcal{S}^d$ be the set of symbols of degree d in the variable p with values in $\mathrm{End}(V)$, with parameters x, ξ, h, τ, λ which are associated to the weight J_τ. Let η be the variable dual to p. If $a(x,\xi,h,\tau,\lambda,p,\eta) \in \mathcal{S}^d$, for any multiindices α, β,

$$\left|\partial_p^\alpha \partial_\eta^\beta a(x,\xi,h,\tau,\lambda,p,\eta)\right| \le C\left(|\eta|^2 + A_\tau^2(\xi,\lambda)\right)^{\frac{d-|\beta|}{2}}. \tag{17.9.57}$$

The class of symbols $\mathcal{S}^d$ coincides with the class $S_{1,0}^d$ which was considered in [Hör85, section 18.1] with large parameter $A_\tau(\xi,\lambda)$.

Let $\widehat{u}(\eta)$ be the Fourier transform in the variable p of $u \in \mathcal{S}$. The quantification $A = \mathrm{Op}(a)$ of a is such that

$$A(x,\xi,h,\tau,\lambda,p,D_p)u(q) = (2\pi)^{-n}\int_{\mathbf{R}^n} e^{i\langle p,\eta\rangle} a(x,\xi,h,\tau,\lambda,p,\eta)\widehat{u}(\eta)\,d\eta. \tag{17.9.58}$$

Let $\mathcal{E}^d$ be the corresponding class of pseudodifferential operators over $\mathbf{R}^n$. Then $\widehat{\nabla}, A_\tau(\xi,\lambda) \in \mathcal{E}^1$.

Let $\theta_0(p)$ be a smooth cutoff function which is equal to 1 on the ball $\mathcal{B}_0$. Set

$$R_{h,\tau} = \theta_0(Q_{h,\tau} - \lambda)\theta_0. \tag{17.9.59}$$

If the principal symbol of $E_d \in \mathcal{E}_d$ is scalar, as in (17.5.19), we get

$$[R_{h,\tau}, E_d] \in \tau^2\mathcal{E}^d\widehat{\nabla} + \tau^{-1}\mathcal{E}^d. \tag{17.9.60}$$

In the sequel, we omit the index j, and we replace 2^{-j} by τ. Set

$$Y = \left(Q_{h,\tau}^0 - \lambda\right)U. \tag{17.9.61}$$

We denote by $|\,|$ the standard L^2 norm. By equation (17.9.7) in Theorem 17.9.1, for any U with support in $\mathcal{B}$ and in the annulus $\mathcal{R}$ for $j \ge 1$,

$$\begin{aligned} &\left|\widehat{\nabla} U\right| + \tau^{-2}|U_j| \le C|Y|, \\ &\tau^2\left|\Delta^V U\right| + |B_\tau U| \le C|(1 + h < p/\tau >)Y|. \end{aligned} \tag{17.9.62}$$

Let $C_0 > 0$ be a large constant. We first observe that if $\tau^{-1} + B_\tau \le C_0$, by (17.8.8), the parameters $\tau^{-1}, < \xi >, |\lambda'|$ are bounded. Therefore for $s \in \mathbf{R}$, in the above range of parameters, $\tau^{n/2}|U|_{p,s}$ is uniformly equivalent to the usual s Sobolev norm U. In this range of parameters, we claim that the inequalities in (17.9.54) follow from (17.9.62). Indeed let $\theta(p) \in C_0^\infty(\mathbf{R}^n)$ be a radial cutoff function which is equal to 1 near $\mathcal{B}$ for $j = 0$, and near the annulus $\mathcal{R}$ for $j \ge 1$, and vanishes near 0 also in that case.

Equation (17.9.61) is equivalent to the equation

$$(Q_{h,\tau} - \lambda)U = \theta(Y - P_{\pm,\tau}U). \tag{17.9.63}$$

Note that $Q_{h,\tau} - \lambda$ is a second order elliptic operator in the variable $p \in \mathbf{R}^n$ with uniformly bounded coefficients (indeed $|\xi|$ is uniformly bounded), and that $\Pi_{0,\tau}$ is a smoothing operator. From (17.9.63), we find that given $s, s' \in \mathbf{R}$,

$$|U|_{p,s+2} \le C_{s,s'} \left(|\theta Y|_{p,s} + |U|_{p,s'} \right). \tag{17.9.64}$$

Note that in (17.9.64), $C_{s,s'}$ depends implicitly on C_0.

Since τ^{-1} remains uniformly bounded, we have the obvious inequality

$$|\theta Y|_{p,s} \le C \left| K_\tau Y \right|_{p,s}. \tag{17.9.65}$$

By (17.9.64) and (17.9.65), we obtain

$$|U|_{p,s+2} \le C_{s,s'} \left(|K_\tau Y|_{p,s} + |U|_{p,s'} \right). \tag{17.9.66}$$

Using the injectivity of $Q^0_{h,\tau}$, (17.9.62), (17.9.66) and the familiar contradiction argument, we get

$$|U|_{p,s+2} \le C \left| K_\tau Y \right|_{p,s}. \tag{17.9.67}$$

Therefore we have established (17.9.54) under the conditions which were given above.

Now we will consider the case where τ is small. Let ψ be a radial cutoff function which has the same properties as θ before, and which also vanishes near 0 for $j \ge 1$. If $E_d \in \mathcal{E}^d, N \in \mathbf{N}$, using (17.9.57), (17.9.58) and integration by parts, we get

$$\begin{aligned} \left| < p >^N (1 - \psi) E_d U \right|_{p,0} &\le C_N \left| U \right|_{p,-N}, \\ \left| [\psi, E_d] U \right|_{p,0} &\le C_N \left| U \right|_{p,-N}. \end{aligned} \tag{17.9.68}$$

Recall that J_τ was defined in (17.8.10). By (17.9.60), replacing U by $\psi J^s_\tau U$ in (17.9.62), and using (17.9.68), we get

$$\begin{aligned} \left| \widehat{\nabla} U \right|_{p,s} + \tau^{-2} |U|_{p,s} \le C_s \Bigg(& \left| \psi J^s_\tau Y \right|_{p,0} + \tau^2 \left| \widehat{\nabla} U \right|_{p,s} + \tau^{-1} |U|_{p,s} \\ & + \left| [P_{\pm,\tau}, \psi J^s] U \right|_{p,0} \Bigg), \end{aligned} \tag{17.9.69}$$

$$\begin{aligned} \tau^2 \left| \Delta^V U \right|_{p,s} + B_\tau |U|_{p,s} \le C_s (1 + h/\tau) \Bigg(& \left| \psi J^s_\tau Y \right|_{p,0} + \tau^2 \left| \widehat{\nabla} U \right|_{p,s} \\ & + \tau^{-1} |U|_{p,s} \Bigg) + C_s \left| (1 + h < p/\tau >) [P_{\pm,\tau}, \psi J^s_\tau] U \right|_{p,0}. \end{aligned}$$

Since the operator $< p > P_\pm$ acts as a bounded operator on L^2,

$$\left| (1 + h < p/\tau >) P_{\pm,\tau} \psi J^s_\tau U \right|_{p,0} \le C |U|_{p,s}. \tag{17.9.70}$$

Set

$$e_\tau (p) = \frac{1}{\pi^{n/4} \tau^{n/2}} e^{-|p|^2/2\tau^2}. \tag{17.9.71}$$

By (17.9.52), (17.9.71), we get

$$J^s_\tau P_{\pm,\tau} u = J^s_\tau e_\tau(p) \int_{\mathbf{R}^n} e_\tau(q) Q_\pm u(q)\, dq. \tag{17.9.72}$$

By (17.9.72), using the fact that p remains bounded on the support of ψ,

$$\left|(1+h<p/\tau>)\psi J^s P_{0,\tau} U\right|_{p,0} \leq C(1+h/\tau)\tau^n \left|J^s_\tau e_\tau\right|_{p,0} \left|J^{-s}_\tau e_\tau\right|_{p,0} |U|_{p,s}. \tag{17.9.73}$$

Note that the factor τ^n in the right-hand side of (17.9.73) comes from the conventions used in (17.8.11). Using the Fourier transform, we get

$$\tau^n \left|J^s_\tau e_\tau\right|^2_{p,0} = 2^n \pi^{n/2} \int_{\mathbf{R}^n} \left(4\pi^2 \frac{|p|^2}{\tau^2} + A^2_\tau\right)^s \exp\left(-4\pi^2 |p|^2\right) dp. \tag{17.9.74}$$

Now by noting that $A_\tau \geq \tau^{-2}$, and using the fact that

$$4\pi^2 \frac{|p|^2}{\tau^2} + A^2_\tau = \left(4\pi^2 \frac{|p|^2}{A^2_\tau \tau^2} + 1\right) A^2_\tau, \tag{17.9.75}$$

we find that given $s \in \mathbf{R}$, as $\tau \to 0$,

$$\tau^n \left|J^s_\tau e_\tau\right|^2_{p,0} \simeq C_s A^{2s}_\tau. \tag{17.9.76}$$

By (17.9.70), (17.9.73), (17.9.76), we get

$$\begin{aligned} &\left|\left[P_{\pm,\tau}, \psi J^s_\tau\right] U\right|_{p,0} \leq C_s |U|_{p,s}, \\ &\left|(1+h<p/\tau>)\left[P_{\pm,\tau}, \psi J^s_\tau\right] U\right|_{p,0} \leq C_s((1+h/\tau)|U|_{p,s}. \end{aligned} \tag{17.9.77}$$

By combining (17.9.69) and (17.9.77), we obtain

$$\begin{aligned} &\left|\widehat{\nabla} U\right|_{p,s} + \tau^{-2}|U|_{p,s} \leq C_{p,s}\left(\left|\psi J^s_\tau V\right|_{p,0} + \tau^2 \left|\widehat{\nabla} U\right|_{p,s} + \tau^{-1}|U|_{p,s}\right), \\ &\tau^2 \left|\Delta^V U\right|_{p,s} + B_\tau |U|_{p,s} \leq C_s(1+h/\tau) \\ &\qquad \left(\left|\psi J^s_\tau V\right|_{p,0} + \tau^2 \left|\widehat{\nabla} U\right|_{p,s} + \tau^{-1}|U|_{p,s}\right). \end{aligned} \tag{17.9.78}$$

Given $s \in \mathbf{R}$, if $\tau_0 \in]0,1]$ is small enough, if $\tau \in]0,\tau_0]$, by (17.9.78), we get

$$\begin{aligned} &\left|\widehat{\nabla} U\right|_{p,s} + \tau^{-2}|U|_{p,s} \leq C_{p,s}\left|\psi J^s_\tau Y\right|_{p,0}, \\ &\tau^2 \left|\Delta^V U\right|_{p,s} + B_\tau |U|_{p,s} \leq C_s(1+h/\tau)\left|\psi J^s_\tau Y\right|_{p,0}. \end{aligned} \tag{17.9.79}$$

Moreover,

$$\left|\psi J^s_\tau Y\right|_{p,0} \leq C_s \left\|K_\tau Y\right\|_{p,s}. \tag{17.9.80}$$

From (17.9.79), (17.9.80), we deduce that for $\tau \in]0,\tau_0]$, (17.9.54) holds.

Observe that there exists $C_s > 0$ such that for $\varepsilon > 0$, $s \in \mathbf{R}$,

$$\left|\widehat{\nabla} U\right|_{p,s} \leq C\left(\varepsilon\left|\Delta^V U\right|_{p,s} + \frac{1}{\varepsilon}|U|_{p,s}\right). \tag{17.9.81}$$

For $\tau \geq \tau_0$, and $B_\tau \geq C_0 > 0$, if C_0 is large enough, from the second inequality in (17.9.78) and from (17.9.81), we get

$$\left|\Delta^V U\right|_{p,s} + B_\tau \left|U\right|_{p,s} \leq C_s \left|\psi J_\tau^s V_j\right|_{p,0}. \tag{17.9.82}$$

Using (17.9.80) and (17.9.82), we find that if $C_0 > 0$ is large enough, the second inequality in (17.9.79) still holds. Using again (17.9.81) and this second inequality, we also get the first inequality in (17.9.79). Combining with (17.9.80), we get again the estimates in (17.9.54). This concludes the proof of these inequalities in full generality.

Now we will establish (17.9.55). We use the notation

$$\chi_\tau(p) = \chi\left(\tau < p/\tau >\right). \tag{17.9.83}$$

We will proceed as in (15.5.5)-(15.5.11). Recall that Y was defined in (17.9.61). Set

$$W = K_{\tau^{-1}}\left[Q_h^0, \chi\left(\tau < p >\right)\right] u. \tag{17.9.84}$$

Since $\left(Q_h^0 - \lambda\right) u = v$, as in (15.5.6), we get

$$Y = V + W. \tag{17.9.85}$$

Put

$$W' = K_{\tau^{-1}}\left[Q_h, \chi\left(\tau < p >\right)\right] u, \qquad W'' = \left[P_{\pm,\tau}, \chi_\tau\right] K_{\tau^{-1}} u. \tag{17.9.86}$$

By (17.9.84), (17.9.86), we get

$$W = W' + W''. \tag{17.9.87}$$

By (15.1.3), (17.9.2) and (17.9.4), we get

$$\begin{aligned}\left[Q_h, \chi\left(\tau < p >\right)\right] = \frac{1}{2}\left(-\Delta^V \chi\left(\tau < p >\right) - 2\left(\nabla_{\widehat{e}^i} \chi\left(\tau < p >\right)\right) \nabla_{\widehat{e}^i}\right) \\ - \frac{h}{2}\omega\left(\nabla^F, g^F\right)\left(e_i\right) \nabla_{\widehat{e}^i} \chi\left(\tau < p >\right).\end{aligned} \tag{17.9.88}$$

By (17.9.88), we obtain

$$\begin{aligned}K_{\tau^{-1}}\left[Q_h, \chi\left(\tau < p >\right)\right] K_\tau = \frac{1}{2}\Bigg(-K_{\tau^{-1}}\left(\Delta^V \chi\left(\tau < p >\right)\right) \\ -2K_{\tau^{-1}}\left(\nabla_{\widehat{e}^i} \chi\left(\tau < p >\right)\right) \tau \nabla_{\widehat{e}^i}\Bigg) - \frac{h}{2}\omega\left(\nabla^F, g^F\right)\left(e_i\right) K_{\tau^{-1}} \nabla_{\widehat{e}^i} \chi\left(\tau < p >\right).\end{aligned} \tag{17.9.89}$$

As in (15.5.10), we write (17.4.1) in the form

$$u = \sum K_{\tau'} U'. \tag{17.9.90}$$

From (17.9.86), (17.9.89), (17.9.90), we get, as in (15.5.11),

$$\begin{aligned}W' = \Bigg(\frac{1}{2}\left(-K_{\tau^{-1}}\left(\Delta^V \chi\left(\tau < p >\right)\right) - 2K_{\tau^{-1}}\left(\nabla_{\widehat{e}^i} \chi\left(\tau < p >\right)\right) \tau \nabla_{\widehat{e}^i}\right) \\ - \frac{h}{2}\omega\left(\nabla^F, g^F\right)\left(e_i\right) K_{\tau^{-1}} \nabla_{\widehat{e}^i} \chi\left(\tau < p >\right)\Bigg) \sum K_{\tau'/\tau} U'.\end{aligned} \tag{17.9.91}$$

Now the considerations we made after (15.5.11) show that if $\tau = 2^{-j}, \tau' = 2^{-j'}$, given $j \in \mathbf{N}$, the nonzero terms in (17.9.91) appear only with $|j' - j| \leq 1$.

By (17.9.49), (17.9.54), (17.9.85), (17.9.87), (17.9.91), we get

$$D_{s,j}(U_j) \leq C_s \Bigg(|V_j|_{p,s,j} + \sum_{|j'-j|\leq 1} \left(\left| 2^{-2j} \widehat{\nabla} U_{j'} \right|_{p,s,j'} + \left(2^{-2j} + h 2^{-j}\right) |U_{j'}|_{p,s,j'} \right) + \left\| \left[P_\pm, \chi\left(2^{-j} < p >\right) \right] u \right\|_{p,s} \Bigg). \quad (17.9.92)$$

We claim that there exist $C > 0, \sigma > 0$ such that in the given range of parameters,

$$\left(|\eta|^2 + A_\tau^2 \right)^{\frac{\sigma}{2}} \left(1 + \tau^2 < \eta > + h\tau\right) \leq C \left(\tau^{-2} + |\eta| + \frac{\tau^2 |\eta|^2 + B_\tau}{1 + h/\tau} \right). \quad (17.9.93)$$

Indeed,

$$\tau^{-2} + |\eta| + \frac{\tau^2 |\eta|^2 + B_\tau}{1 + h/\tau} \geq \tau^{-2} + |\eta| + \frac{1}{2} \left(\tau^3 |\eta|^2 + \tau B_\tau \right). \quad (17.9.94)$$

Also there is $C > 0$ such that

$$\tau^{-2} + \frac{1}{2} \tau^3 |\eta|^2 \geq C |\eta|^{6/5}, \qquad \tau^{-2} + \frac{1}{2} \tau B_\tau \geq C B_\tau^{2/3}. \quad (17.9.95)$$

By (17.9.94), (17.9.95), we obtain

$$\tau^{-2} + |\eta| + \frac{\tau^2 |\eta|^2 + B_\tau}{1 + h/\tau} \geq C \left(\tau^{-2} + B_\tau^{2/3} + |\eta|^{6/5} \right). \quad (17.9.96)$$

Using equations (17.8.8) for A_τ, B_τ, we deduce from (17.9.96) that there exists $C > 0, \sigma > 0$ such that

$$\left(|\eta|^2 + A_\tau^2 \right)^{\sigma/2} (1 + < \eta >) \leq C \left(\tau^{-2} + |\eta| + \frac{\tau^2 |\eta|^2 + B_\tau}{1 + h/\tau} \right). \quad (17.9.97)$$

Since $\tau, h\tau \in]0,1]$, (17.9.93) follows from (17.9.97).

By making $\tau = 2^{-j}$ in the right-hand side of (17.9.93), we obtain precisely the weight that was used in the definition of $D_{s,j}(U_j)$ in (17.9.49).

Using (17.9.93), we get

$$\begin{gathered} |U_j|_{p,s,j} \leq C D_{s-\sigma,j}(U_j), \\ \sum_{|j'-j|\leq 1} \left(2^{-2j} \left| \widehat{\nabla} U_{j'} \right|_{p,s,j'} + \left(2^{-2j} + h 2^{-j}\right) |U_{j'}|_{p,s,j'} \right) \\ \leq C \sum_{|j'-j|\leq 1} D_{s-\sigma,j'}(U_{j'}). \end{gathered} \quad (17.9.98)$$

By (15.2.14), we get

$$\left\|\chi\left(2^{-j}<p>\right)u\right\|_{p,s}\leq\sum_{|j'-j|\leq 1}|U_{j'}|_{p,s,j'}\,. \tag{17.9.99}$$

By (17.9.98) and (17.9.99), we get

$$\left\|\chi\left(2^{-j}<p>\right)u\right\|_{p,s}\leq C_s\sum_{|j'-j|\leq 1}D_{s-\sigma,j'}\left(U_{j'}\right). \tag{17.9.100}$$

Using an obvious modification of inequality (17.5.1) in Lemma 17.5.1 and (17.9.100), we obtain

$$\left\|P_{\pm}\chi\left(2^{-j}<p>\right)u\right\|_{p,s}\leq C_s\sum_{|j'-j|\leq 1}D_{s-\sigma,j'}\left(U_{j'}\right). \tag{17.9.101}$$

Let $\theta(p)\in C_0^{\infty}\left(\mathbf{R}^n\right)$ be a radial cutoff function which has the same properties as the function θ which was considered after (17.9.62). Since the choice of θ depends on whether $j=0$ or $j\geq 1$, we will write θ_j instead of θ, even though we only pick θ_0 and θ_1. In particular there is $\epsilon>0$ such that if $|p|\leq\epsilon$, then $\theta_1\left(p\right)=0$. By (17.2.14) and using the notation in (17.9.72), we get

$$\left\|\chi\left(2^{-j}<p>\right)P_{\pm}u\right\|_{p,s}\leq 2^{-jn/2}\left|\theta_j e_{2^{-j}}\right|_{p,s,j}\left|\int_{\mathbf{R}^n}e_1(p)u(p)dp\right|. \tag{17.9.102}$$

Moreover, using (17.4.1), we have

$$\left|\int_{\mathbf{R}^n}e_1(p)u(p)dp\right|\leq\sum_{k=0}^{+\infty}2^{kn/2}\left|\int_{\mathbf{R}^n}e_{2^{-k}}(p)U_k(p)dp\right|. \tag{17.9.103}$$

By (17.9.102), (17.9.103), we obtain

$$\begin{aligned}&\left\|\chi\left(2^{-j}<p>\right)P_{\pm}u\right\|_{p,s}\\&\leq C_s 2^{-jn/2}|\theta_j e_{2^{-j}}|_{p,s,j}\sum_{k=0}^{+\infty}2^{-kn/2}\left|\theta_k e_{2^{-k}}\right|_{p,-s,k}\left|U_k\right|_{p,s,k}\,.\end{aligned} \tag{17.9.104}$$

Clearly,

$$2^{-jn}\left|\theta_j e_{2^{-j}}\right|^2_{p,s,j}\leq C\int_{\mathbf{R}^n}\left(|\eta|^2+A^2_{2^{-j}}\right)^s\left|\widehat{\theta_j e_{2^{-j}}}\right|^2(\eta)d\eta. \tag{17.9.105}$$

By construction, there is $\epsilon>0$ such that $\theta_1\left(p\right)$ vanishes for $|p|\leq\varepsilon$. Therefore,

$$\begin{aligned}\widehat{\theta_1 e_\tau}\left(\eta\right)&=\pi^{-n/4}\tau^{-n/2}\exp\left(-\varepsilon^2/4\tau^2\right)\\&\int_{\mathbf{R}^n}e^{-i\langle\eta,p\rangle}\theta_1\left(p\right)\exp\left(-\frac{1}{2\tau^2}\left(|p|^2-\varepsilon^2/2\right)\right)dp.\end{aligned} \tag{17.9.106}$$

We claim that there is a rapidly decreasing function $\Gamma\left(\eta\right)$ such that

$$\left|\int_{\mathbf{R}^n}e^{-i\langle\eta,p\rangle}\theta_1\left(p\right)\exp\left(-\frac{1}{2\tau^2}\left(|p|^2-\varepsilon^2/2\right)\right)dp\right|\leq\Gamma\left(\eta\right). \tag{17.9.107}$$

This is indeed obtained using the support property of θ_1 and integration by parts.

By (17.9.106), (17.9.107), we find that there exists $c > 0$ such that

$$\left|\widehat{\theta_1 e_\tau}\right|(\eta) \leq \Gamma(\eta) e^{-c\tau^{-2}}. \tag{17.9.108}$$

By (17.9.105), (17.9.108), we get for any $s \in \mathbf{R}$,

$$2^{-jn/2} \left|\theta_j e_{2^{-j}}\right|_{p,s,j} \leq C_s A^s_{2^{-j}} e^{-c2^{2j}}. \tag{17.9.109}$$

Also given $s \in \mathbf{R}, c > 0$, there exists $C > 0, c' > 0$ such that for $\tau, \tau' \in]0,1]$,

$$\left(\frac{A_\tau}{A_{\tau'}}\right)^s \exp\left(-c\left(\tau^{-2} + \tau'^{-2}\right)\right) \leq C \exp\left(-c'\left(\tau^{-2} + \tau'^{-2}\right)\right). \tag{17.9.110}$$

By (17.9.98), (17.9.104), (17.9.109) and (17.9.110), we obtain

$$\left\|\chi(2^{-j} < p >) P_\pm u\right\|_{p,s} \leq C_s \sum_{k=0}^{+\infty} e^{-c(2^{2j}+2^{2k})} D_{s-\sigma,k}(U_k). \tag{17.9.111}$$

By (17.9.92), (17.9.98), (17.9.101), (17.9.111), we find that for any s, there exists $C_s > 0$ such that

$$D_{s,j}(U_j) \leq C_s \Bigg(|V_j|_{p,s,j} + \sum_{|j'-j| \leq 1} D_{s-\sigma,j'}\left(U_{j'}\right) + e^{-c2^{2j}} \sum_{k=0}^{+\infty} e^{-c2^{2k}} D_{s-\sigma,k}\left(U_k\right) \Bigg). \tag{17.9.112}$$

For $N \in \mathbf{Z}$, put

$$|||u|||^2_{s,N} = \sum_{j=0}^{+\infty} 2^{2jN} D^2_{s,j}(U_j), \quad \|v\|^2_{s,N} = \sum_{j=0}^{+\infty} 2^{2jN} \left|V_j\right|^2_{p,s,j}. \tag{17.9.113}$$

By (17.9.112), we get

$$|||u|||_{s,N} \leq C_{s,N} \left(\|v\|_{s,N} + |||u|||_{s-\sigma,N}\right). \tag{17.9.114}$$

By Theorem 17.9.1, $Q^0_h(x,\xi) - \lambda$ is injective. By the familiar contradiction argument, we get, from (17.9.114),

$$|||u|||_{s,N} \leq C_{s,N} \|v\|_{s,N}. \tag{17.9.115}$$

By (17.9.115), we get (17.9.55). The proof of Lemma 17.9.5 is complete. □

Now we will finish the proof of equation (17.9.48) in the case $k = 0$. For $N \in \mathbf{N}$, put

$$\gamma_j = 2^{jN} D_{s,j}(U_j), \qquad \delta_j = 2^{jN} |V_j|_{p,s,j}. \tag{17.9.116}$$

Using (17.8.23) and the comments we made after equation (17.9.49), we get

$$M_{s,N}(u) \sim \|\gamma\|_{\ell^2}, \qquad \left\|< p >^N v\right\|_{p,s} \sim \|\delta\|_{\ell^2}. \tag{17.9.117}$$

Moreover, since Γ_s is rapidly decreasing, given $N \in \mathbf{N}$, there is a rapidly decreasing function $\Gamma_{s,N}$ such that for any $j, k \in \mathbf{N}$,

$$2^{jN}\Gamma_s(2^{|j-k|}) \leq \Gamma_{s,N}(2^{|j-k|})2^{kN}. \tag{17.9.118}$$

By (17.9.55) and (17.9.118), one gets

$$\|\gamma\|_{\ell^2} \leq 2\,\|\Gamma_{s,N}(2^{\cdot})\|_{\ell^1}\,\|\delta\|_{\ell^2}. \tag{17.9.119}$$

Note that the factor 2 in (17.9.119) comes from the absolute value in $2^{|j-k|}$. By (17.9.117), (17.9.119), we get (17.9.48) for $\alpha = 0, \beta = 0$, that is, for $k = 0$.

We establish inequality (17.9.48) for arbitrary $k \in \mathbf{N}$ by recursion. We will assume that it has been proved for $k' \leq k-1$, and we will establish it for k.

We will use the notation

$$u_{\alpha,\beta} = \partial_x^\alpha \partial_\xi^\beta u. \tag{17.9.120}$$

Recall that v depends only on p and not on (x, ξ). We will take derivatives in x, ξ of the equation $\left(Q_h^0 - \lambda\right) u = v$. For $k = |\alpha| + |\beta| \geq 1$, and taking into account the fact that Q_h^0 is an affine function of ξ, we get

$$\left(Q_h^0 - \lambda\right) u_{\alpha,\beta} = \sum S_{\alpha_1} u_{\alpha_2,\beta} + \sum T_{\nu,\beta_2} u_{\nu,\beta_2}. \tag{17.9.121}$$

In (17.9.121), the operators S_{α_1} are proportional to $\partial_x^{\alpha_1} Q_h^0$, the operators T_{ν,β_2} involve just one derivative in ξ of Q_h^0, and derivatives of Q_h^0 in X of order lower than α, and moreover

$$\alpha_1 + \alpha_2 = \alpha, \alpha_2 < \alpha, \qquad \nu \leq \alpha, \qquad |\beta_2| + 1 = |\beta|. \tag{17.9.122}$$

In particular, in (17.9.122),

$$|\nu| + |\beta_2| \leq k - 1. \tag{17.9.123}$$

By (17.9.2), for $1 \leq i \leq n$,

$$T_{0,i} = \pm i \left\langle \left(g^{TX}\right)^{-1} \sigma p, e^i \right\rangle. \tag{17.9.124}$$

By the already proved equation (17.9.48) with $k = 0$, we get

$$M_{s,N}(u_{\alpha,\beta}) \leq C_{s,N} \left\| < p >^N \left(Q_h^0(x,\xi) - \lambda\right) u_{\alpha,\beta} \right\|_{p,s}. \tag{17.9.125}$$

Observe that for $\rho \in]1/3, 2/3]$ and with $c_1(\rho) = 9\rho/2 - 1$, there exist $C > 0$ such that

$$\tau^{-1}\left(|\eta|^2 + A_\tau^2\right)^{\rho/2} \leq C\tau^{-c_1(\rho)}\left(\tau^{-2} + |\eta| + \frac{B_\tau + \tau^2|\eta|^2}{1 + h/\tau}\right). \tag{17.9.126}$$

In fact we will prove the stronger inequality,

$$\tau^{-1}\left(|\eta|^2 + A_\tau^2\right)^{\rho/2} \leq C\tau^{-c_1(\rho)}\left(\tau^{-2} + |\eta| + \tau B_\tau\right). \tag{17.9.127}$$

Now using the definition A_τ, B_τ in (17.8.8), we get

$$\left(|\eta|^2 + A_\tau^2\right)^{\rho/2} \simeq |\eta|^\rho + \tau^{-2\rho} + B_\tau^{3\rho/2}. \tag{17.9.128}$$

Therefore to establish (17.9.127), we only need to verify the inequalities

$$\tau^{-1}|\eta|^{\rho} + \tau^{-(1+2\rho)} \le C\tau^{-c_1(\rho)}\left(\tau^{-2} + |\eta|\right), \tag{17.9.129}$$
$$\tau^{-1}B_{\tau}^{3\rho/2} \le C\tau^{-c_1(\rho)}\left(\tau^{-2} + \tau B_{\tau}\right).$$

Now the first inequality in (17.9.129) is trivial since for $|\eta| \le \tau^{-2}$, it reduces to $1 + 2\rho \le c_1(\rho) + 2 = \frac{9}{2}\rho + 1$, and for $|\eta| \ge \tau^{-2}$, it comes from the fact that $\tau^{-1}|\eta|^{\rho} \le \tau^{-c_1(\rho)}|\eta|$, which is itself a consequence of the inequality $2(1-\rho) \ge 1 - c_1(\rho) = 2 - \frac{9}{2}\rho$.

Also since $c_1(\rho) = \frac{9}{2}\rho - 1$, if $x = \tau^{-1}$, we can write the second inequality in (17.9.129) in the form

$$B_{\tau}^{3\rho/2} \le C\left(x^{9\rho/2} + x^{9\rho/2-3}B_{\tau}\right). \tag{17.9.130}$$

Now (17.9.130) is obvious for $\rho = 2/3$. Moreover, for $\rho \in]1/3, 2/3[$, the minimum in $x \in \mathbf{R}_+$ of the right-hand side of (17.9.130) is attained for $x \simeq B_{\tau}^{1/3}$, for which we get again (17.9.130). This completes the proof of (17.9.126).

Comparing (17.9.46) with (17.9.126), we get

$$\left\|<p>^{N+1}u\right\|_{p,s+\rho} \le CM_{s,N+c_1(\rho)}u. \tag{17.9.131}$$

So using (17.9.48) for $k-1$, (17.9.122)-(17.9.125), and (17.9.131), we get

$$\begin{aligned} \|<p>^{N}T_{\nu,\beta_2}u_{\nu,\beta_2}\|_{p,s} &\le C_{s,\alpha}\|<p>^{N+1}u_{\nu,\beta_2}\|_{p,s} \\ &\le C_{s,N,\alpha}M_{s-\rho,N+c_1(\rho)}(u_{\nu,\beta_2}) \\ &\le C_{s,N,\alpha,\beta}\|<p>^{N+2|\alpha|+c_1(\rho)|\beta|}v\|_{p,s+|\alpha|/3-\rho|\beta|}. \end{aligned} \tag{17.9.132}$$

As was already pointed out, the operators $\alpha_{\pm}, P_{\pm}$ do not depend on x, and so they do not contribute to S_{α_1}. Therefore S_{α_1} is essentially of the same type as the contribution to Q_h^0 of $h\beta_{\pm} + h^2\gamma_{\pm}$.

We claim that

$$\begin{aligned} &\left(\tau^{-1}<\xi> + h\tau^{-1}|\eta| + h^2\tau^{-2}\right) \\ &\quad \le C\left(|\eta|^2 + A_{\tau}^2\right)^{1/6}\tau^{-2}\left(\tau^{-2} + |\eta| + \frac{B_{\tau} + \tau^2|\eta|^2}{1 + h/\tau}\right). \end{aligned} \tag{17.9.133}$$

The last two terms in the left-hand side of (17.9.133) are easy to control. To control the first term, it is enough to show that

$$<\xi> \le A_{\tau}^{1/3}B_{\tau}, \tag{17.9.134}$$

which is obvious since $<\xi> \le A_{\tau}, <\xi>^{2/3} \le B_{\tau}$.

By (17.9.133), we get

$$\|<p>^{N}S_{\alpha_1}u\|_{p,s} \le CM_{s+1/3,N+2}(u). \tag{17.9.135}$$

Using the estimate (17.9.48) for $k' < k$, (17.9.122), and (17.9.135),

$$\begin{aligned} &\left\|<p>^{N}S_{\alpha_1}(u_{\alpha_2,\beta})\right\|_{p,s} \le CM_{s+1/3,N+2}(u_{\alpha_2,\beta}) \\ &\qquad \le C_{s,N,\alpha,\beta}\left\|<p>^{N+2|\alpha|+c_1(\rho)|\beta|}v\right\|_{p,s+|\alpha|/3-\rho|\beta|}. \end{aligned} \tag{17.9.136}$$

By (17.9.121)-(17.9.125), (17.9.135), and (17.9.136), we get (17.9.48) for arbitrary $k \in \mathbf{N}$.

The proof of Theorem 17.9.4 is complete. □

17.10 THE PARAMETRIX FOR $\mathbf{S_{h,\lambda}}$

Recall that $S_{h,\lambda}$ and $e_{0,h,\lambda}(x,\xi)$ were respectively defined in Definitions 17.6.2 and 17.9.3. In the sequel, we will write e_0 instead of $e_{0,h,\lambda}(x,\xi)$.

We define $c_1(\rho)$ as in (17.9.44). If $x_0 \in X$, we still use the coordinate system and the trivializations which were defined in sections 17.8 and 17.9. The class of symbols and operators which were defined in section 17.8 refer to this particular trivialization.

Now we will obtain a parametrix for $S_{h,\lambda}$.

Theorem 17.10.1. *Take $\rho \in]1/3, 2/3]$. Let $x_0 \in X$, let $K, U \subset X$ be small neighborhoods of x_0, with $K \subset U$, K compact and U open, and let $\varphi(x) \in C_0^\infty(U)$ be a cutoff function which is equal to 1 near K. There exists $E_0 \in \mathcal{P}^{-2/3,1}_{\rho,1/3,(2,c_1(\rho))}$, whose symbol is exactly $e_0 \in \mathcal{S}^{-2/3,1}_{\rho,1/3,(2,c_1(\rho))}$, and $E_1 \in \mathcal{P}^{-1/3,4}_{\rho,1/3,(2,c_1(\rho))}$ with symbol $e_1 \in \mathcal{S}^{-1/3,4}_{\rho,1/3,(2,c_1(\rho))}$ compactly supported in $x \in U$, such that for $u \in \mathcal{S}^{\cdot}(T^*X, \pi^*F)$ with support in $\pi^{-1}K$, and $M \in \mathbf{N}$, then*

$$S_{h,\lambda} u = \sum_{0 \le j < M} h^j \varphi E_0 E_1^j u + h^M S_{h,\lambda} E_1^M u. \tag{17.10.1}$$

In particular,

$$E_0 \in \mathcal{P}^{-2/3,1}, \qquad E_1 \in \mathcal{P}^{-1/3,4}. \tag{17.10.2}$$

Proof. Let $\widehat{u}(\xi, p)$ be the Fourier transform of $u(x,p)$ in the variable x. Let $E_0(x, hD_x, h, \nu)$ be the pseudodifferential operator associated to $e_{0,h,\lambda}(x,\xi)$ as in (17.8.26), i.e.,

$$E_0(x, hD_x, h, \lambda)\, u(x,p) = (2\pi h)^{-n} \int_{\mathbf{R}^n} e^{\frac{i}{h}\langle x,\xi\rangle} e_{0,h,\lambda}(x,\xi) \widehat{u}\left(\frac{\xi}{h}, p\right) d\xi. \tag{17.10.3}$$

By Theorem 17.9.4, we know that $E_0 \in \mathcal{P}^{-2/3,1}_{\rho,1/3,(2,c_1(\rho))}$.

Let U, K be taken as in Theorem 17.10.1, and let $\varphi(x) \in C_0^\infty(U)$ be equal to 1 near K. By (15.1.3), (17.6.2), and (17.8.6),

$$\left[P_h^0, \varphi\right] = \mp h \left\langle \left(g^{TX}\right)^{-1} \sigma p, d_x \varphi \right\rangle. \tag{17.10.4}$$

Assume that the support of $u(x,\cdot)$ is included in K. Set

$$\begin{aligned} &E_1(x, hD_x, h, \lambda) u(x,p) \\ &= \pm (2\pi h)^{-n} \int_{\mathbf{R}^n} e^{\frac{i}{h}\langle x,y\rangle} \left\langle \left(g^{TX}\right)^{-1} \sigma p, d_x\left(\varphi e_{0,h,\lambda}\right)(x,\xi)\, \widehat{u}\left(\frac{\xi}{h}, p\right)\right\rangle d\xi. \end{aligned} \tag{17.10.5}$$

Using the fact that $e_{0,h,\lambda}(x,\xi)$ is the inverse of $Q_h^0(x,\xi) - \lambda$ and (17.10.4), we get

$$\left(P_h^0 - \lambda\right) \varphi E_0 u = u - h E_1 u. \tag{17.10.6}$$

By the considerations we made after Definition 17.8.1, by Theorem 17.9.4, and by (17.10.5), noting that $-2/3+1/3=-1/3, 1+2=3$, we get

$$E_1 \in \mathcal{P}^{-1/3,4}_{\rho,1/3,(2,c_1(\rho))}. \tag{17.10.7}$$

Note that the fact that 3 is changed into 4 in (17.10.7) reflects the extra linear dependence on p in the right-hand side of (17.10.5).

Recall that $S_{h,\lambda} = \left(P_h^0 - \lambda\right)^{-1}$. Replacing u by $\sum_{0\leq j<M} h^j E_1^j u$ in equation (17.10.6), for $M \geq 1$, we get

$$S_{h,\lambda}u = \sum_{0\leq j<M} h^j \varphi E_0 E_1^j u + h^M S_{h,\lambda} E_1^M u, \tag{17.10.8}$$

which is just (17.10.1). By using the same argument as in the beginning of the proof of Theorem 17.9.4, we get (17.10.2). The proof of our theorem is completed. □

17.11 A LOCALIZATION PROPERTY FOR $\mathbf{E_0, E_1}$

We now state a lemma on the local properties in the variable x of the operators E_0, E_1 which appear in (17.10.1).

Lemma 17.11.1. *Let K a compact subset of U, and let $\varphi(x) \in C_0^\infty(U\setminus K)$. Then for any $s \in \mathbf{R}, t \in \mathbf{R}, N \in \mathbf{N}$, there exists $C_{s,t,N} > 0$ such that for $h \in]0,h_0], \lambda \in \mathbf{C}, \operatorname{Re}\lambda \leq \lambda_1$, and $u \in \mathcal{S}^{\cdot}(T^*X, \pi^*F)$ whose support is included in $\pi^{-1}K$, for $j=0,1$,*

$$\left\|\varphi E_j u\right\|_{\lambda,\mathrm{sc},t} \leq C_{s,t,N} h^N \left\|u\right\|_{\lambda,\mathrm{sc},s}. \tag{17.11.1}$$

Proof. For $m \in \mathbf{N}$, and $j=0,1$, we have

$$\begin{aligned}&\varphi E_j(x,hD_x,h,\lambda)u(x,p)\\&\quad = h^{2m}(2\pi h)^{-n}\int_{\mathbf{R}^n} e^{\frac{i}{h}\langle x-y,\xi\rangle}\left(-\Delta_\xi^V\right)^m e_j(x,\xi,h,\lambda)\frac{\varphi(x)u(y,.)}{|x-y|^{2m}}\,dyd\xi.\end{aligned} \tag{17.11.2}$$

Using the properties listed after (17.8.24) and also Theorem 17.10.1, for $\rho \in]1/3,2/3]$,

$$\begin{aligned}\left(-\Delta_\xi^V\right)^m(e_0) &\in \mathcal{S}^{-2/3-2m\rho,1+2mc_1(\rho)}_{\rho,1/3,(2,c_1(\rho))},\\ \left(-\Delta_\xi^V\right)^m(e_1) &\in \mathcal{S}^{-1/3-2m\rho,4+2mc_1(\rho)}_{\rho,1/3,(2,c_1(\rho))}.\end{aligned} \tag{17.11.3}$$

By (17.8.8) and (17.8.24), for $a>0$,

$$\mathcal{S}^{d,k}_{\rho,\delta,(2,c_1(\rho))} \subset \mathcal{S}^{d+a,k-2a}_{\rho,\delta,(2,c_1(\rho))}. \tag{17.11.4}$$

Given $\rho \in]1/3,2/5[$, then $c_1(\rho) = 9\rho/2-1 < 2\rho$. Take $\sigma \in]c_1(\rho),2\rho[$. By (17.11.4), for $m \in \mathbf{N}$,

$$\mathcal{S}^{d-2m\rho,k+2mc_1(\rho)}_{\rho,\delta,(2,c_1(\rho))} \subset \mathcal{S}^{d-2m\rho+m\sigma,k+2mc_1(\rho)-2m\sigma}_{\rho,\delta,(2,c_1(\rho))}. \tag{17.11.5}$$

Now observe that given d,k,ρ,σ taken as before, for $m \in \mathbf{N}$ large enough, $d-2m\rho+m\sigma$ and $k+2mc_1(\rho)-2m\sigma$ become arbitrarily negative. Using equation (17.8.33) in Lemma 17.8.4, (17.11.2),(17.11.3), and the above, we obtain our lemma. □

17.12 THE OPERATOR $\mathbf{P_{\pm}S_{h,\lambda}}$

Let L^2_X (resp. $\mathcal{D}_X$, resp. $\mathcal{D}'_X$) be the space of square integrable (resp. smooth, resp. distribution) sections over X of $\Lambda^{\cdot}(T^*X)\widehat{\otimes}F$ in the $+$ case, and of $\Lambda^{\cdot}(T^*X)\widehat{\otimes}F\widehat{\otimes}o(TX)$ in the $-$ case. We identify L^2_X to $\ker\alpha_{\pm}\in H^0$ by the isometric embedding $i_{\pm}$ described before Theorem 2.3.2. In fact, if $u\in L^2_X$, then

$$i_+u=\pi^*u\exp\left(-|p|^2/2\right)/\pi^{n/4},\quad i_-u=\pi^*s\exp\left(-|p|^2/2\right)\wedge\eta/\pi^{n/4}. \tag{17.12.1}$$

In (17.12.1), η is a unit volume form in T^*X.

For $s\in\mathbf{R}$, let H^s_X be the Sobolev space of sections of $\Lambda^{\cdot}(T^*X)\widehat{\otimes}F$ on X in the $+$ case, of $\Lambda^{\cdot}(T^*X)\widehat{\otimes}F\widehat{\otimes}o(TX)$ in the $-$ case. For any $s\in\mathbf{R}$, the map $i_{\pm}$ maps H^s_X into $\mathcal{H}_s\cap\ker\alpha_{\pm}$.

Theorem 17.12.1. *There exists $\lambda_1>0$ such that if $s\in\mathbf{R}$, there is $C_s>0$ for which if $\lambda\in\mathbf{C},\operatorname{Re}\lambda\le\lambda_1$, for $u\in\mathcal{S}^{\cdot}(T^*X,\pi^*F)\cap\ker\alpha_{\pm}$, then*

$$\|S_{h,\lambda}u\|_{\lambda,\mathrm{sc},s+5/6}\le C_s\|u\|_{\lambda,\mathrm{sc},s}, \tag{17.12.2}$$

and for $u\in\mathcal{S}^{\cdot}(T^*X,\pi^*F)$,

$$\|P_{\pm}S_{h,\lambda}u\|_{\lambda,\mathrm{sc},s+5/6}\le C_s\|u\|_{\lambda,\mathrm{sc},s}. \tag{17.12.3}$$

Proof. Notice that the structure of $S^*_{h,\lambda}=((P^0_h)^*-\overline{\lambda})^{-1}$ is similar to the one of $S_{h,\lambda}$. By duality and interpolation, we only need to establish (17.12.3). Using partition of unity on X, we may and we will assume that the support of u is included in $\pi^{-1}K$, where the open set U and the compact subset $K\subset U$ are taken as in Theorem 17.10.1. Let $\psi(x)\in C^{\infty}_0(U)$ be a cutoff function which is equal to 1 near K. We will show that

$$\|\psi P_{\pm}S_{h,\lambda}u\|_{\lambda,\mathrm{sc},s+5/6}\le C_s\|u\|_{\lambda,\mathrm{sc},s}. \tag{17.12.4}$$

Combining Lemma 17.6.4 with the above estimate then leads to a proof of (17.12.3).

We use the coordinate system and the trivializations of vector bundles which were described at the beginning in section 17.8. Also the notation will be the same as in the proof of Theorem 17.10.1. We may and we will assume that $\psi\varphi=\psi$.

By (17.2.14), given N,L,α,β, there exists $C_{s,N,L,\alpha,\beta}>0$ such that in the given range of parameters,

$$\left\|<p>^N\widehat{\nabla}^{\alpha}P_{\pm}<p>^L\widehat{\nabla}^{\beta}u\right\|_{\lambda,\mathrm{sc},s}\le C_{s,N,L,\alpha,\beta}\|u\|_{\lambda,\mathrm{sc},s}. \tag{17.12.5}$$

Note that (17.12.5) gives an extension of (17.5.1).

By (17.10.1) in Theorem 17.10.1, we get

$$\psi P_{\pm}S_{h,\lambda}u=\sum_{0\le j<M}h^j\psi P_{\pm}E_0E^j_1u+h^M\psi P_{\pm}S_{h,\lambda}E^M_1u. \tag{17.12.6}$$

By (17.6.14) in Theorem 17.6.3, by (17.8.33) in Lemma 17.8.4, by (17.10.2) in Theorem 17.10.1, and by (17.12.5), if $M' \in \mathbf{N}$ is the integer associated to E_1^M and to $s - 1/4$ as in (17.8.33), we get

$$\left\|\psi P_\pm S_{h,\lambda} E_1^M u\right\|_{\lambda,\mathrm{sc},s} \leq C_s \left\|< p >^{-M'} S_{h,\lambda} E_1^M u\right\|_{\lambda,\mathrm{sc},s}$$
$$\leq C_s \left\|< p >^{-M'} E_1^M u\right\|_{\lambda,\mathrm{sc},s-1/4} \leq C_s \left\|u\right\|_{\lambda,\mathrm{sc},s-1/4-M/3}. \quad (17.12.7)$$

Using again Lemma 17.8.4 and (17.10.2) in Theorem 17.10.1, and proceeding as in (17.12.7), we find that for $0 \leq j < M$,

$$\left\|\psi P_\pm E_0 E_1^j u\right\|_{\lambda,\mathrm{sc},s} \leq C \left\|u\right\|_{\lambda,\mathrm{sc},s-2/3-j/3}. \quad (17.12.8)$$

Therefore in order to prove (17.12.4), we just have to prove the stronger inequality than (17.12.8) for $j = 0$,

$$\left\|\psi P_\pm E_0 u\right\|_{\lambda,\mathrm{sc},s+5/6} \leq C \left\|u\right\|_{\lambda,\mathrm{sc},s}. \quad (17.12.9)$$

Remark 17.12.2. The fact that we obtain the better $5/6$ instead of $2/3$ in (17.12.2) and (17.12.3) will play a crucial role in the sequel. Indeed otherwise, the proof of Theorem 17.16.3 would lead to replacing $-1/6$ in (17.16.10) by $-1/3$, which would not be enough to establish the crucial Theorem 17.21.3.

17.13 A PROOF OF EQUATION (17.12.9)

By (17.10.3), we have the identity

$$P_\pm E_0(x, hD_x, h, \lambda) u(x, p) = (2\pi h)^{-n} \int_{\mathbf{R}^n} e^{\frac{i}{h}\langle x,\xi\rangle} P_\pm e_{0,h,\lambda}(x,\xi) \widehat{u}\left(\frac{\xi}{h}, p\right) d\xi. \quad (17.13.1)$$

Equation (17.13.1) just says that the operator $P_\pm E_0$ is associated to the symbol $P_\pm e_0$.

To establish (17.12.9), we will show in Proposition 17.13.4 that for any $L \in \mathbf{N}$, $< p >^L P_\pm e_0 < p >^L \in \mathcal{S}^{-5/6,0}$, and then we will use Lemma 17.8.4 to conclude.

First, we give a refinement of the estimate (17.9.20) in the proof of Theorem 17.9.1.

If $\theta(p) \in \mathcal{S}, M \in \mathbf{N}$, set

$$\left\|\theta\right\|_M = \sum_{|\alpha|+|\beta|\leq M} \left\|p^\alpha \partial_p^\beta \theta\right\|_{L^\infty}. \quad (17.13.2)$$

Lemma 17.13.1. *There exists $C > 0$ for which if $a > 0, u \in \mathcal{S}$ are such that*

$$\left\|\Delta^V u\right\|_{L^2} + a \left\|p_1 u\right\|_{L^2} \leq 1, \quad (17.13.3)$$

then

$$\left\|u\right\|_{L^2} \leq C a^{-2/3}, \quad \left|\int_{\mathbf{R}^n} \theta(p) u(p) dp\right| \leq C a^{-5/6} \int_{\mathbf{R}} \left\|\widehat{\theta}(\eta^1, .)\right\|_{L^2(\eta')} d\eta^1. \quad (17.13.4)$$

There exist $C > 0, M \in \mathbf{N}$ for which if $b \in \mathbf{R}, k \in \mathbf{R}^n, u \in \mathcal{S}$ are such that

$$\|u\|_{L^2} + \left\|\Delta^V u\right\|_{L^2} + \|(b - \langle k, p\rangle)\, u\|_{L^2} \leq 1, \tag{17.13.5}$$

then

$$\left\|\left(< k > + \frac{|b|}{< p >}\right)^{2/3} u\right\|_{L^2} \leq C, \tag{17.13.6}$$

$$\left|\int_{\mathbf{R}^n} \theta(p)u(p)dp\right| \leq C\,(< k > + |b|)^{-5/6}\, \|\theta\|_M\,.$$

Proof. Take $u, v \in \mathcal{S}$. Consider the equation

$$v = \left(-\frac{1}{2}\Delta^V - iap_1\right) u. \tag{17.13.7}$$

We denote by $\widehat{u}$ the Fourier transform of u in the variable p. Then (17.13.7) is equivalent to

$$\widehat{v} = \left(a\frac{\partial}{\partial \eta^1} + \frac{1}{2}\,|\eta|^2\right)\widehat{u}. \tag{17.13.8}$$

By proceeding as in (17.9.24), (17.9.27), we get

$$\widehat{u}(\eta^1, \eta') = \frac{1}{a}\int_0^{+\infty} e^{-\frac{1}{a}\left(|\eta|^2 s/2 - \eta^1 s^2/2 + s^3/6\right)}\widehat{v}(\eta^1 - s, \eta')\, ds. \tag{17.13.9}$$

Observe that for $s \geq 0, \eta^1 \in \mathbf{R}$,

$$\eta^{1,2} s/2 - \eta^1 s^2/2 + s^3/6 \geq s^3/24. \tag{17.13.10}$$

For $a > 0, \eta' \in \mathbf{R}^{n-1}$, set

$$\varphi_{a,\eta'}(s) = 1_{s\geq 0}\frac{1}{a}e^{-\frac{1}{a}\left(|\eta'|^2 s/2 + s^3/24\right)}. \tag{17.13.11}$$

There exists $C > 0$ such that for any $a > 0, \eta' \in \mathbf{R}^{n-1}$,

$$\|\varphi_{a,\eta'}\|_{L^1} \leq Ca^{-2/3}, \qquad \|\varphi_{a,\eta'}\|_{L^2} \leq Ca^{-5/6}. \tag{17.13.12}$$

By (17.13.9)-(17.13.11),

$$\left|\widehat{u}\left(\eta^1, \eta'\right)\right| \leq \int_0^{+\infty} \varphi_{a,\eta'}(s)\left|\widehat{v}\left(\eta^1 - s, \eta'\right)\right| ds. \tag{17.13.13}$$

By (17.13.13), we get

$$\|\widehat{u}(., \eta')\|_{L^2(\eta^1)} \leq \|\varphi_{a,\eta'}\|_{L^1}\, \|\widehat{v}(., \eta')\|_{L^2(\eta^1)}\,. \tag{17.13.14}$$

When (17.13.3) holds, by (17.13.12), (17.13.14), we get the first inequality in (17.13.4).

Clearly,

$$\int_{\mathbf{R}^n} \theta\,(p)\, u\,(p)\, dp = (2\pi)^n \int_{\mathbf{R}^n} \widehat{u}\,(\eta)\,\overline{\widehat{v}}\,(\eta)\, d\eta. \tag{17.13.15}$$

By (17.13.13), (17.13.15), we get

$$\begin{aligned}\left|\int_{\mathbf{R}^n} \theta(p)\, u(p)\, dp\right| &\le C \int_{\mathbf{R}^n} \left|\widehat{\theta}(\eta)\right| \left(\int_0^{+\infty} \varphi_{a,\eta'}(s) \left|\widehat{v}\left(\eta^1 - s, \eta'\right)\right| ds\right) d\eta \\ &\le C \int \left|\widehat{\theta}(\eta)\right| \|\varphi_{a,\eta'}\|_{L^2} \|\widehat{v}(.,\eta')\|_{L^2(\eta^1)}\, d\eta \\ &\le C a^{-5/6} \|v\|_{L^2} \int \left\|\widehat{\theta}(\eta^1, .)\right\|_{L^2(\eta')} d\eta^1, \quad (17.13.16)\end{aligned}$$

which gives the second inequality in (17.13.4) under (17.13.3).

Now we establish (17.13.6). These inequalities are obvious for $< k > + |b| \le 1$ and also for $k = 0$.

For $k \ne 0$, by a rotation in the variable p, we may and we will assume that $k = (|k|, 0, ..., 0)$, so $\langle k, p\rangle - b = |k|(p_1 - \frac{b}{|k|})$. Since the Fourier transform of $\theta(p_1 - \frac{b}{|k|}, p')$ is equal to $e^{-i\frac{b\eta^1}{|k|}}\widehat{\theta}(\eta)$, by (17.13.4), we get

$$\left\| < k >^{2/3} u\right\|_{L^2} \le C, \quad (17.13.17)$$
$$\left|\int_{\mathbf{R}^n} \theta(p) u(p) dp\right| \le C < k >^{-5/6} \int \left\|\widehat{\theta}(\eta^1, .)\right\|_{L^2(\eta')} d\eta^1.$$

We claim that there exists $C > 0$ such that

$$< k >^{2/3} + |b - \langle k, p\rangle| \ge C\left(< k > + \frac{|b|}{< p >}\right)^{2/3}. \quad (17.13.18)$$

The proof is similar to the proof of (17.9.41). Indeed (17.13.18) is true if $|b - \langle k, p\rangle| \ge \frac{1}{2}|b|$. If not, then $|b| / < p > \le 2|k|$, and (17.13.18) still holds.

By using (17.13.5), (17.13.17), and (17.13.18), we get the first inequality in (17.13.6). For a given $C' > 0$, if $|b| \le C' < k >$, the second estimate in (17.13.6) also follows from (17.13.17). To establish this second estimate in full generality, we may and we will assume $b \ge C' < k >$, where C' is a large positive constant.

Let $\psi \in C_0^\infty(]-1/2, 1/2[)$ be equal to 1 on $[-1/4, 1/4]$. Set

$$u_1 = \psi\left(1 - \frac{\langle k, p\rangle}{b}\right) u, \qquad u_2 = \left(1 - \psi\left(1 - \frac{\langle k, p\rangle}{b}\right)\right) u. \quad (17.13.19)$$

Since $b \ge C' < k >$, the functions u_1, u_2 verify bounds similar to the bounds in (17.13.5) for u, possibly with a bound which is larger than 1.

First we study the contribution of u_2 to the second inequality in (17.13.6). Observe that $|b - \langle k, p\rangle| \ge \frac{b}{4}$ on the support of u_2, and so by (17.13.5), we get

$$\|u_2\|_{L^2} \le \frac{C}{|b|}. \quad (17.13.20)$$

Since $b \ge C' < k >$, by (17.13.20), we derive a bound on $\left|\int_{\mathbf{R}^n} \theta(p)\, u_2(p)\right| dp$ which is compatible with the second estimate in (17.13.6).

We still make a rotation on p so that $\langle k,p\rangle = |k|\,p_1$. Set

$$v = K_{b/|k|}u_1. \tag{17.13.21}$$

Then the support of v is included in the set of p such that $|1-p_1|\leq 1/2$. As we saw before, u_1 verifies an estimate similar to (17.13.5). This estimate can be written in the form

$$\|v\|_{L^2} + \frac{|k|^2}{b^2}\left\|\Delta^V v\right\|_{L^2} + b\,\|(1-p_1)v\|_{L^2} \leq \left(\frac{|k|}{b}\right)^{n/2}. \tag{17.13.22}$$

Let $\varphi \in C_0^\infty([1/4,5/4])$ be equal to 1 on $[1/2,3/2]$. Set

$$\theta_1(p) = \left(\frac{b}{|k|}\right)^{n/2}\varphi(p_1)\,\theta\left(\frac{bp}{|k|}\right), \quad w(p) = \left(\frac{b}{|k|}\right)^{n/2-2} v(p). \tag{17.13.23}$$

Clearly,

$$\int_{\mathbf{R}^n}\theta(p)\,u_1(p)\,dp = \frac{b^2}{|k|^2}\int_{\mathbf{R}^n}\theta_1(p)\,w(p)\,dp. \tag{17.13.24}$$

Now by (17.13.22), (17.13.23), we get

$$\left\|\Delta^V w\right\|_{L^2} + \frac{b^2}{|k|^2}b\,\|(1-p_1)\,w\|_{L^2} \leq 1. \tag{17.13.25}$$

Using (17.13.3), the second estimate in (17.13.4), (17.13.24), (17.13.25), we obtain

$$\left|\int_{\mathbf{R}^n}\theta(p)\,u_1(p)\,dp\right| \leq C\left(\frac{b}{|k|}\right)^{1/3} b^{-5/6}\int_{\mathbf{R}}\left\|\widehat{\theta}_1\left(\eta^1,\cdot\right)\right\|_{L^2(\eta')}d\eta^1. \tag{17.13.26}$$

Now observe that since the support of φ is included in $[1/4,5/4]$, given $M, M' \in \mathbf{N}$, there exists $C_{M,M'}>0$ such that

$$\|\theta_1\|_{M'} \leq C_{M,M'}\left|\frac{|k|}{b}\right|^M. \tag{17.13.27}$$

By (17.13.26), (17.13.27), we find that the contribution of u_1 to the integral in the second inequality in (17.13.6) is also compatible with the corresponding estimate. The proof of our lemma is completed. □

A smooth function $\varphi(x,p)$ with values in $\mathrm{End}(V)$ is said to be a symbol if there is $d\in\mathbf{R}$ such that

$$\left|\partial_x^\alpha\partial_p^\beta\varphi(x,p)\right| \leq C_{\alpha,\beta} < p >^{d-|\beta|}. \tag{17.13.28}$$

If e, e' are symbols, we denote by ee' their pointwise product, i.e., we use the same notation as after (17.8.24).

Lemma 17.13.2. *Given a multiindex α, there is $c_{\alpha,\beta}\in\mathcal{S}^{\frac{|\alpha-\beta|}{3}+2/3,|\alpha-\beta|-1}$ and $c'_{\alpha,\beta}\in\mathcal{S}^{\frac{|\alpha-\beta|}{3},|\alpha-\beta|}$ such that*

$$\left[e_0,\partial_p^\alpha\right] = \sum_{\beta<\alpha}\partial_p^\beta e_0 c_{\alpha,\beta}e_0, \qquad e_0\partial_p^\alpha = \sum_{\beta\leq\alpha}\partial_p^\beta e_0 c'_{\alpha,\beta}. \tag{17.13.29}$$

If $k \in \mathbf{N}$,

$$e_0 <p>^k \in \sum_{0\le j\le k} <p>^j e_0 \mathcal{S}^{0,0}. \tag{17.13.30}$$

Given multiindices α, β, *there exist* $a_{\alpha,\beta,\gamma} \in \mathcal{S}^{\frac{|\alpha|-|\gamma|}{3}-2/3|\beta|,0}$ *and symbols* $\varphi_{\alpha,\beta,\gamma}$ *such that*

$$\partial_x^\alpha \partial_\xi^\beta e_0 = \sum_{|\gamma|\le|\alpha|} \partial_p^\gamma \varphi_{\alpha,\beta,\gamma} e_0 a_{\alpha,\beta,\gamma}. \tag{17.13.31}$$

Proof. As we saw after (17.8.24), with respect to the pointwise product of symbols, $\mathcal{S}^{d,k}\mathcal{S}^{d',k'} \subset \mathcal{S}^{d+d',k+k'}$. Since $e_0 \in \mathcal{S}^{-2/3,1}$, by taking $c'_{\alpha,\beta} = c_{\alpha,\beta} e_0$ for $|\beta| < |\alpha|$ and $c'_{\alpha,\alpha} = 1$, the second identity in (17.13.29) follows from the first one. By multiplication on the left and on the right by Q_h^0, this first identity is equivalent to

$$\partial_p^\alpha \left(Q_h^0 - \lambda\right) = \left(Q_h^0 - \lambda\right) \left(\partial_p^\alpha + \sum_{\beta<\alpha} \partial_p^\beta e_0 c_{\alpha,\beta}\right). \tag{17.13.32}$$

To establish (17.13.32), we will argue by induction on $k = |\alpha|$, the case $k = 0$ being obvious. By (15.1.3), (17.9.2), and (17.9.3), for $\beta \neq 0$, the iterated commutator $\mathrm{Ad}_{\partial_p}^\beta Q_h^0$ is a linear combination with coefficients smooth functions in x of the operators

$$\xi_i, \quad p_i \frac{\partial}{\partial p_j}, \quad \frac{\partial}{\partial p_i}, \quad p_i M(x), \quad M(x), \quad \mathrm{Ad}_{\partial p}^\beta (P_\pm), \tag{17.13.33}$$

where $M(x)$ denotes a smooth matrix operator. Incidentally note that for $|\beta| \ge 3$, only the last operator in (17.13.33) still appears. The operators in (17.13.33) lie in $\mathcal{S}^{1,0}$. Note in particular that this is the case for $p_i \frac{\partial}{\partial p_j}$ since this operator is scale-invariant, so that we may ultimately exploit the fact that $|p|$ is bounded on the ball $\mathcal{B}_0$.

For multiindices α, γ such that $\beta < \alpha$, let $g_{\alpha,\beta}$ be a symbol which is a linear combination with integer coefficients of the operators $\mathrm{Ad}_{\partial_p}^\mu Q_h^0$ with $|\mu| = |\alpha - \beta|$. By the above, $g_{\alpha,\beta} \in \mathcal{S}^{1,0}$. By recursion, we see that given α, there are $g_{\alpha,\beta}$ such that

$$\partial_p^\alpha \left(Q_h^0 - \lambda\right) = \left(Q_h^0 - \lambda\right) \partial_p^\alpha + \sum_{\beta<\alpha} \partial_p^\beta g_{\alpha,\beta}. \tag{17.13.34}$$

To the left of the second sum in the right-hand side of (17.13.34), we may as well introduce the factor $\left(Q_h^0 - \lambda\right) e_0 = 1$. Also since $|\beta| < |\alpha|$, by recursion we can replace $e_0 \partial_p^\beta$ by the expression in the right-hand side of the first identity in (17.13.29). We get

$$\partial_p^\alpha \left(Q_h^0 - \lambda\right) = \left(Q_h^0 - \lambda\right) \left(\partial_p^\alpha + \sum_{\beta<\alpha} \partial_p^\beta e_0 \, g_{\alpha,\beta} + \sum_{\gamma<\beta<\alpha} \partial_p^\gamma \, e_0 c_{\beta,\gamma} e_0 g_{\alpha,\beta}\right). \tag{17.13.35}$$

By (17.13.35), we find that (17.13.32) holds with

$$c_{\alpha,\beta} = g_{\alpha,\beta} + \sum_{\beta<\gamma<\alpha} c_{\gamma,\beta}\, e_0\, g_{\alpha,\gamma}. \tag{17.13.36}$$

From known results on $e_0, g_{\alpha,\beta}$ and using recursion on the $c_{\beta,\gamma}$, we obtain the required result on $c_{\alpha,\beta}$.

By (17.9.46), (17.9.48), we get

$$<p>^2\ e_0 \in \mathcal{S}^{0,0}, \qquad <p>\ \partial_p\, e_0 \in \mathcal{S}^{0,0}, \qquad \partial_p^2\, e_0 \in \mathcal{S}^{0,1}. \tag{17.13.37}$$

Let $\varphi(x,p)$ be a symbol of degree d. By (15.1.3), (17.9.2), and (17.9.3), we get

$$[Q_h, \varphi] \in \mathcal{O}\left(\partial_p\varphi\left(\partial_p + <p>^2\right) + \partial_p^2\varphi\right). \tag{17.13.38}$$

Moreover,

$$[e_0, \varphi] = -e_0\left[Q_h^0, \varphi\right] e_0. \tag{17.13.39}$$

By (17.13.37), (17.13.39), we find that $[e_0, \varphi] \in e_0\mathcal{S}^{0,d-1}$. If $E \in \mathcal{S}^{0,d-1}$, then $<p>^{1-d} E \in \mathcal{S}^{0,0}$. We can then proceed by recursion and obtain (17.13.30).

We will establish (17.13.31) by recursion on $k = |\alpha| + |\beta|$. Equation (17.13.31) is obvious for $k=0$. For any $L \in \mathbf{R}$,

$$\mathcal{S}^{0,0}\ <p>^L = <p>^L\ \mathcal{S}^{0,0} = \mathcal{S}^{0,L}. \tag{17.13.40}$$

By (17.13.40), it is enough to establish (17.13.31) with $a_{\alpha,\beta,\gamma}(x,\xi,h,\lambda) \in \mathcal{S}^{\frac{|\alpha-\gamma|}{3}-2/3|\beta|,*}$, where $*$ denotes an unspecified real number.

Put

$$e_0^{\alpha,\beta} = \partial_x^\alpha \partial_\xi^\beta e_0. \tag{17.13.41}$$

By (17.9.121), we get

$$e_0^{\alpha,\beta} = \sum e_0\ S_{\alpha_1} e_0^{\alpha_2,\beta} + \sum\ e_0 T_{\nu,\beta_2}\ e_0^{\nu,\beta_2}. \tag{17.13.42}$$

The sum in (17.13.42) is submitted to the conditions in (17.9.122) and (17.9.123). Moreover, by (17.9.124), T_{ν,β_2} depends linearly on p, and so it is a symbol of degree 1.

For a symbol $\varphi^{(d)}(x,p)$ of degree d,

$$\partial_p^\alpha \varphi^{(d)} = \sum_{\beta\leq\alpha} \psi_{\alpha,\beta}^{(d-|\alpha-\beta|)} \partial_p^\beta, \tag{17.13.43}$$

where $\psi_{\alpha,\beta}^{(d-|\alpha-\beta|)}$ is a symbol of degree $d - |\alpha-\beta|$. In what follows we will often use the same notation for different symbols.

Using (17.13.29), (17.13.31), (17.13.43), and also recursion, we get

$$\begin{aligned} e_0 T_{\nu,\beta_2} e_0^{\nu,\beta_2} &= \sum_{|\gamma|\leq|\nu|} e_0 \varphi \partial_p^\gamma \varphi_{\nu,\beta_2,\gamma} e_0 a_{\nu,\beta_2,\gamma} \\ &= \sum_{\substack{|\gamma|\leq|\nu| \\ \gamma_1\leq\gamma}} e_0 \partial_p^{\gamma_1} \varphi_{\nu,\beta_2,\gamma} e_0 a_{\nu,\beta_2,\gamma} \subset \sum_{\substack{|\gamma|\leq|\nu| \\ \sigma\leq\gamma_1\leq\gamma}} \partial_p^\sigma e_0 \mathcal{S}^{\frac{|\gamma_1-\sigma|}{3},*} \varphi e_0 a_{\nu,\beta_2,\gamma}. \end{aligned} \tag{17.13.44}$$

The last term in (17.13.44) is of the required form in the right-hand side of (17.13.31) (with 0 replaced by $*$), because by (17.9.122),

$$\mathcal{S}^{\frac{|\gamma_1-\sigma|}{3},*}\varphi e_0 a_{\nu,\beta_2,\gamma} \subset \mathcal{S}^{\frac{|\gamma_1|-|\sigma|}{3}-2/3+\frac{|\nu|-|\gamma|}{3}-2/3|\beta_2|,*} \subset \mathcal{S}^{\frac{|\alpha|-|\sigma|}{3}-2/3|\beta|,*}. \tag{17.13.45}$$

We proceed in the same way to deal with the terms containing S_{α_1} in (17.13.42). The structure of S_{α_1} was described after (17.9.132). First we study the contribution of the ξ linear component of S_{α_1} which is bilinear in ξ, p. Using recursion, we get

$$\begin{aligned} e_0\xi\phi e_0^{\alpha_2,\beta} &= \sum_{|\gamma|\le|\alpha_2|} \xi e_0\varphi\partial_p^\gamma \varphi_{\alpha_2,\beta,\gamma} e_0 a_{\alpha_2,\beta,\gamma} \\ &= \sum_{|\gamma|\le|\alpha_2|\gamma_1\le\gamma} \xi e_0 \partial_p^{\gamma_1}\varphi_{\alpha_2,\beta,\gamma} e_0 a_{\alpha_2,\beta,\gamma} \\ &\subset \sum_{\substack{|\gamma|\le|\alpha_2| \\ \sigma\le\gamma_1\le\gamma}} \partial_p^\sigma e_0 \mathcal{S}^{\frac{|\gamma_1-\sigma|}{3},*}\xi\varphi\ e_0\ a_{\alpha_2,\beta,\gamma}. \end{aligned} \tag{17.13.46}$$

Using again (17.9.122), we get

$$\mathcal{S}^{\frac{|\gamma_1|-|\sigma|}{3},*}\xi\varphi e_0 a_{\alpha_2,\beta,\gamma} \subset \mathcal{S}^{\frac{|\gamma_1|-|\sigma|}{3}+1/3+\frac{|\alpha_2|-|\gamma|}{3}-2/3|\beta|,*} \subset \mathcal{S}^{\frac{|\alpha|-|\sigma|}{3}-2/3|\beta|,*}, \tag{17.13.47}$$

which takes care of the term in (17.13.46).

What remains in S_{α_1} are either matrix values symbols, which can be dealt with as above, or a term containing just one differential in the p variable, with a matrix coefficient. Now note that

$$\begin{aligned} e_0\partial_p e_0^{\alpha_2,\beta} &= \sum_{|\gamma|\le|\alpha_2|+1} e_0\partial_p^\gamma\varphi_{\alpha_2,\beta,\gamma} e_0 a_{\alpha_2,\beta,\gamma} \\ &\subset \sum_{\substack{|\gamma|\le|\alpha_2|+1 \\ \sigma\le\gamma}} \partial_p^\sigma e_0 \mathcal{S}^{\frac{|\gamma-\sigma|}{3},*}\varphi e_0 a_{\alpha_2,\beta,\gamma}. \end{aligned} \tag{17.13.48}$$

Using again (17.9.122), we get

$$\mathcal{S}^{\frac{|\gamma-\sigma|}{3},*}\varphi e_0 a_{\alpha_2,\beta,\gamma} \in \mathcal{S}^{\frac{|\gamma-\sigma|}{3}-2/3+\frac{|\alpha_2|-|\gamma|}{3}-2/3|\beta|,*} \subset \mathcal{S}^{\frac{|\alpha|-|\sigma|}{3}-1-2/3|\beta|,*}. \tag{17.13.49}$$

We find that (17.13.48), (17.13.49) are compatible with (17.13.31).

The proof of our lemma is complete. □

Remark 17.13.3. Since $e_0 \in \mathcal{S}^{-2/3,1}$, by the considerations which follow (17.8.24), we get $\partial_x e_0 \in \mathcal{S}^{-1/3,3}$. We will briefly show that

$$[p\partial_p, e_0] \in \mathcal{S}^{-1/3,3}. \tag{17.13.50}$$

By (17.13.50), we find that the estimates on $\partial_x e_0$ and $[p\partial_p, e_0]$ are compatible. Now e_0 is a symbol which is globally defined. As explained in Remark 17.8.3, the above compatibility indicates that our computations are indeed consistent.

To establish (17.13.50), note that

$$[p\partial_p, e_0] = p\,[\partial_p, e_0] + [p, e_0]\,\partial_p. \tag{17.13.51}$$

By (17.13.29) in Lemma 17.13.2, $[\partial_p, e_0] \in \mathcal{S}^{-1/3,2}$, and so $p\,[\partial_p, e_0] \in \mathcal{S}^{-1/3,3}$. Moreover, by (17.13.39),

$$[p, e_0] = e_0\left[Q_h^0, p\right] e_0. \tag{17.13.52}$$

The term $P_\pm$ in the expression (17.9.3) for Q_h^0 is easily dealt with. By (17.13.38),

$$[Q_h, p] \in \mathcal{O}\left(\partial_p + < p >^2\right). \tag{17.13.53}$$

Now note that $\partial_p \in \mathcal{S}^{1,-1}$, the -1 coming from the presence of $< p >^2$ in the expression for $A(\xi, p, \lambda)$ in (17.8.7). Therefore

$$e_0 < p >^2 e_0\partial_p \in \mathcal{S}^{-1/3,3}. \tag{17.13.54}$$

Also

$$e_0\partial_p e_0\partial_p = e_0\partial_p^2 e_0 + e_0\partial_p\,[e_0, \partial_p]\,. \tag{17.13.55}$$

By (17.13.37), $\partial_p^2 e_0 \in \mathcal{S}^{0,1}$, and so $e_0\partial_p^2 e_0 \in \mathcal{S}^{-2/3,3}$. Also by using (17.13.37) again, $\partial_p e_0 \in \mathcal{S}^{0,-1}$. By noting that our classes of operators are invariant when taking adjoints, we thus find that $e_0\partial_p \in \mathcal{S}^{0,-1}$. Moreover, $[e_0, \partial_p] \in \mathcal{S}^{-1/3,2}$, and so

$$e_0\partial_p\,[e_0, \partial_p] \in \mathcal{S}^{-1/3,1} \subset \mathcal{S}^{-1/3,3}. \tag{17.13.56}$$

By (17.13.51)-(17.13.56), we find that indeed $[p\partial_p, e_0] \in \mathcal{S}^{-1/3,3}$.

Proposition 17.13.4. *For any $L \in \mathbf{N}$, then*

$$< p >^L P_\pm e_0 < p >^L \in \mathcal{S}^{-5/6,0} \tag{17.13.57}$$

Proof. In our trivialization, $P_\pm$ and $< p >$ do not depend on x, and so

$$\partial_x^\alpha\partial_\xi^\beta < p >^L P_\pm e_0 < p >^L = < p >^L P_\pm \partial_x^\alpha\partial_\xi^\beta e_0 < p >^L\,. \tag{17.13.58}$$

Using the defining equation (17.8.24) and (17.13.58), our proposition will be a consequence of the inequality

$$\left\| < p >^N P_\pm \partial_x^\alpha\partial_\xi^\beta e_0 < p >^L u \right\|_{p,s+5/6} \leq \|u\|_{p,s+1/3|\alpha|-2/3|\beta|}\,. \tag{17.13.59}$$

We may temporarily assume that $\dim F = 1$. Observe that the operator in the left-hand side of (17.13.59) has rank 1. By (17.13.30) and (17.13.31) in Lemma 17.13.2, and integration by part in p, we see that it is sufficient to prove that for $\theta \in \mathcal{S}, s \in \mathbf{R}$, there exists $C > 0$ such that

$$\left|(< \xi > + |\lambda|)^{s+5/6} \int_{\mathbf{R}^n} \theta(p)\, e_0 u dp\right| \leq C \|u\|_{p,s}\,. \tag{17.13.60}$$

By interpolation, we just have to verify (17.13.60) for $s \in \mathbf{Z}$.

First we prove (17.13.60) when $s \in \mathbf{N}$. By (17.8.8), for $\tau \in]0,1]$,

$$A_\tau \geq \tau\,(< \xi > + |2\lambda_1 - \lambda|)\,. \tag{17.13.61}$$

By (17.13.61), we find that if $s \in \mathbf{N}$,

$$\left\| < p >^{-s} u \right\|_{L^2} \leq C_s \left(< \xi > + |\lambda| \right)^{-s} \|u\|_{p,s} . \tag{17.13.62}$$

By (17.13.30) and (17.13.62), we find that to establish (17.13.60) for $s \in \mathbf{N}$, we only need to show that

$$\left(< \xi > + |\lambda| \right)^{5/6} \left| \int_{\mathbf{R}^n} \theta(p) e_0 u dp \right| \leq \| < p > u \|_{L^2} . \tag{17.13.63}$$

Using the inequalities (17.9.7) in Theorem 17.9.1 with $a = 2, b = 0$ allows us to dominate the norms $\|u\|_{L^2}$, $\left\| \Delta^V e_0 u \right\|_{L^2}$ in terms of $\| < p > u \|_{L^2}$. Recall that $Q_h(x, \xi)$ is given by (17.9.2). Using this formula, we can dominate $|\mathrm{Re}\,\lambda| \, \|u\|_{L^2}$ and $\|(\mathrm{Im}\,\lambda \pm \langle k, p \rangle) u\|_{L^2}$, with $k = \widetilde{\sigma} \left(g^{TX} \right)^{-1} \xi$ in terms of $\| < p > u \|_{L^2}$. By using Lemma 17.13.1, we get (17.13.63).

To establish (17.13.60) for $s \in \mathbf{Z}, s \leq 0$, we just have to show that for $|\alpha| \leq |s|$,

$$\left| \left(< \xi > + |\lambda| \right)^{5/6 - |\alpha|} \int_{\mathbf{R}^n} \theta e_0 \partial_p^\alpha u dp \right| \leq C_s \|u\|_{L^2} . \tag{17.13.64}$$

Using (17.13.29), we find that to establish (17.13.64), we only need to show that for $k \in \mathbf{N}, a \in \mathcal{S}^{k/3,*}$ (with $* \in \mathbf{R}$),

$$\left| \left(< \xi > + |\lambda| \right)^{5/6 - k} \int_{\mathbf{R}^n} \theta e_0 a u dp \right| \leq C \|u\|_{L^2} . \tag{17.13.65}$$

We argue by induction on k. Indeed the case $k = 0$ was already considered. For $a \in \mathcal{S}^{\frac{k+1}{3},*}$, there exist $a_1 \in S^{(k-2)/3,*}, a_2 \in S^{k/3,*}$ such that

$$a = \partial_p a_1 + \left(< \xi > + |\lambda| \right)^{1/3} a_2. \tag{17.13.66}$$

Indeed if $c = - < p > \Delta^V < p > + A_\tau^2 \in \mathcal{S}^{2,0}$, then $c^{-1} a \in \mathcal{S}^{\frac{k-5}{3},*}$, and moreover,

$$a = - < p > \Delta^V < p > c^{-1} a + \left(< \xi > + |\lambda| \right)^{1/3} \left(< \xi > + |\lambda| \right)^{-1/3} A_\tau^2 c^{-1} a. \tag{17.13.67}$$

Then (17.13.67) is a form of (17.13.66).

Using (17.13.66), we get

$$e_0 a = \partial_p e_0 a_1 + e_0 \left[\partial_p, Q_h^0 \right] e_0 a_1 + e_0 \left(< \xi > + |\lambda| \right)^{1/3} a_2. \tag{17.13.68}$$

Recursion and integration by parts takes care of the contribution of the first and last terms in (17.13.68) to (17.13.65). Moreover, using (15.1.3) and equation (17.9.2) for Q_h, we find that the ξ-linear part of $\left[\partial_p, Q_h^0 \right]$ is compatible with (17.13.65). The only terms of $\left[\partial_p, Q_h^0 \right]$ which remain to be controlled are of the form p or $p\partial_p$. Using (17.13.30), we may as well replace $p\partial_p$ by ∂_p. Now note that

$$e_0 \partial_p e_0 a_1 = \partial_p e_0 e_0 a_1 + e_0 \left[\partial_p, Q_h^0 \right] e_0 e_0 a_1. \tag{17.13.69}$$

Using the fact that $e_0 \in \mathcal{S}^{-2/3,1}$ and the form of Q_h^0, we get $\left[\partial_p, Q_h^0 \right] e_0 e_0 \in \mathcal{S}^{-1/3,*}$. By recursion, we find that the contribution of (17.13.69) is also compatible with (17.13.64). As to the terms of the form p, they can be handled using (17.13.30). The proof of Proposition 17.13.4 is complete. □

Now we establish (17.12.9). As we saw in (17.13.1), the symbol of ψE_0 is just ψe_0. Then (17.12.9) follows from Lemma 17.8.4 and Proposition 17.13.4. This completes the proof of Theorem 17.12.1. □

17.14 AN EXTENSION OF THE PARAMETRIX TO $\lambda \in \mathcal{V}$

Finally, we investigate the holomorphic extension of the resolvent $S_{h,\lambda}$ to $\lambda \in \mathcal{V}$, where $\mathcal{V}$ is defined in (17.7.1). Recall that in Theorem 17.7.2, we showed that the resolvent $S_{h,\lambda}$ extends to $\lambda \in \mathcal{V}$ and verifies corresponding uniform estimates. Now we will extend the results we obtained using the construction of the parametrix to such λ. Recall that $\mathcal{V}$ depends on $\lambda_1 > 0, c_0 > 0$.

Theorem 17.14.1. *There exist $\lambda_1 \in]0, 1/2[, c_0 > 0$, such that $S_{h,\lambda}$ extends as a holomorphic function of $\lambda \in \mathcal{V}$, and Theorems 17.9.1, 17.9.4, 17.10.1, and 17.12.1 as well as Lemma 17.13.2 and Proposition 17.13.4 still hold.*

Proof. We use the notation $\lambda = \mu + \nu$ as in equation (17.7.1). As we already observed in the proof of Theorem 17.7.2, in the given range of parameters, the norms $\| \; \|_{\lambda,\mathrm{sc},s}$ and $\| \; \|_{\mu,\mathrm{sc},s}$ are equivalent. Using Theorem 17.7.2, and following the same strategy as we did before, the only point to verify is that Theorem 17.9.1 extends to $\lambda \in \mathcal{V}$. Thus we must show that if $\left(Q_h^0 - \lambda\right) u = v$, the estimates in (17.9.7) still hold. Note that

$$\left(Q_h^0 - \mu\right) u = v + \nu u. \tag{17.14.1}$$

By (17.14.1), we get the inequalities (17.9.7) in Theorem 17.9.1 in which λ is replaced by μ and v is replaced by $v + \nu u$.

Moreover, for $x > 0$,

$$x + \frac{|\mu|^{2/3}}{x^{5/6}} \geq C \, |\mu|^{4/11}, \tag{17.14.2}$$

so that when $c_0 \in]0, 1]$,

$$< p >^2 + \left(\frac{|\mu|}{< p >} \right)^{2/3} \frac{1}{< p >} \geq C < \mu >^{4/11} \geq C < \lambda >^{1/6}. \tag{17.14.3}$$

Using the form of the inequalities (17.9.7) which was described before and (17.14.3), we get

$$|\lambda|^{1/6} \left\| < p >^a u \right\|_{L^2} \leq C_a \left\| < p >^a (v + \nu u) \right\|_{L^2}. \tag{17.14.4}$$

Now recall that if $\lambda \in \mathcal{V}$, then $|\nu| \leq c_0 \, |\mu|^{1/6}$. The constants C_a are uniformly bounded as long as a varies in a compact set of $\mathbf{R}$. From (17.14.4), we find that when a varies in such a compact domain, we can choose $c_0 \in]0, 1]$ small enough so that

$$\left\| < p >^a \nu u \right\|_{L^2} \leq C \left\| < p >^a v \right\|_{L^2}. \tag{17.14.5}$$

From (17.14.5), we get the inequalities in (17.9.7) when a remains bounded. However, by proceeding as in the proof of Theorem 17.9.1, we find that once our inequalities have been established bounded a, b, they extend to arbitrary a, b.

The proof of Theorem 17.14.1 is complete. □

17.15 PSEUDODIFFERENTIAL ESTIMATES FOR $\mathbf{P}_\pm \mathbf{S}_{\mathbf{h},\lambda}\mathbf{i}_\pm$

We still use the embedding of L^2_X into H^0 which was described at the beginning of section 17.12. Other spaces of distributions are embedded as well.

Recall that $\square^X = \left(d^X + d^{X*}\right)^2$ is the Hodge Laplacian, which acts on $\mathcal{D}_X$ and on $\mathcal{D}'_X$.

Definition 17.15.1. If $s\in\mathbf{R}, \lambda\in\mathbf{C}, u\in H^s_X$, set

$$\begin{aligned} \|u\|_{X,s} &= \left\|\left(1+\square^X\right)^{s/2} u\right\|_{L^2_X},\\ \|u\|_{X,\lambda^{1/2},s} &= \left\|\left(1+|\lambda|+\square^X\right)^{s/2} u\right\|_{L^2_X},\\ \|u\|_{X,h,\lambda,s} &= \left\|\left(1+|\lambda|^2+h^2\square^X\right)^{s/2} u\right\|_{L^2_X}. \end{aligned} \tag{17.15.1}$$

By comparing (17.4.3) and (17.15.1), one finds easily that given $s\in\mathbf{R}$, the norms $\|u\|_{X,h,\lambda,s}$ and $\|u\|_{\lambda,\mathrm{sc},s}$ are uniformly equivalent, the constants in the equivalence not depending on h, λ.

Recall that $\mathcal{W}_\delta$ was defined in (16.4.1). Namely,

$$\mathcal{W}_\delta = \left\{\lambda\in\mathbf{C}, \operatorname{Re}\lambda\le \delta_0+\delta_1\left|\operatorname{Im}\lambda\right|^{\delta_2}\right\}. \tag{17.15.2}$$

Moreover, $\mathcal{V}$ was defined in (17.7.1) and depends on $\lambda_1>0, c_0>0$. We claim that when taking $\delta_0=\lambda_1, \delta_1=c_0, \delta_2=1/6$, then

$$\mathcal{W}_\delta\subset\mathcal{V}. \tag{17.15.3}$$

Indeed if $\lambda=a+ib, \lambda\in\mathcal{W}_\delta$, if $a\le\lambda_1$, then $\lambda\in\mathcal{V}$. If $a>\lambda_1$, then

$$\lambda=\lambda_1+ib+a-\lambda_1, \tag{17.15.4}$$

and moreover $0\le a-\lambda_1\le\delta_1\left|b\right|^{1/6}$, so that again $\lambda\in\mathcal{V}$.

To simplify the exposition, many statements will only be given in the $+$ case. However, the corresponding statement in the $-$ case will be obtained simply by replacing F by $F\otimes o(TX)$.

We fix temporarily $\delta=(\delta_0,\delta_1,\delta_2)$. The precise value of δ will be determined later.

Take $x_0\in X$. Let $U\subset X$ be a small open neighborhood of x_0, and let $x^1,\dots,x^n$ be a coordinate system on U. Consider a trivialization of $\Lambda^\cdot(T^*X)\widehat{\otimes}F$ on U. In the $+$ case, a symbol $a(x,\xi,h,\lambda)$ of degree d is a smooth function of (x,ξ) with values in $\operatorname{End}\left(\Lambda^\cdot(T^*X)\widehat{\otimes}\Lambda^\cdot(TX)\widehat{\otimes}F\right)_{x_0}$, which is holomorphic in the parameter $\lambda\in\mathcal{W}_\delta$ and also depends on $h\in$

$]0, h_0]$, such that for any α, β, there exists $C_{\alpha,\beta} > 0$ such that for $x \in \mathbf{R}^n, h \in]0, h_0], \lambda \in \mathcal{W}_\delta$,

$$\left|\partial_x^\alpha \partial_\xi^\beta a(x, \xi, h, \lambda)\right| \leq C_{\alpha,\beta} \left(1 + |\lambda| + |\xi|\right)^{d+1/3|\alpha|-2/3|\beta|}. \tag{17.15.5}$$

We denote by $\mathbb{S}^d_{\delta,h}$ the set of symbols of degree d. Note that $\mathbb{S}^d_{\delta,h}$ is simply a semiclassical version of the Hörmander class $S^d_{2/3,1/3}$ defined in [Hör85, section 18.1].

A smoothing operator on X is a family of operators $B(h, \lambda)$, depending holomorphically on $\lambda \in \mathcal{W}_\delta$, and also depending on $h \in]0, h_0]$, such that if $s \in \mathbf{R}, t \in \mathbf{R}, N \in \mathbf{N}$, there exists $C_{s,t,N} > 0$ such that if $u \in \mathcal{D}_X$,

$$\|B(h, \lambda)u\|_{X,h,\lambda,s} \leq C_{s,t,N}\ h^N \|u\|_{X,h,\lambda,t}\,. \tag{17.15.6}$$

Note in particular that because N is arbitrary in (17.15.6), if $B(h, \lambda)$ is smoothing, it is also a uniformly regularizing family of operators in the classical sense, which converges to 0 as $h \to 0$.

In the above coordinate system, if u is a smooth section of $\Lambda^{\cdot}(T^*X) \widehat{\otimes} F$ with support included in U, let $\widehat{u}(\xi)$ denote its Fourier transform. We quantify a symbol a into an operator $A = \operatorname{Op}(a)$ by the usual formula

$$A(x, hD_x, h, \lambda)\, u(x) = (2\pi h)^{-n} \int_{\mathbf{R}^n} e^{\frac{i}{h}\langle x,\xi\rangle} a(x, \xi, h, \lambda) \widehat{u}\left(\frac{\xi}{h}\right) d\xi. \tag{17.15.7}$$

Let $\mathbb{E}^d_{\delta,h}$ be the associated set of pseudodifferential operators of degree d on X. If $A \in \mathbb{E}^d_{\delta,h}$, if $K \subset X$ is a small compact set, if $\varphi(x)$ is a cutoff function with support in a small neighborhood of K, there is a cutoff function φ' equal to 1 near the support of φ, a symbol $a \in \mathbb{S}^d_{\delta,h}$ and a smoothing operator $B(h, \lambda)$ such that in the given local coordinates and trivializations,

$$A\varphi = \varphi' \operatorname{Op}(a) \varphi + B(h, \lambda). \tag{17.15.8}$$

For $A \in \mathbb{E}^d_{\delta,h}$, we denote by $\sigma_d(A)$ the semiclassical principal symbol of A. Namely, if $A = \operatorname{Op}(a)$, $\sigma_d(A)$ is the class of a in the quotient space $\mathbb{S}^d_{\delta,h}/h\mathbb{S}^d_{\delta,h}$. If $E_d \in \mathbb{E}^d_{\delta,h}, E_{d'} \in \mathbb{E}^{d'}_{\delta,h}$, then $E_d E_{d'} \in \mathbb{E}^{d+d'}_{\delta,h}, \sigma(E_d E_{d'}) = \sigma(E_d)\sigma(E_{d'})$. Moreover, if $E_d = \operatorname{Op}(a), E_{d'} = \operatorname{Op}(a')$, then

$$[E_d, E_{d'}] - \operatorname{Op}([a, a'] + \frac{h}{i}\{a, a'\}) \in h^2 \mathcal{E}^{d+d'-2/3}_{\delta,h}. \tag{17.15.9}$$

Operators in $\mathbb{E}^0_{\delta,h}$ act as a family of uniformly bounded operators on L^2_X.

We will denote by $\mathbb{S}^d_{\delta,h,0}$ and $\mathbb{E}^d_{\delta,h,0}$ the classes of symbols and corresponding operators, where the estimates in (17.15.5) are replaced by the stronger estimates

$$\left|\partial_x^\alpha \partial_\xi^\beta a(x, \xi, h, \lambda)\right| \leq C_{\alpha,\beta} \left(1 + |\lambda| + |\xi|\right)^{d-|\beta|}. \tag{17.15.10}$$

An operator $A \in \mathbb{E}^d_{\delta,h}$ is said to be elliptic if there exist $h_1 \in]0, h_0]$ and $C > 0$ such that if $h \in]0, h_1], \lambda \in \mathcal{W}_\delta$ such that

$$|\sigma_d(A)(x, \xi, h, \lambda)| \geq C \left(1 + |\lambda| + |\xi|\right)^d. \tag{17.15.11}$$

If $A \in \mathbb{E}^d_{\delta,h}$ is elliptic, then for h_0 small enough, there exist $B \in \mathbb{E}^{-d}_{\delta,h}$ such that $AB = 1, BA = 1$.

Recall that $e_{0,h,\lambda}(x,\xi)$ was defined in Definition 17.9.3 as the inverse of $Q^0_h(x,\xi) - \lambda$.

Definition 17.15.2. For $\lambda \in \mathbf{R}, \lambda < 1$ or $\lambda \notin \mathbf{R}$, let $e^0_{0,\lambda}(x,\xi)$ be the inverse of $Q^0_0(x,\xi) - \lambda$. In the sequel, we will often use the notation e^0_0 instead of $e^0_{0,\lambda}$.

We have the obvious

$$e^0_{0,\lambda} = e_{0,0,\lambda}. \tag{17.15.12}$$

Therefore the estimates which were proved for e_0 are also valid for e^0_0.

In the sequel we will often write e^0_0 instead of $e^0_{0,\lambda}$.

Observe that $P_\pm S_{h,\lambda} i_\pm$ maps $\mathcal{D}_X$ into itself. Note that $a^0_0(\xi)$ was defined in (16.5.1). An explicit formula for $a^0_0(\xi)$ was given in (16.5.11). In particular $a^0_0(\xi)$ depends only on $|\xi|$.

Theorem 17.15.3. *There exists $\delta' = (\delta'_0, \delta'_1, \delta'_2)$ with $\delta'_0 \in]0,1[, \delta'_1 > 0, \delta'_2 = 1/6$ such that $P_\pm S_{h,\lambda} i_\pm \in \mathbb{E}^{-1}_{\delta',h}$. Moreover, this operator is elliptic, and its principal symbol is given by*

$$\sigma_{-1}(P_\pm S_{h,\lambda} i_\pm) = a^0_0. \tag{17.15.13}$$

Given $a,b,c,d \in \mathbf{R}$, then

$$P_\pm \partial^a_p p^b S_{h,\lambda} p^c \partial^d_p i_\pm \in \mathbb{E}^{-1}_{\delta',h}. \tag{17.15.14}$$

Proof. We take $\delta' = \delta$, with δ as in (17.15.3). By Lemma 17.6.4, by Theorem 17.7.2, and by (17.15.3), our problem is local on X. As in the proof of Theorem 17.12.1, we will work in a small open neighborhood U of a given point $x_0 \in X$, and we use the corresponding coordinate system and trivialization.

Let K be a compact subset of U, and let $\varphi, \varphi' \in C^\infty_0(U)$ be cutoff functions with φ equal to 1 near K and φ' equal to 1 near the support of φ. By equation (17.10.1) in Theorem 17.10.1, by Theorem 17.14.1, and by (17.15.3), if $\lambda \in \mathcal{W}_{\delta'}$,

$$\varphi' P_\pm S_{h,\lambda} i_\pm \varphi u = \sum_{0 \le j < M} h^j \varphi P_\pm \, E_0 E^j_1 i_\pm \varphi u + h^M \varphi' P_\pm S_{h,\lambda} E^M_1 i_\pm \varphi u. \tag{17.15.15}$$

Moreover, by (17.10.2) in Theorem 17.10.1 and by proceeding as in the proof of Lemma 17.8.4, if $L \in \mathbf{N}$, we get

$$E^j_1 P_\pm \varphi \in \mathcal{P}^{-j/3,-L}, \qquad E_0 E^j_1 P_\pm \varphi \in \mathcal{P}^{-2/3-j/3,-L}. \tag{17.15.16}$$

Note here that the fact that $P_\pm$ appears in (17.15.16) overcomes the fact that the $\mathcal{P}^{d,k}$ do not form an algebra under composition.

We claim that

$$\varphi' P_\pm S_{h,\lambda} i_\pm \varphi \in \varphi' P_\pm E_0 i_\pm \varphi + h \mathbb{E}^{-1}_{\delta',h}. \tag{17.15.17}$$

In fact by using Theorems 17.6.3 and 17.7.2, Lemma 17.8.4, Theorem 17.10.1, and the considerations we made after (17.15.1), we get

$$\left\|\varphi' P_\pm S_{h,\lambda} E_1^M i_\pm \varphi u\right\|_{X,h,\lambda,s+1/4+M/3} \leq C_{M,s} \|u\|_{X,h,\lambda,s} . \tag{17.15.18}$$

Since (17.15.15) and (17.15.18) are valid for any $M \in \mathbf{N}$, it is a classical result that $\varphi'\ P_\pm S_{h,\lambda} i_\pm \varphi \in \mathbb{E}^{-1}_{\delta',h}$. By (17.15.15), we obtain (17.15.17).

We already know that $\varphi P_\pm E_0 i_\pm \varphi \in \mathbb{E}^{-1}_{\delta',h}$. By (17.15.17), to establish our theorem, we only need to prove that this is an elliptic operator on K, whose principal symbol is given by (17.15.13).

Set

$$\mathbb{Q}_h(x,\xi) = \frac{1}{h}\left(Q_h - Q_0\right)(x,\xi) . \tag{17.15.19}$$

By (17.9.2), (17.9.3),

$$Q_h^0 = Q_0^0 + h\mathbb{Q}_h. \tag{17.15.20}$$

Moreover, $\mathbb{Q}_h$ is a differential operator of degree 1 along the fibers T^*X, whose coefficients are polynomials of degree at most 2 in p.

Clearly,

$$e_0 = e_0^0 - he_0\mathbb{Q}_h\, e_0^0. \tag{17.15.21}$$

By equation (17.13.29) in Lemma 17.13.2, $\left[\partial_p, e_0^0\right] \in \mathcal{S}^{-1/3,1}$. Using the considerations we made on $\mathbb{Q}_h$, we find that for any $L \in \mathbf{N}$, $\mathbb{Q}_h e_0^0 P_\pm \in \mathcal{S}^{-1/3,-L}$. Then the same arguments show that for any $L \in \mathbf{N}$,

$$P_\pm e_0 \mathbb{Q}_h e_0^0 i_\pm \in \mathcal{S}^{-1,-L}. \tag{17.15.22}$$

Let $E_0^0 \in \mathcal{P}^{-2/3,1}$ be associated to e_0^0 as in (17.8.26). By (17.15.21), (17.15.22), and replacing the symbols by the corresponding operators, we get

$$\varphi P_\pm E_0 i_\pm \varphi \in \varphi P_\pm E_0^0 i_\pm \varphi + h\mathbb{E}^{-1}_{\delta',h}. \tag{17.15.23}$$

To establish the first part of our theorem, what remains to prove is that $P_\pm e_0^0 i_\pm$ is an elliptic symbol of degree -1. By using the notation of sections 16.1 and 16.4, and comparing equations (15.1.3) and (17.9.2) with (16.2.1), we get in the $+$ case,

$$e_0^0 = \left(B\left(i\widetilde{\sigma}_x\left(g_x^{TX}\right)^{-1}\xi\right) + P_+ + N^V - \lambda\right)^{-1}. \tag{17.15.24}$$

In the $-$ case, N^V should be replaced by $n - N^V$, P_+ by P_-, and ξ by $-\xi$. Now observe that $\ker \alpha_+$ is concentrated in vertical degree 0 and $\ker \alpha_-$ in vertical degree n. Moreover, the operator $B(i\xi)$ is scalar, so that it does not change the vertical degree. Recall that the operator P was defined in section 16.5. By (17.15.24), we obtain

$$P_\pm e_0^0 P_\pm = P\left(B\left(\pm i\widetilde{\sigma}_x\left(g_x^{TX}\right)^{-1}\xi\right) + P - \lambda\right)^{-1} P. \tag{17.15.25}$$

By (16.5.1), and taking into account the fact that a_0^0 is a radial function of ξ, we get

$$P_\pm e_0^0 P_\pm = a_0^0\left(\widetilde{\sigma}_x\left(g_x^{TX}\right)^{-1}\xi, \lambda\right). \tag{17.15.26}$$

Now we can use Proposition 16.5.1 to control the derivatives in the ξ variable of the right-hand side of (17.15.26), and we obtain the estimates in (17.15.10). Incidentally note that since we take $\delta_2 = 1$ in Theorem 16.4.1, the domain in which these final estimates are valid is bigger than our $\mathcal{W}_{\delta'}$. The derivatives in the variable x are also easy to control by using Proposition 16.5.1 and by (17.15.24). Finally, since a_0^0 is a globally defined symbol, we may as well evaluate (17.15.26) at $x = x_0$, so that we get equation (17.15.13). This completes the proof of the first part of our theorem.

To establish the second part of our theorem, note that $P_\pm \partial_p^a p^b S_{h,\lambda} p^c \partial_p^d i_\pm$ is a linear combination of operators of the form $P_\pm p^c S_{h,\lambda} p^{c'} i_\pm$. Using the same arguments as above, we see that to establish the second part of our theorem, we must show that $P_\pm p^b e_0^0 p^c i_\pm$ is a symbol of degree -1. Of course we can instead replace the p^a by corresponding Hermite polynomials. Then we use Proposition 16.5.1 for the a_α^β and (17.15.24) to complete the proof of our theorem. □

17.16 THE OPERATOR $\Theta_{h,\lambda}$

In the sequel, if an operator acts on $\ker \alpha_\pm^\perp$, we extend it to an operator acting on H by making it act like the 0 map on $\ker \alpha_\pm$. Recall that $P_\pm^\perp = 1 - P_\pm$.

Definition 17.16.1. Set

$$\Theta_{h,\lambda} = P_\pm^\perp (P_h - \lambda) P_\pm^\perp. \tag{17.16.1}$$

Then $\Theta_{h,\lambda}$ acts on $\ker \alpha_\pm^\perp$.

Recall that the orthogonal projection operator P was defined in section 16.5 and that $P^\perp = 1 - P$.

Theorem 17.16.2. *There exists $\delta' = (\delta_0', \delta_1', \delta_2')$ with $\delta_0' \in]0,1[, \delta_1' > 0, \delta_2' = 1/6$, such that for $h_0 > 0$ small enough and $h \in]0, h_0]$, $\lambda \in \mathcal{W}_{\delta'}$, the operator $\Theta_{h,\lambda}$ is one to one from $\ker \alpha_\pm^\perp$ into itself. Moreover, we have the identity of operators acting on H,*

$$\Theta_{h,\lambda}^{-1} = S_{h,\lambda} - S_{h,\lambda} (P_\pm S_{h,\lambda} i_\pm)^{-1} S_{h,\lambda}. \tag{17.16.2}$$

If $a, b, c, d \in \mathbf{R}$, then

$$P_\pm p^a \partial_p^b P_\pm^\perp \Theta_{h,\lambda}^{-1} P_\pm^\perp p^c \partial_p^d i_\pm \in \mathbb{E}_{\delta',h}^{-1}. \tag{17.16.3}$$

Moreover,

$$\sigma_{-1} \left(P_\pm p^a \partial_p^b P_\pm^\perp \Theta_{h,\lambda}^{-1} P_\pm^\perp p^c \partial_p^d i_\pm \right) = P p^a \partial_p^b \left(P^\perp (B(i\xi) - \lambda) P^\perp \right)^{-1} p^c \partial_p^d i_\pm. \tag{17.16.4}$$

Proof. By Theorems 17.6.1 and 17.7.2 and by choosing $\delta' = \delta$ as in (17.15.3), the operator $P_h^0 - \lambda$ is invertible with inverse $S_{h,\lambda}$. By Theorem 17.15.3, for h small enough and $\lambda \in \mathcal{W}_{\delta'}$, the operator $P_\pm S_{h,\lambda} i_\pm$ acts as an invertible operator on $\ker \alpha_\pm$. Equation (17.16.2) now follows from (17.1.4).

By Theorem 17.15.3, $P_\pm S_{h,\lambda} i_\pm \in \mathbb{S}^{-1}_{\delta',h}$. Therefore for h small enough, $(P_\pm S_{h,\lambda} i_\pm)^{-1} \in \mathbb{E}^{1}_{\delta',h}$. Using (17.16.2), we find that to establish (17.16.3), we only need to show that the operators

$$P_\pm p^a \partial_p^b P_\pm^\perp S_{h,\lambda} i_\pm, \quad P_\pm S_{h,\lambda} P_\pm^\perp p^c \partial_p^d i_\pm, \quad P_\pm p^a \partial_p^b P_\pm^\perp S_{h,\lambda} P_\pm^\perp p^c \partial_p^d i_\pm \tag{17.16.5}$$

lie in $\mathbb{E}^{-1}_{\delta',h}$. Since $P_\pm^\perp = 1 - P_\pm$, this is a consequence of Theorem 17.15.3.

By equation (17.10.1) for $S_{h,\lambda}$ in Theorem 17.10.1 and by (17.16.2), we get

$$\begin{aligned} & P_\pm p^a \partial_p^b P_\pm^\perp \Theta^{-1}_{h,\lambda} P_\pm^\perp p^c \partial_p^d i_\pm \\ &= P_\pm p^a \partial_p^b P_\pm^\perp \left(E_0 - E_0 \left(P_\pm E_0 i_\pm \right)^{-1} E_0 \right) P_\pm^\perp p^c \partial_p^d i_\pm \bmod h\mathbb{E}^{-1}_{\delta',h}. \end{aligned} \tag{17.16.6}$$

By Theorem 17.10.1, e_0 is the principal symbol of E_0, and by Theorem 17.15.3, $P_\pm e_0^0 i_\pm$ is the principal symbol of $P_\pm S i_\pm$. To find the principal symbol of the operator in the left-hand side of (17.16.6), we only need to evaluate

$$P_\pm p^a \partial_p^b P_\pm^\perp \left(e_0 - e_0 \left(P_\pm e_0^0 i_\pm \right)^{-1} e_0 \right) P_\pm^\perp p^c \partial_p^d i_\pm. \tag{17.16.7}$$

Now we use the notation in (17.15.19), the identity (17.15.21) By the above, we find that the principal symbol of the operator in (17.16.3) is given by

$$P_\pm p^a \partial_p^b P_\pm^\perp \left(e_0^0 - e_0^0 \left(P_\pm e_0^0 i_\pm \right)^{-1} e_0^0 \right) P_\pm^\perp p^c \partial_p^d i_\pm. \tag{17.16.8}$$

Equation (17.1.4) shows that (17.16.8) is just the right-hand side of (17.16.4) with $B(i\xi)$ replaced by $B(i\xi) + P$. Since $Q_0^0 = Q_0 + P_\pm$, we may indeed replace Q_0^0 by Q_0. The proof of our theorem is completed. □

Theorem 17.16.3. *For $s \in \mathbf{R}$, there exists $C_s > 0$ such that for $h \in]0, h_0], \lambda \in \mathcal{W}_{\delta'}, u \in \mathcal{H}^s$,*

$$\left\| \Theta^{-1}_{h,\lambda} u \right\|_{\lambda,\mathrm{sc},s+1/4} \le C_s \left\| u \right\|_{\lambda,\mathrm{sc},s}. \tag{17.16.9}$$

For $s \in \mathbf{R}$, there exists $C_s > 0$ such that if $h \in]0, h_0], \lambda \in \mathcal{W}_{\delta'}$, if $u \in \ker \alpha_\pm \cap \mathcal{H}^s, v \in \mathcal{H}^s$,

$$\begin{aligned} \left\| \Theta^{-1}_{h,\lambda} h L_3 u \right\|_{\lambda,\mathrm{sc},s-1/6} &\le C_s \left\| u \right\|_{\lambda,\mathrm{sc},s}, \\ \left\| h L_2 \Theta^{-1}_{h,\lambda} v \right\|_{\lambda,\mathrm{sc},s-1/6} &\le C_s \left\| v \right\|_{\lambda,\mathrm{sc},s}. \end{aligned} \tag{17.16.10}$$

Proof. By Theorem 17.15.3, for h small enough, $P_\pm S_{h,\lambda} i_\pm \in \mathbb{E}^{-1}_{\delta',h}$ is invertible, and so $(P_\pm S_{h,\lambda} i_\pm)^{-1} \in \mathbb{E}^{1}_{\delta',h}$. By equation (17.16.2) in Theorem 17.16.2, we can express $\Theta^{-1}_{h,\lambda}$ in terms of $S_{h,\lambda}$.

First we establish (17.16.9). Indeed by Theorems 17.6.3 and 17.7.2,

$$\left\| S_{h,\lambda} u \right\|_{\lambda,\mathrm{sc},s+1/4} \le C_s \left\| u \right\|_{\lambda,\mathrm{sc},s}. \tag{17.16.11}$$

Moreover, by Theorems 17.12.1, 17.14.1, and 17.15.3, we obtain

$$\left\| S_{h,\lambda} \left(P_\pm S_{h,\lambda} i_\pm \right)^{-1} S_{h,\lambda} u \right\|_{\lambda,\mathrm{sc},s+2/3} \leq C \left\| u \right\|_{\lambda,\mathrm{sc},s}. \tag{17.16.12}$$

By (17.16.2), (17.16.11), (17.16.12) we get (17.16.9).

Now we establish the second equation in (17.16.10). By (17.2.8),

$$hL_2 \Theta^{-1}_{h,\lambda} = P_\pm \left(h\beta_\pm + h^2 \gamma_\pm \right) P^\perp_\pm \Theta^{-1}_{h,\lambda}. \tag{17.16.13}$$

Also we have the trivial

$$P_\pm \left(\alpha_\pm + P_\pm - \lambda \right) P^\perp_\pm = 0. \tag{17.16.14}$$

By (17.2.3), (17.6.1), (17.16.13), (17.16.14), we obtain

$$hL_2 \Theta^{-1}_{h,\lambda} = P_\pm \left(P^0_h - \lambda \right) \Theta^{-1}_{h,\lambda}. \tag{17.16.15}$$

We write $\Theta^{-1}_{h,\lambda}$ in terms of $S_{h,\lambda}$ as in (17.16.2) and we use (17.6.10) and (17.16.15). We obtain

$$hL_2 \Theta^{-1}_{h,\lambda} = P_\pm - P_\pm \left(P_\pm S_{h,\lambda} i_\pm \right)^{-1} S_{h,\lambda}. \tag{17.16.16}$$

Using again Theorems 17.12.1, 17.14.1, and 17.15.3, we get

$$\left\| \left(P_\pm S_{h,\lambda} i_\pm \right)^{-1} S_{h,\lambda} v \right\|_{\lambda,\mathrm{sc},s-1/6} \leq C_s \left\| v \right\|_{\lambda,\mathrm{sc},s}. \tag{17.16.17}$$

The second inequality in (17.16.10) follows from equation (17.5.1) in Lemma 17.5.1, from (17.16.16), and from (17.16.17). The first inequality in (17.16.10) can be established by the same method. The proof of our theorem is completed. □

Remark 17.16.4. The inequalities in (17.16.12) can be established in a different way, in which (17.16.14), (17.16.15) are not used. By (15.1.3) and (17.16.2), the only possible difficulty comes from

$$P_\pm h \nabla^{\Lambda^\cdot(T^*T^*X)\widehat{\otimes}F,u}_{Y^{\mathcal{H}}} S_{h,\lambda} \left(P_\pm S_{h,\lambda} i_\pm \right)^{-1} S_{h,\lambda} u.$$

Clearly

$$\begin{aligned} P_\pm h \nabla^{\Lambda^\cdot(T^*T^*X)\widehat{\otimes}F,u}_{Y^{\mathcal{H}}} & S_{h,\lambda} \left(P_\pm S_{h,\lambda} i_\pm \right)^{-1} S_{h,\lambda} u \\ &= \nabla^{\Lambda^\cdot(T^*T^*X)\widehat{\otimes}F,u}_{e_i} P_\pm p_i S_{h,\lambda} P_\pm \left(P_\pm S_{h,\lambda} i_\pm \right)^{-1} S_{h,\lambda} u. \end{aligned} \tag{17.16.18}$$

By Theorem 17.15.3,

$$P_\pm p_i S_{h,\lambda} P_\pm \left(P_\pm S_{h,\lambda} i_\pm \right)^{-1} \in \mathbb{E}^0_{\delta',h}. \tag{17.16.19}$$

By (17.16.18) and (17.16.19), we get the second inequality in (17.16.10). We can prove the first inequality along the same lines.

17.17 THE OPERATOR $\mathbf{T}_{\mathbf{h},\lambda}$

By [B05, eq. (3.67)],

$$P_+\gamma_+P_+ = -\frac{1}{4}\left\langle R^{TX}\left(e_i,e_j\right)e_k,e_l\right\rangle e^ie^ji_{e_k}i_{e_l} + \frac{1}{2}\left(\left\langle S^Xe_i,e_j\right\rangle - \nabla^F_{e_i}\omega\left(\nabla^F,g^F\right)\left(e_j\right)\right)e^ii_{e_j}. \quad (17.17.1)$$

By proceeding as in [B05], $P_-\gamma_-P_-$ is obtained from the right-hand side of (17.17.1) by adding the term $\frac{1}{2}\nabla_{e_i}\omega\left(\nabla^F,g^F\right)(e_i)$.

Definition 17.17.1. For $h\in]0,h_0]$, $\lambda\in\mathcal{W}_{\delta'}$, set

$$T_{h,\lambda} = P_\pm\gamma_\pm P_\pm - P_\pm\left(\beta_\pm + h\gamma_\pm\right)\Theta^{-1}_{h,\lambda}\left(\beta_\pm + h\gamma_\pm\right)P_\pm. \quad (17.17.2)$$

With the notation in (17.2.8), we get

$$T_{h,\lambda} = L_1 - L_2\left(L_4-\lambda\right)^{-1}L_3. \quad (17.17.3)$$

The operator $T_{h,\lambda}$ acts on $\mathcal{D}_X$. Let $T^*_{h,\lambda}$ denote the formal adjoint of $T_{h,\lambda}$.

Theorem 17.17.2. *The following identity holds:*

$$T^*_{h,\lambda} = T_{h,\overline{\lambda}}. \quad (17.17.4)$$

Proof. As we saw after (2.1.25), by [B05, Theorem 2.30], the operator $\mathfrak{A}'_{\phi,\mathcal{H}^c}$ is self-adjoint with respect to the Hermitian form $h^{\Omega^{\cdot}(T^*X,\pi^*F)}$ defined in (2.1.24). By (15.1.2)-(15.1.4), it follows that the operators $\alpha_\pm,\beta_\pm,\gamma_\pm$ are self-adjoint with respect to $h^{\Omega^{\cdot}(T^*X,\pi^*F)}$. By (17.2.3) or (17.2.4), P_h is also $h^{\Omega^{\cdot}(T^*X,\pi^*F)}$ self-adjoint. Also observe that $\ker\alpha_\pm$ is r^*-invariant. More precisely, r^* acts like the identity on $\ker\alpha_\pm$ for $c>0$ or when $c<0$ and n is even, and like -1 for $c<0$ and n is odd.

If C is an operator, we denote by $C^\dagger$ its $h^{\Omega^{\cdot}(T^*X,\pi^*F)}$ adjoint. By the above we get

$$\Theta^\dagger_{h,\lambda} = \Theta_{h,\overline{\lambda}}. \quad (17.17.5)$$

By (17.17.2), (17.17.5), we obtain

$$T^\dagger_{h,\lambda} = T_{h,\overline{\lambda}}. \quad (17.17.6)$$

When restricted to $\ker\alpha_\pm$, $g^{\Omega^{\cdot}(T^*X,\pi^*F)}$ and $h^{\Omega^{\cdot}(T^*X,\pi^*F)}$ are proportional, the constant of proportionality being ±1. Also the restriction of $g^{\Omega^{\cdot}(T^*X,\pi^*F)}$ to $\ker\alpha_\pm\simeq\mathcal{D}_X$ is just the L^2 Hermitian product of $\mathcal{D}_X$. Since $T_{h,\lambda}$ acts on $\ker\alpha_\pm$, (17.17.4) and (17.17.6) are equivalent. The proof of our theorem is completed. □

Definition 17.17.3. For $h \in [0, h_0], \lambda \in \mathcal{W}_{\delta'}$, set

$$\begin{aligned}
&\mathbb{A}_{h,\lambda} = P_\pm p \Theta_{h,\lambda}^{-1} p i_\pm, \\
&\mathbb{B}_{h,\lambda} = \pm P_\pm p \Theta_{h,\lambda}^{-1} \left(\frac{1}{2}\omega\left(\nabla^F, g^F\right)(e_i)\nabla_{\widehat{e}^i} - h\gamma_\pm\right) i_\pm, \\
&\mathbb{B}'_{h,\lambda} = \pm P_\pm \left(\frac{1}{2}\omega\left(\nabla^F, g^F\right)(e_i)\nabla_{\widehat{e}^i} - h\gamma_\pm\right) \Theta_{h,\lambda}^{-1} p i_\pm, \\
&\mathbb{C}_{h,\lambda} = -P_\pm \left(\frac{1}{2}\omega\left(\nabla^F, g^F\right)(e_i)\nabla_{\widehat{e}^i} - h\gamma_\pm\right) \Theta_{h,\lambda}^{-1} \\
&\qquad\qquad \left(\frac{1}{2}\omega\left(\nabla^F, g^F\right)(e_i)\nabla_{\widehat{e}^i} - h\gamma_\pm\right) i_\pm.
\end{aligned} \tag{17.17.7}$$

By Theorem 17.16.2, for $h \in]0, h_0], \lambda \in \mathcal{W}_{\delta'}$, $\Theta_{h,\lambda}$ is indeed well-defined. Including $h = 0$ in the definition is harmless.

Let π_1, π_2 be the first and second projections of $T^*X \times T^*X$ on X, or from $X \times X$ on X. By Lemma 17.8.4 and by Theorem 17.16.2, $\Theta_{h,\lambda}^{-1}$ is an operator on sections of $\pi^*\left(\Lambda^\cdot(T^*X)\widehat{\otimes}\Lambda^\cdot(TX)\widehat{\otimes}F\right)$ over the total space of T^*X. The kernel $\Theta_{h,\lambda}^{-1}(\cdot,\cdot)$ is a distribution on $T^*X \times T^*X$ with values in

$$\pi_1^*\left(\Lambda^\cdot(T^*X)\widehat{\otimes}\Lambda^\cdot(TX)\widehat{\otimes}F\right)\widehat{\otimes}\pi_2^*\left(\Lambda^\cdot(T^*X)\widehat{\otimes}\Lambda^\cdot(TX)\widehat{\otimes}F\right)^*.$$

Recall that we identified TX and T^*X by the metric g^{TX}. By Theorem (17.16.2), we find that $\mathbb{A}_{h,\lambda}$ is a distribution on $X \times X$.

In the sequel, we use the notation A^* to denote the adjoint of an operator A acting on $\Omega^\cdot(X, F)$ or on $\Omega^\cdot(X, F \otimes o(TX))$ with respect to the standard L^2 Hermitian product.

Given a smooth section Y of TX over X, one associates the operator $\nabla_Y^{\Lambda^\cdot(T^*X)\widehat{\otimes}F,u}$. It is such that $\nabla_Y^{\Lambda^\cdot(T^*X)\widehat{\otimes}F,u*} = -\nabla_Y^{\Lambda^\cdot(T^*X)\widehat{\otimes}F,u} - \operatorname{div}(Y)$.

In the + case, the operator $\nabla^{\Lambda^\cdot(T^*X)\widehat{\otimes}Fu,*}\mathbb{A}_+\nabla^{\Lambda^\cdot(T^*X)\widehat{\otimes}F,u}$ is a well defined pseudodifferential operator. A more explicit expression for this operator can be obtained as follows. Take $x_0, y_0 \in X$. Let $e_1, \ldots, e_n$ be smooth orthonormal basis of TX near x_0, let $e'_1, \ldots, e'_n$ be a smooth orthonormal basis of TX near y_0. We denote with a superscript the corresponding dual bases. The distribution kernel for $\nabla^{\Lambda^\cdot(T^*X)\widehat{\otimes}Fu,*}\mathbb{A}_+\nabla^{\Lambda^\cdot(T^*X)\widehat{\otimes}F,u}$ near x_0, y_0 is given by

$$\begin{aligned}
&\nabla^{\Lambda^\cdot(T^*X)\widehat{\otimes}F,u*}\mathbb{A}_+\nabla^{\Lambda^\cdot(T^*X)\widehat{\otimes}F,u} \\
&\qquad = \nabla_{e_i}^{\Lambda^\cdot(T^*X)\widehat{\otimes}F,u*} P_+\left\langle p, e^i\right\rangle \Theta_{h,\lambda}^{-1}\left\langle p, e^j\right\rangle P_+ \nabla_{e_j}^{\Lambda^\cdot(T^*X)\widehat{\otimes}F,u}.
\end{aligned} \tag{17.17.8}$$

Similar formulas also hold in the $-$ case.

We will use the notation in sections 16.1 and 16.4. In particular 1 denotes the canonical generator of $\ker \mathcal{N}$, P is the orthogonal projection on $\ker \mathcal{N}$, and $P^\perp = 1 - P$. Also recall that $\mathbb{S}_\delta^d$ was defined in section 16.5.

Theorem 17.17.4. *We have*

$$\begin{aligned}
&\mathbb{A}_{h,\lambda}, \mathbb{B}_{h,\lambda}, \mathbb{B}'_{h,\lambda}, \mathbb{C}_{h,\lambda} \in \mathbb{E}_{\delta',h}^{-1}, \\
&\mathbb{A}_{h,\lambda}^* = \mathbb{A}_{h,\overline{\lambda}}, \qquad \mathbb{B}_{h,\lambda}^* = -\mathbb{B}'_{h,\overline{\lambda}}, \qquad \mathbb{C}_{h,\lambda}^* = \mathbb{C}_{h,\overline{\lambda}}.
\end{aligned} \tag{17.17.9}$$

Moreover,

$$T_{h,\lambda} = \nabla^{\Lambda^{\cdot}(T^*X)\widehat{\otimes}F,u*}\mathbb{A}_{h,\lambda}\nabla^{\Lambda^{\cdot}(T^*X)\widehat{\otimes}F,u} + \nabla^{\Lambda^{\cdot}(T^*X)\widehat{\otimes}F,u*}\mathbb{B}_{h,\lambda} - \mathbb{B}'_{h,\lambda}\nabla^{\Lambda^{\cdot}(T^*X)\widehat{\otimes}F,u} + P_{\pm}\gamma_{\pm}P_{\pm} + \mathbb{C}_{h,\lambda}. \quad (17.17.10)$$

The principal symbol $\mathfrak{a}(\xi,\lambda)$ *of* $\mathbb{A}_{h,\lambda}$ *is such that if* $\zeta \in T^*X$*,*

$$\begin{aligned}\langle \mathfrak{a}\zeta, \zeta\rangle &= \frac{1}{2}\left\langle \left(P^{\perp}\left(B\left(i\xi\right)-\lambda\right)P^{\perp}\right)^{-1} a^*\left(\zeta\right)1, a^*\left(\zeta\right)1\right\rangle \\ &= \frac{1}{2}\left(J_0 - J_1\right)\left(\left|\xi\right|/\sqrt{2},\lambda\right)\left|\zeta\right|^2 + \frac{1}{4}\left(\frac{J_1^2}{J_0} - J_2\right)\left(\left|\xi\right|/\sqrt{2},\lambda\right)\langle\xi,\zeta\rangle^2.\end{aligned} \quad (17.17.11)$$

Moreover,

$$\langle \mathfrak{a}\xi, \xi\rangle - \lambda = \left\langle \left(B\left(i\xi\right)-\lambda\right)^{-1}1, 1\right\rangle^{-1} = J_0^{-1}\left(\left|\xi\right|/\sqrt{2},\lambda\right), \quad (17.17.12)$$

$$\langle \mathfrak{a}\xi, \xi\rangle = \frac{\left|\xi\right|^2}{2}\frac{J_1}{J_0}\left(\left|\xi\right|/\sqrt{2},\lambda\right).$$

Finally,

$$T_{0,0} = \frac{1}{2}\square^X. \quad (17.17.13)$$

Proof. The inclusions in (17.17.9) follow from (17.16.3) in Theorem 17.16.2. The same arguments as in the proof of Theorem 17.17.2 lead immediately to the identities in (17.17.9). Note in particular that the fact that p is an odd function explains the minus sign in the identity $\mathbb{B}^*_{h,\lambda} = -\mathbb{B}'_{h,\overline{\lambda}}$.

We claim that (17.17.10) follows from (15.1.3) and from (17.17.2). The main point is to explain the contribution of $\nabla^{\Lambda^{\cdot}(T^*T^*X)\widehat{\otimes}F,u}_{Y^{\mathcal{H}}}$ to the right-hand side of (17.17.10). If $e_1,\ldots,e_n$ is a locally defined orthonormal basis of TX, then

$$\nabla^{\Lambda^{\cdot}(T^*T^*X)\widehat{\otimes}F,u}_{Y^{\mathcal{H}}} = \left\langle p, e^i\right\rangle \nabla^{\Lambda^{\cdot}(T^*T^*X)\widehat{\otimes}F,u}_{e_i}. \quad (17.17.14)$$

Since $Y^{\mathcal{H}}$ is divergence free,

$$\nabla^{\Lambda^{\cdot}(T^*X)\widehat{\otimes}F,u*}_{Y^{\mathcal{H}}} = -\nabla^{\Lambda^{\cdot}(T^*T^*X)\widehat{\otimes}F,u}_{Y^{\mathcal{H}}}. \quad (17.17.15)$$

From (17.17.14), (17.17.15), we get

$$\nabla^{\Lambda^{\cdot}(T^*T^*X)\widehat{\otimes}F,u}_{Y^{\mathcal{H}}} = -\nabla^{\Lambda^{\cdot}(T^*T^*X)\widehat{\otimes}F,u*}_{e_i}\left\langle p, e^i\right\rangle. \quad (17.17.16)$$

Using (17.17.14), (17.17.16), we get (17.17.10).

By equation (17.16.4) in Theorem 17.16.2, we get

$$\langle \mathfrak{a}\zeta, \zeta\rangle = P\left\langle p,\zeta\right\rangle\left(P^{\perp}\left(B\left(i\xi\right)-\lambda\right)P^{\perp}\right)^{-1}\left\langle p,\zeta\right\rangle P. \quad (17.17.17)$$

Using (16.1.5) and (17.17.17), we get the first equality in (17.17.11).

Set

$$\mathfrak{S} = \left(B\left(i\xi\right)-\lambda\right)^{-1}. \quad (17.17.18)$$

By (17.1.4), we get

$$\left(P^{\perp}\left(B\left(i\xi\right)-\lambda\right)P^{\perp}\right)^{-1}=\mathfrak{S}-\mathfrak{S}\left(P\mathfrak{S}P\right)^{-1}\mathfrak{S}. \tag{17.17.19}$$

Using (16.4.66), the first identity in (17.17.11), and (17.17.19), we get easily

$$\langle\mathfrak{a}\zeta,\zeta\rangle=\frac{1}{2}\left(\psi^i_j\zeta_i\zeta_j-\frac{1}{\psi^0_0}\psi^0_i\psi^j_0\zeta_i\zeta_j\right). \tag{17.17.20}$$

By (16.4.67)-(16.4.69), we get

$$\begin{aligned}\psi^0_0&=J_0\left(\left|\xi\right|/\sqrt{2},\lambda\right),\\ \psi^0_i&=J_1\left(\left|\xi\right|/\sqrt{2},\lambda\right)i\xi_i/\sqrt{2},\\ \psi^i_0&=J_1\left(\left|\xi\right|/\sqrt{2},\lambda\right)i\xi_i/\sqrt{2},\\ \psi^i_j&=\left(J_0-J_1\right)\left(\left|\xi\right|\sqrt{2},\lambda\right)\delta'_{ij}-\frac{1}{2}J_2\left(\left|\xi\right|/\sqrt{2},\lambda\right)\xi_i\xi_j.\end{aligned} \tag{17.17.21}$$

By (17.17.20), (17.17.21), we obtain the second part of (17.17.11).

By (17.17.11), we get

$$\langle\mathfrak{a}\xi,\xi\rangle=\frac{1}{2}\left\langle a\left(\xi\right)\left(P^{\perp}\left(B\left(i\xi\right)-\lambda\right)P^{\perp}\right)^{-1}a^*\left(\xi\right)1,1\right\rangle. \tag{17.17.22}$$

Comparing (17.17.22) with (16.5.12), we get

$$\langle\mathfrak{a}\xi,\xi\rangle=\mathfrak{T}_{\xi,\lambda}. \tag{17.17.23}$$

The first line in (17.17.12) follows from (16.3.14), from equation (16.5.13) in Proposition 16.5.4, and from (17.17.23). The second line in (17.17.12) also follows from (17.17.23) and (16.5.13).

By the result in [B05, Theorem 3.14], which was stated in Theorem 2.3.2, we get (17.17.13). The proof of our theorem is completed. □

Remark 17.17.5. More generally, one can give an explicit formula for $T_{0,\lambda}$ which extends (17.17.13). Indeed $P_\pm\gamma_\pm P_\pm$ was evaluated in (17.17.1). Moreover, inspection of the proof of [B05, Theorem 3.14] shows that with respect to $T_{0,0}$, the contribution of the second term in the right-hand side of (17.17.2) will be simply scaled by $1-\lambda$. So we get

$$T_{0,\lambda}=\frac{1}{1-\lambda}\left(-\lambda P_\pm\gamma_\pm P_\pm+\frac{1}{2}\square^X\right). \tag{17.17.24}$$

17.18 THE OPERATOR $(\mathbf{J_1}/\mathbf{J_0})\left(\mathbf{hD^X}/\sqrt{\mathbf{2}},\lambda\right)$

We take $\delta=(\delta_0,\delta_1,\delta_2)$ as in Theorem 16.4.1.

By Theorem 17.9.5, given $y\in\mathbf{R}$, the poles of $J_0(y,\cdot)$ are simple and given by $y^2+\mathbf{N}$. Moreover, by (16.4.4) and (16.4.5) in Theorem 17.9.5, we get

$$\begin{aligned}&\left|\partial^k_y J_0\left(y,\lambda\right)\right|\le C_k\left(1+|y|+|\lambda|\right)^{-k-1},\ (y,\lambda)\in\mathbf{R}\times\mathcal{W}_\delta,\ |y|+|\lambda|\ge 1,\\ &\left|J_0^{-1}\left(y,\lambda\right)\right|\le C\left(1+|y|+|\lambda|\right),\ (y,\lambda)\in\mathbf{R}\times\mathcal{W}_\delta.\end{aligned} \tag{17.18.1}$$

Recall that the class of symbols $\mathbb{S}^d_\delta$ was defined in section 16.5. By (17.18.1), $J_0^{-1} \in \mathbb{S}^1_\delta$ and by (16.5.15), $\frac{J_1}{J_0} \in \mathbb{S}^{-1}_\delta$. Note if $y^2 < \delta_0$, y^2 is a pole of $J_0(y, \lambda)$ in $\mathcal{W}_\delta$. Therefore, we will not write that $J_0 \in \mathbb{S}^{-1}_\delta$.

By (1.2.2),

$$D^X = d^X + d^{X,*}. \tag{17.18.2}$$

By the above, if $h \in]0,1], \lambda \in \mathcal{W}_\delta/h^2 \backslash \mathrm{Sp}\,\square^X/2$, $J_0\left(hD^X/\sqrt{2}, h^2\lambda\right)$ is a pseudodifferential operator of order -1, and this operator is invertible. Moreover,

$$J_0\left(hD^X/\sqrt{2}, h^2\lambda\right)^* = J_0\left(hD^X/\sqrt{2}, h^2\overline{\lambda}\right). \tag{17.18.3}$$

Also note that by (16.3.16) or (16.4.3), we may as well replace D^X by $\sqrt{\square^X}$.

Now we fix $h \in]0,1]$. If $\lambda \in \mathcal{W}_\delta/h^2$, $\left(h^2 J_0\right)^{-1}\left(hD^X/\sqrt{2}, h^2\lambda\right)$ is a pseudodifferential operator of order 1, and $\frac{J_1}{J_0}\left(hD^X/\sqrt{2}, h^2\lambda\right)$ is a pseudodifferential operator of order -1. By (16.4.63), if $\lambda \in \mathcal{W}_\delta$,

$$\left(h^2 J_0\right)^{-1}\left(hD^X/\sqrt{2}, h^2\lambda\right) + \lambda = \frac{\square^X}{2}\frac{J_1}{J_0}\left(hD^X/\sqrt{2}, h^2\lambda\right). \tag{17.18.4}$$

Also since $\frac{J_1}{J_0} \in \mathbb{S}^{-1}_\delta$, $\frac{J_1}{J_0}\left(hD^X/\sqrt{2}, \lambda\right) \in \mathbb{E}^{-1}_{\delta,h,0}$.

In the next proposition, we consider the operators $B(i\xi)$ in the case where $n = 1$.

Proposition 17.18.1. *If* $h \in]0,1], \lambda \in \mathcal{W}_\delta$,

$$\frac{J_1}{J_0}\left(hD^X/\sqrt{2}, \lambda\right) = Pa(1)\left(P^\perp\left(B\left(ihD^X\right) - \lambda\right)P^\perp\right)^{-1} a^*(1)P. \tag{17.18.5}$$

Moreover,

$$\sigma_{-1}\left(\frac{J_1}{J_0}\left(hD^X/\sqrt{2}, \lambda\right)\right) = \frac{J_1}{J_0}\left(|\xi|/\sqrt{2}, \lambda\right). \tag{17.18.6}$$

Proof. By (16.5.12), (16.5.13),

$$\frac{J_1}{J_0}(y, \lambda) = Pa(1)\left(P^\perp\left(B\left(i\sqrt{2}y\right) - \lambda\right)P^\perp\right)^{-1} a^*(1)P. \tag{17.18.7}$$

By (17.18.7), we get (17.18.5).

First we consider the + case. To evaluate the principal symbol of the operator in (17.18.5), we take a coordinate system on an open neighborhood U of $x_0 \in X$ as in section 17.8 and we use the corresponding notation. In particular $\Lambda^{\cdot}(T^*X)\widehat{\otimes}F$ is trivialized as indicated there. Let $\Gamma^{\Lambda^{\cdot}(T^*X)\widehat{\otimes}F,u}$ be the connection form for $\nabla^{\Lambda^{\cdot}(T^*X)\widehat{\otimes}F,u}$ in this trivialization. Let $e_1, \ldots, e_n$ be an orthonormal basis of $T_{x_0}X$. By (1.2.11), in the above trivialization, D^X can be written in the form

$$D^X = \sum_1^n c(e_i)\left(\nabla_{\widetilde{\sigma}_x^{-1}e_i} + \Gamma^{\Lambda^{\cdot}(T^*X)\widehat{\otimes}F,u}\left(\widetilde{\sigma}_x^{-1}e_i\right)\right) - \frac{1}{2}\sum_1^n \widehat{c}(e_i)\,\omega\left(\nabla^F, g^F\right)\left(\widetilde{\sigma}_x^{-1}e_i\right). \tag{17.18.8}$$

By (17.18.8), we find that D^X is a first order differential operator whose principal symbol $\sigma\left(D^X\right)$ is given by

$$\sigma\left(D^X\right) = ic\left(\sigma_x^{-1}\xi\right). \tag{17.18.9}$$

The semiclassical symbol $h\mathfrak{D}^X$ of hD^X is obtained from the right-hand side of (17.18.8) by multiplication by h and by replacing ∂_{x^j} by $i\xi_j/h$. The semiclassical symbol of $\frac{J_1}{J_0}\left(hD^X/\sqrt{2},\lambda\right)$ is then just $\frac{J_1}{J_0}\left(h\mathfrak{D}^X,\lambda\right)$. It follows from the above that

$$\sigma_{-1}\left(\frac{J_1}{J_0}\left(hD^X/\sqrt{2},\lambda\right)\right) = \frac{J_1}{J_0}\left(ihc\left(\sigma_x^{-1}\xi\right)/\sqrt{2},\lambda\right). \tag{17.18.10}$$

Also $\frac{J_1}{J_0}(y,\lambda)$ is a function of y^2. Moreover, since $\sigma_x : T^*_{x_0}X \to T^*_xX$ is an isometry,

$$c\left(i\sigma_x^{-1}\xi\right)^2 = |\xi|_x^2. \tag{17.18.11}$$

By (17.18.10), (17.18.11), we get (17.18.6). The proof of our proposition is completed. □

17.19 THE OPERATOR $\mathrm{U}_{h,\lambda}$

If $h \in]0,h_0]$, $h^2\lambda \in \mathcal{W}_\delta$, using (17.18.4), we get

$$T_{h,h^2\lambda} - \lambda = \left(h^2J_0\right)^{-1}\left(hD^X/\sqrt{2},h^2\lambda\right) + T_{h,h^2\lambda} - \frac{\square^X}{2}\frac{J_1}{J_0}\left(hD^X/\sqrt{2},h^2\lambda\right). \tag{17.19.1}$$

If $\lambda \in \mathcal{W}_\delta \setminus h^2\mathrm{Sp}\,\square^X/2$, as we saw before, for a given $h > 0$, the operator $h^2J_0\left(hD^X/\sqrt{2},\lambda\right)$ is a well-defined pseudodifferential operator of order -1. Set

$$U_{h,\lambda} = h^2J_0\left(hD^X/\sqrt{2},\lambda\right)\left(T_{h,\lambda} - \frac{\square^X}{2}\frac{J_1}{J_0}\left(hD^X/\sqrt{2},\lambda\right)\right). \tag{17.19.2}$$

We can then rewrite (17.19.1) in the form

$$T_{h,h^2\lambda} - \lambda = \left(h^2J_0\right)^{-1}\left(hD^X/\sqrt{2},h^2\lambda\right)\left(1 + U_{h,h^2\lambda}\right). \tag{17.19.3}$$

We will use the notation in (16.4.41) with respect to $S = \mathrm{Sp}\,\square^X/2$. In particular the function $\rho(\lambda)$ is defined as in (16.4.41).

Theorem 17.19.1. *There exists $\delta' = (\delta'_0,\delta'_1,\delta'_2)$, with $\delta'_0 \in]0,1[$, $\delta'_1 > 0$, $\delta'_2 = 1/6$ and there exists $C > 0$ such that for $h \in]0,h_0]$, $\lambda \in \mathcal{W}_{\delta'}/h^2$, if $u \in \Omega^{\cdot}(X,F)$, then*

$$\left\|U_{h,h^2\lambda}u\right\|_{L^2} \leq Ch\rho(\lambda)\left\|u\right\|_{L^2}. \tag{17.19.4}$$

Proof. We may and we will assume that $\|u\|_{L^2} = 1$. Set

$$V_{h,\lambda} = T_{h,\lambda} - \frac{\square^X}{2}\frac{J_1}{J_0}\left(hD^X/\sqrt{2},\lambda\right). \tag{17.19.5}$$

We use Theorem 17.17.4, particularly equations (17.17.12) and (17.17.13), and also equation (17.18.6) in Proposition 17.18.1. We find in particular that for $h \in]0, h_0]$, the operator $V_{h,h^2\lambda}$ is a pseudodifferential operator of order 1.

We consider the first term in the right-hand side of (17.17.10). First, as we explained after (17.15.6), the contribution of a semiclassical smoothing operator in $\mathbb{A}_{h,h^2\lambda}$ will be uniformly smoothing, and its corresponding norm will decay faster as $h \to 0$ than any $h^N, N \in \mathbf{N}$. Moreover, since $\mathbb{A}_{h,\lambda} \in \mathbb{E}^{-1}_{h,\delta'}$, when replacing $\mathbb{A}_{h,h^2\lambda}$ by the operator associated to the semiclassical principal symbol $\mathfrak{a}(\xi, \lambda)$ of $\mathbb{A}_{h,\lambda}$, this introduces at the level of operators an error whose classical symbol can be dominated by $Ch/\left(1 + h|\xi| + h^2|\lambda|\right)$. Using the bound (16.4.45) in Proposition 16.4.2, we find that when estimating (17.19.4), this replacement introduces an error which can be dominated at the level of symbols by

$$C\rho(\lambda) \frac{1 + h^2|\lambda| + h|\xi|}{1 + |\lambda| + |\xi|^2} \frac{h}{1 + h^2|\lambda| + h|\xi|} \left(1 + |\xi|^2\right) \leq Ch\rho(\lambda), \quad (17.19.6)$$

which is compatible with (17.19.4).

We will then replace $\mathbb{A}_{h,h^2\lambda}$ by $\mathfrak{a}\left(h\xi, h^2\lambda\right)$, which we identify to the corresponding operator as in (17.15.7). In what follows, it should be clear that in the local coordinates centered at x_0, $\partial_{x^j} = i\xi_j, 1 \leq j \leq n$. Using the notation in Theorem 17.17.4, the operator to be considered is

$$\nabla^{\Lambda^{\cdot}(T^*X)\widehat{\otimes}F,u*}\mathfrak{a}\left(h\xi, h^2\lambda\right) \nabla^{\Lambda^{\cdot}(T^*T^*X)\widehat{\otimes}F,u}. \quad (17.19.7)$$

Equation (17.19.7) is an expression for the composition of several operators. A priori we cannot commute $\nabla^{\Lambda^{\cdot}(T^*X)\widehat{\otimes}F,u*}$ with $\mathfrak{a}\left(h\xi, h^2\lambda\right)$. Now we explain how to handle such commutations. Incidentally observe that although the expression (17.19.7) seems to be coordinate-invariant, the underlying operator is not.

We use equation (17.17.11) for the semiclassical principal symbol $\mathfrak{a}(\xi, \lambda)$ of $\mathbb{A}_{h,\lambda}$. By (16.4.62), the functional equation (16.4.39) for J_0 extends to the equation

$$(J_0 - J_1)(y, \lambda) = J_0(y, \lambda - 1). \quad (17.19.8)$$

By Theorem 16.4.1, $J_0(y, \lambda - 1) \in \mathbb{S}^{-1}_{\delta}$. From (17.19.8), we find that $J_0 - J_1 \in \mathbb{S}^{-1}_{\delta}$.

By (16.4.63) with $k = 0$ and $k = 1$, we get

$$\left(\frac{J_1^2}{J_0} - J_2\right) = \frac{1}{y^2}\left(\frac{J_1}{J_0} - J_0 + J_1\right). \quad (17.19.9)$$

Moreover, $\frac{J_1}{J_0} - J_0 + J_1 \in \mathbb{S}^{-1}_{\delta}$.

We will now use the above considerations to write (17.19.7) in a more explicit form. Namely, we will push the operators ξ to the very left of the considered expression. Incidentally note that $\mathfrak{a}\left(h\xi, h^2\lambda\right)$ is normally ordered, that is, the operators ξ should be thought of as being to the right of whatever function of x appears. Using (17.17.11), we find that a first contribution to (17.19.7) is given by

$$-\frac{1}{2}\nabla^{\Lambda^{\cdot}(T^*X)\widehat{\otimes}F,u,*}(J_0 - J_1)\left(h|\xi|/\sqrt{2}, h^2\lambda\right) \nabla^{\Lambda^{\cdot}(T^*T^*X)\widehat{\otimes}F,u}. \quad (17.19.10)$$

In (17.19.10), there is an implicit trace which is taken via the metric g^{TX}. Now using standard pseudodifferential calculus, we find that if $K \in \mathbb{S}_\delta^{-1}$, then

$$\left[\nabla^{\Lambda^\cdot(T^*X)\widehat{\otimes}F,u,*}, K\left(h\left|\xi\right|/\sqrt{2}, h^2\lambda\right)\right] \simeq h\left(1+h^2\left|\lambda\right|+h\left|\xi\right|\right)^{-2}\left(1+\left|\xi\right|\right). \tag{17.19.11}$$

The sign $\simeq$ means here that at the level of symbols, the operator in the right-hand side is of the order of the right-hand side. Note that the first term in the right-hand side of (17.19.11) appears because of differentiation in the ξ variable of $K\left(\left|\xi\right|/\sqrt{2}, h^2\lambda\right)$, the other two terms appearing because of the Lie bracket. Ultimately the contribution of the commutator to (17.19.4) can be dominated as in (17.19.6), that is, it is compatible with the estimate we want to prove. Therefore we can replace the expression in (17.19.10) by

$$-\frac{1}{2}\left(J_0 - J_1\right)\left(h\left|\xi\right|/\sqrt{2}, h^2\lambda\right)\Delta^{H,u}. \tag{17.19.12}$$

Using (17.19.9), we find that the contribution of the second term in the right-hand side of (17.19.7) is given by

$$\frac{1}{2}\nabla^{\Lambda^\cdot(T^*X)\widehat{\otimes}F,u,*}_{\xi^*}\left(\frac{J_1}{J_0} - J_0 + J_1\right)\left(h\left|\xi\right|/\sqrt{2}, h^2\lambda\right)\frac{1}{\left|\xi\right|^2}\nabla^{\Lambda^\cdot(T^*T^*X)\widehat{\otimes}F,u}_{\xi^*}. \tag{17.19.13}$$

In (17.19.13), ξ^* is dual to ξ by the metric g^{TX}. The expression involving the ξ is again normally ordered. Besides, in $\nabla^{\Lambda^\cdot(T^*X)\widehat{\otimes}F,u,*}_{\xi^*}$, ξ^* should be understood as being to the right of $\nabla^{\Lambda^\cdot(T^*X)\widehat{\otimes}F,u,*}$, while in $\nabla^{\Lambda^\cdot(T^*X)\widehat{\otimes}F,u}_{\xi^*}$, it is to the left of $\nabla^{\Lambda^\cdot(T^*X)\widehat{\otimes}F,u}$.

Take again $K \in \mathbb{S}_\delta^{-1}$. By using the same notation as in (17.19.11), we find that if $K \in \mathbb{S}_\delta^{-1}$,

$$\left[\nabla^{\Lambda^\cdot(T^*X)\widehat{\otimes}F,u,*}_{\xi^*}, K\left(h\left|\xi\right|/\sqrt{2}, h^2\lambda\right)\right] \simeq h\left(1+h^2\left|\lambda\right|+h\left|\xi\right|\right)^{-2}\left(1+\left|\xi\right|^2\right). \tag{17.19.14}$$

Still using the bound (16.4.45), we find that the contribution of the commutator (17.19.14) to the estimation of (17.19.4) can be dominated for $\left|\xi\right| \geq 1$ by

$$C\rho\left(\lambda\right)\frac{1+h^2\left|\lambda\right|+h\left|\xi\right|}{1+\left|\lambda\right|+\left|\xi\right|^2}\frac{h\left(1+\left|\xi\right|^2\right)}{\left(1+h^2\left|\lambda\right|+h\left|\xi\right|\right)^2} \leq Ch\rho\left(\lambda\right). \tag{17.19.15}$$

By (17.17.11) and by (17.19.10), (17.19.12), (17.19.13), (17.19.15), we get

$$\begin{aligned}\nabla^{\Lambda^\cdot(T^*X)\widehat{\otimes}F,u*}\mathfrak{a}\left(h\xi, h^2\lambda\right)\nabla^{\Lambda^\cdot(T^*T^*X)\widehat{\otimes}F,u} &\\ \simeq -\frac{1}{2}\left(J_0 - J_1\right)\left(h\left|\xi\right|/\sqrt{2}, h^2\lambda\right)\Delta^{H,u} &\\ + \frac{1}{2}\left(\frac{J_1}{J_0} - J_0 + J_1\right)\left(h\left|\xi\right|/\sqrt{2}, h^2\lambda\right)\nabla^{\Lambda^\cdot(T^*X)\widehat{\otimes}F,u^*}_{\xi^*}\frac{1}{\left|\xi\right|^2}\nabla^{\Lambda^\cdot(T^*T^*X)\widehat{\otimes}F,u}_{\xi^*}.&\end{aligned} \tag{17.19.16}$$

Since $\frac{J_1}{J_0} \in \mathbb{S}_\delta^{-1}$, by the same arguments as before, we find that

$$\frac{1}{2}\frac{J_1}{J_0}\left(hD^X/\sqrt{2}, h^2\lambda\right)\Delta^{H,u} \simeq \frac{1}{2}\frac{J_1}{J_0}\left(h\left|\xi\right|/\sqrt{2}, h^2\lambda\right)\Delta^{H,u}. \tag{17.19.17}$$

By (17.19.16), (17.19.17), (17.19.9), we obtain

$$\begin{aligned}\nabla^{\Lambda^{\cdot}(T^*X)\widehat{\otimes}F,u*}\mathfrak{a}\left(h\xi, h^2\lambda\right)\nabla^{\Lambda^{\cdot}(T^*T^*X)\widehat{\otimes}F,u} + \frac{1}{2}\frac{J_1}{J_0}\left(hD^X/\sqrt{2}, h^2\lambda\right)\Delta^{H,u}\\ \simeq \frac{1}{2}\left(\frac{J_1}{J_0} - J_0 + J_1\right)\left(h\left|\xi\right|/\sqrt{2}, h^2\lambda\right)\\ \left(\Delta^{H,u} + \nabla_{\xi^*}^{\Lambda^{\cdot}(T^*X)\widehat{\otimes}F,u*}\frac{1}{\left|\xi\right|^2}\nabla_{\xi^*}^{\Lambda^{\cdot}(T^*T^*X)\widehat{\otimes}F,u}\right).\end{aligned} \tag{17.19.18}$$

Set

$$A = \Delta^{H,u} + \nabla_{\xi^*}^{\Lambda^{\cdot}(T^*X)\widehat{\otimes}F,u*}\frac{1}{\left|\xi\right|^2}\nabla_{\xi^*}^{\Lambda^{\cdot}(T^*T^*X)\widehat{\otimes}F,u}. \tag{17.19.19}$$

We claim that A is a classical pseudodifferential operator of order 0. To prove this, we necessarily have to use local coordinates. To do this, we will just take $X = \mathbf{R}^n$. Let $e_1, \ldots, e_n$ be the canonical basis of $\mathbf{R}^n$. The metric g^{TX} is then given by g_x which is an (n,n) self-adjoint matrix which depends smoothly on x. The corresponding matrix elements are denoted by $g_{i,j}, 1 \le i,j \le n$. The matrix elements of the dual metric $g^{T^*X} = g_x^{-1}$ are denoted $g^{i,j}, 1 \le i,j \le n$.

Let Γ^{TX} be the connection form of the Levi-Civita connection ∇^{TX} with respect to the trivialization $TX \simeq \mathbf{R}^n$, and let $\Gamma^{\Lambda^{\cdot}(T^*X)}$ be the corresponding connection form on $\Lambda^{\cdot}(T^*X)$. Given a trivialization of F, let $\Gamma^{F,u}$ be the connection form for $\nabla^{F,u}$. Let Γ be the corresponding connection form on $\Lambda^{\cdot}(T^*X)\widehat{\otimes}F$. We denote by $\Gamma_i, 1 \le i \le n$ the components of Γ. Let η be the 1-form

$$\eta = \frac{1}{2}d\log\det g. \tag{17.19.20}$$

We will now write the full symbol of the operator, so that in particular $\xi_j = -i\partial_{x^j}$. Also we will use the notation $*$ to indicate a product in the algebra of pseudodifferential operators, as opposed to the pointwise product. Clearly,

$$\Delta^{H,u} = -\left(\xi_i - i\eta_i - i\Gamma_i\right) * g^{ij} * \left(\xi_j - i\Gamma_j\right). \tag{17.19.21}$$

By definition,

$$\begin{aligned}\nabla_{\xi^*}^{\Lambda^{\cdot}(T^*X)\widehat{\otimes}F,u,*} &= -i\left(\xi_i - i\eta_i - i\Gamma_i\right) * \left(g^{ik}\xi_k\right),\\ \nabla_{\xi^*}^{\Lambda^{\cdot}(T^*X)\widehat{\otimes}F,u} &= i\left(g^{jl}\xi_l\right) * \left(\xi_j - i\Gamma_j\right).\end{aligned} \tag{17.19.22}$$

Finally,

$$\left|\xi\right|^2 = g^{pq}\xi_p\xi_q. \tag{17.19.23}$$

By (17.19.19), (17.19.21)-(17.19.23),

$$A = (\xi_i - i\eta_i - i\Gamma_i) * \left(-g^{ij} + \left(g^{ik}\xi_k\right)\left(g^{pq}\xi_p\xi_q\right)^{-1}\left(g^{jl}\xi_l\right)\right) * \left(\xi_j - i\Gamma_j\right). \quad (17.19.24)$$

Now observe that

$$\left(g^{j,l}\xi_l\right) * \xi_j = g^{j,l}\xi_l\xi_j. \quad (17.19.25)$$

By (17.19.24), (17.19.25), we get

$$A = (\eta_i + \Gamma_i)\, g^{ij}\Gamma_j - (\eta_i + \Gamma_i)\left(g^{ik}\xi_k\right)\left(g^{pq}\xi_p\xi_q\right)^{-1}\left(g^{jl}\xi_l\right) * \Gamma_j + i\xi_i * \left(g^{ij}\Gamma_j\right) - i\xi_i * \left(g^{ik}\xi_k\right)\left(g^{pq}\xi_p\xi_q\right)^{-1}\left(g^{jl}\xi_l\right) * \Gamma_j. \quad (17.19.26)$$

Now the operator in the first line of (17.19.26) is a classical pseudodifferential operator of order 0. Let B be the operator in the second line of (17.19.26). Then B is a pseudodifferential operator of order 1. Its classical principal symbol is obtained by deleting the $*$. When doing this, we find that the principal symbol of B vanishes. Therefore B is also a classical pseudodifferential operator of order 0.

We conclude from the above that A is indeed a classical pseudodifferential operator of order 0.

By (16.4.39),

$$J_0(y, \lambda - 1) = \frac{1}{y^2 - \lambda + 1} + \frac{y^2}{y^2 - \lambda + 1} J_0(y, \lambda - 2). \quad (17.19.27)$$

By (17.19.27), we deduce that

$$J_0(y, \lambda - 1) - 1 = -\frac{y^2 - \lambda}{y^2 - \lambda + 1} + \frac{y^2}{y^2 - \lambda + 1} J_0(y, \lambda - 2). \quad (17.19.28)$$

We claim that by taking $\delta'_0 \in]0,1[$, $\delta'_1 > 0$ small enough, there exists $C > 0$ such that for $\lambda \in \mathcal{W}_{\delta'}$,

$$\left|y^2 - \lambda + 1\right| \geq C\left(1 + |\lambda| + y^2\right). \quad (17.19.29)$$

Set $\lambda = a + ib, a, b \in \mathbf{R}$. For $a \leq 0$, (17.19.29) is trivial. For $a \geq 0$,

$$\left|y^2 - \lambda + 1\right| \geq y^2 - a + 1 + |b|. \quad (17.19.30)$$

Moreover, for $0 < C < 1$,

$$y^2 - a + 1 + |b| - C\left(1 + y^2 + a + |b|\right) = (1 - C)\left(1 + y^2\right) + (1 - C)|b| - (1 + C)\, a \geq 1 - C + (1 - C)|b| - (1 + C)\left(\delta'_0 + \delta'_1 |b|^{1/6}\right). \quad (17.19.31)$$

By taking δ'_0, δ'_1, C small enough, for any $b \in \mathbf{R}$, the right-hand side of (17.19.31) is nonnegative. Therefore we have established (17.19.29).

Note that if $\lambda \in \mathcal{W}_\delta$, then $\lambda - 1 \in \mathcal{W}_\delta$. Using (16.4.40), if $\lambda \in \mathcal{W}_\delta$,

$$\left|J_0(y, \lambda - 2)\right| \leq C\left(1 + |y| + |\lambda|\right)^{-1}. \quad (17.19.32)$$

By (17.19.8), (17.19.28), (17.19.29) and (17.19.32), if $\lambda \in \mathcal{W}_{\delta'}$,

$$\left|\left(J_0 - J_1\right)(y, \lambda) - 1\right| \le C \frac{y^2 + |\lambda|}{1 + |\lambda| + y^2}. \tag{17.19.33}$$

By (16.4.45), (17.19.33), we find that if $h \in]0, h_0], y^2 \in S, h^2\lambda \in \mathcal{W}_{\delta'}$,

$$\begin{aligned} &\left|h^2 J_0\left(hy, h^2\lambda\right)\left(\left(J_0 - J_1\right)\left(hy, h^2\lambda\right) - 1\right)\right| \\ &\le C\rho(\lambda) \frac{1 + h^2|\lambda| + h|y|}{1 + |\lambda| + y^2} h^2 \frac{y^2 + \lambda}{1 + h^2|\lambda| + h^2 y^2} \le C\rho(\lambda) h^2. \end{aligned} \tag{17.19.34}$$

The estimate (17.19.34) indicates that in the right-hand side of (17.19.18), we may as well replace $J_1 - J_0$ by 1.

In the right-hand side of (17.19.18), we should then estimate the contribution of $\frac{J_1}{J_0}\left(h|\xi|/\sqrt{2}, h^2\lambda\right) - 1$.

By (16.4.40) and (17.19.8), if $\lambda \in \mathcal{W}_\delta$,

$$\left|\left(J_1 - J_0\right)(y, \lambda)\right| \le \frac{C}{1 + |y| + |\lambda|}. \tag{17.19.35}$$

By (17.19.35), we find that if $y^2 \in S, h^2\lambda \in \mathcal{W}_\delta$,

$$\left|h^2 J_0\left(hy, h^2\lambda\right)\left(\frac{J_1}{J_0}\left(hy, h^2\lambda\right) - 1\right)\right| \le Ch^2. \tag{17.19.36}$$

The estimate (17.19.36) takes care of the difference $\frac{J_1}{J_0} - 1$ in the right-hand side of (17.19.18).

Now we inspect the other terms in the right-hand side of (17.17.10). First we replace $\mathbb{B}_{h,h^2\lambda}, \mathbb{B}'_{h,h^2\lambda}, \mathbb{C}_{h,h^2\lambda}$ by their semiclassical principal symbol. Indeed the same argument as before shows that the contribution of the difference to $U_{h,h^2\lambda}$ can be dominated by $C\rho(\lambda) h \frac{1+|\xi|}{1+|\lambda|+|\xi|^2}$. In the sequel, we will also use without further mention the same commutation arguments as the ones outlined after (17.19.7).

As we saw in equation (17.16.4) in Theorem 17.16.2, the semiclassical symbols of $\mathbb{B}_{h,\lambda}, \mathbb{B}'_{h,\lambda}, \mathbb{C}_{h,\lambda}$ are given by

$$\begin{aligned} \overline{\mathbb{B}}_{\xi,\lambda} &= \pm P p \left(P^\perp \left(B(i\xi) - \lambda\right) P^\perp\right)^{-1} \frac{1}{2}\omega\left(\nabla^F, g^F\right)(e_i) \nabla_{\widehat{e}^i} P, \\ \overline{\mathbb{B}}'_{\xi,\lambda} &= \pm P \frac{1}{2}\omega\left(\nabla^F, g^F\right)(e_i) \nabla_{\widehat{e}^i} \left(P^\perp \left(B(i\xi) - \lambda\right) P^\perp\right)^{-1} pP, \\ \overline{\mathbb{C}}_{\xi,\lambda} &= -P \frac{1}{2}\omega\left(\nabla^F, g^F\right)(e_i) \nabla_{\widehat{e}^i} \left(P^\perp \left(B(i\xi) - \lambda\right) P^\perp\right)^{-1} \\ &\qquad \frac{1}{2}\omega\left(\nabla^F, g^F\right)(e_i) \nabla_{\widehat{e}^i} P. \end{aligned} \tag{17.19.37}$$

Using the identity (17.17.11) in Theorem 17.17.4, we can rewrite (17.19.37) in the form

$$\begin{aligned} \overline{\mathbb{B}}_{\xi,\lambda} &= \mp \frac{1}{2}\mathfrak{a}(\xi, \lambda)\, \omega\left(\nabla^F, g^F\right), \\ \overline{\mathbb{B}}'_{\xi,\lambda} &= \pm \frac{1}{2}\mathfrak{a}(\xi, \lambda)\, \omega\left(\nabla^F, g^F\right), \\ \overline{\mathbb{C}}_{\xi,\lambda} &= \frac{1}{4}\left\langle \mathfrak{a}(\xi, \lambda)\, \omega\left(\nabla^F, g^F\right), \omega\left(\nabla^F, g^F\right)\right\rangle. \end{aligned} \tag{17.19.38}$$

When F is of dimension 1, the interpretation of (17.19.38) is clear. In general $\omega\left(\nabla^F, g^F\right)$ is a section of $T^*X \otimes \operatorname{End}(F)$. If $e_1, \ldots, e_n$ is an orthonormal basis of TX, the interpretation of the last identity is that

$$\overline{\mathbb{C}}_{\xi,\lambda} = \frac{1}{4}\left\langle \mathfrak{a}(\xi,\lambda) e_i, e_j\right\rangle \omega\left(\nabla^F, g^F\right)(e_i)\, \omega\left(\nabla^F, g^F\right)(e_j). \tag{17.19.39}$$

We already know $\frac{J_1}{J_0}$ and $J_0 - J_1$ lie in $\mathbb{S}^{-1}_{\delta}$. By (17.17.11) and (17.19.9), we conclude that for any ζ, $\left\langle \mathfrak{a}(\xi,\lambda)\zeta,\zeta\right\rangle \in \mathbb{S}^{-1}_{\delta}$.

We take a system of local coordinates on X and trivializations as we did after (17.19.19). We identify $\mathfrak{a}\left(h\xi, h^2\lambda\right)$ to the corresponding classical pseudodifferential operator. By the above, it follows that

$$\left[\nabla^{\Lambda^{\cdot}(T^*T^*X)\widehat{\otimes}F,u}, \mathfrak{a}\left(h\xi, h^2\lambda\right)\right] \simeq \frac{h\xi}{\left(1 + h^2|\lambda| + h|\xi|\right)^2}. \tag{17.19.40}$$

By proceeding as in (17.19.6), we find that (17.19.40) is irrelevant in our estimates.

We also identify $\overline{\mathbb{B}}_{h\xi,h^2\lambda}, \overline{\mathbb{B}}'_{h\xi,h^2\lambda}$ to the associated classical pseudodifferential operators. By (17.19.38)-(17.19.40), we get

$$\begin{aligned}\nabla^{\Lambda^{\cdot}(T^*X)\otimes F,u*}&\overline{\mathbb{B}}_{h\xi,h^2\lambda} - \overline{\mathbb{B}}'_{h\xi,h^2\lambda}\nabla^{\Lambda^{\cdot}(T^*X)\otimes F,u}\\ &\simeq \pm\frac{1}{2}\left\langle \mathfrak{a}\left(h\xi, h^2\lambda\right) e_i, e_j\right\rangle \nabla^{F,u}_{e_i}\omega\left(\nabla^F, g^F\right)(e_j), \qquad (17.19.41)\\ \overline{\mathbb{C}}_{h\xi,h^2\lambda} &\simeq \left\langle \mathfrak{a}\left(h\xi, h^2\lambda\right) e_i, e_j\right\rangle \frac{1}{4}\omega\left(\nabla^F, g^F\right)(e_i)\,\omega\left(\nabla^F, g^F\right)(e_j).\end{aligned}$$

Now recall that by (1.2.10), $\nabla^{F,u}_{U}\omega\left(\nabla^F, g^F\right)(V)$ is a symmetric tensor. Using equation (17.17.11) in Theorem 17.17.4 and (17.19.41), we obtain

$$\begin{aligned}&\nabla^{\Lambda^{\cdot}(T^*X)\otimes F,u*}\overline{\mathbb{B}}_{h\xi,h^2\lambda} - \overline{\mathbb{B}}'_{h\xi,h^2\lambda}\nabla^{\Lambda^{\cdot}(T^*X)\otimes F,u}\\ &= \pm\frac{1}{4}\left(J_0 - J_1\right)\left(h|\xi|/\sqrt{2}, h^2\lambda\right)\nabla^{F,u}_{e_i}\omega\left(\nabla^F, g^F\right)(e_i)\\ &\pm\frac{1}{4}\left(\frac{J_1}{J_0} - J_0 + J_1\right)\left(h|\xi|/\sqrt{2}, h^2\lambda\right)\frac{1}{|\xi|^2}\nabla^{F,u}_{\xi}\omega\left(\nabla^F, g^F\right)(\xi), \qquad (17.19.42)\\ &\overline{\mathbb{C}}_{h\xi,h^2\lambda} = \frac{1}{8}\left(J_0 - J_1\right)\left(h|\xi|/\sqrt{2}, h^2\lambda\right)\omega\left(\nabla^F, g^F\right)(e_i)^2\\ &+\frac{1}{8}\left(\frac{J_1}{J_0} - J_0 + J_1\right)\left(h|\xi|/\sqrt{2}, h^2\lambda\right)\frac{\omega\left(\nabla^F, g^F\right)^2(\xi)}{|\xi|^2}.\end{aligned}$$

The same arguments as in (17.19.18)-(17.19.36) show that in (17.19.42), we can replace $J_1 - J_0$ and $\frac{J_1}{J_0}$ by 1.

Using now equation (1.2.14) for $\square^X$, equation (17.17.1) for $P_+\gamma_+P_+$, and the considerations which follow, we finally get (17.19.4). The proof of our theorem is completed. □

17.20 ESTIMATES ON THE RESOLVENT OF $\mathrm{T}_{\mathbf{h},\mathbf{h^2}\lambda}$

For $r > 0, h > 0$, set

$$\mathcal{W}_{\delta',h,r} = \left\{\lambda \in \mathcal{W}_{\delta'}/h^2, r\left(\operatorname{Re}\lambda + 1\right) \leq |\operatorname{Im}\lambda|\right\}. \tag{17.20.1}$$

In the sequel we take δ' as in Theorem 17.19.1.

Theorem 17.20.1. *There exists $C>0$ such that if $h\in]0,h_0],\lambda\in\mathbf{C},\lambda\in\mathcal{W}_{\delta'}/h^2\setminus\operatorname{Sp}\square^X/2$ are such that $h\rho(\lambda)\le C$, for any $s\in\mathbf{R}$, the operator $\left(T_{h,h^2\lambda}-\lambda\right)^{-1}$ maps H^s_X into H^{s+1}_X, and moreover there exists $C_s>0$ such that if $u\in\Omega^{\cdot}(X,F)$ in the + case, or $u\in\Omega^{\cdot}(X,F\otimes o(TX))$ in the − case, then*

$$\left\|\left(T_{h,h^2\lambda}-\lambda\right)^{-1}\left(\frac{1}{2}\square^X-\lambda\right)u-u\right\|_{X,\lambda^{1/2},s}\\ \le C_sh\left\|\left(\rho(\lambda)+\square^{X,1/2}\right)u\right\|_{X,\lambda^{1/2},s}. \quad (17.20.2)$$

For any $r>0$, there exists $h_r\in]0,h_0]$ such that for $h\in]0,h_r],\lambda\in\mathcal{W}_{\delta',h,r}$, the operator $\left(T_{h,h^2\lambda}-\lambda\right)^{-1}$ exists. Moreover, given $s\in\mathbf{R}$, there exist $C_r>0,C_s>0$ such that if h,λ are taken as before,

$$\left\|\left(T_{h,h^2\lambda}-\lambda\right)^{-1}u\right\|_{X,\lambda^{1/2},s+1}\le C_s\left(h+C_r\left(1+|\lambda|\right)^{-1/2}\right)\|u\|_{X,\lambda^{1/2},s}. \quad (17.20.3)$$

Proof. If $A\in\operatorname{End}\left(L^2_X\right)$, we denote by $\|A\|$ the norm of A. By Theorem 17.19.1, it is clear that if $h\rho(\lambda)$ is small enough, the operator $1+U_{h,h^2\lambda}$ is invertible when acting on $H^0_X=L^2_X$, and that its inverse is uniformly bounded.

We will show that under the same conditions, given $s\in\mathbf{R}$, there exists $C_s>0$ such that

$$\left\|\left(1+U_{h,h^2\lambda}\right)^{-1}u\right\|_{X,\lambda^{1/2},s}\le C_s\|u\|_{X,\lambda^{1/2},s}, \quad (17.20.4)$$
$$\left\|\left(\left(1+U_{h,h^2\lambda}\right)^{-1}-1\right)u\right\|_{X,\lambda^{1/2},s}\le C_sh\rho(\lambda)\|u\|_{X,\lambda^{1/2},s}.$$

As we just saw, this is true for $s=0$.

Set

$$\Lambda=\left(1+|\lambda|+\square^X\right)^{1/2}. \quad (17.20.5)$$

Then

$$\|u\|_{X,\lambda^{1/2},s}=\|\Lambda^su\|. \quad (17.20.6)$$

If $v=\left(1+U_{h,h^2\lambda}\right)^{-1}u$, then

$$\Lambda^sv=\left(1+U_{h,h^2\lambda}\right)^{-1}\left(\Lambda^su-\left[\Lambda^s,U_{h,h^2\lambda}\right]v\right). \quad (17.20.7)$$

By Theorem 17.19.1, there exists $C>0$ such that for $h\rho(\lambda)$ small enough, for any $s\in\mathbf{R}$,

$$\left\|\left(1+U_{h,h^2\lambda}\right)^{-1}\Lambda^su\right\|\le C\|u\|_{X,\Lambda^{1/2},s}. \quad (17.20.8)$$

We claim that given $s\in\mathbf{R}$, there is $C_s>0$ such that

$$\left\|\left[\Lambda^s,U_{h,h^2\lambda}\right]w\right\|\le C_s\left(h\rho(\lambda)+h^2\rho^2(\lambda)\right)\|w\|_{X,\lambda^{1/2},s-1/3}. \quad (17.20.9)$$

The proof of (17.20.9) will be delayed. By (17.20.7)-(17.20.9), for $h\rho(\lambda)$ small enough, we get

$$\|v\|_{X,\lambda^{1/2},s} \leq C_s \left(\|u\|_{X,\lambda^{1/2},s} + \|v\|_{X,\lambda^{1/2},s-1/3}\right). \tag{17.20.10}$$

By (17.20.10), we deduce that the first equation in (17.20.4) holds for $s \geq 0$. Since the formal adjoint of $U_{h,h^2\lambda}$ has the same structure as $U_{h,h^2\lambda}$, the above proof also leads to a proof of the first equation in (17.20.4) for arbitrary $s \in \mathbf{R}$. A similar argument allows us to also obtain the second equation in (17.20.4) for any $s \in \mathbf{R}$.

Now we concentrate on the proof of (17.20.9). The idea is to go along the proof of Theorem 17.19.1 and check that the corresponding estimates can be safely "commuted" with Λ^s.

Moreover, recall that all the functions of D^X which we considered are in fact functions of $\square^X$. As an aside , let us observe that D^X and $\square^X$ commute anyway. By (17.19.2), (17.19.5), we get

$$\left[\Lambda^s, U_{h,h^2\lambda}\right] = h^2 J_0\left(hD^X/\sqrt{2}, h^2\lambda\right)\left[\Lambda^s, V_{h,h^2\lambda}\right]. \tag{17.20.11}$$

We will evaluate the commutator in the right-hand side of (17.20.9) using equation (17.17.10) in Theorem 17.17.4. We will use the arguments we already gave in the proof of Theorem 17.19.1.

First we consider the contribution of the first term in the right-hand side of (17.17.10) for $T_{h,h^2\lambda}$. As we already explained in (17.19.18)-(17.19.36), once $\mathbb{A}_{h,h^2\lambda}$ is replaced by its semiclassical principal symbol $\mathfrak{a}\left(h\xi, h^2\lambda\right)$, there is an approximate cancellation with the term $-\frac{J_1}{J_0}\left(hD^X/\sqrt{2}, h^2\lambda\right)\frac{\Delta^{H,u}}{2}$ which is compatible with the above estimates. Commuting Λ^s with $\nabla^{\Lambda^\cdot(T^*X)\otimes F,u*}$ or $\nabla^{\Lambda^\cdot(T^*T^*X)\widehat{\otimes}F,u}$ does not raise any special difficulty, the above cancellations still occurring. The only potential difficulty consists in controlling the commutator of Λ^s with an operator of the type $\nabla^{\Lambda^\cdot(T^*X)\otimes F,u*}hB\nabla^{\Lambda^\cdot(T^*X)\otimes F,u}$, where B is itself a semiclassical pseudodifferential operator which lies in $\mathbb{E}^{-1}_{\delta,h}$. Apart from a smoothing semiclassical operator, whose contribution is irrelevant by (17.15.6), we can as well assume that $B = \mathrm{Op}(b)$, where $b \in \mathbb{S}^{-1}_{\delta,h}$. For h small enough, the classical symbol of B is given by $b\left(x, h\xi, h, h^2\lambda\right)$. When commuting Λ^s with B, we can then use the classical rules of composition of pseudodifferential operators, where the parameter λ is incorporated. Using the bounds in (17.15.5) with $d = -1$, we get

$$\begin{aligned}\left[\Lambda^s(x,\xi), hb\left(x, h\xi, h, h^2\lambda\right)\right]&\\ \simeq h\left(1 + |\lambda|^{1/2} + |\xi|\right)^{s-1}&\left(1 + h^2|\lambda| + h|\xi|\right)^{-2/3}\\ - h^2\left(1 + |\lambda|^{1/2} + |\xi|\right)^{s}&\left(1 + h^2|\lambda| + h|\xi|\right)^{-5/3}.\end{aligned} \tag{17.20.12}$$

Incidentally observe that mixing classical and semiclassical pseudodifferential operators produces terms with different homogeneities. By (17.20.12), we get

$$C\rho(\lambda)\frac{1+h^2|\lambda+h|\xi||}{1+|\lambda|+|\xi|^2}\left|\left[\Lambda^s(x,\xi),hb\left(x,h\xi,h,h^2\lambda\right)\right]\right|\left(1+|\xi|^2\right)$$

$$\leq Ch\rho(\lambda)\left(\left(1+|\lambda|^{1/2}+|\xi|\right)^{s-7/3}+h\frac{\left(1+|\lambda|^{1/2}+|\xi|\right)^{s-2}}{\left(1+h^2|\lambda|+h|\xi|\right)^{2/3}}\right)\left(1+|\xi|^2\right)$$

$$\leq Ch\rho(\lambda)\left(1+|\lambda|^{1/2}+|\xi|\right)^{s-1/3}, \quad (17.20.13)$$

which fits with (17.20.9).

Establishing the corresponding bounds for the other terms in the right-hand side of (17.17.10) follows the same principle as before. So we get (17.20.9).

By (16.4.4), (17.19.3), and (17.20.4), the first part of our theorem is now obvious. By (16.4.6), we get

$$\left|h^2J_0\left(hy,h^2\lambda\right)\left(y^2-\lambda\right)-1\right|\leq Ch|y|. \quad (17.20.14)$$

By (17.19.3), (17.20.4), (17.20.14), we get (17.20.2).

Now we will establish (17.20.3). Indeed given $r>0$, there exists $C_r>0$ such that if $\lambda\in\mathcal{W}_{\delta',h,r}$, then $\rho(\lambda)\leq C_r$, and so by the above, there is $h_r\in\,]0,h_0]$ such that for $h\in]0,h_r]$, the operator $\left(T_{h,h^2\lambda}-\lambda\right)^{-1}$ is well defined. Using the first identity in (16.4.44), (17.19.3), and (17.20.4), we get (17.20.3).

The proof of our theorem is completed. □

17.21 THE ASYMPTOTICS OF $(\mathbf{L_c}-\lambda)^{-1}$

Take $h_0>0$ small enough. We take $h\in]0,h_0]$. Here we make $c=\pm1/h^2$.

Put

$$R_{h,\lambda}=\left(T_{h,h^2\lambda}-\lambda\right)^{-1}. \quad (17.21.1)$$

By (17.2.12), we have the formal equality

$$(L_c-\lambda)^{-1}$$
$$=\begin{bmatrix} R_{h,\lambda} & -hR_{h,\lambda}L_2\Theta^{-1}_{h,h^2\lambda} \\ -h\Theta^{-1}_{h,h^2\lambda}L_3R_{h,\lambda} & h^2\Theta^{-1}_{h,h^2\lambda}+\Theta^{-1}_{h,h^2\lambda}hL_3R_{h,\lambda}hL_2\Theta^{-1}_{h,h^2\lambda}\end{bmatrix}. \quad (17.21.2)$$

To analyze the action of pseudodifferential operators which lie in the class $\mathbb{E}^d_{\delta,h}$ on the chain of Hilbert spaces which is associated to the norms $\| \|_{X,\lambda^{1/2},s}$, we introduce a new norm. Indeed set

$$\Lambda=\left(1+|\lambda|+\Box^X\right)^{1/2}, \qquad \Lambda'_h=\left(1+\left|h^2\lambda\right|^2+h^2\Box^X\right)^{1/2}. \quad (17.21.3)$$

Let $H^{s_1,s_2}_{X,h,\lambda}$ be the Sobolev space $H^{s_1+s_2}_X$ equipped with the norm

$$\|u\|_{\lambda,s_1,s_2}=\left\|\Lambda^{s_1}\Lambda'^{s_2}_h u\right\|_{L^2}. \quad (17.21.4)$$

By (17.15.1), (17.21.4), we get

$$\|u\|_{X,\lambda^{1/2},s} = \|u\|_{\lambda,s,0}\,, \qquad \|u\|_{X,h,h^2\lambda,s} = \|u\|_{\lambda,0,s}\,. \tag{17.21.5}$$

From the considerations we made after (17.15.1), using (17.21.5), we find that given $s \in \mathbf{R}$, the norms $\| \ \|_{h^2\lambda,\mathrm{sc},s}$ and $\| \ \|_{\lambda,0,s}$ are uniformly equivalent in the considered range of parameters h, λ.

If $E \in \mathbb{E}^d_{\delta,\lambda}$, we denote by E_h the operator E in which λ has been replaced by $h^2\lambda$, so that E_h is defined for $\lambda \in \mathcal{W}_\delta/h^2$.

Proposition 17.21.1. *If B is an operator which is a polynomial of degree p in $|\lambda|^{1/2}, \nabla_{e_i}, 1 \leq i \leq n$, for $s_1, s_2 \in \mathbf{R}$, there exists $C_{s_1,s_2} > 0$ such that if $u \in \mathcal{D}_X$,*

$$\|Bu\|_{\lambda,s_1,s_2} \leq C_{s_1,s_2} \|u\|_{\lambda,p+s_1,s_2}\,. \tag{17.21.6}$$

Moreover, given $E \in \mathbb{E}^d_{\delta,h}$ and any $s_1, s_2 \in \mathbf{R}$, there exists $C_{s_1,s_2} > 0$ such that if $u \in \mathcal{D}_X$,

$$\|E_h u\|_{\lambda,s_1,s_2} \leq C_{s_1,s_2} \|u\|_{\lambda,s_1,s_2+d}\,. \tag{17.21.7}$$

Proof. Using duality and interpolation, it is enough to establish our result when $s_1, s_2 \in 2\mathbf{N}$.

Equation (17.21.6) is true when $s_2 = 0$. To establish this equation in full generality, it is enough to prove it for $B = \nabla_{e_i}$. If $s_2 = 2\ell_2, \ell_2 \in \mathbf{N}$, there is a partial operator $Q(\partial_x)$ of degree $2\ell_2$ such that

$$\left[\nabla_{e_i}, \Lambda'^{2\ell_2}_h\right] = Q(h\partial_x)\,. \tag{17.21.8}$$

Therefore,

$$\begin{aligned}\|\nabla_{e_i} u\|_{\lambda,2\ell_1,2\ell_2} &= \left\|\Lambda'^{2\ell_2}\nabla_{e_i} u\right\|_{\lambda,2\ell_1,0} \\ &\leq \left\|\nabla_{e_i}\Lambda'^{2\ell_2} u\right\|_{\lambda,2\ell_1,0} + \|Q(h\partial x)\,u\|_{\lambda,2\ell_1,0} \\ &\leq C_{\ell_1,\ell_2}\|u\|_{\lambda,2\ell_1+1,2\ell_2}\,.\end{aligned} \tag{17.21.9}$$

So we have established (17.21.6).

Clearly (17.21.7) holds when $s_1 = 0$. Moreover, by (17.15.5), if $E \in \mathbb{E}^d_{\delta,h}$, then

$$[\nabla_{e_i}, E] \in \mathbb{E}^{d+1/3}_{\delta,h}\,. \tag{17.21.10}$$

So if A is a polynomial in $|\lambda|^{1/2}, \nabla_{e_i}$ of total degree $2\ell_1$,

$$[A, E] \in \sum_{1\leq j\leq 2\ell_1} \mathbb{E}^{d+j/3}_{\delta,h} A_j, \tag{17.21.11}$$

where the A_j are themselves polynomials in $|\lambda|^{1/2}, \nabla_{e_i}$ of degree $\leq 2\ell_1 - j$. Using (17.21.6), (17.21.7) with $s_1 = 0$ and (17.21.11) with $A = \Lambda^{\ell_1}$, we get

$$\|E_h u\|_{\lambda,2\ell_1,s_2} \leq C_{\ell_1,s_2}\left(\|u\|_{\lambda,2\ell_1,s_2+d} + \sum_{1\leq j\leq 2\ell_1}\|u\|_{\lambda,2\ell_1-j,s_2+d+j/3}\right). \tag{17.21.12}$$

Also for $0 \leq j \leq 2\ell_1$, we have the trivial

$$\|u\|_{\lambda,2\ell_1-j,s_2+d+j/3} \leq C_{\ell_1,s_2} \|u\|_{\lambda,2\ell_1,s_2+d}. \tag{17.21.13}$$

Then (17.21.7) follows from (17.21.12) and (17.21.13). The proof of our proposition is completed. □

We will extend the estimate (17.20.3) in Theorem 17.20.1, taking into account the refined norms $\| \|_{\lambda,s_1,s_2}$.

Proposition 17.21.2. *For $r > 0, s_1 \in \mathbf{R}, s_2 \in \mathbf{R}$, there exists $h_r > 0, C_r > 0, C_{r,s_1,s_2} > 0$ such that for $h \in]0, h_r], \lambda \in \mathcal{W}_{\delta',h,r}, u \in \mathcal{D}_X$,*

$$\left\| \left(T_{h,h^2\lambda} - \lambda\right)^{-1} u \right\|_{\lambda,s_1+1,s_2} \leq C_{r,s_1,s_2} \left(h + C_r \left(1 + |\lambda|\right)^{-1/2}\right) \|u\|_{\lambda,s_1,s_2}. \tag{17.21.14}$$

Proof. By equation (17.20.3) in Theorem 17.20.1 and by (17.21.5), we get (17.21.14) when $s_2 = 0$.

By duality and interpolation, it is enough to establish (17.21.14) for $s_2 = d > 0$. We will show that given $a > 0$, if (17.21.14) holds for $d \in [0, a]$, then it still holds for $d \in [0, a + 2/3]$. Recall that $\mathbb{E}^d_{\delta,h,0}$ was defined via the inequalities in (17.15.10). Let $E \in \mathbb{E}^d_{\delta,h,0}$ with symbol $e = e_d + he_{d-1}, e_i \in \mathbb{S}^i_{\delta,h,0}$ for $i = d-1, d$, so that e_d is scalar. If $A \in \mathbb{E}^s_{\delta,h}$, by (17.15.5), (17.15.10), we get

$$[E, A] \in h\mathbb{E}^{d+s-2/3}_{\delta,h}. \tag{17.21.15}$$

Using Theorem 17.17.4 and in particular (17.17.9) and (17.17.10), we get

$$[E, T_{h,\lambda}] \in \sum \mathbb{E}^{d-2/3}_{\delta,h} \nabla_{e_i} + \mathbb{E}^{d-2/3}_{\delta,h}. \tag{17.21.16}$$

Also,

$$\left\| \left(T_{h,h^2\lambda} - \lambda\right)^{-1} u \right\|_{\lambda,s_1,d} = \left\| \Lambda'^d_h \left(T_{h,h^2\lambda} - \lambda\right)^{-1} u \right\|_{\lambda,s_1,0}. \tag{17.21.17}$$

Moreover,

$$\begin{aligned}\Lambda'^d_h \left(T_{h,h^2\lambda} - \lambda\right)^{-1} &= \left(T_{h,h^2\lambda} - \lambda\right)^{-1} \Lambda'^d_h \\ &+ \left(T_{h,h^2\lambda} - \lambda\right)^{-1} \left[T_{h,h^2\lambda}, \Lambda'^d_h\right] \left(T_{h,h^2\lambda} - \lambda\right)^{-1}.\end{aligned} \tag{17.21.18}$$

Put

$$m = h + C_r \left(1 + |\lambda|\right)^{-1/2}. \tag{17.21.19}$$

By Proposition 17.21.1, by (17.21.14) with $s_2 = 0$, and using also (17.21.16)-(17.21.18), we obtain

$$\begin{aligned}\left\| \left(T_{h,h^2\lambda} - \lambda\right)^{-1} u \right\|_{\lambda,s_1+1,d} &\leq C_{s_1} m \Bigg(\left\|\Lambda'^d_h u\right\|_{\lambda,s_1,0} \\ &+ \left\| \left[T_{h,h^2\lambda}, \Lambda'^d_h\right] \left(T_{h,h^2\lambda} - \lambda\right)^{-1} u \right\|_{\lambda,s_1,0} \Bigg) \\ &\leq C_{s_1,d} m \left(\|u\|_{\lambda,s_1,d} + \left\| \left(T_{h,h^2\lambda} - \lambda\right)^{-1} u \right\|_{\lambda,s_1+1,d-2/3} \right).\end{aligned} \tag{17.21.20}$$

By (17.21.20), we obtain the announced recursion on d. The proof of our proposition is completed. □

Let $\mathcal{Q}^\ell$ be the set of differential operators with smooth coefficients in $x \in X$, which are polynomials in $|\lambda|^{1/2}, \nabla_{e_i}, \nabla_{\widehat{e}^i}, p_j \nabla_{\widehat{e}^i}, p_i, p_i p_j$ which are of total degree at most ℓ. Let $\mathcal{R}^\ell$ be a finite family of operators which generate $\mathcal{Q}^\ell$ over $C^\infty(X, \mathbf{R})$.

We will denote by $H_{X,\ell}$ the associated Sobolev space with the norm

$$\|u\|_\ell^2 = \sum_{Q \in \mathcal{R}^\ell} \|Qu\|_{L_X^2}^2 . \tag{17.21.21}$$

We denote by $\mathcal{H}_{X,-\ell}$ the vector space which is dual to $\mathcal{H}_{X,\ell}$.

If $A \in \mathcal{L}\left(L_X^2\right)$, let $\|A\|$ be the norm of A. In the case where A extends to a bounded operator from $\mathcal{H}_{-\ell}$ into $\mathcal{H}_\ell$, set

$$\|A\|_\ell = \sum_{Q, Q' \in \mathcal{R}^\ell} \|QAQ'\| . \tag{17.21.22}$$

Recall that H is the standard L^2 space over T^*X. If $B \in \operatorname{End}(H)$, we still denote by $\|B\|$ the norm of B.

Theorem 17.21.3. *There exist $h_0 > 0, \delta' = (\delta'_0, \delta'_1, \delta'_2)$ with $\delta'_0 \in]0,1[, \delta'_1 > 0, \delta'_2 = \frac{1}{6}$, such that for any $r > 0$,*

- *For $h \in]0, h_0], \lambda \in \mathcal{W}_{\delta',h,r}$, the resolvent $(L_c - \lambda)^{-1}$ exists and is given by equation (17.21.2). There exists $C > 0$ such that if b, λ are taken as before, then*

$$\begin{aligned} &\left\|(L_c - \lambda)^{-1}\right\| \leq C, \\ &\left\|i_\pm \left(T_{h,h^2\lambda} - \lambda\right)^{-1} P_\pm\right\| \leq C, \\ &\left\|(L_c - \lambda)^{-1} - i_\pm \left(T_{h,h^2\lambda} - \lambda\right)^{-1} P_\pm\right\| \leq Ch. \end{aligned} \tag{17.21.23}$$

- *Let $v \in]0,1[$. For any $\ell \in \mathbf{N}$, for $N \in \mathbf{N}^*$ large enough, there exists $C_N > 0$ such that if h, λ are taken as before, then*

$$\begin{aligned} &\left\|(L_c - \lambda)^{-N}\right\|_\ell \leq C_N, \\ &\left\|i_\pm \left(T_{h,h^2\lambda} - \lambda\right)^{-N} P_\pm\right\|_\ell \leq C_N, \\ &\left\|(L_c - \lambda)^{-N} - i_\pm \left(T_{h,h^2\lambda} - \lambda\right)^{-N} P_\pm\right\|_\ell \leq C_N h^v. \end{aligned} \tag{17.21.24}$$

Proof. We take δ' as in Theorem 17.20.1. If $\lambda \in \mathcal{W}_{\delta',h,r}$, by equation (17.20.3) in Theorem 17.20.1, we get the second estimate in (17.21.23), and we also find that the operators $L_2 \left(T_{h,h^2\lambda} - \lambda\right)^{-1}, \left(T_{h,h^2\lambda} - \lambda\right)^{-1} L_3$ are uniformly bounded when acting on $H = L^2$. By (17.16.9),

$$\left\|\Theta_{h,h^2\lambda}^{-1} u\right\|_{h^2\lambda,\mathrm{sc},s+1/4} \leq C_s \|u\|_{h^2\lambda,\mathrm{sc},s} . \tag{17.21.25}$$

In particular $\Theta_{h,h^2\lambda}^{-1}$ is uniformly bounded as an operator acting on L^2.

By Proposition 17.21.2, we find that given $s_1, s_2 \in \mathbf{R}$,

$$\left\| \left(T_{h,h^2\lambda} - \lambda\right)^{-1} L_2 u \right\|_{\lambda,s_1,s_2} \leq C_{s_1,s_2} \left\| L_2 u \right\|_{\lambda,s_1-1,s_2}. \tag{17.21.26}$$

Recall that L_2 is given by (17.2.8). Also in the $+$ case,

$$P_+ \nabla_{Y^{\mathcal{H}}}^{\Lambda^{\cdot}(T^*T^*X)\widehat{\otimes}F,u} u = \nabla_{e_i}^{\Lambda^{\cdot}(T^*T^*X)\widehat{\otimes}F,u} P_+ p_i u. \tag{17.21.27}$$

By (17.21.27), we find easily that

$$\left\| L_2 u \right\|_{\lambda,s_1-1,s_2} \leq C_{s_1,s_2} \left\| P_\pm \left(1 + |p|^2\right) u \right\|_{\lambda,s_1,s_2}. \tag{17.21.28}$$

Using the considerations after (17.21.5), and also (17.21.26), (17.21.28) with $s_1 = 0, s_2 = s$, we get

$$\left\| \left(T_{h,h^2\lambda} - \lambda\right)^{-1} L_2 u \right\|_{h^2\lambda,\mathrm{sc},s} \leq C_s \left\| u \right\|_{h^2\lambda,\mathrm{sc},s}. \tag{17.21.29}$$

The same arguments as in the proof of (17.21.29) also show that

$$\left\| L_3 \left(T_{h,h^2\lambda} - \lambda\right)^{-1} u \right\|_{h^2\lambda,\mathrm{sc},s} \leq C_s \left\| u \right\|_{h^2\lambda,\mathrm{sc},s}. \tag{17.21.30}$$

Set

$$A = \Theta_{h,h^2\lambda}^{-1} h L_3 \left(T_{h,h^2\lambda} - \lambda\right)^{-1} L_2 \Theta_{h,h^2\lambda}^{-1}. \tag{17.21.31}$$

We claim that A acts on L^2 as a uniformly bounded operator. Indeed by Theorem 17.16.3 and by (17.21.29) with $s = 1/6$, since $1/6 - 1/4 = -1/12$, we get

$$\begin{aligned} \left\| Au \right\| &\leq C \left\| \left(T_{h,h^2\lambda} - \lambda\right)^{-1} L_2 \Theta_{h,h^2\lambda}^{-1} u \right\|_{h^2\lambda,\mathrm{sc},1/6} \\ &\leq C \left\| \Theta_{h,h^2\lambda}^{-1} u \right\|_{h^2\lambda,\mathrm{sc},1/6} \leq C \left\| u \right\|_{h^2\lambda,\mathrm{sc},-1/12} \leq C \left\| u \right\|. \end{aligned} \tag{17.21.32}$$

Incidentally note that it is here that the critical $-1/6$ is used in our proof.

Using (17.21.2) and the above estimates, we deduce that the resolvent $(L_c - \lambda)^{-1}$ is indeed given by (17.21.2) and moreover that (17.21.23) holds.

The second equation in (17.21.24) follows from equation (17.20.3) in Theorem 17.20.1. By (17.21.23), we find that for any $N \in \mathbf{N}$,

$$\left\| (L_c - \lambda)^{-N} - i_\pm \left(T_{h,h^2\lambda} - \lambda\right)^{-N} P_\pm \right\| \leq C_N h. \tag{17.21.33}$$

We will now show that the third estimate in (17.21.24) follows from the first two. Take $v \in]0, 1[$, and let $p \in \mathbf{N}$ such that $1 - 1/p \geq v$. Put $\ell' = p\ell$. Assume that $N \in \mathbf{N}$ is such that the first two identities in (17.21.24) hold with respect to ℓ'. Using these two estimates, (17.21.33), and classical interpolation, we get

$$\left\| (L_c - \lambda)^{-N} - i_\pm \left(T_{h,h^2\lambda} - \lambda\right)^{-N} P_\pm \right\|_\ell \leq C h^{1-1/p}, \tag{17.21.34}$$

which implies the third identity in (17.21.24).

Therefore, we only need to establish the first estimate in (17.21.24). Recall that $R_{h,\lambda}$ was defined in (17.21.1). Set

$$R = \begin{bmatrix} 0 & -R_{h,\lambda} L_2 \Theta^{-1}_{h,h^2\lambda} \\ -\Theta^{-1}_{h,h^2\lambda} L_3 R_{h,\lambda} & h\Theta^{-1}_{h,h^2\lambda}\left(1 + L_3 R_{h,\lambda} L_2 \Theta^{-1}_{h,h^2\lambda}\right) \end{bmatrix}, \tag{17.21.35}$$

$$J = i_\pm \left(T_{h,h^2\lambda} - \lambda\right)^{-1} P_\pm .$$

By (17.21.2),

$$\left(L_c - \lambda\right)^{-1} = J + hR. \tag{17.21.36}$$

Take $\sigma = 1/12$. We claim that for $s \in \mathbf{R}$, there exists $C_s > 0$ such that if $h \in]0, h_0], \lambda \in \mathcal{W}_{\delta',h,r}$,

$$\|Ru\|_{h^2\lambda,\mathrm{sc},s+\sigma} \le C_s \|u\|_{h^2\lambda,\mathrm{sc},s}, \qquad \|Ju\|_{h^2\lambda,\mathrm{sc},s+\sigma} \le C_s \|u\|_{h^2\lambda,\mathrm{sc},s}. \tag{17.21.37}$$

Indeed since $\sigma \le 1/2$, by proceeding as in (17.21.13) and (17.21.14), we get

$$\left\|\left(T_{h,h^2\lambda} - \lambda\right)^{-1} u\right\|_{\lambda,0,s+\sigma} \le C_s \|u\|_{\lambda,-1,s+\sigma} \le C'_s \|u\|_{\lambda,0,s}. \tag{17.21.38}$$

The second inequality in (17.21.37) follows the considerations we made after (17.21.5) and from (17.21.38).

Since $\sigma \le 1/4$, we can use (17.21.29), (17.21.30), and the same arguments as in the proof of (17.21.32) to show that

$$\begin{gathered} \left\|\left(T_{h,h^2\lambda} - \lambda\right)^{-1} L_2 \Theta^{-1}_{h,h^2\lambda} u\right\|_{h^2\lambda,\mathrm{sc},s+\sigma} \le C_s \|u\|_{h^2\lambda,\mathrm{sc},s}, \\ \left\|\Theta^{-1}_{h,h^2\lambda} L_3 \left(T_{h,h^2\lambda} - \lambda\right)^{-1} u\right\|_{h^2\lambda,\mathrm{sc},s} \le C_s \|u\|_{h^2\lambda,\mathrm{sc},s}. \end{gathered} \tag{17.21.39}$$

Similarly, since $\sigma = 1/12$, the same arguments as in (17.21.32) show that

$$\begin{aligned} &\left\|\Theta^{-1}_{h,h^2\lambda} h L_3 \left(T_{h,h^2\lambda} - \lambda\right)^{-1} L_2 \Theta^{-1}_{h,h^2\lambda} u\right\|_{h^2\lambda,\mathrm{sc},s+\sigma} \\ &\qquad \le C_s \left\|\left(T_{h,h^2\lambda} - \lambda\right)^{-1} L_2 \Theta^{-1}_{h,h^2\lambda} u\right\|_{h^2\lambda,\mathrm{sc},s+\sigma+1/6} \le C_s \|u\|_{h^2\lambda,\mathrm{sc},s}. \end{aligned} \tag{17.21.40}$$

By (17.16.9), (17.21.35), (17.21.39), (17.21.40), we get the first inequality of (17.21.37), which completes the proof of (17.21.37).

By (17.21.36), we obtain

$$\left(L_c - \lambda\right)^{-N} = \sum h^{N_2} J^{i_1} R^{j_1} \dots J^{i_q} R^{j_q} \tag{17.21.41}$$

where the nonzero indices are such that $\sum_{k=1}^q i_k = N_1, \sum_{k=1}^q j_k = N_2, N_1 + N_2 = N$. By (17.21.37), (17.21.41), we get

$$\left\|J^{i_1} R^{j_1} \dots J^{i_q} R^{j_q}\right\|_{h^2\lambda,\mathrm{sc},-l+N\sigma} \le C \|u\|_{h^2\lambda,\mathrm{sc},-l}. \tag{17.21.42}$$

Clearly, for $j \in \mathbf{N}$,

$$|\lambda|^{1/2} \le \frac{1}{2}\left(2^{-j}|\lambda| + 2^j\right). \tag{17.21.43}$$

By (17.4.3), (17.4.4), and (17.21.43), we get for $\ell \in \mathbf{N}$,

$$\|u\|_{\ell} \leq C_{\ell} h^{-\ell} \|u\|_{h^2\lambda,\mathrm{sc},\ell}, \tag{17.21.44}$$

and so

$$\|u\|_{h^2\lambda,\mathrm{sc},-\ell} \leq C_{\ell} h^{-\ell} \|u\|_{-\ell}. \tag{17.21.45}$$

Using (17.21.37) and (17.21.41)-(17.21.45), we find that to prove the first estimate in (17.21.24), the only potentially annoying terms are the ones with $N_2 < 2\ell$. Since $N_2 \geq q$, we may assume that $q < 2\ell$. This is what we will do now.

We claim that there exists $M_\ell \in \mathbf{N}^*$ such that for any $Q \in \mathcal{R}^\ell$, then

$$JQ = \sum_{Q' \in \mathcal{R}^\ell} Q' J_{Q'}, \qquad QJ = \sum_{Q' \in \mathcal{R}^\ell} J'_{Q'} Q', \tag{17.21.46}$$
$$RQ = \sum_{Q' \in \mathcal{R}^\ell} Q' R_{Q'}, \qquad QR = \sum_{Q' \in \mathcal{R}^\ell} R'_{Q'} Q'.$$

so that in (17.21.46), the sums are finite, and moreover we should have the estimates

$$\|J_{Q'}u\|_{h^2\lambda,\mathrm{sc},s} + \left\|J'_{Q'}u\right\|_{h^2\lambda,\mathrm{sc},s} \leq C_s \|u\|_{h^2\lambda,\mathrm{sc},s}, \tag{17.21.47}$$
$$\|R_{Q'}u\|_{h^2\lambda,\mathrm{sc},s} + \left\|R'_{Q'}u\right\|_{h^2\lambda,\mathrm{sc},s} \leq C_s \|u\|_{h^2\lambda,\mathrm{sc},s+M_\ell}.$$

First we establish the required properties for J. Recall that Λ was defined in (17.21.3). If $Q \in \mathcal{R}_\ell$,

$$QJ = \left(QJ\Lambda^{-\ell}\right)\Lambda^{\ell}. \tag{17.21.48}$$

By Proposition 17.21.2,

$$\left\|\Lambda^{\ell} J \Lambda^{-\ell} u\right\|_{\lambda,s_1,s_2} \leq C_{s_1,s_2} \|u\|_{\lambda,s_1-1,s_2}. \tag{17.21.49}$$

By (17.21.49), we obtain

$$\left\|QJ\Lambda^{-\ell}u\right\|_{\lambda,s_1,s_2} \leq C_{s_1,s_2} \|u\|_{\lambda,s_1-1,s_2}. \tag{17.21.50}$$

Using the considerations we made after (17.21.5) and by making $s_1 = 0$ in (17.21.50), we get

$$\left\|QJ\Lambda^{-\ell}u\right\|_{h^2\lambda,\mathrm{sc},s} \leq C_s \|u\|_{h^2\lambda,\mathrm{sc},s}. \tag{17.21.51}$$

By (17.21.48), (17.21.51), we get the second identity in (17.21.46). To obtain the first identity, instead of (17.21.48), we write

$$JQ = \Lambda^{\ell}\left(\Lambda^{-\ell}JQ\right), \tag{17.21.52}$$

and we proceed as before.

Now we establish the commutation relations in (17.21.46), (17.21.47) for R. Note that if R^1, R^2 are operators such that the estimates in (17.21.46), (17.21.47) hold with $R = R^1$ and $R = R^2$ with a given M, then $R^1 + R^2$ verifies similar estimates with the same M, and R^1R^2 with M replaced by $2M$. Using equation (17.21.35) for R and also the estimates for J in

(17.21.47), we need to prove only estimates similar to (17.21.46), (17.21.47) with R replaced by $\Theta^{-1}_{h,h^2\lambda}$. To do this we will use again equation (17.16.2) for $\Theta^{-1}_{h,\lambda}$.

By Theorem 17.15.3, we know that $(P_\pm S_{h,\lambda} i_\pm)^{-1} \in \mathbb{E}^1_{\delta',h}$. Therefore if $\lambda \in \mathcal{W}_{\delta',h,r}$, the commutators of $\left(P_\pm S_{h,h^2\lambda} i_\pm\right)^{-1}$ with Q can be easily evaluated using the pseudodifferential calculus. So we only need to prove the relevant estimates for $S_{h,h^2\lambda}$. By Lemma 17.6.4 and by Theorem 17.7.2, the evaluation of the commutators $\left[Q, S_{h,h^2\lambda}\right]$ can be reduced to the evaluation of the commutators $[E_0, Q], [E_1, Q]$, where E_0, E_1 appear in the parametrix formula (17.10.1) in Theorem 17.10.1.

By Theorem 17.10.1, the symbol of E_0 is e_0, where e_0 was defined in Definition 17.9.3. We use equation (17.10.5) for E_1 together with Lemma 17.13.2, in order to evaluate the commutators of E_0, E_1 with $p_i, \nabla_{\widehat{e}_j}, \nabla_{e_i}$. The results on the commutators with the Q follow easily.

Now we come back to the estimation of the finite number of terms in the right-hand side of (17.21.41) such that $N_2, q < 2\ell$, which is the final point needed in the proof of the first estimate in (17.21.24). We only need to show that if $Q, Q' \in \mathcal{R}^\ell$,

$$\left\|QJ^{i_1}R^{j_1}\dots J^{i_q}R^{j_q}Q'u\right\|_{L^2} \le C\left\|u\right\|_{L^2}. \tag{17.21.53}$$

Now $N_1 \ge N - 2\ell$, and so at least one of the i_j is such that

$$i_j \ge \frac{N_1}{q} \ge \frac{N}{2\ell} - 1. \tag{17.21.54}$$

By (17.21.46), (17.21.47), we only need to check that given $l, M' \in \mathbf{N}$, for $N' \in \mathbf{N}$ large enough,

$$\left\|QJ^{N'}Q'\right\|_{h^2\lambda,\mathrm{sc},M'} \le C_{N'}\left\|u\right\|_{h^2\lambda,\mathrm{sc},-M'}, \tag{17.21.55}$$

which is itself a consequence of equation (17.20.3) in Theorem 17.20.1. The proof of our theorem is completed. □

Let $K\left((x,p),(x',p')\right)$ be the kernel of an operator K. For $\ell \in \mathbf{N}$, let $|||K|||_\ell$ be the least upper bound of the norms of the kernels $QKQ'\left((x,p),(x',p')\right)$, with $Q, Q' \in \mathcal{R}^\ell$.

By Sobolev's inequalities, if $n = \dim X$,

$$|||K|||_\ell \le C_\ell \left\|K\right\|_{\ell+3n+1}. \tag{17.21.56}$$

where the right-hand side is defined as in (17.21.22).

Proposition 17.21.4. *Take $v \in]0,1[, \ell \in \mathbf{N}$. For $N \in \mathbf{N}^*$ large enough, there exists $C_N > 0$ such that for $h \in]0, h_0], \lambda \in \mathcal{W}_{\delta',h,r}$, then*

$$\begin{aligned} &\left|\left|\left|(L_c - \lambda)^{-N}\right|\right|\right|_\ell \le C_N, \\ &\left|\left|\left|(L_c - \lambda)^{-N} - i_\pm\left(T_{h,h^2\lambda} - \lambda\right)^{-N} P_\pm\right|\right|\right|_\ell \le C_N h^v. \end{aligned} \tag{17.21.57}$$

Proof. This is an obvious consequence of (17.21.24) and of (17.21.56). □

Theorem 17.21.5. *Take $v \in]0,1[, \ell \in \mathbf{N}$. For $N \in \mathbf{N}^*$ large enough, there exists $C_N > 0$ such that for $h \in]0, h_0], \lambda \in \mathcal{W}_{\delta',h,r}$, then*

$$\left|\left|\left|(L_c - \lambda)^{-N} - i_\pm \left(\frac{1}{2}\square^X - \lambda\right)^{-N} P_\pm\right|\right|\right|_\ell \leq C_N h^v. \tag{17.21.58}$$

Proof. Set

$$A = \left(T_{h,h^2\lambda} - \lambda\right)^{-1}, \qquad B = \left(\frac{1}{2}\square^X - \lambda\right)^{-1}. \tag{17.21.59}$$

As we saw after (17.20.14), given $r > 0$, there is $C_r > 0$ such that if $\lambda \in \mathcal{W}_{\delta',h,r}$, then $\rho(\lambda) \leq C_r$.

By (17.20.2) in Theorem 17.20.1 and by (17.21.5),

$$\|(A - B)\, u\|_{\lambda,s,0} \leq C_s h \|u\|_{\lambda,s-1,0}\,. \tag{17.21.60}$$

Moreover,

$$\|Bu\|_{\lambda,s,0} \leq C \|u\|_{\lambda,s-2,0}\,, \tag{17.21.61}$$

By (17.21.60), (17.21.61), we obtain

$$\left\|\left(A^N - B^N\right) u\right\|_{\lambda,s,0} \leq C_{s,N} h \|u\|_{\lambda,s-N,0}\,. \tag{17.21.62}$$

By (17.21.57) and (17.21.62), we get (17.21.58). The proof of our theorem is completed. □

Remark 17.21.6. Recall that the function $\rho(\lambda)$ defined in (16.4.41) is associated here to $S = \mathrm{Sp}\,\square^X/2$. Let K be a compact subset of $\mathbf{C}$, not containing 0, and such that

$$K \cap \mathrm{Sp}\,\square^X/2 = \emptyset. \tag{17.21.63}$$

Then the function $\rho(\lambda)$ is bounded on S. The results of Theorem 17.20.1 are obviously valid for $\lambda \in K$, as long as $h > 0$ is small enough. It follows that the results contained in Propositions 17.21.1, 17.21.2, in Theorem 17.21.3, in Proposition 17.21.4, and in Theorem 17.21.5 are also valid when $\lambda \in K$. In particular, by Theorem 17.21.3, we find that for $h > 0$ small enough,

$$\mathrm{Sp}\, L_c \cap K = \emptyset. \tag{17.21.64}$$

17.22 A LOCALIZATION PROPERTY

Observe that by (13.2.4), for $\varepsilon \in]0,1[, h > 0$, for $c = \pm 1/h^2$,

$$\epsilon^2 L_{c/\epsilon^2} = \frac{\alpha_\pm}{h^2} + \frac{\epsilon\beta_\pm}{h} + \epsilon^2\gamma_\pm. \tag{17.22.1}$$

Comparing with (13.2.4), we see that $\varepsilon^2 L_{c/\epsilon^2}$ is obtained from L_c by scaling $\beta_\pm, \gamma_\pm$ by the factors ϵ, ϵ^2 respectively.

Let us denote explicitly the dependence of L_c on the metric g^{TX}, i.e., we write $L_c^{g^{TX}}$ instead of L_c. Then one verifies easily that up to a trivial conjugation,

$$L_c^{g^{TX}/\varepsilon^2} = \varepsilon^2 L^{g^{TX}}_{c/\varepsilon^2}. \tag{17.22.2}$$

Incidentally observe (17.22.2) is also a consequence of the identities in (2.8.8) in degree 0.

In the sequel, it will often be convenient to use the notation

$$h = \epsilon b, \qquad c = \pm 1/b^2. \tag{17.22.3}$$

Also for notational convenience, we set $b_0 = h_0$.

By (17.21.1), (17.21.2), we have the formal equality

$$\left(\epsilon^2 L_{c/\epsilon^2} - \lambda\right)^{-1} = \begin{bmatrix} \epsilon^{-2} R_{\epsilon b,\lambda/\epsilon^2} & -b\epsilon^{-2} R_{\epsilon b,\lambda/\epsilon^2} \epsilon L_2 \Theta^{-1}_{\epsilon b,b^2\lambda} \\ -b\Theta^{-1}_{\epsilon b,b^2\lambda} \epsilon L_3 \epsilon^{-2} R_{\epsilon b,\lambda/\epsilon^2} & b^2\left(\Theta^{-1}_{\epsilon b,b^2\lambda} + \Theta^{-1}_{\epsilon b,b^2\lambda} L_3 R_{\epsilon b,\lambda/\epsilon^2} L_2 \Theta^{-1}_{\epsilon b,b^2\lambda}\right)\end{bmatrix}. \tag{17.22.4}$$

Proposition 17.22.1. *Let $v \in]0,1[$. For any $\ell \in \mathbf{N}$, for $N \in \mathbf{N}^*$ large enough, there exists $C_N > 0$ such that for $b \in]0,b_0], \epsilon \in]0,1], \lambda \in \mathcal{W}_{\delta',b,r}$, then*

$$\left|\left|\left|\left(\epsilon^2 L_{c/\epsilon} - \lambda\right)^{-N} - i_\pm \left(\frac{\epsilon^2}{2}\square^X - \lambda\right)^{-N} P_\pm \right|\right|\right|_\ell \le C\epsilon^{-N} h^v. \tag{17.22.5}$$

Proof. Note that if $\lambda \in \mathcal{W}_{\delta',b,r}$, then $\lambda/\epsilon^2 \in \mathcal{W}_{\delta',h,r}$. Our proposition now follows from Theorem 17.21.5. □

Given $\ell \in \mathbf{N}$, we define the family of operators $\mathcal{Q}^\ell_\epsilon$ as the family $\mathcal{Q}^\ell$, by simply replacing the ∇_{e_i} by $\epsilon\nabla_{e_i}$. We define corresponding norms $\| \|_{\epsilon,\ell}$ as in (17.21.21).

Theorem 17.22.2. *Let K be a compact subset of X, and let $\phi(x) \in C_0^\infty(X \setminus K)$. Given $N \in \mathbf{N}, M \in \mathbf{N}$, there exists $b_0 \in]0,1], \epsilon_0 \in]0,1], C_{M,N} > 0$ such that for $b \in]0,b_0], \epsilon \in]0,\epsilon_0], \lambda \in \mathcal{W}_{\delta',b,r}$, if the support of u is included in $\pi^{-1}(K)$, then*

$$\left\|\phi(x)\left(\epsilon^2 L_{c/\epsilon^2} - \lambda\right)^{-1} u\right\|_M \le C_{M,N}\epsilon^N \|u\|_{-M}. \tag{17.22.6}$$

Proof. For the moment, we only consider the parameter h as in the previous sections, and we take $\lambda \in \mathcal{W}_{\delta',h,r}$. Take $\phi_1(x), \phi_2(x) \in C_0^\infty(X)$ with disjoint supports. By Lemma 17.6.4 and by Theorems 17.7.2, 17.15.3, and 17.16.2, we find that for any $s \in \mathbf{R}, t \in \mathbf{R}, N \in \mathbf{N}$, there exists $C > 0$ such that

$$\left\|\phi_1 \Theta^{-1}_{h,h^2\lambda} \phi_2 u\right\|_{h^2\lambda,\mathrm{sc},t} \le C_{s,t,N} h^N \|u\|_{h^2\lambda,\mathrm{sc},s}. \tag{17.22.7}$$

Since s, t, N are arbitrary, from (17.22.7), we deduce that for any $M \in \mathbf{N}, N \in \mathbf{N}$, there exists $C_{M,N} > 0$ such that

$$\left\| \phi_1 \Theta_{h,h^2\lambda}^{-1} \phi_2 u \right\|_M \leq C_{M,N} h^N \|u\|_{-M} . \tag{17.22.8}$$

Moreover, by commuting $\Theta_{\epsilon b, b^2\lambda}^{-1}$ with operators in $\mathcal{Q}^\ell$ as in the proof of Theorem 17.21.3, from (17.21.46), (17.21.47), we find that given $\ell \in \mathbf{Z}$, there exists $\ell' \in \mathbf{Z}, C > 0$ such that

$$\left\| \Theta_{h,h^2\lambda}^{-1} u \right\|_\ell \leq C \|u\|_{\ell'} . \tag{17.22.9}$$

We claim that if $M \in \mathbf{N}, N \in \mathbf{N}$, there exists $C_{M,N} > 0$ such that under the conditions stated in our theorem,

$$\left\| \phi_1 i_\pm \left(\epsilon^2 T_{\epsilon b, b^2\lambda} - \lambda \right)^{-1} P_\pm \phi_2 u \right\|_M \leq C_{M,N} \epsilon^N \|u\|_{-M} . \tag{17.22.10}$$

By (17.21.56), (17.22.4), (17.22.8)-(17.22.10), we get (17.22.6), i.e., we get a proof of Theorem 17.22.2.

Now we concentrate on the proof of (17.22.10). If $\theta = (\theta_1(x), \dots, \theta_N(x))$ is a family of smooth real functions on X, we use the notation $\mathrm{Ad}_\theta \mathfrak{T}$ as in (17.6.11).

Assume that $\theta_1, \dots, \theta_N$ are smooth real functions which are equal to 1 on the support of φ_1 and equal to 0 on the support of φ_2. Then if $\lambda \in \mathcal{W}_{\delta', h, r}$,

$$\varphi_1 i_\pm \left(T_{h,\lambda} - \lambda \right)^{-1} P_\pm \varphi_2 = i_\pm \varphi_1 \mathrm{Ad}_\theta^N \left(T_{h,\lambda} - \lambda \right)^{-1} \phi_2 P_\pm . \tag{17.22.11}$$

Also we have the identity

$$\begin{aligned} &\mathrm{Ad}_\theta^N \left(T_{h,\lambda} - \lambda \right)^{-1} \\ &\quad = (-1)^N \sum \left(T_{h,\lambda} - \lambda \right)^{-1} \mathrm{Ad}_\theta^{i_1} \left(T_{h,\lambda} \right) \left(T_{h,\lambda} - \lambda \right)^{-1} \dots \\ &\qquad\qquad \mathrm{Ad}_\theta^{i_p} \left(T_{h,\lambda} \right) \left(T_{h,\lambda} - \lambda \right)^{-1} , \end{aligned} \tag{17.22.12}$$

with $i_j \geq 1, i_1 \dots + i_p = N$.

We claim that for $i \geq 1$,

$$\mathrm{Ad}_\theta^i T_{h,\lambda} \in \sum_{1 \leq i \leq n} \mathbb{E}_{\delta',h}^{-2/3} \nabla_{e_i} + \mathbb{E}_{\delta',h}^{-2/3} . \tag{17.22.13}$$

To establish (17.22.13), we use equation (17.17.10) for $T_{h,\lambda}$. Observe that in the right-hand side of (17.17.10), $P_\pm \gamma_\pm P_\pm$ does not contribute to the commutator in (17.22.13). By Theorem 17.17.4, $\mathbb{A}_{h,\lambda} \in \mathbb{E}_{\delta,h}^{-1}$. By (17.15.5) and (17.15.9), we find that

$$[\theta_1, \mathbb{A}_{h,\lambda}] \in h \mathbb{E}_{\delta',h}^{-5/3} . \tag{17.22.14}$$

By (17.22.14), we get

$$\nabla^{\Lambda^\cdot (T^*X) \widehat{\otimes} F, u, *} [\theta_1, \mathbb{A}_{h,\lambda}] \in h \mathbb{E}_{\delta',h}^{-2/3} . \tag{17.22.15}$$

The other commutators with θ_1 in the right-hand side of (17.17.10) can be handled in the same way. A recursion argument then leads to the proof of (17.22.13).

Observe that given $r > 0$, for $h \in]0,1], y \in \mathbf{R}, \lambda \in \mathcal{W}_{\delta',h,r}$,

$$\left|\lambda - y^2\right| \geq C\left(1 + |\lambda| + y^2\right). \tag{17.22.16}$$

By (17.22.16), if $\lambda \in \mathcal{W}_{\delta',h,r}$,

$$\left\|\left(\frac{\square^X}{2} - \lambda\right)^{-1} u\right\|_{\lambda,s,0} \leq C_s \|u\|_{\lambda,s-2,0}. \tag{17.22.17}$$

By equation (17.20.2) in Theorem 17.20.1, by (17.21.5), and by (17.22.17), for $\lambda \in \mathcal{W}_{\delta',h,r}$, we get

$$\left\|\left(T_{h,h^2\lambda} - \lambda\right)^{-1} u\right\|_{\lambda,s,0} \leq C_s\left(\|u\|_{\lambda,s-2,0} + h\|u\|_{\lambda,s-1,0}\right). \tag{17.22.18}$$

So using equation (17.21.7) in Proposition 17.21.1, (17.22.13), and (17.22.18), we get

$$\left\|\left(T_{h,h^2\lambda} - \lambda\right)^{-1}\left(\mathrm{Ad}^i_\theta T_{h,h^2\lambda}\right) u\right\|_{\lambda,s,0} \leq C_s\Bigg(\left\|\left(\mathrm{Ad}^i_\theta T_{h,h^2\lambda}\right) u\right\|_{\lambda,s-2,0}$$
$$+ h\left\|\left(\mathrm{Ad}^i_\theta T_{h,h^2\lambda}\right)\right\|_{\lambda,s-1,0}\Bigg) \leq C_s\left(\|u\|_{\lambda,s-1,-2/3} + h\|u\|_{\lambda,s,-2/3}\right). \tag{17.22.19}$$

Now we use the notation in (17.21.3). Clearly,

$$\Lambda \leq \frac{\Lambda'_h}{h}, \tag{17.22.20}$$

so that

$$\|u\|_{s,-2/3} \leq h^{-2/3}\|u\|_{s-2/3,0}. \tag{17.22.21}$$

By (17.22.19), (17.22.21), we obtain

$$\left\|\left(T_{h,h^2\lambda} - \lambda\right)^{-1} \mathrm{Ad}^i_\theta T_{h,h^2\lambda} u\right\|_{s,0} \leq C_s\|u\|_{s-2/3,0}. \tag{17.22.22}$$

Then inequality (17.22.10) with $\epsilon = 1$ follows from (17.22.11), (17.22.12) and (17.22.22).

Recall that by (17.22.3), $h = \epsilon b$. Take $\lambda \in \mathcal{W}_{\delta',b,r}$. Set $\mu = \lambda/\epsilon^2 \in \mathcal{W}_{\delta',h,r}$. Then

$$\left(\epsilon^2 T_{h,h^2\mu} - \lambda\right)^{-1} = \epsilon^{-2}\left(T_{h,h^2\mu} - \mu\right)^{-1}. \tag{17.22.23}$$

By (17.22.10) with $\epsilon = 1$, we get

$$\left\|\left(1 + |\mu| + \square^X/2\right)^{M/2} \phi_1 i_\pm \left(T_{h,h^2\mu} - \mu\right)^{-1} P_\pm \phi_2 u\right\|_{L^2}$$
$$\leq C_M \left\|\left(1 + |\mu| + \square^X/2\right)^{-M/2} u\right\|_{L^2}. \tag{17.22.24}$$

By (17.22.24), we obtain

$$\left\| \left(\epsilon^2 + |\lambda| + \epsilon^2 \square^X/2\right)^{M/2} \phi_1 i_\pm \left(\epsilon^2 T_{h,h^2\mu} - \lambda\right)^{-1} P_\pm \phi_2 u \right\|_{L^2} \\ \leq C_M \epsilon^{2M-2} \left\| \left(\epsilon^2 + |\lambda| + \epsilon^2 \square^X/2\right)^{-M/2} u \right\|_{L^2}. \quad (17.22.25)$$

Moreover, there exists $c > 0$ such that for $b \in]0, b_0], \lambda \in \mathcal{W}_{\delta',b,r}$, then $|\lambda| \geq c$. By (17.22.25) we obtain

$$\left\| \left(1 + |\lambda| + \epsilon^2 \square^X/2\right)^{M/2} \phi_1 i_\pm \left(\epsilon^2 T_{h,h^2\mu} - \mu\right)^{-1} P_\pm \phi_2 u \right\|_{L^2} \leq \\ C_M \epsilon^{2M-2} \left\| \left(1 + |\lambda| + \epsilon^2 \square^X/2\right)^{-M/2} u \right\|_{L^2}. \quad (17.22.26)$$

By (17.22.26), if $\lambda \in \mathcal{W}_{\delta',b,r}$, we get

$$\left\| \phi_1 \left(\epsilon^2 T_{h,h^2\mu} - \lambda\right)^{-1} P_\pm \phi_2 u \right\|_{\epsilon,M} \leq C_M \epsilon^{2M-2} \|u\|_{\epsilon,-M}. \quad (17.22.27)$$

By (17.22.27), we get (17.22.10).

This completes the proof of our theorem. □

Remark 17.22.3. Given $N \in \mathbf{N}^*$, it is possible to replace $\left(\epsilon^2 L_{c/\epsilon^2 - \lambda}\right)^{-1}$ by $\left(\epsilon^2 L_{c/\epsilon^2 - \lambda}\right)^{-N}$. We still get the obvious analogue of Theorem 17.22.2. Indeed by proceeding as in the proof of (17.22.9), we find that given $\ell \in \mathbf{Z}$, there exists $\ell' \in \mathbf{Z}, C$ such that

$$\left\| \left(\epsilon^2 L_{c/\epsilon^2 - \lambda}\right)^{-1} u \right\|_\ell \leq C \|u\|_{\ell'}. \quad (17.22.28)$$

By combining (17.22.6) in Theorem 17.22.2 with (17.22.28), we get the corresponding statement for $\left(\epsilon^2 L_{c/\epsilon^2 - \lambda}\right)^{-N}$.

Bibliography

[ABP73] M. Atiyah, R. Bott, and V. K. Patodi. On the heat equation and the index theorem. *Invent. Math.*, 19:279–330, 1973.

[BeB94] A. Berthomieu and J.-M. Bismut. Quillen metrics and higher analytic torsion forms. *J. Reine Angew. Math.*, 457:85–184, 1994.

[B81a] J.-M. Bismut. Martingales, the Malliavin calculus and Hörmander's theorem. In *Stochastic integrals (Proc. Sympos., Univ. Durham, Durham, 1980)*, volume 851 of *Lecture Notes in Math.*, pages 85–109. Springer, Berlin, 1981.

[B81b] J.-M. Bismut. Martingales, the Malliavin calculus and hypoellipticity under general Hörmander's conditions. *Z. Wahrsch. Verw. Gebiete*, 56(4):469–505, 1981.

[B84] J.-M. Bismut. *Large deviations and the Malliavin calculus.* Birkhäuser Boston, Boston, MA, 1984.

[B86] J.-M. Bismut. The Atiyah-Singer index theorem for families of Dirac operators: two heat equation proofs. *Invent. Math.*, 83(1):91–151, 1986.

[B90] J.-M. Bismut. Koszul complexes, harmonic oscillators, and the Todd class. *J. Amer. Math. Soc.*, 3(1):159–256, 1990. With an appendix by the author and C. Soulé.

[B94] J.-M. Bismut. Equivariant short exact sequences of vector bundles and their analytic torsion forms. *Compositio Math.*, 93(3):291–354, 1994.

[B95] J.-M. Bismut. Equivariant immersions and Quillen metrics. *J. Differential Geom.*, 41(1):53–157, 1995.

[B97] J.-M. Bismut. Holomorphic families of immersions and higher analytic torsion forms. *Astérisque*, (244):viii+275, 1997.

[B04] J.-M. Bismut. Le Laplacien hypoelliptique. In *Séminaire: Équations aux Dérivées Partielles, 2003–2004*, Sémin. Équ. Dériv. Partielles, pages Exp. No. XXII, 15. École Polytech., Palaiseau, 2004.

[B05] J.-M. Bismut. The hypoelliptic Laplacian on the cotangent bundle. *J. Amer. Math. Soc.*, 18(2):379–476 (electronic), 2005.

[B06] J-.M. Bismut. Loop spaces and the hypoelliptic Laplacian. *Comm. Pure Appl. Math.*, 61(4):559–593, 2008.

[BG01] J.-M. Bismut and S. Goette. Families torsion and Morse functions. *Astérisque*, (275):x+293, 2001.

[BG04] J.-M. Bismut and S. Goette. Equivariant de Rham torsions. *Ann. Math.*, 159:53–216, 2004.

[BL91] J.-M. Bismut and G. Lebeau. Complex immersions and Quillen metrics. *Inst. Hautes Études Sci. Publ. Math.*, (74):ii+298 pp. (1992), 1991.

[BL05] J.-M. Bismut and G. Lebeau. Laplacien hypoelliptique et torsion analytique. *C. R. Math. Acad. Sci. Paris*, 341(2):113–118, 2005.

[BLo95] J.-M. Bismut and J. Lott. Flat vector bundles, direct images and higher real analytic torsion. *J. Amer. Math. Soc.*, 8(2):291–363, 1995.

[BZ92] J.-M. Bismut and W. Zhang. An extension of a theorem by Cheeger and Müller. *Astérisque*, (205):235, 1992. With an appendix by François Laudenbach.

[BZ94] J.-M. Bismut and W. Zhang. Milnor and Ray-Singer metrics on the equivariant determinant of a flat vector bundle. *Geom. Funct. Anal.*, 4(2):136–212, 1994.

[C79] J. Cheeger. Analytic torsion and the heat equation. *Ann. Math. (2)*, 109(2):259–322, 1979.

[F86] D. Fried. The zeta functions of Ruelle and Selberg. I. *Ann. Sci. École Norm. Sup. (4)*, 19(4):491–517, 1986.

[F88] D. Fried. Torsion and closed geodesics on complex hyperbolic manifolds. *Invent. Math.*, 91(1):31–51, 1988.

[G86] E. Getzler. A short proof of the local Atiyah-Singer index theorem. *Topology*, 25(1):111–117, 1986.

[Gi84] P. B. Gilkey. *Invariance theory, the heat equation, and the Atiyah-Singer index theorem*, volume 11 of *Mathematics Lecture Series*. Publish or Perish, Wilmington, DE, 1984.

[Ha79] U. G. Haussmann. On the integral representation of functionals of Itô processes. *Stochastics*, 3(1):17–27, 1979.

[HeN05] B. Helffer and F. Nier. *Hypoelliptic estimates and spectral theory for Fokker-Planck operators and Witten Laplacians*, volume 1862 of *Lecture Notes in Mathematics*. Springer-Verlag, Berlin, 2005.

[HeSj85] B. Helffer and J. Sjöstrand. Puits multiples en mécanique semi-classique. IV. Étude du complexe de Witten. *Comm. Partial Differential Equations*, 10(3):245–340, 1985.

[HN04] F. Hérau and F. Nier. Isotropic hypoellipticity and trend to equilibrium for the Fokker-Planck equation with a high-degree potential. *Arch. Ration. Mech. Anal.*, 171(2):151–218, 2004.

[Hör67] L. Hörmander. Hypoelliptic second order differential equations. *Acta Math.*, 119:147–171, 1967.

[Hör85] L. Hörmander. *The analysis of linear partial differential operators. III.* Pseudodifferential operators. Volume 274 of Die Grundlehren der mathematischen Wissenschaften. Springer-Verlag, Berlin, 1985.

[IM74] K. Itô and H. P. McKean Jr. *Diffusion processes and their sample paths.* Second printing, corrected, volume 125 of Die Grundlehren der mathematischen Wissenschaften. Springer-Verlag, Berlin, 1974.

[KMu76] F. F. Knudsen and D. Mumford. The projectivity of the moduli space of stable curves. I. Preliminaries on "det" and "Div". *Math. Scand.*, 39(1):19–55, 1976.

[Ko73] J. J. Kohn. Pseudo-differential operators and hypoellipticity. In *Partial differential equations (Proc. Sympos. Pure Math., Vol. XXIII, Univ. California, Berkeley, 1971)*, pages 61–69. Amer. Math. Soc., Providence, R.I., 1973.

[Kol34] A. Kolmogoroff. Zufällige Bewegungen (zur Theorie der Brownschen Bewegung). *Ann. Math. (2)*, 35(1):116–117, 1934.

[L05] G. Lebeau. Geometric Fokker-Planck equations. *Port. Math.*, 62(4), 2005.

[L06] G. Lebeau. Equations de Fokker-Planck géométriques ii:: Estimations hypoelliptiques maximales. *Ann. Inst. Fourier (Grenoble)*, 2006.

[Le88] M. Lerch. Note sur la fonction $\mathfrak{r}(w,x,s) = \sum_0^\infty \frac{e^{2i\pi kx}}{(w+k)^s}$. *Acta Math.*, 11:19–24, 1887-1888.

[M78] P. Malliavin. Stochastic calculus of variation and hypoelliptic operators. In *Proceedings of the International Symposium on Stochastic Differential Equations (Res. Inst. Math. Sci., Kyoto Univ., Kyoto, 1976)*, pages 195–263. Wiley, New York, 1978.

[MatQ86] V. Mathai and D. Quillen. Superconnections, Thom classes, and equivariant differential forms. *Topology*, 25(1):85–110, 1986.

[MaMe90] R. R. Mazzeo and R. B. Melrose. The adiabatic limit, Hodge cohomology and Leray's spectral sequence for a fibration. *J. Differential Geom.*, 31(1):185–213, 1990.

[MoSta91] H. Moscovici and R. J. Stanton. R-torsion and zeta functions for locally symmetric manifolds. *Invent. Math.*, 105(1):185–216, 1991.

[Mül78] W. Müller. Analytic torsion and R-torsion of Riemannian manifolds. *Adv. Math.*, 28(3):233–305, 1978.

[P71] V. K. Patodi. Curvature and the eigenforms of the Laplace operator. *J. Differential Geometry*, 5:233–249, 1971.

[Q85a] D. Quillen. Determinants of Cauchy-Riemann operators on Riemann surfaces. *Functional Anal. Appl.*, 19(1):31–34, 1985.

[Q85b] D. Quillen. Superconnections and the Chern character. *Topology*, 24(1):89–95, 1985.

[RS71] D. B. Ray and I. M. Singer. R-torsion and the Laplacian on Riemannian manifolds. *Adv. Math.*, 7:145–210, 1971.

[ReSi78] M. Reed and B. Simon. *Methods of modern mathematical physics. IV. Analysis of operators.* Academic Press [Harcourt Brace Jovanovich Publishers], New York, 1978.

[Re35] K. Reidemeister. Homotopieringe und Linsenraüm. *Hamburger Abhandl.*, pages 102–109, 1935.

[Sm61] S. Smale. On gradient dynamical systems. *Ann. Math. (2)*, 74:199–206, 1961.

[St81a] D. W. Stroock. The Malliavin calculus, a functional analytic approach. *J. Funct. Anal.*, 44(2):212–257, 1981.

[St81b] D. W. Stroock. The Malliavin calculus and its applications. In *Stochastic integrals (Proc. Sympos., Univ. Durham, Durham, 1980)*, volume 851 of *Lecture Notes in Math.*, pages 394–432. Springer, Berlin, 1981.

[StV72] D.W. Stroock and S. R. S. Varadhan. On the support of diffusion processes with applications to the strong maximum principle. In *Proc. of the Sixth Berkeley Symp. on Math. Stat. and Prob. (Univ. California, Berkeley, 1970/1971), Vol. III: Probability theory*, pages 333–359. Univ. California Press, Berkeley, 1972.

[T49] R. Thom. Sur une partition en cellules associée à une fonction sur une variété. *C. R. Acad. Sci. Paris*, 228:973–975, 1949.

[W76] A. Weil. *Elliptic functions according to Eisenstein and Kronecker*. Volume 88 of Ergebnisse der Mathematik und ihrer Grenzgebiete. Springer-Verlag, Berlin, 1976.

[Wi82] E. Witten. Supersymmetry and Morse theory. *J. Differential Geom.*, 17(4):661–692 (1983), 1982.

[Wi85] E. Witten. Global anomalies in string theory. In *Symposium on anomalies, geometry, topology (Chicago, Ill., 1985)*, pages 61–99. World Sci. Publishing, Singapore, 1985.

[Y68] K. Yosida. *Functional analysis.* Second edition. Volume 123 of Die Grundlehren der mathematischen Wissenschaften. Springer-Verlag New York, New York, 1968.

Subject Index

analytic torsion forms, 20

Bargman kernel, 248
Berezin integral, 63
Brownian motion, 72, 215

characteristic spaces, 46
Chern analytic torsion forms, 22
Chern hypoelliptic torsion forms, 114

eigenvalues, 46
elliptic odd Chern forms, 18
equivariant determinant, 24, 123

Feynman-Kac formula, 72
Fredholm determinants, 140

Gauss-Manin connection, 18
generalized Ray-Singer metric, 119
Girsanov formula, 216

harmonic oscillator, 49, 111, 203, 228, 251
Heisenberg algebra, 247
Hodge theory, 13
Hodge type, 58
Horizontal Laplacian, 13
hypoelliptic torsion forms, 114
hypoelliptic estimates, 229
hypoelliptic odd Chern forms, 66

integration by parts, 216
Itô calculus, 72, 222
Itô differential, 215
Itô integral, 73
Itô's formula, 221

Lefschetz formula, 16
Levi-Civita superconnection, 15
Littlewood-Paley decomposition, 226

Malliavin calculus, 73
Malliavin covariance matrix, 217

normal ordering, 249
number operator, 18, 41, 225

parametrix, 306
Pfaffian, 63
Poisson bracket, 271
pseudodifferential operators, 231

Ray-Singer metric, 23

secondary classes, 22
semiclassical Poisson bracket, 271
semiclassical Sobolev norms, 271
stochastic differential equation, 72, 215, 216
Stratonovitch differential, 215
superconnection, 15
supertrace, 12
symbol, 231

Thom form, 31
Thom isomorphism, 45

von Neumann supertrace, 137

Weitzenböck formula, 15, 29
Weitzenböck formula, 14

Index of Notation

A, 15, 334
$\mathcal{A}$, 79
a, 65
$\|A\|_1$, 51, 243
a_α^β, 261
$A_{b,t}^{\mathcal{M}}$, 41
$A_{\phi_b,\mathcal{H}}$, 29
$\mathbb{A}_{h,\lambda}$, 327
$\|A\|_\ell$, 343
$\alpha_\pm$, 32
$\alpha^{\max}$, 91
$\mathfrak{A}'_{\phi_b,\mathcal{H}}$, 29
$A_{\phi,\mathcal{H}}$, 28
$\mathfrak{A}_{\phi,\mathcal{H}}$, 28
$A_{\phi,\mathcal{H}-\omega^H}^{\mathcal{M}}$, 34
$A'^{\mathcal{M}}$, 34
$\overline{A}'^{\mathcal{M}}$, 34
$\mathfrak{a}_\pm$, 30
$\overline{A}_\phi'^{\mathcal{M}}$, 34
$\widehat{A}_\dagger'^{\mathcal{M}}\theta$, 39
$\alpha_{\pm,\tau}$, 229
$\mathfrak{A}'_{\phi,\mathcal{H}}$, 28
$\alpha'_\pm$, 207
$a^*(V)$, 247
A_t, 18
$A_\tau(\xi,\lambda)$, 284
$a(U)$, 247
a_X, 71
$\mathfrak{a}(\xi,\lambda)$, 328
$A(\xi,p,\lambda)$, 284

B, 15
$\mathcal{B}$, 227
$\mathcal{B}_0$, 227
B, 248
$B_{b,t}^{\mathcal{M}}$, 41
$\beta_\pm$, 32
$B_{\phi_b,\mathcal{H}}$, 29
$\mathbb{B}_{h,\lambda}$, 327
$\beta'_\pm$, 40
$\mathfrak{B}'_{\phi_b,\mathcal{H}}$, 29
$B_{\phi,\mathcal{H}}$, 28
$\mathfrak{B}_{\phi,\mathcal{H}}$, 28
$\mathbb{B}'_{h,\lambda}$, 327
$B_{\phi,\mathcal{H}-\omega^H}^{\mathcal{M}}$, 34
$\mathfrak{b}_\pm$, 30
$\beta_{\pm,\tau}$, 229
$\mathfrak{B}'_{\phi,\mathcal{H}}$, 28
$\overline{\mathbb{B}}'_{\xi,\lambda}$, 336
B_t, 18
b_t, 20
b_t^F, 100
$B^{T_xX}(0,\epsilon)$, 76
$B_\tau(\xi,\lambda)$, 284
$B(V^*)$, 247
$B(\xi)$, 249
$B^X(x,\epsilon)$, 76
$\overline{\mathbb{B}}_{\xi,\lambda}$, 336
$B(\xi,p,\lambda)$, 284

$c_1(\rho)$, 295
$c_{\alpha,\beta}$, 312
$\xi_G(F)$, 16
$\mathcal{C}_g^\infty([0,1],E)$, 140
$\widetilde{\mathrm{ch}}_g^\circ(\nabla^F, g_0^F, g_1^F)$, 22
$\chi(c,g,A)$, 141
$\mathrm{ch}_g^\circ(\nabla^F, g^F)$, 21
$\overline{\chi}_g(F)$, 64
$\chi_j(r)$, 226
$\chi'_g(F)$, 19
$\overline{\chi}'_g(F)$, 64
$\chi(r)$, 226
$\mathbb{C}_{h,\lambda}$, 327
C_j, 227
$\mathfrak{C}_{\phi_b,\mathcal{H}-b\omega^H,t}^{\mathcal{M}}$, 42
$\underline{\widehat{\mathfrak{C}}}_{\phi,\mathcal{H}^c-\omega^H}^{\mathcal{M}^E,2}$, 134
$\mathfrak{C}_{\phi_b,\mathcal{H}-b\omega^H}^{\mathcal{M}}$, 35
$\underline{\widehat{\mathfrak{C}}}_{\phi,\mathcal{H}^c-\omega^H}^{\mathcal{M}}$, 38
$c'_{\alpha,\beta}$, 312
$\mathfrak{C}_{\phi,\mathcal{H}-\omega^H}^{\mathcal{M}}$, 34
$\mathfrak{C}'^{\mathcal{M}}_{\mathcal{H}-\omega^H}$, 35
C_t, 18
c_t, 100
$\chi_\tau(p)$, 300
$\widehat{c}(U)$, 12
$c(U)$, 12
$c(V)$, 11

$\underline{\chi}_W$, 23
$\overline{\mathbb{C}}_{\xi,\lambda}$, 336
$\mathcal{C}_z([0,1],\mathbf{C})$, 132

d, 102
D_4, 269
$\mathcal{D}^{a,z}_{c,t,\lambda}$, 210
$\mathcal{D}^{a,z,N}_{c,t,\lambda}$, 211
Δ^V, 29
det (E), 23
$D_{\phi,\mathcal{H}^c}$, 117
Δ^H, 13
$D^0_h(x,\xi)$, 290
$\Delta^{H,u}$, 14, 226
$\Delta^{H,V}$, 87
$\delta_j(u)$, 226
D_μ, 251
$\mathfrak{D}^{\mathcal{M}}_{\phi_b,\mathcal{H}-b\omega^H,t}$, 42
$\mathfrak{D}^{\mathcal{M}}_{\phi_b,\mathcal{H}-b\omega^H}$, 35
$\underline{\widehat{\mathfrak{D}}}^{\mathcal{M}}_{\phi,\mathcal{H}^c-\omega^H}$, 38
dp, 78
$\mathfrak{D}^{\mathcal{M}}_{\phi,\mathcal{H}-\omega^H}$, 34
$\delta_\pm$, 46
$\mathcal{D}'_X$, 308
$D_{s,j}(U_j)$, 296
$D_s(L_c)$, 238
$D_s(P_c)$, 238
D_t, 18
$\widehat{d}^{T^*X}$, 48
$\widehat{d}^{T^*X}_{\mathcal{H}^c}$, 48
$d^{T^*X'}_{\phi,\mathcal{H}}$, 28
$\overline{d}^{T^*X'}_{\phi,\mathcal{H}}$, 28
$\widehat{d}^{T^*X*}$, 48
$\overline{d}^{T^*X}_{\phi}$, 27
$d^{T^*X'}_{\phi_b,\mathcal{H}}$, 29
$\overline{d}^{T^*X}_{\phi,\mathcal{H}}$, 27
$d^{T^*X}_{\mathcal{H}}$, 27
$dv_{N_{X_g/X}}$, 71
$dv_{\mathbf{P}(T^*X\oplus\mathbf{R})}$, 228
dv_{T^*X}, 25
dv_X, 12
dv_{X_g}, 71
D^X, 12
Δ^X, 215
$\mathcal{D}_X$, 308
d^X, 12
δ_X, 175
d_X, 71
d^{X*}, 12
Δ^Y, 228
dy, 78

E, 41
$E_{-1/2}$, 234
E_0, 306
e_0, 306
e^0_0, 321
$e^0_{0,\lambda}(x,\xi)$, 321
$e^{\alpha,\beta}_0$, 314
$e_{0,h,\lambda}(x,\xi)$, 295
E_1, 306
$:e^{\langle\alpha,p\rangle}:$, 249
$\mathcal{E}^d$, 231, 274
$\mathcal{E}^d$, 297
$\mathbb{E}^d_{\delta,h}$, 320
$\mathbb{E}^d_{\delta,h,0}$, 320
$\mathcal{E}^d_\lambda$, 242
$e(E)$, 64
$e(E,\nabla^E)$, 64
$\mathfrak{E}^{\mathcal{M}}_{\phi,\mathcal{H}-\omega^H}$, 34
$\mathfrak{e}^i$, 79
$\mathfrak{e}_i$, 79
$\widehat{\mathfrak{e}}_i$, 79
ℓ, 16
$\mathfrak{E}^{\mathcal{M}}_{\phi_b,\mathcal{H}-b\omega^H,t}$, 42
$\mathfrak{E}^{\mathcal{M}}_{\phi_b,\mathcal{H}-b\omega^H}$, 35
E^P, 72
$\epsilon\left(|\,|^2_\lambda\right)$, 117
$E^{\cdot}_r$, 59
$E^{S(y_0,p_0)}$, 138
η_0, 71
$e_\tau(p)$, 298
$\eta(\theta,s)$, 155
$e\left(TX_g,\nabla^{TX_g}\right)$, 17
$e(TX_g)$, 16, 17
$\epsilon\left(\|\,\|^2_\lambda\right)$, 119
$\exp\left(-\underline{\widehat{\mathfrak{M}}}_{b,t}\right)(z,z')$, 70

F, 12, 26
$\mathfrak{f}$, 26
$\mathfrak{f}^{TT^*X}$, 27
f, 26
$f^{(>0)}(x)$, 156
F_b, 28
$\mathfrak{f}_b$, 28
f_b, 28
$Ff(x)$, 21
$\mathfrak{F}^{\mathcal{M}}_{\phi,\mathcal{H}-\omega^H}$, 34
$\mathfrak{L}^{E'}_c$, 138
$\mathfrak{F}^{\mathcal{M}}_{\phi_b,\mathcal{H}-b\omega^H,t}$, 42
$\mathfrak{M}^E_c$, 152
$\mathfrak{F}^{\mathcal{M}}_{\phi_b,\mathcal{H}-b\omega^H}$, 35
$\mathfrak{F}^{\mathcal{M},(i)}_{\phi_b,\pm\mathcal{H}-b\omega^H}$, 104
$\mathfrak{f}^{\Omega^{\cdot}(T^*X,\pi^*F)}$, 27
$\phi(r)$, 226

$\Phi(t)$, 254

Γ, 164
γ, 46, 195
γ_b, 53
$\gamma_\pm$, 32
g^F, 12
$\widehat{g}$, 23
$g^{H^\cdot(X,F)}$, 13
$\widehat{\mathfrak{G}}^{\mathcal{M}^E}_{\phi,\mathcal{H}^c-\omega^H}$, 134
$\mathfrak{G}^{\mathcal{M}}_{\phi,\mathcal{H}^c-\omega^H}$, 36
$\mathbf{g}^{\Omega^\cdot(T^*X,\pi^*F)}$, 27
$g^{\Omega^\cdot(T^*X,\pi^*F)}$, 28
$g^{\Omega^\cdot(X,F)}$, 12
$g_t^{\Omega^\cdot(X,F)}$, 18
$\gamma'_\pm$, 40
$\gamma_{\pm,\tau}$, 229
$\gamma(s)$, 76
$g^{T\mathbf{P}(T^*X\oplus\mathbf{R})}$, 228
Γ^{T^*X}, 283
$\mathbf{g}^{TT^*X}$, 26
g^{TT^*X}, 28
g^{TX}, 12
g^{TX_g}, 16
g_t^{TX}, 18, 41
$g_x^{F_x}$, 77
$g_x^{T_xX}$, 77

H, 32, 226, 269
$\mathcal{H}$, 29, 228
H, 228
$\mathcal{H}_a$, 29
$\mathbb{H}_b^\cdot(X,F)$, 99
$\mathcal{H}^c$, 29
$H^{c,\cdot}(T^*X,\pi^*F)$, 44
$h_g\left(A', g^{\Omega^\cdot(X,F)}\right)$, 18
$h_g\left(A'^{\mathcal{M}\times\mathbf{R}_+^{*2}}, \mathfrak{h}^{\Omega^\cdot(T^*X,\pi^*F)}_{\overline{\mathcal{H}}-\omega^H}\right)$, 66
$h_g\left(\nabla^F, g_\ell^F\right)$, 22
$h_g\left(\nabla^F, g^F\right)$, 17
$h_g^\wedge\left(A', g^{\Omega^\cdot(X,F)}\right)$, 19
$\mathfrak{h}_b^{\mathfrak{H}^\cdot(X,F)}$, 99
H^∞, 226
$\mathcal{H}^\infty$, 228
$H_{X,\ell}$, 343
$\overline{\mathcal{H}}$, 41
$\widehat{\omega}\left(\nabla^F, g^F\right)$, 39
$\mathfrak{h}^{\Omega^\cdot(T^*X,\pi^*F)}_{\mathcal{H}-\omega^H}$, 34
$\mathfrak{h}^{\Omega^\cdot(T^*X,\pi^*F)}$, 27
$\mathfrak{h}^{\Omega^\cdot(T^*X,\pi^*F)_0}$, 116
$\mathfrak{h}^{\Omega^\cdot(T^*X,\pi^*F)_{<0}}_{\mathcal{H}^c}$, 117
$\mathfrak{h}^{\Omega^\cdot(T^*X,\pi^*F)}_{\mathcal{H}}$, 27
$\mathfrak{h}^{\Omega^\cdot(T^*X,\pi^*F)_{<r}}_{\mathcal{H}^c}$, 121
$h^{\Omega^\cdot(T^*X,\pi^*F)}$, 28
$\mathfrak{H}_\pm$, 40
$\mathcal{H}_{p,s}$, 285
H^s, 51, 226
$\mathcal{H}^s$, 51, 228
$\mathfrak{h}^{\mathcal{S}^\cdot(T^*X,\pi^*F)}$, 45
H_X^s, 308
$H_{X,h,\lambda}^{s_1,s_2}$, 340
$\widetilde{h}_g\left(\nabla^F, g_0^F, g_1^F\right)$, 22
$H^\cdot(T^*X,\pi^*F)$, 44
$\mathcal{H}^X$, 13
$h(x)$, 17
$H^\cdot(X,F)$, 12
$\mathbf{H}^\cdot(X,F)$, 173
$\mathfrak{H}^\cdot(X,F)$, 45
$\mathcal{H}_z$, 248

i, 80
I_a, 78
$\int^{\widehat{B}}$, 63
$\mathbf{I}_g\left(E,\nabla^E\right)$, 160
$i_\pm$, 32
$i_{\widehat{R^{TX}p}}$, 26
$I^\theta(x)$, 157
$\mathbf{I}(\theta,x)$, 157
${}^0I(\theta,x)$, 158
$\mathbf{I}_x$, 209
$\mathbf{I}_x^{\sigma,\sigma'}$, 207

J, 32, 132, 345
j, 31
$J_0(y,\lambda)$, 252
J_g, 140
$\mathbf{J}_g\left(E,\nabla^E\right)$, 160
$J_k(y,\lambda)$, 252
$\mathfrak{I}_\pm$, 40
$J^\theta(x)$, 157
$\mathbf{J}(\theta,x)$, 157
${}^0J(\theta,x)$, 158
$J_\tau(\xi,\lambda)$, 284
J_z, 132

k, 44
$K_{0,t,\lambda,N}(x,x')$, 197
K_a, 28
$K_{b,t,\lambda,N}((x,p),(x',p'))$, 197
κ^F, 40
$K^\theta(c,x)$, 156
$k(x)$, 283
$k(x,y)$, 71

L, 290
Λ, 231, 273, 338, 340

λ, 116
$\underline{\lambda_0}$, 38
L^2_X, 308
λ_0, 26
Λ, 242
$\underline{\widehat{\mathfrak{L}}}_{b,t}$, 69
$\mathfrak{L}_{b,t}$, 69
L_c, 225, 264
$\mathcal{L}_c$, 217
$L_{c,\tau}$, 229
$\mathfrak{L}^E_c$, 135
$\lambda(F)$, 23, 24
$L(g)$, 16
L_i, 269
Λ_j, 228
$\Lambda_{\lambda,j}$, 241
$\Lambda_{\lambda,\mathrm{sc},j}$, 271
λ', 290
$L'_{c,\tau}$, 229
$L\Lambda'_h$, 340
$L_\pm(g)$, 16
$L''_{c,\tau}$, 229
$L^\theta(c,x)$, 156
$|\,|^2_{\lambda(F)}$, 23
$\|\,\|^2_{\lambda(F)}$, 23
$L_{Y^\mathcal{H}}$, 29
$\mathcal{L}_{Y^\mathcal{H}}$, 38

$\mathcal{M}$, 33
$m_{a,z}(y)$, 208
$\widehat{\mathfrak{M}}_{b,t}$, 70
$\widehat{\mathfrak{M}}'_{b,t}$, 71
M_c, 59
$\underline{\widehat{\mathfrak{M}}}_{c,t}$, 182
$\underline{\widehat{\mathcal{M}}}_{c,t}$, 197
M^E, 133
M_g, 16
$\mathfrak{M}_{r,c,i}$, 59
$M_{s,N}(u)$, 295
$m_\tau(\xi,h,\lambda)$, 285
μ_0, 26

N, 18, 102
$\|\,\|$, 290
$\mathcal{N}$, 102, 248
$\|A\|$, 243
$\widehat{\nabla}$, 234
${}^1\nabla^{\Lambda^\cdot(T^*X)\widehat{\otimes}F,u}$, 15
${}^1\nabla^{\Lambda^\cdot(T^*X)\widehat{\otimes}F}$, 15
$\underline{\widehat{\mathfrak{N}}}_{b,t}$, 77
$\underline{\widehat{\mathcal{N}}}_{c,t}$, 198
∇^F, 12
$\nabla^{F,u}$, 13
N^H, 41
$\nabla^{H^\cdot(X,F)}$, 18
$\nabla^{\mathfrak{H}^\cdot(X,F)}$, 100
N_∞, 103
$\underline{\widehat{N}}_{\infty,t}$, 203
${}^1\nabla_t^{\Lambda^\cdot(T^*X)\widehat{\otimes}\Lambda^\cdot(T^*X)\widehat{\otimes}\Lambda^n(TX)\widehat{\otimes}F,u}$, 182
$\|\,\|_{b^2\lambda,\mathrm{sc},s}$, 209
N', 225
$\underline{\widehat{\mathfrak{N}}}'_{b,t}$, 77
N'', 225
N^{T^*X}, 41
∇^{TX}, 14
$\widehat{\nabla}U$, 229
ν_c, 36
N^V, 41
$N_{X_g/X}$, 71

$\underline{\widehat{\mathfrak{O}}}_{b,t}$, 78
$\underline{\widehat{\mathfrak{O}}}_{c,t}$, 183
$\underline{\widehat{\mathcal{O}}}_{c,t}$, 198
$o(E)$, 63, 136
$\Omega^\cdot(E\oplus E^*)$, 134
$\Omega^\cdot(X,F)$, 13
$O_{\ell,m,m'}(b)$, 106
$\Omega^\cdot(E\oplus E^*)$, 135
ω, 25, 33
ω^H, 33
ω^V, 33
$\mathrm{Op}(a)$, 320
$\mathrm{Op}(a)$, 231, 274
$\Omega^\cdot(T^*X,\pi^*F)$, 26
$\Omega^\cdot(T^*X,\pi^*F)_*$, 99
$\Omega^\cdot(T^*X,\pi^*F)^0$, 46
$\Omega^\cdot(T^*X,\pi^*F)_{<0}$, 117
$\Omega^\cdot(T^*X,\pi^*F)_\lambda$, 99
$\Omega^\cdot(T^*X,\pi^*F)_{<r}$, 121
$o(TX)$, 15
$\Omega^\cdot(X,F)$, 12

P, 72, 261
$\underline{\widehat{\mathfrak{P}}}_c$, 185
$\underline{\mathfrak{P}}$, 47
$\underline{\mathfrak{P}}$, 80
$\mathfrak{p}$, 26
$< p >$, 226
p, 14
$P_{<0}$, 117
$P^0_{b,t}$, 202
P_b, 99
$\{\}^{T^*Y}$, 271
$\mathcal{P}_b$, 180
$\mathfrak{P}'_b$, 102
$\{\}^{T^*Y_{\mathrm{sc}}}$, 271
$\{\}_{\mathrm{sc}}$, 271
$P_{b,t}$, 202
$\widehat{\mathfrak{P}}_{b,t}$, 79
$\underline{\widehat{\mathfrak{P}}}^\kappa_{b,t}$, 85

$\{\}^{T^{V}*Y}$, 271
P_c, 229
$\underline{\widehat{\mathcal{P}}}_c$, 200
$\underline{\widehat{\mathcal{P}}}_{c,0}$, 202
$P_{c,\lambda}$, 241
$P_{c,\lambda,\tau}$, 241
$P_{c,\tau}$, 229
$\underline{\widehat{\mathfrak{P}}}_{c,t}$, 183
$\underline{\widehat{\mathcal{P}}}_{c,t}$, 198
$\underline{\widehat{\mathcal{P}}}_{c,t}^{a,z}$, 208
$P_{c,x}(\lambda)$, 131
$\mathcal{P}^{d,k}$, 288
$\mathcal{P}^{d,k}_{\rho,\delta,c}$, 267
$\mathcal{P}^{d,k}_{\rho,\delta,c}$, 287
$\mathcal{P}^{d}_{\rho,\delta}$, 288
Pf, 63
$P_{g,p}$, 150
P_h, 264, 269
$\underline{\widehat{\mathcal{P}}}$, 201
$\widehat{p}$, 29
P_h^0, 265
P_h^0, 277
φ, 17
Φ^{T^*X}, 31
$\left[\Phi^{T^*X}\right]$, 45
$P_{h,\lambda}$, 272
$P_{h,\lambda,\tau}$, 273
π, 25, 33
$\mathfrak{P}^{\lambda}$, 56
P_{λ}, 56
$\mathfrak{P}_{\lambda}$, 49, 246
$\mathfrak{P}_{\lambda,\overline{\lambda}}$, 50
$\mathcal{P}_{\mu}$, 52
$P(p)$, 248
P_p, 148
$P'_{c,\tau}$, 229
$P^{\perp}$, 262
$P_{\pm}$, 32, 270
$P_{\pm}^{\perp}$, 270
$P_{\pm,\tau}$, 296
$\mathfrak{P}'_r$, 59
$\mathfrak{P}_{<r}$, 120
$P_{r,b,i}(z)$, 59
$P''_{c,\tau}$, 229
ψ_a, 18
ψ_{α}^{β}, 260
$\psi(t)$, 254
P^{TX}, 14
$P_{x,y}^t$, 216
$p_t(x,y)$, 216
$P_{z,z'}^t$, 221
P_x, 215
P_z, 218

Q, 21
$\mathfrak{Q}$, 47
q, 33
$Q_{h,\tau}^0(x,\xi)$, 296
$Q_c((y,p),(y',p'))$, 135
$Qf(x)$, 21
$\underline{\widehat{\mathcal{Q}}}$, 201
$Q_h^0(x,\xi)$, 289
$Q_{h,\tau}(x,\xi)$, 296
$Q_h(x,\xi)$, 289
$\mathcal{Q}^{\ell}$, 343
$\mathfrak{Q}_{\lambda}$, 49, 246
$\mathcal{Q}^{\ell}_{\epsilon}$, 349
$\mathfrak{Q}_{\lambda,\overline{\lambda}}$, 50
Q_-, 31
$Q_+^{T^*X}$, 31
$Q'_c((y,p),(y',p'))$, 138
$Q_{\pm}$, 270
$Q_{r,b,i}$, 59
$q_t(z,z')$, 220
Q_u, 148
$Q_{u,g}$, 150
$q((y,p),(y',p'))$, 138
$\mathbb{Q}_h(x,\xi)$, 322

R, 232, 345
$\mathcal{R}$, 227
r, 27
$r_1(\lambda)$, 257
r_a, 28
R^F, 13
$R_{h,\lambda}$, 340
$\rho(\lambda)$, 257
$\rho(y)$, 76
R_{λ}, 282
$\mathcal{R}^{\ell}$, 343
$r(\lambda)$, 102
R_{λ}, 242
$r_N(-\lambda)$, 195
R^{TX}, 13, 14
$R_{(y_0,p_0)}$, 138

$\mathfrak{S}$, 328
S, 51, 226, 234, 331
$\mathbb{S}$, 228
$\mathcal{S}$, 228, 284
$|s|_{0,t}$, 207
$\sigma(A)$, 231, 274
$S_b(g^{TX},\nabla^F,g^F)$, 119
$S_b(g^{TX},\nabla^F,g^F)_{>r}$, 121
$S_{b,\lambda}$, 203
$S_{b,t}$, 72
$\mathfrak{S}_{b,t}$, 83
S^d, 231
S^d, 274
$\mathcal{S}^d$, 297
$\sigma_d(A)$, 320
$\mathbb{S}_{\delta}^d$, 261

$\mathbb{S}^d_{\delta,h}$, 320
$\mathbb{S}^d_{\delta,h,0}$, 320
$\mathcal{S}^{d,k}$, 288
$\mathcal{S}^{d,k}_{\rho,\delta,c}$, 267
$\mathcal{S}^{d,k}_{\rho,\delta,c}$, 286
S^d_λ, 242
$S^d_{\rho,\delta}$, 267
$\mathcal{S}^d_{\rho,\delta}$, 288
$S_{g,b}\left(g^{TX}, \nabla^F, g^F\right)$, 124
$S_{h,\lambda}$, 265, 278
$S^m_{\rho,\delta}$, 286
$\mathrm{Sp}\, A \subset \mathbf{C}$, 244
$\mathcal{S}'^{\cdot}\left(T^*X, \pi^*F^*\right)$, 226
$\widehat{\square}^{T^*X}_{\mathcal{H}^c}$, 49
$\square^X$, 13
$\mathbb{S}_{\mathrm{sc}}$, 271
$\mathcal{S}^{\cdot}\left(T^*X, \pi^*F\right)$, 45
$\mathcal{S}^{\cdot}\left(T^*X, \pi^*F\right)_*$, 47
$\mathcal{S}^{\cdot}\left(T^*X, \pi^*F\right)_\lambda$, 47
$\mathcal{S}^{\cdot}\left(T^*X, \pi^*F\right)_\lambda$, 55
$\mathcal{S}^{\cdot}\left(T^*X, \pi^*F\right)_{\lambda,\overline{\lambda*}}$, 50
$\mathcal{S}^{\cdot}\left(T^*X, \pi^*F\right)_{\lambda,\overline{\lambda}}$, 50
$\mathcal{S}^{\cdot}\left(T^*X, \pi^*F\right)_{\lambda,*}$, 49
$\mathcal{S}^{\cdot}\left(T^*X, \pi^*F\right)_{<r}$, 120
$\mathcal{S}^{\cdot}\left(T^*X, \pi^*F\right)^\lambda$, 56
$\sigma(u,\eta,x)$, 155
S^X, 14
$*^X$, 15
σ_x, 283
$S_{(y_0,p_0)}$, 138

T, 14
τ_b, 42
$\mathcal{T}_{\mathrm{ch},g}\left(T^HM, g^{TX}, \nabla^F, g^F\right)$, 21
$\mathcal{T}_{\mathrm{ch},g,b}\left(T^HM, g^{TX}, \nabla^F, g^F\right)$, 114
$\tau\left(c,\eta,x\right)$, 132
T^H, 33
$\mathcal{T}^H$, 33
θ_0, 232
$T_{h,b}\left(g^{TX}, \nabla^F, g^F\right)_{>0}$, 120
$T_{h,b}\left(g^{TX}, \nabla^F, g^F\right)$, 119
$T_{h,b}\left(g^{TX}, \nabla^F, g^F\right)_{>r}$, 121
$\vartheta_g\left(s\right)$, 117
$\vartheta_{>r}\left(s\right)$, 121
θ, 25
$\vartheta_{<0}\left(s\right)$, 117
$\vartheta_{>0}\left(s\right)$, 117
$\vartheta_g\left(s\right)$, 20
$\vartheta\left(s\right)$, 117
$T_{h,g,b}\left(g^{TX}, \nabla^F, g^F\right)$, 124
$\mathcal{T}_{h,g,b}\left(T^HM, g^{TX}, \nabla^F, g^F\right)$, 114
$\mathcal{T}_{h,g}\left(T^HM, g^{TX}, \nabla^F, g^F\right)$, 19
$T_{h,\lambda}$, 326
$\Theta_{h,\lambda}$, 264, 323

T^HM, 14
$T^H\mathcal{M}$, 33
T^HM_g, 16
$T^{H\prime}\mathcal{M}$, 38
$\vartheta\left(s\right)$, 20
T^HT^*X, 26
$\|\theta\|_M$, 309
$\langle T^0, p\rangle$, 38
$T'X$, 225
$\mathrm{Tr_s}$, 12
$\widehat{\mathrm{Tr_s}}$, 79
$\widehat{\mathrm{Tr_s}}$, 184
$\mathrm{Tr_s}^{\mathrm{even}}$, 139
$\mathrm{Tr_s}^{\mathrm{odd}}$, 140
$T''X$, 225
T^*Y_{sc}, 270
T^VY, 270
$\mathfrak{T}_{\xi,\lambda}$, 262
T_{y_0}, 134

U, 283
$\mathcal{U}$, 244
u, 27, 253
$u_{0,t}$, 100
$u_{\alpha,\beta}$, 304
$u_{b,\infty}$, 100
$U_{b,t}$, 41
$u_{b,t}$, 65
$\mathcal{U}_\eta$, 71
$\widehat{U}$, 231
$U_{h,\lambda}$, 331
$\widehat{u}\left(\xi,p\right)$, 287
U_j, 227
$|U|_{j,s}$, 228
$\|u\|_\ell$, 343
$|U|_{\lambda,j,s}$, 241
$\|u\|_{\lambda,s}$, 241
$|U|_{\lambda,\mathrm{sc},j,s}$, 272
$||||u||||_{\lambda,\mathrm{sc},s}$, 279
$|U|_{\lambda,\mathrm{sc},s}$, 273
$\|u\|_{\lambda,\mathrm{sc},s}$, 272
$|||u|||_{\lambda,\mathrm{sc},s}$, 276
$\|u\|_{\lambda,s_1,s_2}$, 340
$\|u\|_{p,s}$, 284
$|U|_{p,s,j}$, 284
$|U|_s$, 231
$|u|_s$, 226
$\|u\|_s$, 228
$|||U|||_{\mathrm{sc},s}$, 276
$\|u\|_{\mathrm{sc},s}$, 272
$|||u|||_{s,N}$, 303
$\|u\|_{X,h,\lambda,s}$, 319
$\|u\|_{X,\lambda^{1/2},s}$, 319
$\|u\|_{X,s}$, 319

V, 283

$\mathcal{V}$, 281
v, 27
$\overline{v}_{0,N}(x, y, \lambda)$, 205
$v_{0,t}$, 100
$v_{0,t,N}(x, \lambda)$, 197
$v_{b,\infty}$, 100
$V_{b,t}$, 102
$v_{b,t}$, 65
$\mathcal{U}_\eta$, 71
$V_{h,\lambda}$, 331
V_λ, 244
$\| \|_\lambda^2$, 119
$| |_\lambda^2$, 116
$| |_{\lambda,<r}^2$, 121
V_r, 122
$\|s\|_{\pm\mathcal{H}}$, 181
$\|v\|_{s,N}$, 303
$v_{\sqrt{t}b,t,N}(x, \lambda)$, 197

$w_{0,0}$, 165
$w_{0,\infty}$, 165
$\overline{w}_{0,N}(x, \lambda)$, 205
$\underline{w}_{0,t}$, 100
$w_{0,t}$, 100
$w_{0,t,N}(x, \lambda)$, 198
$w_{b,\infty}$, 100
$W_{b,t}$, 102
$w_{b,t}$, 65
$\mathcal{W}_\delta$, 53, 253, 319
$\mathcal{W}_{\delta',b,r}$, 53
$\mathcal{W}_{\delta',h,r}$, 337
W^s, 236
$\mathcal{W}^s$, 240
$w_{\sqrt{t}b,t,N}(x, \lambda)$, 198
$\underline{w}_{b,\infty}$, 100
$\underline{w}_{b,t}$, 66

X_g, 16

Y, 227
$Y^{\mathcal{H}}$, 25

$\zeta(\theta, s)$, 155

Milton Keynes UK
Ingram Content Group UK Ltd.
UKHW020740180824
447095UK00005B/344

9 780691 137322